Materials Selection in Mechanical Design

Third Edition

Michael F. Ashby

ELSEVIER
BUTTERWORTH
HEINEMANN

AMSTERDAM • BOSTON • HEIDELBERG • LONDON • NEW YORK • OXFORD
PARIS • SAN DIEGO • SAN FRANCISCO • SINGAPORE • SYDNEY • TOKYO

Butterworth-Heinemann
Linacre House, Jordan Hill, Oxford OX2 8DP
30 Corporate Drive, Burlington, MA 01803

First published by Pergamon Press 1992
Second edition 1999
Third edition 2005

British Library Cataloguing in Publication Data
A catalogue record for this book is available from the British Library

Library of Congress Cataloging in Publication Data
A catalog record for this book is available from the Library of Congress

ISBN 0 7506 6168 2

For information on all Elsevier Butterworth-Heinemann
publications visit our website at http://books.elsevier.com

Typeset by Newgen Imaging Systems (P) Ltd, Chennai, India
Printed and bound in Italy

Contents

Preface

> Materials, of themselves, affect us little; it is the way we use them which influences our lives.
> Epictetus, AD 50–100, *Discourses* Book 2, Chapter 5.

New materials advanced engineering design in Epictetus' time. Today, with more materials than ever before, the opportunities for innovation are immense. But advance is possible only if a procedure exists for making a rational choice. This book develops a systematic procedure for selecting materials and processes, leading to the subset which best matches the requirements of a design. It is unique in the way the information it contains has been structured. The structure gives rapid access to data and allows the user great freedom in exploring the potential of choice. The method is available as software,[1] giving greater flexibility.

The approach emphasizes design with materials rather than materials "science", although the underlying science is used, whenever possible, to help with the structuring of criteria for selection. The first eight chapters require little prior knowledge: a first-year grasp of materials and mechanics is enough. The chapters dealing with shape and multi-objective selection are a little more advanced but can be omitted on a first reading. As far as possible the book integrates materials selection with other aspects of design; the relationship with the stages of design and optimization and with the mechanics of materials, are developed throughout. At the teaching level, the book is intended as the text for 3rd and 4th year engineering courses on Materials for Design: a 6–10 lecture unit can be based on Chapters 1–6; a full 20+ lecture course, with associated project work with the associated software, uses the entire book.

Beyond this, the book is intended as a reference text of lasting value. The method, the charts and tables of performance indices have application in real problems of materials and process selection; and the catalogue of "useful solutions" is particularly helpful in modelling—an essential ingredient of optimal design. The reader can use the book (and the software) at increasing levels of sophistication as his or her experience grows, starting with the material indices developed in the case studies of the text, and graduating to the modelling of new design problems, leading to new material indices and penalty functions, and new—and perhaps novel—choices of material. This continuing education aspect is helped by a list of Further reading at the end of most chapters, and by a set of exercises in Appendix E covering all aspects of the text. Useful reference material is assembled in appendices at the end of the book.

Like any other book, the contents of this one are protected by copyright. Generally, it is an infringement to copy and distribute materials from a copyrighted source. But the best way to use the charts that are a central feature of the book is to have a clean copy on which you can draw, try out alternative selection criteria, write comments, and so forth; and presenting the conclusion of a selection exercise is often most easily done in the same way. Although the book itself is copyrighted, the reader is authorized to make unlimited copies of the charts, and to reproduce these, with proper reference to their source, as he or she wishes.

M.F. Ashby
Cambridge, July 2004

---

1 The *CES materials and process selection platform*, available from Granta Design Ltd, Rustat House, 62 Clifton Road, Cambridge CB1 7EG, UK (www.grantadesign.com).

Acknowledgements

Many colleagues have been generous in discussion, criticism, and constructive suggestions. I particularly wish to thank Professor Yves Bréchet of the University of Grenoble; Professor Anthony Evans of the University of California at Santa Barbara; Professor John Hutchinson of Harvard University; Dr David Cebon; Professor Norman Fleck; Professor Ken Wallace; Dr. John Clarkson; Dr. Hugh Shercliff of the Engineering Department, Cambridge University; Dr Amal Esawi of the American University in Cairo, Egypt; Dr Ulrike Wegst of the Max Planck Institute for Materials Research in Stuttgart, Germany; Dr Paul Weaver of the Department of Aeronautical Engineering at the University of Bristol; Professor Michael Brown of the Cavendish Laboratory, Cambridge, UK, and the staff of Granta Design Ltd, Cambridge, UK.

Features of the Third Edition

Since publication of the Second Edition, changes have occurred in the fields of materials and mechanical design, as well as in the way that these and related subjects are taught within a variety of curricula and courses. This new edition has been comprehensively revised and reorganized to address these. Enhancements have been made to presentation, including a new layout and two-colour design, and to the features and supplements that accompany the text. The key changes are outlined below.

Key changes

New and fully revised chapters:

- Processes and process selection (Chapter 7)
- Process selection case studies (Chapter 8)
- Selection of material and shape (Chapter 11)
- Selection of material and shape: case studies (Chapter 12)
- Designing hybrid materials (Chapter 13)
- Hybrid case studies (Chapter 14)
- Information and knowledge sources for design (Chapter 15)
- Materials and the environment (Chapter 16)
- Materials and industrial design (Chapter 17)
- Comprehensive appendices listing useful formulae; data for material properties; material indices; and information sources for materials and processes.

Supplements to the Third Edition

Material selection charts

Full color versions of the material selection charts presented in the book are available from the following website. Although the charts remain copyright of the author, users of this book are authorized to download, print and make unlimited copies of these charts, and to reproduce these for teaching and learning purposes only, but not for publication, with proper reference to their owner-ship and source. To access the charts and other teaching resources, visit www.grantadesign.com/ashbycharts.htm

Instructor's manual

The book itself contains a comprehensive set of exercises. Worked-out solutions to the exercises are freely available to teachers and lecturers who adopt this book. To access this material online please visit http://books.elsevier.com/manuals and follow the instructions on screen.

Image bank

The Image Bank provides adopting tutors and lecturers with PDF versions of the figures from the book that may be used in lecture slides and class presentations. To access this material please visit http://books.elsevier.com/manuals and follow the instructions on screen.

The CES EduPack

CES EduPack is the software-based package to accompany this book, developed by Michael Ashby and Granta Design. Used together, *Materials Selection in Mechanical Design* and CES EduPack provide a complete materials, manufacturing and design course. For further information please see the last page of this book, or visit www.grantadesign.com.

Chapter 1

Introduction

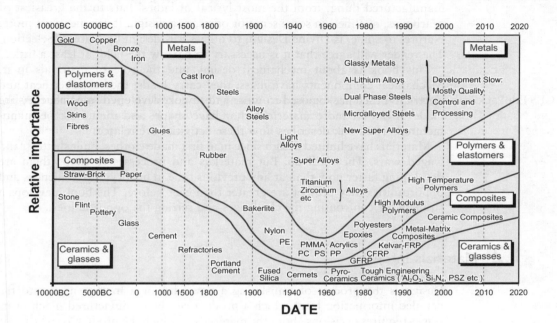

1.1 Introduction and synopsis

"Design" is one of those words that means all things to all people. Every manufactured thing, from the most lyrical of ladies' hats to the greasiest of gearboxes, qualifies, in some sense or other, as a design. It can mean yet more. Nature, to some, is Divine Design; to others it is design by Natural Selection. The reader will agree that it is necessary to narrow the field, at least a little.

This book is about mechanical design, and the role of materials in it. Mechanical components have mass; they carry loads; they conduct heat and electricity; they are exposed to wear and to corrosive environments; they are made of one or more materials; they have shape; and they must be manufactured. The book describes how these activities are related.

Materials have limited design since man first made clothes, built shelters, and waged wars. They still do. But materials and processes to shape them are developing faster now than at any previous time in history; the challenges and opportunities they present are greater than ever before. The book develops a strategy for confronting the challenges and seizing the opportunities.

1.2 Materials in design

Design is the process of translating a new idea or a market need into the detailed information from which a product can be manufactured. Each of its stages requires decisions about the materials of which the product is to be made and the process for making it. Normally, the choice of material is dictated by the design. But sometimes it is the other way round: the new product, or the evolution of the existing one, was suggested or made possible by the new material. The number of materials available to the engineer is vast: something over 120,000 are at his or her (from here on "his" means both) disposal. And although standardization strives to reduce the number, the continuing appearance of new materials with novel, exploitable, properties expands the options further.

How, then, does the engineer choose, from this vast menu, the material best suited to his purpose? Must he rely on experience? In the past he did, passing on this precious commodity to apprentices who, much later in their lives, might assume his role as the in-house materials guru who knows all about the things the company makes. But many things have changed in the world of engineering design, and all of them work against the success of this model. There is the drawn-out time scale of apprentice-based learning. There is job mobility, meaning that the guru who is here today is gone tomorrow. And there is the rapid evolution of materials information, already mentioned.

There is no question of the value of experience. But a strategy relying on experience-based learning is not in tune with the pace and re-dispersion of talent that is part of the age of information technology. We need a *systematic* procedure—one with steps that can be taught quickly, that is robust in the decisions it reaches, that allows of computer implementation, and with the ability to interface with the other established tools of engineering design. The question has to be addressed at a number of levels, corresponding to the stage the design has reached. At the beginning the design is fluid and the options are wide; all materials must be considered. As the design becomes more focused and takes shape, the selection criteria sharpen and the short-list of materials that can satisfy them narrows. Then more accurate data are required (though for a lesser number of materials) and a different way of analyzing the choice must be used. In the final stages of design, precise data are needed, but for still fewer materials—perhaps only one. The procedure must recognize the initial richness of choice, and at the same time provide the precision and detail on which final design calculations can be based.

The choice of material cannot be made independently of the choice of process by which the material is to be formed, joined, finished, and otherwise treated. Cost enters, both in the choice of material and in the way the material is processed. So, too, does the influence material usage on the environment in which we live. And it must be recognized that good engineering design alone is not enough to sell products. In almost everything from home appliances through automobiles to aircraft, the form, texture, feel, color, decoration of the product—the satisfaction it gives the person who owns or uses it—are important. This aspect, known confusingly as "industrial design", is one that, if neglected, can lose the manufacturer his market. Good designs work; excellent designs also give pleasure.

Design problems, almost always, are open-ended. They do not have a unique or "correct" solution, though some solutions will clearly be better than others. They differ from the analytical problems used in teaching mechanics, or structures, or thermodynamics, which generally do have single, correct answers. So the first tool a designer needs is an open mind: the willingness to consider all possibilities. But a net cast widely draws in many fish. A procedure is necessary for selecting the excellent from the merely good.

This book deals with the materials aspects of the design process. It develops a methodology that, properly applied, gives guidance through the forest of complex choices the designer faces. The ideas of *material and process attributes* are introduced. They are mapped on material and process *selection charts* that show the lay of the land, so to speak, and simplify the initial survey for potential candidate-materials. Real life always involves *conflicting objectives*—minimizing mass while at the same time minimizing cost is an example—requiring the use of *trade-off methods*. The interaction between *material and shape* can be built into the method. Taken together, these suggest schemes for expanding the boundaries of material performance by creating *hybrids*—combinations of two or more materials, shapes and configurations with unique property profiles. None of this can be implemented without *data* for material properties and process attributes: ways to find them are described. The role of *aesthetics* in engineering design is discussed. *The forces driving change* in the materials-world are surveyed, the most obvious of which is that dealing with environmental concerns. The appendices contain useful information.

The methods lend themselves readily to implementation as computer-based tools; one, *The CES materials and process selection platform*,[1] has been used for the case studies and many of the figures in this book. They offer, too, potential for interfacing with other computer-aided design, function modeling, optimization routines, but this degree of integration, though under development, is not yet commercially available.

All this will be found in the following chapters, with case studies illustrating applications. But first, a little history.

1.3 The evolution of engineering materials

Throughout history, materials have limited design. The ages in which man has lived are named for the materials he used: stone, bronze, iron. And when he died, the materials he treasured were buried with him: Tutankhamen in his enameled sarcophagus, Agamemnon with his bronze sword and mask of gold, each representing the high technology of their day.

If they had lived and died today, what would they have taken with them? Their titanium watch, perhaps; their carbon-fiber reinforced tennis racquet, their metal-matrix composite mountain bike, their shape-memory alloy eye-glass frames with diamond-like carbon coated lenses, their polyether–ethyl–ketone crash helmet. This is not the age of one material, it is the age of an immense range of materials. There has never been an era in which their evolution was faster and the range of their properties more varied. The menu of materials has expanded so rapidly that designers who left college 20 years ago can be forgiven for not knowing that half of them exist. But not-to-know is, for the designer, to risk disaster. Innovative design, often, means the imaginative exploitation of the properties offered by new or improved materials. And for the man in the street, the schoolboy even, not-to-know is to miss one of the great developments of our age: the age of advanced materials.

This evolution and its increasing pace are illustrated in Figure 1.1. The materials of pre-history (>10,000 BC, the Stone Age) were ceramics and glasses, natural polymers, and composites. Weapons — always the peak of technology — were made of wood and flint; buildings and bridges of stone and wood. Naturally occurring gold and silver were available locally and, through their rarity, assumed great influence as currency, but their role in technology was small. The development of rudimentary thermo-chemistry allowed the

---

[1] Granta Design Ltd, Rustat House, 62 Clifton Road, Cambridge CB1 7EG, UK (www.grantadesign.com).

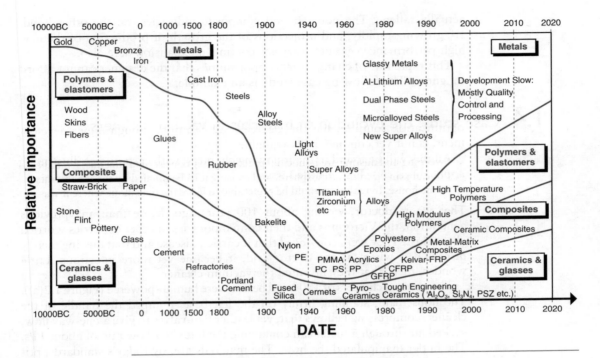

Figure 1.1 The evolution of engineering materials with time. "Relative importance" is based on information contained in the books listed under "Further reading", plus, from 1960 onwards, data for the teaching hours allocated to each material family in UK and US Universities. The projections to 2020 rely on estimates of material usage in automobiles and aircraft by manufacturers. The time scale is non-linear. The rate of change is far faster today than at any previous time in history.

extraction of, first, copper and bronze, then iron (the Bronze Age, 4000–1000 BC and the Iron Age, 1000 BC–1620 AD) stimulating enormous advances, in technology. (There is a cartoon on my office door, put there by a student, showing an aggrieved Celt confronting a sword-smith with the words: "You sold me this bronze sword last week and now I'm supposed to upgrade to iron!") Cast iron technology (1620s) established the dominance of metals in engineering; and since then the evolution of steels (1850 onward), light alloys (1940s) and special alloys, has consolidated their position. By the 1960s, "engineering materials" meant "metals". Engineers were given courses in metallurgy; other materials were barely mentioned.

There had, of course, been developments in the other classes of material. Improved cements, refractories, and glasses, and rubber, bakelite, and polyethylene among polymers, but their share of the total materials market was small. Since 1960 all that has changed. The rate of development of new metallic alloys is now slow; demand for steel and cast iron has in some countries actually fallen.[2] The polymer and composite industries, on the other hand, are growing rapidly, and projections of the growth of production of the new high-performance ceramics suggests continued expansion here also.

This rapid rate of change offers opportunities that the designer cannot afford to ignore. The following case study is an example.

1.4 Case study: the evolution of materials in vacuum cleaners

> Sweeping and dusting are homicidal practices: they consist of taking dust from the floor, mixing it in the atmosphere, and causing it to be inhaled by the inhabitants of the house. In reality it would be preferable to leave the dust alone where it was.

That was a doctor, writing about 100 years ago. More than any previous generation, the Victorians and their contemporaries in other countries worried about dust. They were convinced that it carried disease and that dusting merely dispersed it when, as the doctor said, it became yet more infectious. Little wonder, then, that they invented the vacuum cleaner.

The vacuum cleaners of 1900 and before were human-powered (Figure 1.2(a)). The housemaid, standing firmly on the flat base, pumped the handle of the cleaner, compressing bellows that, via leather flap-valves to give a one-way flow, sucked air through a metal can containing the filter at a flow rate of about 1 l/s. The butler manipulated the hose. The materials are, by today's standards, primitive: the cleaner is made almost entirely from natural materials: wood, canvas, leather and rubber. The only metal is the straps that link the bellows (soft iron) and the can containing the filter (mild steel sheet, rolled to make a cylinder). It reflects the use of materials in 1900. Even a car, in 1900, was mostly made of wood, leather, and rubber; only the engine and drive train had to be metal.

The electric vacuum cleaner first appeared around 1908.[3] By 1950 the design had evolved into the cylinder cleaner shown in Figure 1.2(b) (flow rate about 10 l/s). Air flow is axial, drawn through the cylinder by an electric fan. The fan occupies about half the length of the cylinder; the rest holds the filter. One advance in design is, of course, the electrically driven air pump. The motor, it is true, is bulky and of low power, but it can function continuously without tea breaks or housemaid's elbow. But there are others: this cleaner is almost entirely made of metal: the case, the end-caps, the runners, even the tube to suck up the dust are mild steel: metals have entirely replaced natural materials.

Developments since then have been rapid, driven by the innovative use of new materials. The 1985 vacuum cleaner of Figure 1.2(c) has the power of roughly 16 housemaids working flat out (800 W) and a corresponding air

---

[2] Do not, however, imagine that the days of steel are over. Steel production accounts for 90% of all world metal output, and its unique combination of strength, ductility, toughness, and low price makes steel irreplaceable.

[3] Inventors: Murray Spengler and William B. Hoover. The second name has become part of the English language, along with those of such luminaries as John B. Stetson (the hat), S.F.B. Morse (the code), Leo Henrik Baikeland (Bakelite), and Thomas Crapper (the flush toilet).

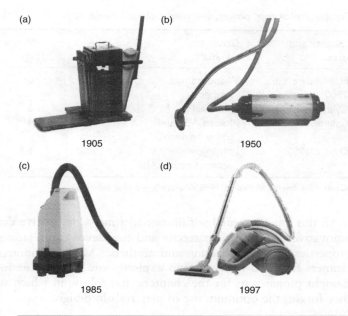

Figure 1.2 Vacuum cleaners: (a) the hand-powered bellows cleaner of 1900, largely made of wood and leather; (b) the cylinder cleaner of 1950; (c) the lightweight cleaner of 1985, almost entirely made of polymer; and (d) a centrifugal dust-extraction cleaner of 1997.

flow-rate; cleaners with twice that power are now available. Air flow is still axial and dust-removal by filtration, but the unit is smaller than the old cylinder cleaners. This is made possible by a higher power-density in the motor, reflecting better magnetic materials, and higher operating temperatures (heat-resistant insulation, windings, and bearings). The casing is entirely polymeric, and is an example of good design with plastics. The upper part is a single molding, with all additional bits attached by snap fasteners molded into the original component. No metal is visible anywhere; even the straight part of the suction tube, metal in all earlier models, is now polypropylene. The number of components is dramatically reduced: the casing has just 4 parts, held together by just 1 fastener, compared with 11 parts and 28 fasteners for the 1950 cleaner. The saving on weight and cost is enormous, as the comparison in Table 1.1 shows. It is arguable that this design (and its many variants) is near-optimal for today's needs; that a change of working principle, material or process could increase performance but at a cost-penalty unacceptable to the consumer. We will leave the discussion of balancing performance against cost to a later chapter, and merely note here that one manufacturer disagrees. The cleaner shown in Figure 1.2(d) exploits a different concept: that of inertial separation rather than filtration. For this to work, the power and rotation speed have to be high; the product is larger, heavier and more expensive than the competition. Yet it sells — a testament to good industrial design and imaginative marketing.

Table 1.1   Comparison of cost, power, and weight of vacuum cleaners

Cleaner and date	Dominant materials	Power (W)	Weight (kg)	Approximate cost*
Hand powered, 1900	Wood, canvas, leather	50	10	£240–$380
Cylinder, 1950	Mild steel	300	6	£96–$150
Cylinder, 1985	Molded ABS and polypropylene	800	4	£60–$95
Dyson, 1995	Polypropylene, polycarbonate, ABS	1200	6.3	£190–$300

*Costs have been adjusted to 1998 values, allowing for inflation.

All this has happened within one lifetime. Competitive design requires the innovative use of new materials and the clever exploitation of their special properties, both engineering and aesthetic. Many manufacturers of vacuum cleaners failed to innovate and exploit; now they are extinct. That sombre thought prepares us for the chapters that follow in which we consider what they forgot: the optimum use of materials in design.

1.5 Summary and conclusions

The number of engineering materials is large: tens of thousands, at a conservative estimate. The designer must select, from this vast menu, the few best suited to his task. This, without guidance, can be a difficult and haphazard business, so there is a temptation to choose the material that is "traditional" for the application: glass for bottles; steel cans. That choice may be safely conservative, but it rejects the opportunity for innovation. Engineering materials are evolving faster, and the choice is wider than ever before. Examples of products in which a new material has captured a market are as common as — well — as plastic bottles. Or aluminium cans. Or polycarbonate eyeglass lenses. Or carbon-fiber golf club shafts. It is important in the early stage of design, or of re-design, to examine the full materials menu, not rejecting options merely because they are unfamiliar. That is what this book is about.

1.6 Further reading

The history and evolution of materials

A History of Technology (21 volumes), edited by Singer, C., Holmyard, E.J., Hall, A.R., Williams, T.I., and Hollister-Short, G. Oxford University Press (1954–2001)

Oxford, UK. ISSN 0307–5451. *(A compilation of essays on aspects of technology, including materials.)*

Delmonte, J. (1985) *Origins of Materials and Processes*, Technomic Publishing Company, Pennsylvania, USA. ISBN 87762-420-8. *(A compendium of information on when materials were first used, any by whom.)*

Dowson, D. (1998) *History of Tribology*, Professional Engineering Publishing Ltd., London, UK. ISBN 1-86058-070-X. *(A monumental work detailing the history of devices limited by friction and wear, and the development of an understanding of these phenomena.)*

Emsley, J. (1998), *Molecules at an Exhibition*, Oxford University Press, Oxford, UK. ISBN 0-19-286206-5. *(Popular science writing at its best: intelligible, accurate, simple and clear. The book is exceptional for its range. The message is that molecules, often meaning materials, influence our health, our lives, the things we make and the things we use.)*

Michaelis, R.R. (1992) editor "Gold: art, science and technology", and "Focus on gold", *Interdisciplinary Science Reviews*, volume 17 numbers 3 and 4. ISSN 0308–0188. *(A comprehensive survey of the history, mystique, associations and uses of gold.)*

The *Encyclopaedia Britannica*, 11th edition (1910). The Encyclopaedia Britannica Company, New York, USA. *(Connoisseurs will tell you that in its 11th edition the Encyclopaedia Britannica reached a peak of excellence which has not since been equalled, though subsequent editions are still usable.)*

Tylecoate, R.F. (1992) *A History of Metallurgy*, 2nd edition, The Institute of Materials, London, UK. ISBN 0-904357-066. *(A total-immersion course in the history of the extraction and use of metals from 6000BC to 1976, told by an author with forensic talent and love of detail.)*

And on vacuum cleaners

Forty, A. (1986) *Objects of Desire — design in society since 1750*, Thames and Hudson, London, UK, p. 174 *et seq*. ISBN 0-500-27412-6. *(A refreshing survey of the design history of printed fabrics, domestic products, office equipment and transport system. The book is mercifully free of eulogies about designers, and focuses on what industrial design does, rather than who did it. The black and white illustrations are disappointing, mostly drawn from the late 19th or early 20th centuries, with few examples of contemporary design.)*

Chapter 2

The design process

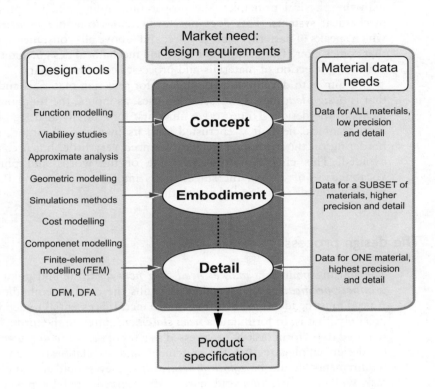

2.1   Introduction and synopsis

It is *mechanical design* with which we are primarily concerned here; it deals with the physical principles, the proper functioning and the production of mechanical systems. This does not mean that we ignore *industrial design*, which speaks of pattern, color, texture, and (above all) consumer appeal — but that comes later. The starting point is good mechanical design, and the ways in which the selection of materials and processes contribute to it.

Our aim is to develop a methodology for selecting materials and processes that is *design-led*; that is, the selection uses, as inputs, the functional requirements of the design. To do so we must first look briefly at design itself. Like most technical fields it is encrusted with its own special jargon, some of it bordering on the incomprehensible. We need very little, but it cannot all be avoided. This chapter introduces some of the words and phrases — the vocabulary — of design, the stages in its implementation, and the ways in which materials selection links with these.

2.2   The design process

The starting point is a *market need* or a *new idea*; the end point is the full *product specification* of a product that fills the need or embodies the idea. A need must be identified before it can be met. It is essential to define the need precisely, that is, to formulate a *need statement*, often in the form: "a device is required to perform task X", expressed as a set of *design requirements*. Writers on design emphasize that the statement and its elaboration in the design requirements should be *solution-neutral* (i.e. they should not imply how the task will be done), to avoid narrow thinking limited by pre-conceptions. Between the need statement and the product specification lie the set of stages shown in Figure 2.1: the stages of *conceptual*, *embodiment* and *detailed designs*, explained in a moment.

The product itself is called a *technical system*. A technical system consists of *sub-assemblies* and *components*, put together in a way that performs the required task, as in the breakdown of Figure 2.2. It is like describing a cat (the system) as made up of one head, one body, one tail, four legs, etc. (the sub-assemblies), each composed of components — femurs, quadriceps, claws, fur. This decomposition is a useful way to analyze an existing design, but it is not of much help in the design process itself, that is, in the synthesis of new designs. Better, for this purpose, is one based on the ideas of systems analysis. It thinks of the inputs, flows and outputs of information, energy, and materials, as in Figure 2.3. The design converts the inputs into the outputs. An electric motor converts electrical into mechanical energy; a forging press takes and reshapes material; a burglar alarm collects information and converts it to noise. In this approach, the system is broken down into connected sub-systems each of

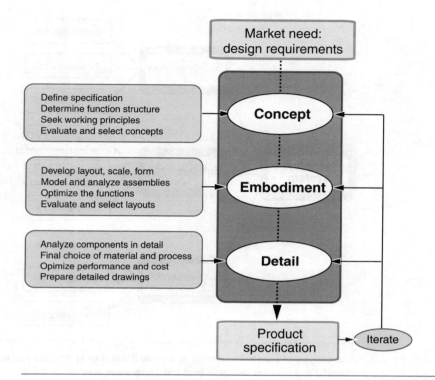

Figure 2.1   The design flow chart. The design proceeds from the identification of a *market need*, clarified as a set of *design requirements*, through *concept, embodiment* and *detailed* analysis to a *product specification*.

which performs a specific function, as in Figure 2.3; the resulting arrangement is called the *function-structure* or *function decomposition* of the system. It is like describing a cat as an appropriate linkage of a respiratory system, a cardio-vascular system, a nervous system, a digestive system and so on. Alternative designs link the unit functions in alternative ways, combine functions, or split them. The function-structure gives a systematic way of assessing design options.

The design proceeds by developing concepts to perform the functions in the function structure, each based on a *working principle*. At this, the conceptual design stage, all options are open: the designer considers alternative concepts and the ways in which these might be separated or combined. The next stage, embodiment, takes the promising concepts and seeks to analyze their operation at an approximate level. This involves sizing the components, and selecting materials that will perform properly in the ranges of stress, temperature, and environment suggested by the design requirements, examining the implications for performance and cost. The embodiment stage ends with a feasible layout, which is then passed to the detailed design stage. Here specifications for each

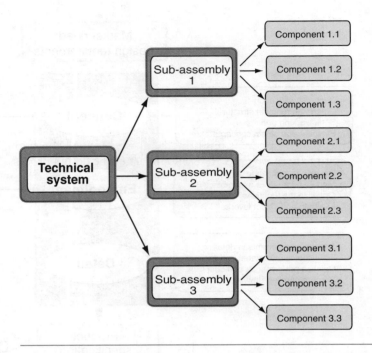

Figure 2.2 The analysis of a technical system as a breakdown into assemblies and components. Material and process selection is at the component level.

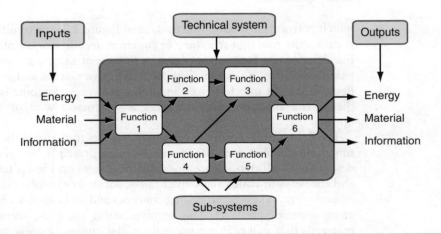

Figure 2.3 The systems approach to the analysis of a technical system, seen as transformation of energy, materials and information (signals). This approach, when elaborated, helps structure thinking about alternative designs.

component are drawn up. Critical components may be subjected to precise mechanical or thermal analysis. Optimization methods are applied to components and groups of components to maximize performance. A final choice of geometry and material is made and the methods of production are analyzed and costed. The stage ends with a detailed production specification.

All that sounds well and good. If only it were so simple. The linear process suggested by Figure 2.1 obscures the strong coupling between the three stages. The consequences of choices made at the concept or embodiment stages may not become apparent until the detail is examined. Iteration, looping back to explore alternatives, is an essential part of the design process. Think of each of the many possible choices that *could* be made as an array of blobs in design space as suggested by Figure 2.4. Here C1, C2,...are possible concepts, and E1, E2,..., and D1, D2,... are possible embodiments and detailed

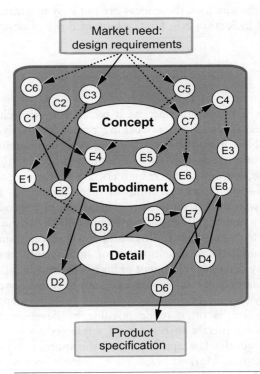

Figure 2.4   The previous figure suggests that the design process is logical and linear. The reality is otherwise. Here the C-blobs represent possible concepts, the E-blobs possible embodiments of the Cs, and the D-blobs possible detailed realizations of the Es. The process is complete when a compatible path form "Need" to "Specification" can be identified. The extreme coupling between the idealized design "stages" leads to a devious path (the full line) and many dead-ends (the broken lines). This creates the need for tools that allow fluid access to materials information at differing levels of breadth and detail.

elaborations of them. Then the design process becomes one of creating paths, linking compatible blobs, until a connection is made from the top ("market need") to the bottom ("product specification"). The trial paths have dead-ends, and they loop back. It is like finding a track across difficult terrain — it may be necessary to go back many times if, in the end, we are to go forward. Once a path is found, it is always possible to make it look linear and logical (and many books do this), but the reality is more like Figure 2.4, not Figure 2.1. Thus a key part of design, and of selecting materials for it, is *flexibility*, the ability to explore alternatives quickly, keeping the big picture as well as the details in focus. Our focus in later chapters is on the selection of materials and processes, where exactly the same need arises. This requires simple mappings of the "kingdoms" of materials and processes that allow quick surveys of alternatives while still providing detail when it is needed. The selection charts of Chapter 4 and the methods of Chapter 5 help do this.

Described in the abstract, these ideas are not easy to grasp. An example will help — it comes in Section 2.6. First, a look at types of design.

2.3 Types of design

It is not always necessary to start, as it were, from scratch. *Original design* does: it involves a new idea or working principle (the ball-point pen, the compact disc). New materials can offer new, unique combinations of properties that enable original design. Thus high-purity silicon enabled the transistor; high-purity glass, the optical fiber; high coercive-force magnets, the miniature earphone, solid-state lasers the compact disc. Sometimes the new material suggests the new product; sometimes instead the new product demands the development of a new material: nuclear technology drove the development of a series of new zirconium-based alloys and low-carbon stainless steels; space technology stimulated the development of light-weight composites; turbine technology today drives development of high-temperature alloys and ceramics.

Adaptive or *developmental design* takes an existing concept and seeks an incremental advance in performance through a refinement of the working principle. This, too, is often made possible by developments in materials: polymers replacing metals in household appliances; carbon fiber replacing wood in sports goods. The appliance and the sports-goods market are both large and competitive. Markets here have frequently been won (and lost) by the way in which the manufacturer has adapted the product by exploiting new materials.

Variant design involves a change of scale or dimension or detailing without change of function or the method of achieving it: the scaling up of boilers, or of pressure vessels, or of turbines, for instance. Change of scale or circumstances of use may require change of material: small boats are made of fiberglass, large ships are made of steel; small boilers are made of copper, large ones of steel; subsonic planes are made of one alloy, supersonic of another; and for good reasons, detailed in later chapters.

2.4 Design tools and materials data

To implement the steps of Figure 2.1, use is made of *design tools*. They are shown as inputs, attached to the left of the main backbone of the design methodology in Figure 2.5. The tools enable the modeling and optimization of a design, easing the routine aspects of each phase. Function-modelers suggest viable function structures. Configuration optimizers suggest or refine shapes. Geometric and 3D solid modeling packages allow visualization and create files that can be down-loaded to numerically controlled prototyping and manufacturing systems. Optimization, DFM, DFA,[1] and cost-estimation

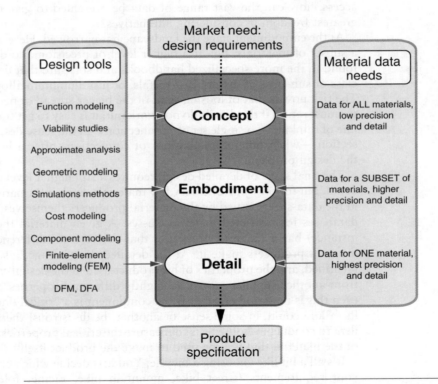

Figure 2.5   The design flow chart, showing how design tools and materials selection enter the procedure. Information about materials is needed at each stage, but at very different levels of breadth and precision.

[1] Design for Manufacture and Design for Assembly.

software allows manufacturing aspects to be refined. Finite element (FE) and Computational Fluid Dynamics (CFD) packages allow precise mechanical and thermal analysis even when the geometry is complex and the deformations are large. There is a natural progression in the use of the tools as the design evolves: approximate analysis and modeling at the conceptual stage; more sophisticated modeling and optimization at the embodiment stage; and precise ("exact"—but nothing is ever that) analysis at the detailed design stage.

Materials selection enters each stage of the design. The nature of the data needed in the early stages differs greatly in its level of precision and breadth from that needed later on (Figure 2.5, right-hand side). At the concept-stage, the designer requires approximate property-values, but for the widest possible range of materials. All options are open: a polymer may be the best choice for one concept, a metal for another, even though the function is the same. The problem, at this stage, is not precision and detail; it is breadth and speed of access: how can the vast range of data be presented to give the designer the greatest freedom in considering alternatives?

At the embodiment stage the landscape has narrowed. Here we need data for a subset of materials, but at a higher level of precision and detail. These are found in the more specialized handbooks and software that deal with a single class or sub-class of materials—metals, or just aluminum alloys, for instance. The risk now is that of loosing sight of the bigger spread of materials to which we must return if the details do not work out; it is easy to get trapped in a single line of thinking—a single set of "connections" in the sense described in the last section—when other combinations of connections offer a better solution to the design problem.

The final stage of detailed design requires a still higher level of precision and detail, but for only one or a very few materials. Such information is best found in the data-sheets issued by the material producers themselves, and in detailed databases for restricted material classes. A given material (polyethylene, for instance) has a range of properties that derive from differences in the ways different producers make it. At the detailed design stage, a supplier must be identified, and the properties of his product used in the design calculations; that from another supplier may have slightly different properties. And sometimes even this is not good enough. If the component is a critical one (meaning that its failure could, in some sense or another, be disastrous) then it may be prudent to conduct in-house tests to measure the critical properties, using a sample of the material that will be used to make the product itself.

It's all a bit like choosing a bicycle. You first decide which concept best suits your requirements (street bike, mountain bike, racing, folding, shopping, reclining, . . .), limiting the choice to one subset. Then comes the next level of detail. What frame material? What gears? Which sort of brakes? What shape of handlebars? At this point you consider the trade-off between performance and cost, identifying (usually with some compromise) a small subset that meet both your desires and your budget. Finally, if your bicycle is important to you, you seek further information in bike magazines, manufacturers' literature or the views of enthusiasts, and try out candidate-bikes yourself. And if you do not like them you go back one or more steps. Only when a match between your need and an available product is found do you make a final selection.

The materials input does not end with the establishment of production. Products fail in service, and failures contain information. It is an imprudent manufacturer who does not collect and analyze data on failures. Often this points to the misuse of a material, one that redesign or re-selection can eliminate.

2.5 Function, material, shape, and process

The selection of a material and process cannot be separated from the choice of shape. We use the word "shape" to include the external, *macro-shape*, and — when necessary — the internal, or *micro-shape*, as in a honeycomb or cellular structure. To make the shape, the material is subjected to processes that, collectively, we shall call *manufacture*: they include primary forming processes (like casting and forging), material removal processes (machining, drilling), finishing processes (such as polishing) and joining processes (e.g. welding). Function, material, shape and process interact (Figure 2.6). Function dictates the choice of both material and shape. Process is influenced by the material: by its formability, machinability, weldability, heat-treatability, and so on. Process obviously interacts with shape — the process determines the shape, the size, the precision and, of course, the cost. The interactions are two-way: specification of shape restricts the choice of material and process; but equally the

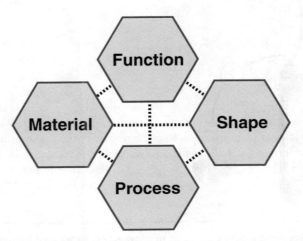

Figure 2.6   The central problem of materials selection in mechanical design: the interaction between function, material, shape and process.

specification of process limits the materials you can use and the shapes they can take. The more sophisticated the design, the tighter the specifications and the greater the interactions. It is like making wine: to make cooking wine, almost any grape and fermentation process will do; to make champagne, both grape and process must be tightly constrained.

The interaction between function, material, shape, and process lies at the heart of the material selection process. But first, a case study to illustrate the design process.

2.6 Case study: devices to open corked bottles

Wine, like cheese, is one of man's improvements on nature. And ever since man has cared about wine, he has cared about cork to keep it safely sealed in flasks and bottles. "Corticum...demovebit amphorae..." — "Uncork the amphora..." sang Horace[2] (27 BC) to celebrate the anniversary of his miraculous escape from death by a falling tree. But how did he do it?

A corked bottle creates a market need: it is the need to gain access to the wine inside. We might state it thus: "A device is required to pull corks from wine bottles." But hold on. The need must be expressed in solution-neutral form, and this is not. The aim is to gain access to the wine; our statement implies that this will be done by removing the cork, and that it will be removed by pulling. There could be other ways. So we will try again: "A device is required to allow access to wine in a corked bottle" (Figure 2.7) and one might add, "with convenience, at modest cost, and without contaminating the wine."

Figure 2.7   The market need: a device is sought to allow access to wine contained in a corked bottle.

---

[2] Horace, Q. 27 BC, *Odes*, Book III, Ode 8, line 10.

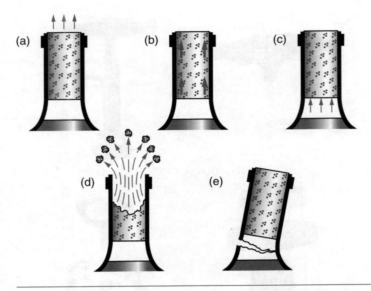

Figure 2.8   Five possible concepts, illustrating physical principles, to fill the need expressed by Figure 2.7.

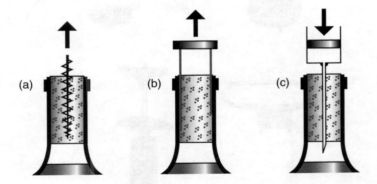

Figure 2.9   Working principles for implementing the first three schemes of Figure 2.8.

Five concepts for doing this are shown in Figure 2.8. In order, they are to remove the cork by axial traction (= pulling); to remove it by shear tractions; to push it out from below; to pulverizing it; and to by-pass it altogether — by knocking the neck off the bottle[3] perhaps.

---

[3] A Victorian invention for opening old port, the cork of which may become brittle with age and alcohol-absorption, involved ring-shaped tongs. The tongs were heated red on an open fire, then clamped onto the cold neck of the bottle. The thermal shock removed the neck cleanly and neatly.

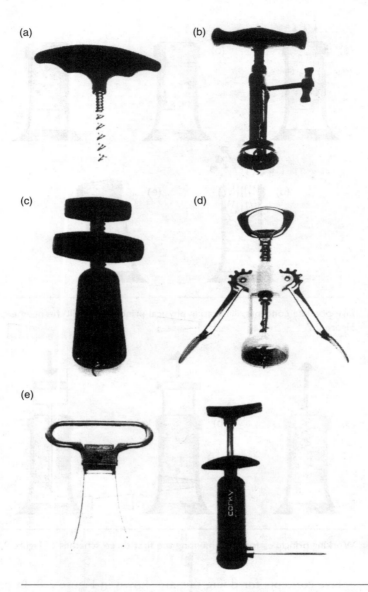

Figure 2.10 Cork removers that employ the working principles of Figure 2.9: (a) direct pull; (b) gear lever, screw-assisted pull; (c) spring-assisted pull (a spring in the body is compressed as the screw is driven into the cork; (d) shear blade systems; (e) pressure-induced removal systems.

Numerous devices exist to achieve the first three of these. The others are used too, though generally only in moments of desperation. We shall eliminate these on the grounds that they might contaminate the wine, and examine the others more closely, exploring working principles. Figure 2.9 shows one for each of

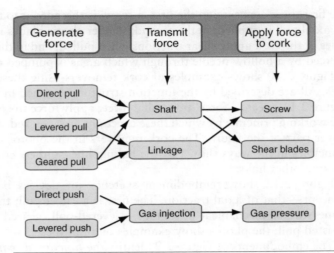

Figure 2.11 The function structure and working principles of cork removers.

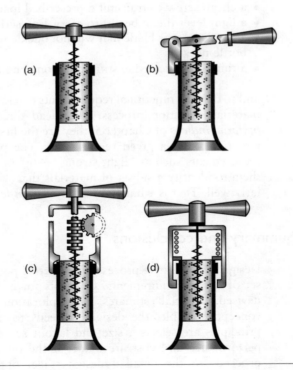

Figure 2.12 Embodiment sketches for four concepts: direct pull, levered pull, geared pull and spring-assisted pull. Each system is made up of components that perform a sub-function. The requirements of these sub-functions are the inputs to the materials selection method.

the first three concepts: in the first, a screw is threaded into the cork to which an axial pull is applied; in the second, slender elastic blades inserted down the sides of the cork apply shear tractions when pulled; and in the third the cork is pierced by a hollow needle through which a gas is pumped to push it out.

Figure 2.10 shows examples of cork removers using these working principles. All are described by the function-structure sketched in the upper part of Figure 2.11: create a force, transmit a force, apply force to cork. They differ in the working principle by which these functions are achieved, as indicated in the lower part of the figure. The cork removers in the photos combine working principles in the ways shown by the linking lines. Others could be devised by making other links.

Figure 2.12 shows embodiment sketches for devices based on just one concept — that of axial traction. The first is a direct pull; the other three use some sort of mechanical advantage — levered-pull, geared pull and spring-assisted pull; the photos show examples of all of these.

The embodiments of Figure 2.9 identify the *functional requirements* of each component of the device, which might be expressed in statements like:

- a cheap screw to transmit a prescribed load to the cork;
- a light lever (i.e. a beam) to carry a prescribed bending moment;
- a slender elastic blade that will not buckle when driven between the cork and bottle-neck;
- a thin, hollow needle, stiff and strong enough to penetrate a cork;

and so on. The functional requirements of each component are the inputs to the materials selection process. They lead directly to the *property limits* and *material indices* of Chapter 5: they are the first step in optimizing the choice of material to fill a given requirement. The procedure developed there takes requirements such as "light strong beam" or "slender elastic blade" and uses them to identify a subset of materials that will perform this function particularly well. That is what is meant by *design-led materials selection*.

2.7 Summary and conclusions

Design is an iterative process. The starting point is a market need captured in a set of design requirements. Concepts for a products that meet the need are devised. If initial estimates and exploration of alternatives suggest that the concept is viable, the design proceeds to the embodiment stage: working principles are selected, size and layout are decided, and initial estimates of performance and cost are made. If the outcome is successful, the designer proceeds to the detailed design stage: optimization of performance, full analysis of critical components, preparation of detailed production drawings (usually as a CAD file), specification of tolerance, precision, joining and finishing methods, and so forth.

Materials selection enters at each stage, but at different levels of breadth and precision. At the conceptual stage all materials and processes are potential candidates, requiring a procedure that allows rapid access to data for a wide range of each, though without the need for great precision. The preliminary selection passes to the embodiment stage, the calculations and optimizations of which require information at a higher level of precision and detail. They eliminate all but a small short-list candidate-materials and processes for the final, detailed stage of the design. For these few, data of the highest quality are necessary.

Data exist at all these levels. Each level requires its own data-management scheme, described in the following chapters. The management is the skill: it must be design-led, yet must recognize the richness of choice and embrace the complex interaction between the material, its shape, the process by which it is given that shape, and the function it is required to perform. And it must allow rapid iteration — back-looping when a particular chain of reasoning proves to be unprofitable. Tools now exist to help with all of this. We will meet one — the CES materials and process selection platform—later in this book.

But given this complexity, why not opt for the safe bet: stick to what you (or others) used before? Many have chosen that option. Few are still in business.

2.8 Further reading

A chasm exists between books on design methodology and those on materials selection: each largely ignores the other. The book by French is remarkable for its insights, but the word 'material' does not appear in its index. Pahl and Beitz has near-biblical standing in the design camp, but is heavy going. Ullman and Cross take a more relaxed approach and are easier to digest. The books by Budinski and Budinski, by Charles, Crane and Furness and by Farag present the materials case well, but are less good on design. Lewis illustrates material selection through case studies, but does not develop a systematic procedure. The best compromise, perhaps, is Dieter.

General texts on design methodology

Cross, N. (2000) *Engineering Design Methods*, 3rd edition, Wiley, Chichester, UK. ISBN 0-471-87250-4. *(A durable text describing the design process, with emphasis on developing and evaluating alternative solutions.)*

French, M.J. (1985) *Conceptual Design for Engineers*, The Design Council, London, UK, and Springer, Berlin, Germany. ISBN 0-85072-155-5 and 3-540-15175-3. *(The origin of the "Concept — Embodiment — Detail" block diagram of the design process. The book focuses on the concept stage, demonstrating how simple physical principles guide the development of solutions to design problems.)*

Pahl, G. and Beitz, W. (1997) *Engineering Design*, 2nd edition, translated by K. Wallace and L. Blessing, The Design Council, London, UK and Springer-Verlag, Berlin,

Germany. ISBN 0-85072-124-5 and 3-540-13601-0. *(The Bible — or perhaps more exactly the Old Testament — of the technical design field, developing formal methods in the rigorous German tradition.)*

Ullman, D.G. (1992) *The Mechanical Design Process*, McGraw-Hill, New York, USA. ISBN 0-07-065739-4. *(An American view of design, developing ways in which an initially ill-defined problem is tackled in a series of steps, much in the way suggested by Figure 2.1 of the present text.)*

Ulrich, K.T. and Eppinger, S.D. (1995) *Product Design and Development*, McGraw-Hill, New York, USA. ISBN 0-07-065811-0. *(A readable, comprehensible text on product design, as taught at MIT. Many helpful examples but almost no mention of materials.)*

General texts on materials selection in design

Budinski, K.G. and Budinski, M.K. (1999) *Engineering Materials, Properties and Selection* 6th edition, Prentice-Hall, Englewood Cliffs, NJ, USA. ISBN 0-13-904715-8. *(A well-established materials text that deals well with both material properties and processes.)*

Charles, J.A., Crane, F.A.A. and Furness, J.A.G. (1997) *Selection and Use of Engineering Materials*, 3rd edition, Butterworth-Heinemann Oxford, UK. ISBN 0-7506-3277-1. *(A materials-science, rather than a design-led, approach to the selection of materials.)*

Dieter, G.E. (1991) *Engineering Design, a Materials and Processing Approach*, 2nd edition, McGraw-Hill, New York, USA. ISBN 0-07-100829-2. *(A well-balanced and respected text focusing on the place of materials and processing in technical design.)*

Farag, M.M. (1989) *Selection of Materials and Manufacturing Processes for Engineering Design*, Prentice-Hall, Englewood Cliffs, NJ, USA. ISBN 0-13-575192-6. *(Like Charles, Crane and Furness, this is Materials-Science approach to the selection of materials.)*

Lewis, G. (1990) *Selection of Engineering Materials*, Prentice-Hall, Englewood Cliffs, N.J., USA. ISBN 0-13-802190-2. *(A text on materials selection for technical design, based largely on case studies.)*

And on corks and corkscrews

McKearin, H. (1973) "On 'stopping', bottling and binning", *International Bottler and Packer*, April issue, pp 47–54.

Perry, E. (1980) *Corkscrews and Bottle Openers*, Shire Publications Ltd, Aylesbury, UK.

The Design Council (1994) Teaching aids program EDTAP DE9, The Design Council, 28 Haymarket, London SW1Y 4SU, UK.

Watney, B.M. and Babbige, H.D. (1981) *Corkscrews*. Sotheby's Publications, London, UK.

Chapter 3

Engineering materials and their properties

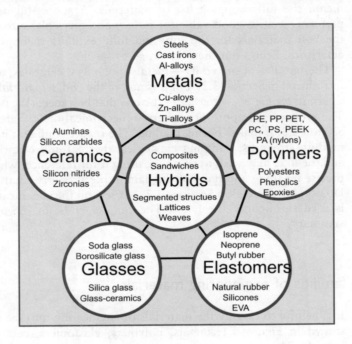

3.1   Introduction and synopsis

Materials, one might say, are the food of design. This chapter presents the menu: the full shopping list of materials. A successful product — one that performs well, is good value for money and gives pleasure to the user — uses the best materials for the job, and fully exploits their potential and characteristics. Brings out their flavor, so to speak.

The families of materials — metals, polymers, ceramics, and so forth — are introduced in Section 3.2. But it is not, in the end, a *material* that we seek; it is a certain *profile of properties* — the one that best meets the needs of the design. The properties, important in thermo-mechanical design, are defined briefly in Section 3.3. It makes boring reading. The reader confident in the definitions of moduli, strengths, damping capacities, thermal and electrical conductivities and the like, may wish to skip this, using it for reference, when needed, for the precise meaning and units of the data in the Property Charts that come later. Do not, however, skip Sections 3.2 — it sets up the classification structure that is used throughout the book. The chapter ends, in the usual way, with a summary.

3.2   The families of engineering materials

It is helpful to classify the materials of engineering into the six broad families shown in Figure 3.1: metals, polymers, elastomers, ceramics, glasses, and hybrids. The members of a family have certain features in common: similar properties, similar processing routes, and, often, similar applications.

Metals have relatively high moduli. Most, when pure, are soft and easily deformed. They can be made strong by alloying and by mechanical and heat treatment, but they remain ductile, allowing them to be formed by deformation processes. Certain high-strength alloys (spring steel, for instance) have ductilities as low as 1 percent, but even this is enough to ensure that the material yields before it fractures and that fracture, when it occurs, is of a tough, ductile type. Partly because of their ductility, metals are prey to fatigue and of all the classes of material, they are the least resistant to corrosion.

Ceramics too, have high moduli, but, unlike metals, they are brittle. Their "strength" in tension means the brittle fracture strength; in compression it is the brittle crushing strength, which is about 15 times larger. And because ceramics have no ductility, they have a low tolerance for stress concentrations (like holes or cracks) or for high-contact stresses (at clamping points, for instance). Ductile materials accommodate stress concentrations by deforming in a way that redistributes the load more evenly, and because of this, they can be used under static loads within a small margin of their yield strength. Ceramics cannot. Brittle materials always have a wide scatter in strength and

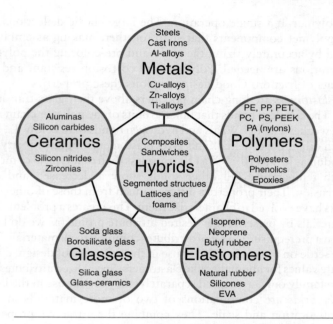

Figure 3.1   The menu of engineering materials. The basic families of metals, ceramics, glasses, polymers, and elastomers can be combined in various geometries to create hybrids.

the strength itself depends on the volume of material under load and the time for which it is applied. So ceramics are not as easy to design with as metals. Despite this, they have attractive features. They are stiff, hard, and abrasion-resistant (hence their use for bearings and cutting tools); they retain their strength to high temperatures; and they resist corrosion well.

Glasses are non-crystalline ("amorphous") solids. The commonest are the soda-lime and boro-silicate glasses familiar as bottles and ovenware, but there are many more. Metals, too, can be made non-crystalline by cooling them sufficiently quickly. The lack of crystal structure suppresses plasticity, so, like ceramics, glasses are hard, brittle and vulnerable to stress concentrations.

Polymers are at the other end of the spectrum. They have moduli that are low, roughly 50 times less than those of metals, but they can be strong — nearly as strong as metals. A consequence of this is that elastic deflections can be large. They creep, even at room temperature, meaning that a polymer component under load may, with time, acquire a permanent set. And their properties depend on temperature so that a polymer that is tough and flexible at 20°C may be brittle at the 4°C of a household refrigerator, yet creep rapidly at the 100°C of boiling water. Few have useful strength above 200°C. If these aspects are allowed for in the design, the advantages of polymers can be exploited. And there are many. When combinations of properties, such as strength-per-unit-weight, are important, polymers are as good as metals. They are easy to shape: complicated parts performing several functions can be molded from a polymer in a single operation. The large elastic deflections allow the design of polymer components that snap together, making assembly fast and cheap. And by accurately sizing the mold and pre-coloring the polymer, no finishing operations are needed. Polymers are corrosion resistant and have low coefficients of friction. Good design exploits these properties.

Elastomers are long-chain polymers above their glass-transition temperature, T_g. The covalent bonds that link the units of the polymer chain remain intact, but the weaker Van der Waals and hydrogen bonds that, below T_g, bind the chains to each other, have melted. This gives elastomers unique property profiles: Young's moduli as low as 10^{-3} GPa (10^5 time less than that typical of metals) that increase with temperature (all other solids show a decrease), and enormous elastic extension. Their properties differ so much from those of other solids that special tests have evolved to characterize them. This creates a problem: if we wish to select materials by prescribing a desired attribute profile (as we do later in this book), then a prerequisite is a set of attributes common to all materials. To overcome this, we settle on a common set for use in the first stage of design, estimating approximate values for anomalies like elastomers. Specialized attributes, representative of one family only, are stored separately; they are for use in the later stages.

Hybrids are combinations of two or more materials in a pre-determined configuration and scale. They combine the attractive properties of the other families of materials while avoiding some of their drawbacks. Their design is the subject of Chapters 13 and 14. The family of hybrids includes fiber and particulate composites, sandwich structures, lattice structures, foams, cables, and laminates. And almost all the materials of nature — wood, bone, skin, leaf — are hybrids. Fiber-reinforced composites are, of course, the most familiar. Most of those at present available to the engineer have a polymer matrix reinforced by fibers of glass, carbon or Kevlar (an aramid). They are light, stiff and strong, and they can be tough. They, and other hybrids using a polymer as one component, cannot be used above 250°C because the polymer softens, but at room temperature their performance can be outstanding. Hybrid components are expensive and they are relatively difficult to form and join. So despite their attractive properties the designer will use them only when the added performance justifies the added cost. Today's growing emphasis on high performance and fuel efficiency provides increasing drivers for their use.

3.3 The definitions of material properties

Each material can be thought of as having a set of attributes: its properties. It is not a material, *per se*, that the designer seeks; it is a specific combination of these attributes: a *property-profile*. The material name is the identifier for a particular property-profile.

The properties themselves are standard: density, modulus, strength, toughness, thermal and electrical conductivities, and so on (Tables 3.1). For

Table 3.1 Basic design-limiting material properties and their usual SI units*

Class	Property	Symbol and units	
General	Density	ρ	(kg/m^3 or Mg/m^3)
	Price	C_m	($/kg)
Mechanical	Elastic moduli (Young's, shear, bulk)	E, G, K	(GPa)
	Yield strength	σ_y	(MPa)
	Ultimate strength	σ_u	(MPa)
	Compressive strength	σ_c	(MPa)
	Failure strength	σ_f	(MPa)
	Hardness	H	(Vickers)
	Elongation	ε	(–)
	Fatigue endurance limit	σ_e	(MPa)
	Fracture toughness	K_{1C}	(MPa.m$^{1/2}$)
	Toughness	G_{1C}	(kJ/m^2)
	Loss coefficient (damping capacity)	η	(–)
Thermal	Melting point	T_m	(C or K)
	Glass temperature	T_g	(C or K)
	Maximum service temperature	T_{max}	(C or K)
	Minimum service temperature	T_{max}	(C or K)
	Thermal conductivity	λ	(W/m.K)
	Specific heat	C_p	(J/kg.K)
	Thermal expansion coefficient	α	(K^{-1})
	Thermal shock resistance	ΔT_s	(C or K)
Electrical	Electrical resistivity	ρ_e	(Ω.m or $\mu\Omega$.cm)
	Dielectric constant	ε_d	(–)
	Breakdown potential	V_b	(10^6 V/m)
	Power factor	P	(–)
Optical	Optical, transparent, translucent, opaque	Yes/No	
	Refractive index	n	(–)
Eco-properties	Energy/kg to extract material	E_f	(MJ/kg)
	CO$_2$/kg to extract material	CO_2	(kg/kg)
Environmental resistance	Oxidation rates	Very low, low, average, high, very high	
	Corrosion rates		
	Wear rate constant	K_A	MPa^{-1}

* Conversion factors to imperial and cgs units appear inside the back and front covers of this book.

completeness and precision, they are defined, with their limits, in this section. If you think you know how properties are defined, you might jump to Section 3.5, returning to this section only if need arises.

General properties

The *density* (units: kg/m^3) is the mass per unit volume. We measure it today as Archimedes did: by weighing in air and in a fluid of known density.

The *price*, C_m (units: $/kg), of materials spans a wide range. Some cost as little as $0.2/kg, others as much as $1000/kg. Prices, of course, fluctuate, and they depend on the quantity you want and on your status as a "preferred customer" or otherwise. Despite this uncertainty, it is useful to have an approximate price, useful in the early stages of selection.

Mechanical properties

The *elastic modulus* (units: GPa or GN/m^2) is defined as the slope of the linear-elastic part of the stress–strain curve (Figure 3.2). Young's modulus, E, describes response to tensile or compressive loading, the shear modulus, G, describes shear loading and the bulk modulus, K, hydrostatic pressure. Poisson's ratio, ν, is dimensionless: it is the negative of the ratio of the lateral strain, ε_2, to the axial strain, ε_1, in axial loading:

$$\nu = -\frac{\varepsilon_2}{\varepsilon_1}$$

In reality, moduli measured as slopes of stress–strain curves are inaccurate, often low by a factor of 2 or more, because of contributions to the strain from

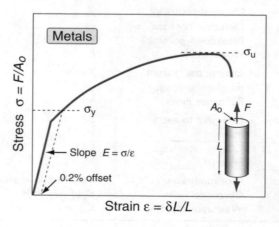

Figure 3.2   The stress–strain curve for a metal, showing the modulus, E, the 0.2 percent yield strength, σ_y, and the ultimate strength, σ_u.

anelasticity, creep and other factors. Accurate moduli are measured dynamically: by exciting the natural vibrations of a beam or wire, or by measuring the velocity of sound waves in the material.

In an isotropic material, the moduli are related in the following ways:

$$E = \frac{3G}{1 + G/3K}; \quad G = \frac{E}{2(1 + \nu)}; \quad K = \frac{E}{3(1 - 2\nu)} \tag{3.1}$$

Commonly $\nu \approx 1/3$ when

$$G \approx \frac{3}{8}E \text{ and } K \approx E \tag{3.2a}$$

Elastomers are exceptional. For these $\nu \approx 1/2$ when

$$G \approx \frac{1}{3}E \text{ and } K \gg E \tag{3.2b}$$

Data sources like those described in Chapter 15 list values for all four moduli. In this book we examine data for E; approximate values for the others can be derived from equation (3.2) when needed.

The *strength* σ_f, of a solid (units: MPa or MN/m^2) requires careful definition. For metals, we identify σ_f with the 0.2 percent offset yield strength σ_y (Figure 3.2), that is, the stress at which the stress–strain curve for axial loading deviates by a strain of 0.2 percent from the linear-elastic line. It is the same in tension and compression. For polymers, σ_f is identified as the stress at which the stress–strain curve becomes markedly non-linear: typically, a strain of 1 percent (Figure 3.3). This may be caused by shear-yielding: the irreversible slipping of molecular chains; or it may be caused by crazing: the formation of low density, crack-like volumes that scatter light, making the polymer look white. Polymers

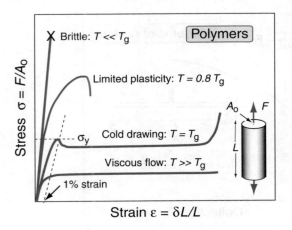

Figure 3.3   Stress–strain curves for a polymer, below, at and above its glass transition temperature, T_g.

are a little stronger (≈ 20 percent) in compression than in tension. Strength, for ceramics and glasses, depends strongly on the mode of loading (Figure 3.4). In tension, "strength" means the fracture strength, σ_t. In compression it means the crushing strength σ_c, which is much larger; typically

$$\sigma_c = 10 \text{ to } 15 \, \sigma_t \tag{3.3}$$

When the material is difficult to grip (as is a ceramic), its strength can be measured in bending. The *modulus of rupture* or MoR (units: MPa) is the maximum surface stress in a bent beam at the instant of failure (Figure 3.5).

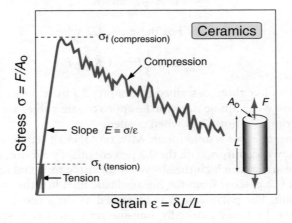

Figure 3.4   Stress–strain curves for a ceramic in tension and in compression. The compressive strength σ_c is 10 to 15 times greater than the tensile strength σ_t.

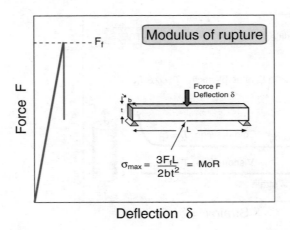

Figure 3.5   The MoR is the surface stress at failure in bending. It is equal to, or slightly larger than the failure stress in tension.

One might expect this to be the same as the strength measured in tension, but for ceramics it is larger (by a factor of about 1.3) because the volume subjected to this maximum stress is small and the probability of a large flaw lying in it is small also; in simple tension all flaws see the maximum stress.

The strength of a composite is best defined by a set deviation from linear-elastic behavior: 0.5 percent is sometimes taken. Composites that contain fibers (and this includes natural composites like wood) are a little weaker (up to 30 percent) in compression than tension because fibers buckle. In subsequent chapters, σ_f for composites means the tensile strength.

Strength, then, depends on material class and on mode of loading. Other modes of loading are possible: shear, for instance. Yield under multi-axial loads is related to that in simple tension by a yield function. For metals, the Von Mises' yield function is a good description:

$$(\sigma_1 - \sigma_2)^2 + (\sigma_2 - \sigma_3)^2 + (\sigma_3 - \sigma_1)^2 = 2\sigma_f^2 \tag{3.4}$$

where σ_1, σ_2, and σ_3 are the principal stresses, positive when tensile; σ_1, by convention, is the largest or most positive, σ_3 the smallest or least. For polymers the yield function is modified to include the effect of pressure:

$$(\sigma_1 - \sigma_2)^2 + (\sigma_2 - \sigma_3)^2 + (\sigma_3 - \sigma_1)^2 = 2\sigma_f^2\left(1 + \frac{\beta p}{K}\right)^2 \tag{3.5}$$

where K is the bulk modulus of the polymer, $\beta \approx 2$ is a numerical coefficient that characterizes the pressure dependence of the flow strength and the pressure p is defined by

$$p = -\frac{1}{3}(\sigma_1 + \sigma_2 + \sigma_3)$$

For ceramics, a Coulomb flow law is used:

$$\sigma_1 - B\sigma_2 = C \tag{3.6}$$

where B and C are constants.

The *ultimate (tensile) strength*, σ_u (units: MPa), is the nominal stress at which a round bar of the material, loaded in tension, separates (see Figure 3.2). For brittle solids — ceramics, glasses, and brittle polymers — it is the same as the failure strength in tension. For metals, ductile polymers and most composites, it is larger than the strength, σ_f, by a factor of between 1.1 and 3 because of work hardening or (in the case of composites) load transfer to the reinforcement.

Cyclic loading not only dissipates energy; it can also cause a crack to nucleate and grow, culminating in fatigue failure. For many materials there exists a fatigue or *endurance limit*, σ_e (units: MPa), illustrated by the $\Delta\sigma - N_f$ curve of Figure 3.6. It is the stress amplitude $\Delta\sigma$ below which fracture does not occur, or occurs only after a very large number ($N_f > 10^7$) of cycles.

The *hardness*, H, of a material is a crude measure of its strength. It is measured by pressing a pointed diamond or hardened steel ball into the surface of the material (Figure 3.7). The hardness is defined as the indenter force

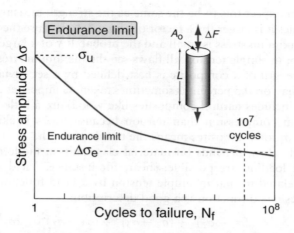

Figure 3.6 The endurance limit, $\Delta\sigma_e$, is the cyclic stress that causes failure in $N_f = 10^7$ cycles.

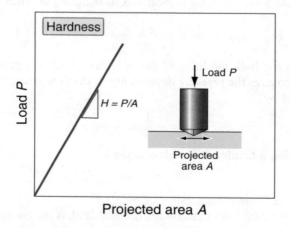

Figure 3.7 Hardness is measured as the load P divided by the projected area of contact, A, when a diamond-shaped indenter is forced into the surface.

divided by the projected area of the indent. It is related to the quantity we have defined as σ_f by

$$H \approx 3\sigma_f \tag{3.7}$$

and this, in the SI system, has units of MPa. Hardness is most usually reported in other units, the commonest of which is the Vickers H_v scale with units of kg/mm^2. It is related to H in the units used here by

$$H_v = \frac{H}{10}$$

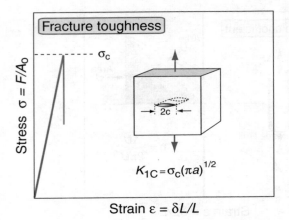

Figure 3.8   The fracture toughness, K_{IC}, measures the resistance to the propagation of a crack. The failure strength of a brittle solid containing a crack of length 2c is $K_{\text{IC}} = Y(\sigma_c / \sqrt{\pi c})$ where Y is a constant near unity.

The *toughness, G_{1C}, (units: kJ/m^2), and the fracture toughness, K_{1C}, (units: $MPa.m^{1/2}$ or $MN/m^{1/2}$), measure the resistance of a material to the propagation of a crack. The fracture toughness is measured by loading a sample containing a deliberately-introduced crack of length 2c (Figure 3.8), recording the tensile stress σ_c at which the crack propagates. The quantity K_{1C} is then calculated from*

$$K_{1C} = Y\sigma_c\sqrt{\pi c} \tag{3.8}$$

and the toughness from

$$G_{1C} = \frac{K_{1C}^2}{E(1+\nu)} \tag{3.9}$$

where Y is a geometric factor, near unity, that depends on details of the sample geometry, E is Young's modulus and ν is Poisson's ratio. Measured in this way K_{1C} and G_{1C} have well-defined values for brittle materials (ceramics, glasses, and many polymers). In ductile materials a plastic zone develops at the crack tip, introducing new features into the way in which cracks propagate that necessitate more involved characterization. Values for K_{1C} and G_{1C} are, nonetheless, cited, and are useful as a way of ranking materials.

The *loss-coefficient, η* (a dimensionless quantity), measures the degree to which a material dissipates vibrational energy (Figure 3.9). If a material is loaded elastically to a stress, σ_{max}, it stores an elastic energy

$$U = \int_0^{\sigma_{\text{max}}} \sigma \, d\varepsilon \approx \frac{1}{2}\frac{\sigma_{\text{max}}^2}{E}$$

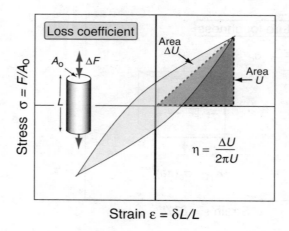

Figure 3.9 The loss coefficient η measures the fractional energy dissipated in a stress–strain cycle.

per unit volume. If it is loaded and then unloaded, it dissipates an energy

$$\Delta U = \oint \sigma \, d\varepsilon$$

The loss coefficient is

$$\eta = \frac{\Delta U}{2\pi U} \tag{3.10}$$

The value of η usually depends on the time-scale or frequency of cycling.

Other measures of damping include the *specific damping capacity*, $D = \Delta U/U$, the *log decrement*, Δ (the log of the ratio of successive amplitudes of natural vibrations), the *phase-lag*, δ, between stress and strain, and the *"Q"-factor* or *resonance factor*, Q. When damping is small ($\eta < 0.01$) these measures are related by

$$\eta = \frac{D}{2\pi} = \frac{\Delta}{\pi} = \tan \delta = \frac{1}{Q} \tag{3.11}$$

but when damping is large, they are no longer equivalent.

Thermal properties

Two temperatures, the *melting temperature*, T_m, and the *glass temperature*, T_g (units for both: K or C) are fundamental because they relate directly to the strength of the bonds in the solid. Crystalline solids have a sharp melting point, T_m. Non-crystalline solids do not; the temperature T_g characterizes the transition from true solid to very viscous liquid. It is helpful, in engineering

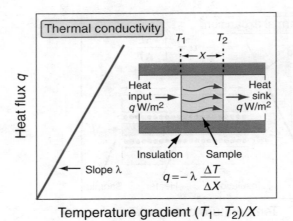

Figure 3.10   The thermal conductivity λ measures the flux of heat driven by a temperature gradient dT/dX.

design, to define two further temperatures: the *maximum* and *minimum service temperatures* T_{max} and T_{min} (both: K or C). The first tells us the highest temperature at which the material can reasonably be used without oxidation, chemical change, or excessive creep becoming a problem. The second is the temperature below which the material becomes brittle or otherwise unsafe to use.

The rate at which heat is conducted through a solid at steady state (meaning that the temperature profile does not change with time) is measured by the *thermal conductivity*, λ (units: W/m.K). Figure 3.10 shows how it is measured: by recording the heat flux q (W/m^2) flowing through the material from a surface at higher temperature T_1 to a lower one at T_2 separated by a distance X. The conductivity is calculated from Fourier's law:

$$q = -\lambda \frac{dT}{dX} = \lambda \frac{(T_1 - T_2)}{X} \qquad (3.12)$$

The measurement is not, in practice, easy (particularly for materials with low conductivities), but reliable data are now generally available.

When heat flow is transient, the flux depends instead on the *thermal diffusivity*, a (units: m^2/s), defined by

$$a = \frac{\lambda}{\rho C_p} \qquad (3.13)$$

where ρ is the density and C_p is the *specific heat at constant pressure* (units: J/kg.K). The thermal diffusivity can be measured directly by measuring the decay of a temperature pulse when a heat source, applied to the material, is switched off; or it can be calculated from λ, via equation (3.13). This requires

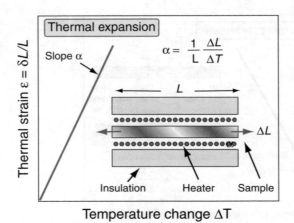

Figure 3.11   The linear-thermal expansion coefficient α measures the change in length, per unit length, when the sample is heated.

values for C_p. It is measured by the technique of calorimetry, which is also the standard way of measuring the glass temperature T_g.

Most materials expand when they are heated (Figure 3.11). The thermal strain per degree of temperature change is measured by the *linear thermal-expansion coefficient*, α (units: K^{-1} or, more conveniently, as "microstrain/C" or $10^{-6}\,C^{-1}$). If the material is thermally isotropic, the volume expansion, per degree, is 3α. If it is anisotropic, two or more coefficients are required, and the volume expansion becomes the sum of the principal thermal strains.

The *thermal shock resistance* ΔT_s (units: K or C) is the maximum temperature difference through which a material can be quenched suddenly without damage. It, and the *creep resistance*, are important in high-temperature design. Creep is the slow, time-dependent deformation that occurs when materials are loaded above about $\frac{1}{3}T_m$ or $\frac{2}{3}T_g$. Design against creep is a specialized subject. Here we rely instead on avoiding the use of a material above its maximum service temperature, T_{max}, or, for polymers, its "heat deflection temperature".

Electrical properties

The *electrical resistivity*, ρ_e (SI units $\Omega.m$, but commonly reported in units of $\mu\Omega.cm$) is the resistance of a unit cube with unit potential difference between a pair of it faces. It is measured in the way shown in Figure 3.12. It has an immense range, from a little more than 10^{-8} in units of $\Omega.m$ (equal to $1\,\mu\Omega.cm$) for good conductors to more than $10^{16}\,\Omega.m$ ($10^{24}\,\mu\Omega.cm$) for the best insulators. The electrical conductivity is simply the reciprocal of the resistivity.

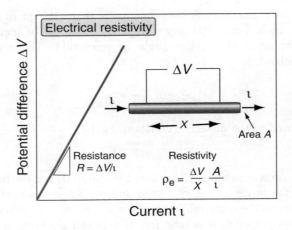

Figure 3.12   Electrical resistivity is measured as the potential gradient $\Delta V/X$ divided by the current density, ι/A.

When an insulator is placed in an electric field, it becomes polarized and charges appear on its surfaces that tend to screen the interior from the electric field. The tendency to polarize is measured by the *dielectric constant*, ε_d (a dimensionless quantity). Its value for free space and, for practical purposes, for most gasses, is 1. Most insulators have values between 2 and 30, though low-density foams approach the value 1 because they are largely air.

The *breakdown potential* (units: MV/m) is the electrical potential gradient at which an insulator breaks down and a damaging surge of current flows through it. It is measured by increasing, at a uniform rate, a 60 Hz alternating potential applied across the faces of a plate of the material until breakdown occurs.

Polarization in an electric field involves the motion of charge particles (electrons, ions, or molecules that carry a dipole moment). In an oscillating field, the charged particles are driven between two alternative configurations. This charge-motion corresponds to an electric current that — if there were no losses — would be 90° out of phase with the voltage. In real dielectrics, the motion of the charged particles dissipates energy and the current leads the voltage by something less that 90°; the loss angle θ is the deviation. The loss tangent is the tangent of this angle. The *power factor* (dimensionless) is the sine of the loss angle, and measures the fraction of the energy stored in the dielectric at peak voltage that is dissipated in a cycle; when small, it is equal to the loss tangent. The *loss factor* is the loss tangent times the dielectric constant.

Optical properties

All materials allow some passage of light, although for metals it is exceedingly small. The speed of light when in the material, v, is always less than that in vacuum, c. A consequence is that a beam of light striking the surface of such a material at an angle α, the angle of incidence, enters the material at an angle β, the angle of refraction. The *refractive index, n* (dimensionless), is

$$n = \frac{c}{\nu} = \frac{\sin \alpha}{\sin \beta} \qquad (3.14)$$

It is related to the dielectric constant, ε_d, by

$$n \approx \sqrt{\varepsilon_d}$$

It depends on wavelength. The denser the material, and the higher its dielectric constant, the larger is the refractive index. When $n = 1$, the entire incident intensity enters the material, but when $n > 1$, some is reflected. If the surface is smooth and polished, it is reflected as a beam; if rough, it is scattered. The percentage reflected, R, is related to the refractive index by

$$R = \left(\frac{n-1}{n+1}\right)^2 \times 100 \qquad (3.15)$$

As n increases, the value of R tends to 100 percent.

Eco properties

The *contained* or *production energy* (units MJ/kg) is the energy required to extract 1 kg of a material from its ores and feedstocks. The associated CO_2 production (units: kg/kg) is the mass of carbon dioxide released into the atmosphere during the production of 1 kg of material. These and other eco-attributes are the subject of Chapter 16.

Environmental resistance

Some material attributes are difficult to quantify, particularly those that involve the interaction of the material within the environments in which it must operate. Environmental resistance is conventionally characterized on a discrete 5-point scale: very good, good, average, poor, very poor. "Very good" means that the material is highly resistant to the environment, "very poor" that it is completely non-resistant or unstable. The categorization is designed to help with initial screening; supporting information should always be sought if environmental attack is a concern. Ways of doing this are described later.

Wear, like the other interactions, is a multi-body problem. None-the-less it can, to a degree, be quantified. When solids slide (Figure 3.13) the volume of material lost from one surface, per unit distance slid, is called the wear rate, W.

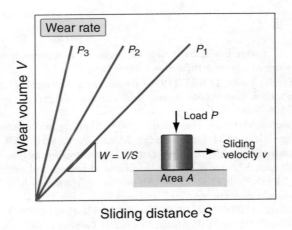

Figure 3.13 Wear is the loss of material from surfaces when they slide. The wear resistance is measured by the Archard wear constant K_A.

The wear resistance of the surface is characterized by the *Archard wear constant*, K_A (units: MPa^{-1}), defined by the equation

$$\frac{W}{A} = K_A P \qquad (3.16)$$

where A is the area of the surface and P the normal force pressing them together. Approximate data for K_A appear in Chapter 4, but must be interpreted as the property of the sliding couple, not of just one member of it.

3.4 Summary and conclusions

There are six important families of materials for mechanical design: metals, ceramics, glasses, polymers, elastomers, and hybrids that combine the properties of two or more of the others. Within a family there is certain common ground: ceramics as a family are hard, brittle, and corrosion resistant; metals are ductile, tough, and good thermal and electrical conductors; polymers are light, easily shaped, and electrical insulators, and so on — that is what makes the classification useful. But in design we wish to escape from the constraints of family, and think, instead, of the material name as an identifier for a certain property-profile — one that will, in later chapters, be compared with an "ideal" profile suggested by the design, guiding our choice. To that end, the properties important in thermo-mechanical design were defined in this chapter. In Chapter 4 we develop a way of displaying these properties so as to maximize the freedom of choice.

3.5 Further reading

Definitions of material properties can be found in numerous general texts on engineering materials, among them those listed here.

Ashby, M.F. and Jones, D.R.H. (1996) *Engineering Materials 1, and Introduction to their Properties and Applications*, 2nd edition, Pergamon Press, Oxford, U.K. ISBN 0–7506–3081–7.

ASM Engineered Materials Handbook (2004) "Testing and characterisation of polymeric materials", ASM International, Metals Park, OH, USA. (*An on-line, subscription-based resource, detailing testing procedures for polymers.*)

ASM Handbooks, Volume 8 (2004) "Mechanical testing and evaluation" ASM International, Metals Park, Ohio, USA. (*An on-line, subscription-based resource, detailing testing procedures for metals and ceramics.*)

ASTM Standards (1988) Vol. 08.01 and 08.02 Plastics; (1989) Vol. 04.02 Concrete; (1990) Vols. 01.01 to 01.05 Steels; Vol. 0201 Copper alloys; Vol. 02.03 Aluminum alloys; Vol. 02.04 Non-ferrous alloys; Vol. 02.05 Coatings; Vol. 03.01 Metals at high and low temperatures; Vol. 04.09 Wood; Vols 09.01 and 09.02 Rubber, American Society for Testing Materials, 1916 Race Street, Philadelphia, PA, USA. ISBN 0–8031–1581–4. (*The ASTM set standards for materials testing.*)

Callister, W.D. (2003) *Materials Science and Engineering, an Introduction*, 6th edition, John Wiley, New York, USA. ISBN 0–471–13576–3. (*A well-respected materials text, now in its 6th edition, widely used for materials teaching in North America.*)

Charles, J.A., Crane, F.A.A. and Furness, J.A.G. (1997) *Selection and Use of Engineering Materials*, 3rd edition, Butterworth-Heinemann, Oxford, UK. ISBN 0–7506–3277–1. (*A materials-science approach to the selection of materials.*)

Dieter, G.E. (1991) *Engineering Design, a Materials and Processing Approach*, 2nd edition, McGraw-Hill, New York, USA. ISBN 0–07–100829–2. (*A well-balanced and respected text focussing on the place of materials and processing in technical design.*)

Farag, M.M. (1989) *Selection of Materials and Manufacturing Processes for Engineering Design*, Prentice-Hall, Englewood Cliffs, NJ, USA. ISBN 0–13–575192–6. (*A materials-science approach to the selection of materials.*)

Chapter 4

Material property charts

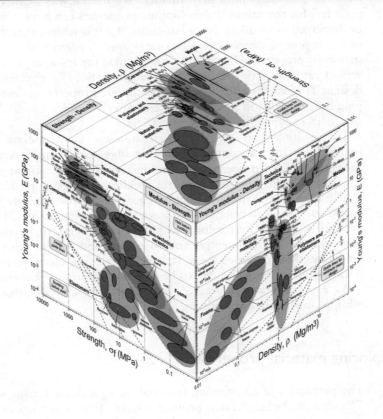

Chapter contents

4.1 Introduction and synopsis

Material properties limit performance. We need a way of surveying them, to get a feel for the values design-limiting properties can have. One property can be displayed as a ranked list or bar-chart. But it is seldom that the performance of a component depends on just one property. Almost always it is a combination of properties that matter: one thinks, for instance, of the strength-to-weight ratio, σ_f/ρ, or the stiffness-to-weight ratio, E/ρ, that enter light-weight design. This suggests the idea of plotting one property against another, mapping out the fields in property-space occupied by each material class, and the sub-fields occupied by individual materials.

The resulting charts are helpful in many ways. They condense a large body of information into a compact but accessible form; they reveal correlations between material properties that aid in checking and estimating data; and in later chapters they become tools for tackling real design problems.

The idea of a materials-selection chart is described briefly in Section 4.2. Section 4.3 is not so brief: it introduces the charts themselves. There is no need to read it all, but it is helpful to persist far enough to be able to read and interpret the charts fluently, and to understand the meaning of the design guidelines that appear on them. If, later, you use one chart, you should read the background to it, given here, to be sure of interpreting it correctly.

As explained in the preface, you may copy and distribute these charts without infringing copyright.[1]

4.2 Exploring material properties

The properties of engineering materials have a characteristic span of values. The span can be large: many properties have values that range over five or more decades. One way of displaying this is as a bar-chart like that of Figure 4.1 for thermal conductivity. Each bar represents a single material. The length of the bar shows the range of conductivity exhibited by that material in its various forms. The materials are segregated by class. Each class shows a characteristic range: metals, have high conductivities; polymers have low; ceramics have a wide range, from low to high.

Much more information is displayed by an alternative way of plotting properties, illustrated in the schematic of Figure 4.2. Here, one property (the modulus, E, in this case) is plotted against another (the density, ρ) on logarithmic scales. The range of the axes is chosen to include all materials, from the

---

[1] A set of the charts in full color (they look much better in color) can be downloaded from www.grantadesign.com. All the charts shown in this chapter were created using Granta Design's CES Materials Selection software.

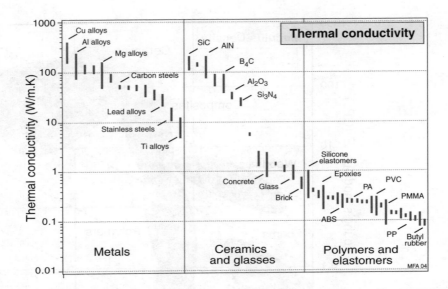

Figure 4.1   A bar-chart showing thermal conductivity for families of solid. Each bar shows the range of conductivity offered by a material, some of which are labeled.

lightest, flimsiest foams to the stiffest, heaviest metals. It is then found that data for a given family of materials (e.g. polymers) cluster together on the chart; the *sub-range* associated with one material family is, in all cases, much smaller than the *full* range of that property. Data for one family can be enclosed in a property-envelope, as Figure 4.2 shows. Within it lie bubbles enclosing classes and sub-classes.

All this is simple enough—just a helpful way of plotting data. But by choosing the axes and scales appropriately, more can be added. The speed of sound in a solid depends on E and ρ; the longitudinal wave speed v, for instance, is

$$v = \left(\frac{E}{\rho}\right)^{1/2}$$

or (taking logs)

$$\log E = \log \rho + 2 \log v$$

For a fixed value of v, this equation plots as a straight line of slope 1 on Figure 4.2. This allows us to add *contours of constant wave velocity* to the chart: they are the family of parallel diagonal lines, linking materials in which longitudinal waves travel with the same speed. All the charts allow additional fundamental relationships of this sort to be displayed. And there is more: design-optimizing parameters called *material indices* also plot as contours on to the charts. But that comes in Chapter 5.

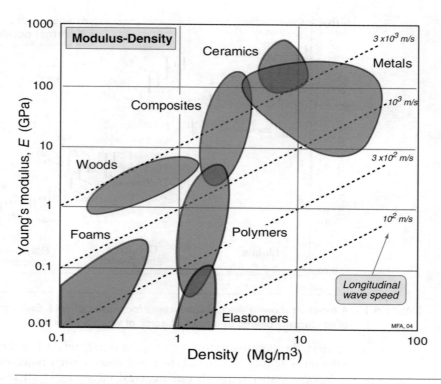

Figure 4.2 The idea of a materials property chart: Young's modulus, *E*, is plotted against the density, *ρ*, on log scales. Each class of material occupies a characteristic part of the chart. The log scales allow the longitudinal elastic wave speed $v = (E/\rho)^{1/2}$ to be plotted as a set of parallel contours.

Among the mechanical and thermal properties, there are 30 or so that are of primary importance, both in characterizing the material, and in engineering design. They were listed in Table 3.1: they include density, moduli, strength, hardness, toughness, thermal and electrical conductivities, expansion coefficient, and specific heat. The charts display data for these properties for the families and classes of materials listed in Table 4.1. The list is expanded from the original six of Figure 3.1 by distinguishing *composites* from *foams* and from *woods* though all are hybrids and by distinguishing the high-strength *engineering ceramics* (like silicon carbide) from the low strength, *porous ceramics* (like brick). Within each family, data are plotted for a representative set of materials, chosen both to span the full range of behavior for the class, and to include the most common and most widely used members of it. In this way the envelope for a family encloses data not only for the materials listed in Table 4.1, but virtually all other members of the family as well.

Table 4.1   Material families and classes

Family	Classes	Short name
Metals (the metals and alloys of engineering)	Aluminum alloys	Al alloys
	Copper alloys	Cu alloys
	Lead alloys	Lead alloys
	Magnesium alloys	Mg alloys
	Nickel alloys	Ni alloys
	Carbon steels	Steels
	Stainless steels	Stainless steels
	Tin alloys	Tin alloys
	Titanium alloys	Ti alloys
	Tungsten alloys	W alloys
	Lead alloys	Pb alloys
	Zinc alloys	Zn alloys
Ceramics	Alumina	Al_2O_3
Technical ceramics (fine ceramics capable of load-bearing application)	Aluminum nitride	AlN
	Boron carbide	B_4C
	Silicon Carbide	SiC
	Silicon Nitride	Si_3N_4
	Tungsten carbide	WC
Non-technical ceramics (porous ceramics of construction)	Brick	Brick
	Concrete	Concrete
	Stone	Stone
Glasses	Soda-lime glass	Soda-lime glass
	Borosilicate glass	Borosilicate glass
	Silica glass	Silica glass
	Glass ceramic	Glass ceramic
Polymers (the thermoplastics and thermosets of engineering)	Acrylonitrile butadiene styrene	ABS
	Cellulose polymers	CA
	Ionomers	Ionomers
	Epoxies	Epoxy
	Phenolics	Phenolics
	Polyamides (nylons)	PA
	Polycarbonate	PC
	Polyesters	Polyester
	Polyetheretherkeytone	PEEK
	Polyethylene	PE
	Polyethylene terephalate	PET or PETE
	Polymethylmethacrylate	PMMA
	Polyoxymethylene (Acetal)	POM
	Polypropylene	PP
	Polystyrene	PS
	Polytetrafluorethylene	PTFE
	Polyvinylchloride	PVC

Table 4.1   (*Continued*)

Family	Classes	Short name
Elastomers	Butyl rubber	Butyl rubber
(engineering rubbers,	EVA	EVA
natural and synthetic)	Isoprene	Isoprene
	Natural rubber	Natural rubber
	Polychloroprene (Neoprene)	Neoprene
	Polyurethane	PU
	Silicone elastomers	Silicones
Hybrids	Carbon-fiber reinforced polymers	CFRP
Composites	Glass-fiber reinforced polymers	GFRP
	SiC reinforced aluminum	Al-SiC
Foams	Flexible polymer foams	Flexible foams
	Rigid polymer foams	Rigid foams
Natural materials	Cork	Cork
	Bamboo	Bamboo
	Wood	Wood

The charts that follow show a *range* of values for each property of each material. Sometimes the range is narrow: the modulus of copper, for instance, varies by only a few percent about its mean value, influenced by purity, texture and such like. Sometimes it is wide: the strength of alumina-ceramic can vary by a factor of 100 or more, influenced by porosity, grain size, and composition. Heat treatment and mechanical working have a profound effect on yield strength and toughness of metals. Crystallinity and degree of cross-linking greatly influence the modulus of polymers. These *structure-sensitive* properties appear as elongated bubbles within the envelopes on the charts. A bubble encloses a typical range for the value of the property for a single material class. Envelopes (heavier lines) enclose the bubbles for a family.

The data plotted on the charts have been assembled from a variety of sources, documented in Chapter 15.

4.3 The material property charts

The Modulus–Density chart

Modulus and density are familiar properties. Steel is stiff, rubber is compliant: these are effects of modulus. Lead is heavy; cork is buoyant: these are effects of density. Figure 4.3 shows the full range of Young's modulus, E, and density, ρ, for engineering materials. Data for members of a particular family of material cluster together and can be enclosed by an envelope (heavy line). The same

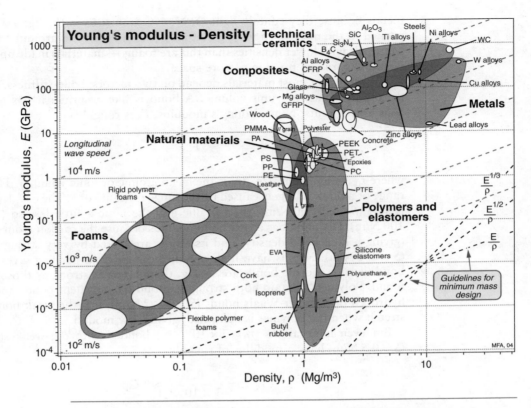

Figure 4.3 Young's modulus, E, plotted against density, ρ. The heavy envelopes enclose data for a given class of material. The diagonal contours show the longitudinal wave velocity. The guidelines of constant E/ρ, $E^{1/2}/\rho$ and $E^{1/3}/\rho$ allow selection of materials for minimum weight, deflection-limited, design.

family-envelopes appear on all the diagrams: they correspond to the main headings in Table 4.1.

The *density* of a solid depends on three factors: the atomic weight of its atoms or ions, their size, and the way they are packed. The size of atoms does not vary much: most have a volume within a factor of two of $2 \times 10^{-29}\,\text{m}^3$. Packing fractions do not vary much either—a factor of two, more or less: close-packing gives a packing fraction of 0.74; open networks (like that of the diamond-cubic structure) give about 0.34. The spread of density comes mainly from that of atomic weight, ranging from 1 for hydrogen to 238 for uranium. Metals are dense because they are made of heavy atoms, packed densely; polymers have low densities because they are largely made of carbon (atomic weight: 12) and hydrogen (atomic weight: 1) in low-density amorphous or crystalline packings. Ceramics, for the most part, have lower densities than metals because they contain light O, N or C atoms. Even the lightest atoms, packed in the most open way, give solids with a density of around $1 \, \text{Mg/m}^3$. Materials with lower densities than this are foams — materials made up of cells containing a large fraction of pore space.

The *moduli* of most materials depend on two factors: bond stiffness, and the density of bonds per unit volume. A bond is like a spring: it has a spring constant, S (units: N/m). Young's modulus, E, is roughly

$$E = \frac{S}{r_0} \tag{4.1}$$

where r_0 is the "atom size" (r_0^3 is the mean atomic or ionic volume). The wide range of moduli is largely caused by the range of values of S. The covalent bond is stiff ($S = 20–200 \, \text{N/m}$); the metallic and the ionic a little less so ($S = 15–100 \, \text{N/m}$). Diamond has a very high modulus because the carbon atom is small (giving a high bond density) and its atoms are linked by very strong springs ($S = 200 \, \text{N/m}$). Metals have high moduli because close-packing gives a high bond density and the bonds are strong, though not as strong as those of diamond. Polymers contain both strong diamond-like covalent bonds and weak hydrogen or Van-der-Waals bonds ($S = 0.5–2 \, \text{N/m}$); it is the weak bonds that stretch when the polymer is deformed, giving low moduli.

But even large atoms ($r_0 = 3 \times 10^{-10} \, \text{m}$) bonded with the weakest bonds ($S = 0.5 \, \text{N/m}$) have a modulus of roughly

$$E = \frac{0.5}{3 \times 10^{-10}} \approx 1 \, \text{Gpa} \tag{4.2}$$

This is the *lower limit* for true solids. The chart shows that many materials have moduli that are lower than this: they are either elastomers or foams. Elastomers have a low E because the weak secondary bonds have melted (their glass temperature, T_g, is below room temperature) leaving only the very weak "entropic" restoring force associated with tangled, long-chain molecules; and foams have low moduli because the cell walls bend easily (allowing large displacements) when the material is loaded.

The chart shows that the modulus of engineering materials spans 7 decades,[2] from 0.0001 GPa (low-density foams) to 1000 GPa (diamond); the density spans a factor of 2000, from less than 0.01 to $20 \, \text{Mg/m}^3$. Ceramics as a family are very stiff, metals a little less so — but none have a modulus less than 10 GPa. Polymers, by contrast, all cluster between 0.8 and 8 GPa. To have a lower modulus than this the material must be either an elastomer or a foam. At the level of approximation of interest here (that required to reveal the relationship between the properties of materials classes) we may approximate

---

[2] Very low density foams and gels (which can be thought of as molecular-scale, fluid-filled, foams) can have lower moduli than this. As an example, gelatin (as in Jello) has a modulus of about 10^{-5} GPa. Their strengths and fracture toughness, too, can be below the lower limit of the charts.

the shear modulus, G, by $3E/8$ and the bulk modulus, K, by E, for all materials except elastomers (for which $G = E/3$ and $K \gg E$) allowing the chart to be used for these also.

The log-scales allow more information to be displayed. As explained in the last section, the velocity of elastic waves in a material, and the natural vibration frequencies of a component made of it, are proportional to $(E/\rho)^{1/2}$. Contours of this quantity are plotted on the chart, labeled with the longitudinal wave speed. It varies from less than 50 m/s (soft elastomers) to a little more than 10^4 m/s (stiff ceramics). We note that aluminum and glass, because of their low densities, transmit waves quickly despite their low moduli. One might have expected the wave velocity in foams to be low because of the low modulus, but the low density almost compensates. That in wood, across the grain, is low; but along the grain, it is high — roughly the same as steel — a fact made use of in the design of musical instruments.

The chart helps in the common problem of material selection for applications in which mass must be minimized. Guidelines corresponding to three common geometries of loading are drawn on the diagram. They are used in the way described in Chapters 5 and 6 to select materials for elastic design at minimum weight.

The strength–density chart

The modulus of a solid is a well-defined quantity with a sharp value. The strength is not. It is shown, plotted against density, ρ, in Figure 4.4.

The word "strength" needs definition (see also Chapter 3, Section 3.3). For metals and polymers, it is the *yield strength*, but since the range of materials includes those that have been worked or hardened in some other way as well as those that have been annealed, the range is large. For brittle ceramics, the strength plotted here is the *modulus of rupture:* the strength in bending. It is slightly greater than the tensile strength, but much less than the compression strength, which, for ceramics is 10 to 15 times larger. For elastomers, strength means the tensile *tear-strength*. For composites, it is the *tensile failure strength* (the compressive strength can be less by up to 30 percent because of fiber buckling). We will use the symbol σ_f for all of these, despite the different failure mechanisms involved to allow a first-order comparison.

The considerable vertical extension of the strength-bubble for an individual material class reflects its wide range, caused by degree-of-alloying, work hardening, grain size, porosity and so forth. As before, members of a family cluster together and can be enclosed in an envelope, each of which occupies a characteristic area of the chart.

The range of strength for engineering materials, like that of the modulus, spans about 6 decades: from less than 0.01 MPa (foams, used in packaging and energy-absorbing systems) to 10^4 MPa (the strength of diamond, exploited in the diamond-anvil press). The single most important concept in understanding this

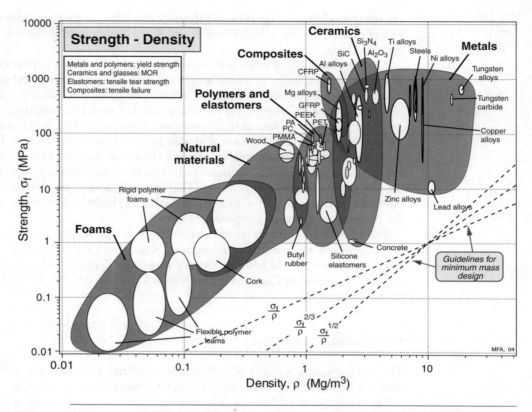

Figure 4.4   Strength, σ_f, plotted against density, ρ (yield strength for metals and polymers, compressive strength for ceramics, tear strength for elastomers and tensile strength for composites). The guidelines of constant σ_f/ρ, $\sigma_f^{2/3}/\rho$ and $\sigma_f^{1/2}/\rho$ are used in minimum weight, yield-limited, design.

wide range is that of the *lattice resistance* or *Peierls stress*: the intrinsic resistance of the structure to plastic shear. Plastic shear in a crystal involves the motion of dislocations. Pure metals are soft because the non-localized metallic bond does little to prevent dislocation motion, whereas ceramics are hard because their more localized covalent and ionic bonds (which must be broken and reformed when the structure is sheared), lock the dislocations in place. In non-crystalline solids we think instead of the energy associated with the unit step of the flow process: the relative slippage of two segments of a polymer chain, or the shear of a small molecular cluster in a glass network. Their strength has the same origin as that underlying the lattice resistance: if the unit step involves breaking strong bonds (as in an inorganic glass), the materials will be strong; if it only involves the rupture of weak bonds (e.g. the Van-der-Waals bonds in polymers), it will be weak. Materials that fail by fracture do so because the lattice resistance or its amorphous equivalent is so large that atomic separation (fracture) happens first.

When the lattice resistance is low, the material can be strengthened by introducing obstacles to slip. In metals this is achieved by adding alloying elements, particles, grain boundaries, and other dislocations ("work hardening"); and in polymers by cross-linking or by orienting the chains so that strong covalent as well as weak Van-der-Waals bonds must be broken when the material deforms. When, on the other hand, the lattice resistance is high, further hardening is superfluous — the problem becomes that of suppressing fracture.

An important use of the chart is in materials selection in light-weight plastic design. Guidelines are shown for materials selection in the minimum-weight design of ties, columns, beams and plates, and for yield-limited design of moving components in which inertial forces are important. Their use is described in Chapters 5 and 6.

The modulus–strength chart

High tensile steel makes good springs. But so does rubber. How is it that two such different materials are both suited for the same task? This and other questions are answered by Figure 4.5, one of the most useful of all the charts.

It shows Young's modulus, E, plotted against strength, σ_f. The qualifications on "strength" are the same as before: it means yield strength for metals and polymers, modulus of rupture for ceramics, tear strength for elastomers, and tensile strength for composite and woods; the symbol σ_f is used for them all. Contours of *yield strain*, σ_f/E (meaning the strain at which the material ceases to be linearly elastic), appear as a family of straight parallel lines.

Examine these first. Engineering polymers have large yield strains of between 0.01 and 0.1; the values for metals are at least a factor of 10 smaller. Composites and woods lie on the 0.01 contour, as good as the best metals. Elastomers, because of their exceptionally low moduli, have values of σ_f/E larger than any other class of material: typically 1 to 10.

The distance over which inter-atomic forces act is small — a bond is broken if it is stretched to more than about 10 percent of its original length. So the force needed to break a bond is roughly

$$F \approx \frac{Sr_0}{10} \tag{4.3}$$

where S, as before, is the bond stiffness. If shear breaks bonds, the strength of a solid should be roughly

$$\sigma_f \approx \frac{F}{r_0^2} = \frac{S}{10r_0} = \frac{E}{10}$$

or

$$\frac{\sigma_f}{E} \approx \frac{1}{10} \tag{4.4}$$

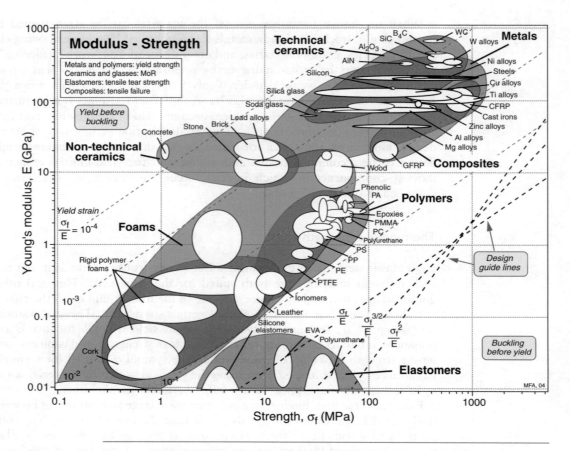

Figure 4.5 Young's modulus, *E*, plotted against strength, σ_f. The design guidelines help with the selection of materials for springs, pivots, knife-edges, diaphragms and hinges; their use is described in Chapters 5 and 6.

The chart shows that, for some polymers, the failure strain approaches this value. For most solids it is less, for two reasons.

First, non-localized bonds (those in which the cohesive energy derives from the interaction of one atom with large number of others, not just with its nearest neighbors) are not broken when the structure is sheared. The metallic bond, and the ionic bond for certain directions of shear, are like this; very pure metals, for example, yield at stresses as low as *E*/10,000, and strengthening mechanisms are needed to make them useful in engineering. The covalent bond *is* localized; and covalent solids do, for this reason, have yield strength that, at low temperatures, are as high as *E*/10. It is hard to measure them (though it can sometimes be done by indentation) because of the second reason for weakness: they generally contain defects — concentrators of stress — from which shear or fracture can propagate, at stresses well below the "ideal" $E/10$. Elastomers are anomalous (they have strengths of about E) because the modulus does not derive from bond-stretching, but from the change in entropy of the tangled molecular chains when the material is deformed.

Materials with high strength and low modulus lie towards the bottom right. Such materials tend to *buckle before they yield* when loaded as panels or columns. Those near the top left have high modulus and low strength: they end to *yield before buckling*.

This has not yet explained how to choose good materials to make springs. This involves the design guidelines shown on the chart. The way to use them is described in Chapter 6, Section 6.7.

The specific stiffness–specific strength chart

Many designs, particularly those for things that move, call for stiffness and strength at minimum weight. To help with this, the data of the previous chart are replotted in Figure 4.6 after dividing, for each material, by the density; it shows E/ρ plotted against σ_f/ρ.

Composites, particularly CFRP, emerge as the material class with the most attractive specific properties, one of the reasons for their increasing use in aerospace. Ceramics have exceptionally high stiffness per unit weight, and the strength per unit weight is as good as metals. Metals are penalized because of their relatively high densities. Polymers, because their densities are low, do better on this chart than on the last one.

The chart has application in selecting materials for light springs and energy-storage devices. But that too has to wait till Section 6.7.

The fracture toughness–modulus chart

Increasing the strength of a material is useful only as long as it remains plastic and does not fail by fast fracture. The resistance to the propagation of a crack is measured by the *fracture toughness*, K_{1C}. It is plotted against modulus E in Figure 4.7. The range is large: from less than 0.01 to over 100 MPa.m$^{1/2}$. At the lower end of this range are brittle materials, which, when loaded, remain elastic until they fracture. For these, linear-elastic fracture mechanics works well, and the fracture toughness itself is a well-defined property. At the upper end lie the super-tough materials, all of which show substantial plasticity before they break. For these the values of K_{1C} are approximate, derived from critical J-integral (J_c) and critical crack-opening displacement (δ_c) measurements (by writing $K_{1C} = (EJ_c)^{1/2}$, for instance). They are helpful in providing a ranking of materials. The figure shows one reason for the dominance of metals in engineering; they almost all have values of K_{1C} above 20 MPa.m$^{1/2}$, a value often quoted as a minimum for conventional design.

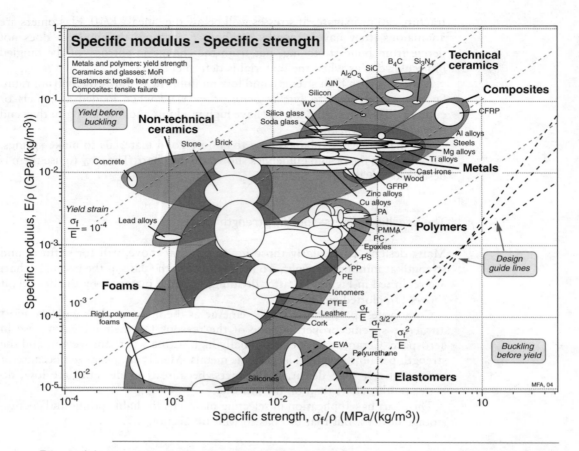

Figure 4.6 Specific modulus, E/ρ plotted against specific strength σ_f/ρ. The design guidelines help with the selection of materials for light-weight springs and energy-storage systems.

As a general rule, the fracture toughness of polymers is less than that of ceramics. Yet polymers are widely used in engineering structures; ceramics, because they are "brittle", are treated with much more caution. Figure 4.7 helps resolve this apparent contradiction. Consider first the question of the *necessary condition for fracture*. It is that sufficient external work be done, or elastic energy released, to supply the surface energy, γ per unit area, of the two new surfaces that are created. We write this as

$$G \geq 2\gamma \tag{4.5}$$

where G is the energy release-rate. Using the standard relation $K = (EG)^{1/2}$ between G and stress intensity K, we find

$$K \geq (2E\gamma)^{1/2} \tag{4.6}$$

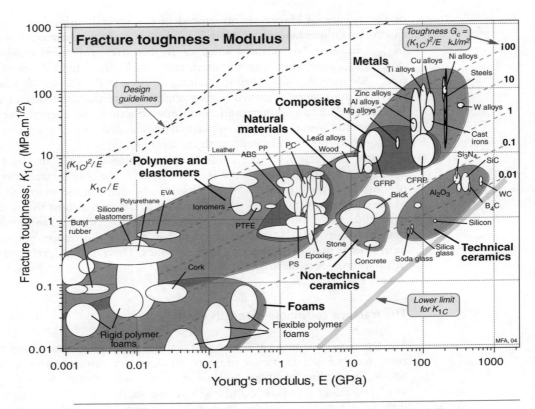

Figure 4.7 Fracture toughness, K_{IC}, plotted against Young's modulus, E. The family of lines are of constant K_{IC}^2/E (approximately G_{IC}, the fracture energy or toughness). These, and the guideline of constant K_{IC}/E, help in design against fracture. The shaded band shows the "necessary condition" for fracture. Fracture can, in fact, occur below this limit under conditions of corrosion, or cyclic loading.

Now the surface energies, γ, of solid materials scale as their moduli; to an adequate approximation $\gamma \approx E r_0/20$ where r_0 is the atom size, giving

$$K \geq E \left(\frac{r_0}{20}\right)^{1/2} \tag{4.7}$$

We identify the right-hand side of this equation with a lower-limiting value of K_{1C}, when, taking as $2 \times 10^{-10}\,\text{m}$,

$$\frac{(K_{1C})_{min}}{E} = \left(\frac{r_0}{20}\right)^{1/2} \approx 3 \times 10^{-6}\,\text{m}^{1/2} \tag{4.8}$$

This criterion is plotted on the chart as a shaded, diagonal band near the lower right corner. It defines a *lower limit* for K_{1C}. The fracture toughness cannot be less than this unless some other source of energy such as a chemical reaction, or the release of elastic energy stored in the special dislocation structures caused by fatigue loading, is available, when it is given a new symbol such as $(K_1)_{scc}$ meaning "the critical value of K_1 for stress-corrosion cracking" or $\Delta(K_1)_{threshold}$ meaning "the minimum range of K_1 for fatigue-crack propagation". We note that the brittlest ceramics lie close to the threshold: when they fracture, the energy absorbed is only slightly more than the surface energy. When metals and polymers and composites fracture, the energy absorbed is vastly greater, usually because of plasticity associated with crack propagation. We come to this in a moment, with the next chart.

Plotted on Figure 4.7 are contours of *toughness*, G_{1C}, a measure of the apparent fracture surface-energy ($G_{1C} \approx K_{1C}^2/E$). The true surface energies, γ, of solids lie in the range 10^{-4} to 10^{-3} kJ/m^2. The diagram shows that the values of the toughness start at 10^{-3} kJ/m^2 and range through almost five decades to over 100 kJ/m^2. On this scale, ceramics (10^{-3}–10^{-1} kJ/m^2) are much lower than polymers (10^{-1}–10 kJ/m^2); and this is part of the reason polymers are more widely used in engineering than ceramics. This point is developed further in Chapter 6, Section 6.10.

The fracture toughness–strength chart

The stress concentration at the tip of a crack generates a *process-zone*: a plastic zone in ductile solids, a zone of micro-cracking in ceramics, a zone of delamination, debonding and fiber pull-out in composites. Within the process zone, work is done against plastic and frictional forces; it is this that accounts for the difference between the measured fracture energy G_{1C} and the true surface energy 2γ. The amount of energy dissipated must scale roughly with the strength of the material within the process zone, and with its size, d_y. This size is found by equating the stress field of the crack ($\sigma = K/\sqrt{2\pi r}$) at $r = d_y/2$ to the strength of the material, σ_f, giving

$$d_y = \frac{K_{1C}^2}{\pi \sigma_f^2} \tag{4.9}$$

Figure 4.8 — fracture toughness against strength — shows that the size of the zone, d_y (broken lines), varies enormously, from atomic dimensions for very brittle ceramics and glasses to almost 1 m for the most ductile of metals. At a constant zone size, fracture toughness tends to increase with strength (as expected): it is this that causes the data plotted in Figure 4.8 to be clustered around the diagonal of the chart.

Materials towards the bottom right have high strength and low toughness; they *fracture before they yield*. Those towards the top left do the opposite: they *yield before they fracture*.

The diagram has application in selecting materials for the safe design of load bearing structures. Examples are given in Sections 6.10 and 6.11.

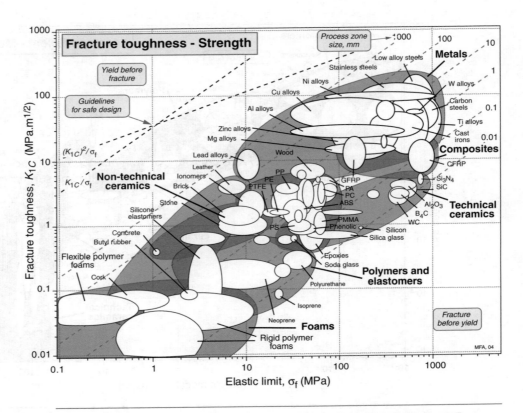

Figure 4.8 Fracture toughness, K_{IC}, plotted against strength, σ_f. The contours show the value of $K_{IC}^2/\pi\sigma_f$ — roughly, the diameter of the process-zone at a crack tip. The design guidelines are used in selecting materials for damage-tolerant design.

The loss coefficient–modulus chart

Bells, traditionally, are made of bronze. They can be (and sometimes are) made of glass; and they could (if you could afford it) be made of silicon carbide. Metals, glasses and ceramics all, under the right circumstances, have low intrinsic damping or "internal friction", an important material property when structures vibrate. Intrinsic damping is measured by the *loss coefficient*, η, which is plotted in Figure 4.9.

There are many mechanisms of intrinsic damping and hysteresis. Some (the "damping" mechanisms) are associated with a process that has a specific time constant; then the energy loss is centered about a characteristic frequency. Others (the "hysteresis" mechanisms) are associated with time-independent mechanisms; they absorb energy at all frequencies. In metals a large part of the loss is hysteretic, caused by dislocation movement: it is high in soft metals like

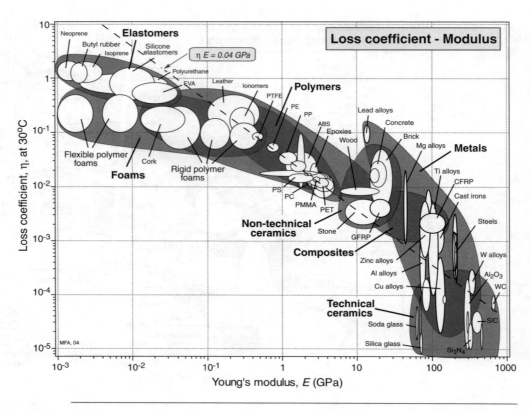

Figure 4.9 The loss coefficient, η, plotted against Young's modulus, E. The guideline corresponds to the condition $\eta = CE$.

lead and pure aluminum. Heavily alloyed metals like bronze and high-carbon steels have low loss because the solute pins the dislocations; these are the materials for bells. Exceptionally high loss is found in the Mn–Cu alloys, because of a strain-induced martensite transformation, and in magnesium, perhaps because of reversible twinning. The elongated bubbles for metals span the large range made accessible by alloying and work hardening. Engineering ceramics have low damping because the enormous lattice resistance pins dislocations in place at room temperature. Porous ceramics, on the other hand, are filled with cracks, the surfaces of which rub, dissipating energy, when the material is loaded; the high damping of some cast irons has a similar origin. In polymers, chain segments slide against each other when loaded; the relative motion dissipates energy. The ease with which they slide depends on the ratio of the temperature T (in this case, room temperature) to the glass temperature, T_g, of the polymer. When $T/T_g < 1$, the secondary bonds are "frozen", the modulus is high and the damping is relatively low. When $T/T_g > 1$, the secondary bonds have melted, allowing easy chain slippage; the modulus is low and the damping is high. This accounts for the obvious inverse dependence of η on E for polymers in Figure 4.9; indeed, to a first approximation,

$$\eta = \frac{4 \times 10^{-2}}{E} \qquad (4.10)$$

(with E in GPa) for polymers, woods and polymer–matrix composites.

The thermal conductivity–electrical resistivity chart

The material property governing the flow of heat through a material at steady-state is the thermal conductivity, λ (units: W/m.K). The valence electrons in metals are "free", moving like a gas within the lattice of the metal. Each electron carries a kinetic energy $\frac{3}{2}kT$, and it is the transmission of this energy, via collisions, that conducts heat. The thermal conductivity is described by

$$\lambda = \frac{1}{3} C_e \bar{c} \lambda \qquad (4.11)$$

where C_e is the electron specific heat per unit volume, $\bar{c}$ is the electron velocity (2×10^5 m/s) and λ the electron mean-free path, typically 10^{-7} m in pure metals. In heavily alloyed solid solution (stainless steels, nickel-based super-alloys, and titanium alloys) the foreign atoms scatter electrons, reducing the mean free path to atomic dimensions ($\approx 10^{-10}$ m), much reducing λ.

These same electrons, when in a potential gradient, drift through the lattice, giving electrical conduction. The electrical conductivity, κ, here measured by its reciprocal, the *resistivity* ρ_e (SI units: Ω.m, units of convenience $\mu\Omega$.cm). The range is enormous: a factor of 10^{28}, far larger than that of any other property. As with heat, the conduction of electricity is proportional to the density of carriers (the electrons) and their mean-free path, leading to the Wiedemann–Franz relation

$$\lambda \propto \kappa = \frac{1}{\rho_e} \qquad (4.12)$$

The quantities λ and ρ_e are the axes of Figure 4.10. Data for metals appear at the top left. The broken line shows that the Wiedemann–Franz relation is well obeyed.

But what of the rest of the chart? Electrons do not contribute to thermal conduction in ceramics and polymers. Heat is carried by phonons–lattice vibrations of short wavelength. They are scattered by each other (through an anharmonic interaction) and by impurities, lattice defects, and surfaces; it is these that determine the phonon mean-free path, λ. The conductivity is still given by equation (4.11), which we write as

$$\lambda = \frac{1}{3} \rho C_p \bar{c} \lambda \qquad (4.13)$$

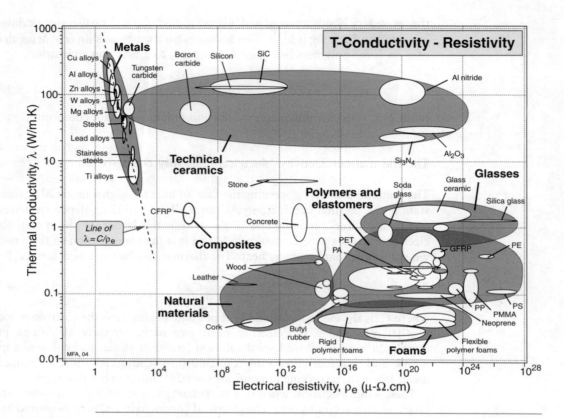

Figure 4.10   Thermal conductivity, λ, plotted against electrical resistivity, ρ_e. For metals the two are related.

but now $\bar{c}$ is the elastic wave speed (around 10^3 m/s — see Figure 4.3), ρ is the density and $C_p{}^3$ is the *specific heat per unit mass* (units: J/kg.K). If the crystal is particularly perfect and the temperature is well below the Debye temperature, as in diamond at room temperature, the phonon conductivity is high: it is for this reason that single crystal silicon carbide and aluminum nitride have thermal conductivities almost as high as copper. The low conductivity of glass is caused by its irregular amorphous structure; the characteristic length of the molecular linkages (about 10^{-9} m) determines the mean free path. Polymers have low conductivities because the elastic wave speed $\bar{c}$ is low (Figure 4.3), and the mean free path in the disordered structure is small. Highly porous materials like firebrick, cork and foams show the lowest thermal conductivities, limited by that of the gas in their cells.

---

[3] The specific heat at constant volume C_v in J/kg.K; for solids, differs only slightly from that at constant pressure, C_p; we will neglect the difference here.

Graphite and many intermetallic compounds such as WC and B_4C, like metals, have free electrons, but the number of carriers is smaller and the resistivity higher. Defects such as vacancies and impurity atoms in ionic solids create positive ions that require balancing electrons. These can jump from ion to ion, conducting charge, but slowly because the carrier density is low. Covalent solids and most polymers have no mobile electrons and are insulators ($\rho_e > 10^{12}$ $\mu\Omega$.cm) — they lie on the right-hand side of Figure 4.10.

Under a sufficiently high potential gradient, anything will conduct. The gradient tears electrons free from even the most possessive atoms, accelerating them into collision with nearby atoms, knocking out more electrons and creating a cascade. The critical gradient is called the breakdown potential V_b (units: MV/m), defined in Chapter 3.

The thermal conductivity–thermal diffusivity chart

Thermal conductivity, as we have said, governs the flow of heat through a material at steady-state. The property governing transient heat flow is the *thermal diffusivity*, a (units: m^2/s). The two are related by

$$a = \frac{\lambda}{\rho C_p} \tag{4.14}$$

where ρ in kg/m^3 is the density. The quantity ρC_p is the *volumetric specific heat* (units: J/m^3.K). Figure 4.11 relates thermal conductivity, diffusivity and volumetric specific heat, at room temperature.

The data span almost five decades in λ and a. Solid materials are strung out along the line[4]

$$\rho C_p \approx 3 \times 10^6 \, J/m^3.K \tag{4.15}$$

As a general rule, then,

$$\lambda = 3 \times 10^6 \, a \tag{4.16}$$

(λ in W/m.K and a in m^2/s). Some materials deviate from this rule: they have lower-than-average volumetric specific heat. The largest deviations are shown

---

4 This can be understood by noting that a solid containing N atoms has 3N vibrational modes. Each (in the classical approximation) absorbs thermal energy kT at the absolute temperature T, and the vibrational specific heat is $C_p \approx C_v = 3N/k$ (J/K) where k is Boltzmann's constant (1.34×10^{-23} J/K). The volume per atom, Ω, for almost all solids lies within a factor of two of 1.4×10^{-29} m^3; thus the volume of N atoms is (NC_p) m^3. The volume specific heat is then (as the Chart shows):

$$\rho C_v \cong 3Nk/N\Omega = \frac{3k}{\Omega} = 3 \times 10^6 \, J/m^3.K$$

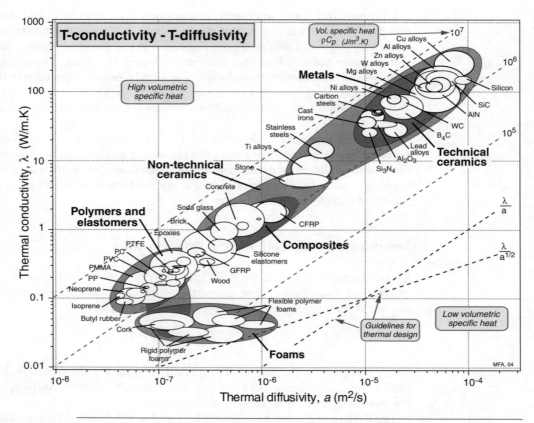

Figure 4.11 Thermal conductivity, λ, plotted against thermal diffusivity, a. The contours show the volume specific heat, ρC_v. All three properties vary with temperature; the data here are for room temperature.

by porous solids: foams, low density firebrick, woods, and the like. Their low density means that they contain fewer atoms per unit volume and, averaged over the volume of the structure, ρC_p is low. The result is that, although foams have low *conductivities* (and are widely used for insulation because of this), their thermal *diffusivities* are not necessarily low: they may not transmit much heat, but they reach a steady-state quickly. This is important in design—a point brought out by the Case Study of Section 6.13.

The thermal expansion–thermal conductivity chart

Almost all solids expand on heating. The bond between a pair of atoms behaves like a linear elastic spring when the relative displacement of the atoms is small, but when it is large, the spring is non-linear. Most bonds become stiffer when the atoms are pushed together, and less stiff when they are pulled apart, and for that reason they are anharmonic. The thermal vibrations of atoms, even at room temperature, involves large displacements; as the temperature is raised, the anharmonicity of the bond pushes the atoms apart, increasing their mean spacing. The effect is measured by the linear *expansion coefficient*

$$\alpha = \frac{1}{\lambda} \frac{d\lambda}{dT} \tag{4.17}$$

where λ is a linear dimension of the body.

The expansion coefficient is plotted against the thermal conductivity in Figure 4.12. It shows that polymers have large values of α, roughly 10 times greater than those of metals and almost 100 times greater than ceramics. This is

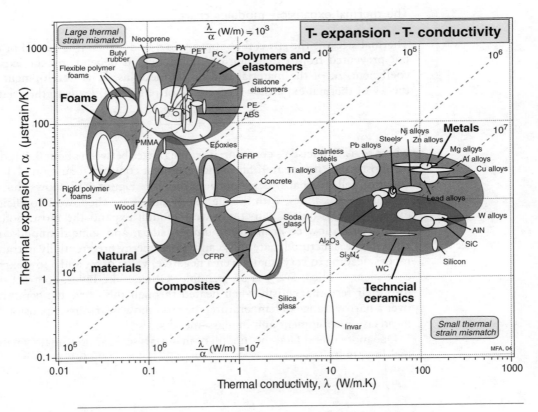

Figure 4.12   The linear expansion coefficient, α, plotted against the thermal conductivity, λ. The contours show the thermal distortion parameter λ/α. An extra material, the nickel alloy Invar, has been added to the chart; it is noted for its exceptionally low expansion at and near room temperature, useful in designing precision equipment that must not distort if the temperature changes.

because the Van-der-Waals bonds of the polymer are very anharmonic. Diamond, silicon, and silica glass (SiO_2) have covalent bonds that have low anharmonicity (i.e. they are almost linear-elastic even at large strains), giving them low expansion coefficients. Composites, even though they have polymer matrices, can have low values of α because the reinforcing fibers — particularly carbon — expand very little.

The chart shows contours of λ/α, a quantity important in designing against thermal distortion. An extra material, Invar (a nickel alloy) has been added to the chart because of its uniquely low expansion coefficient at and near room temperature, an consequence of a trade-off between normal expansion and a contraction associated with a magnetic transformation. An application that uses chart is developed in Chapter 6, Section 6.16.

The thermal expansion–modulus chart

Thermal stress is the stress that appears in a body when it is heated or cooled but prevented from expanding or contracting. It depends on the expansion coefficient, α, of the material and on its modulus, E. A development of the theory of thermal expansion (see, e.g., Cottrell, 1964) leads to the relation

$$\alpha = \frac{\gamma_G \rho C_p}{3E} \tag{4.18}$$

where γ_G is Gruneisen's constant; its value ranges between about 0.4 and 4, but for most solids it is near 1. Since ρC_p is almost constant (equation (4.15)), the equation tells us that α is proportional to $1/E$. Figure 4.13 shows that this is broadly so. Ceramics, with the highest moduli, have the lowest coefficients of expansion; elastomers with the lowest moduli expand the most. Some materials with a low co-ordination number (silica, and some diamond-cubic or zinc-blende structured materials) can absorb energy preferentially in transverse modes, leading to very small (even a negative) value of γ_G and a low expansion coefficient — silica, SiO_2, is an example. Others, like Invar, contract as they lose their ferromagnetism when heated through the Curie temperature and, over a narrow range of temperature, they too show near-zero expansion, useful in precision equipment and in glass–metal seals.

One more useful fact: the moduli of materials scale approximately with their melting point, T_m:

$$E \approx \frac{100kT_m}{\Omega} \tag{4.19}$$

where k is Boltzmann's constant and Ω the volume-per-atom in the structure. Substituting this and equation (4.15) for ρC_p into equation (4.18) for α gives

$$\alpha = \frac{\gamma_G}{100T_m} \tag{4.20}$$

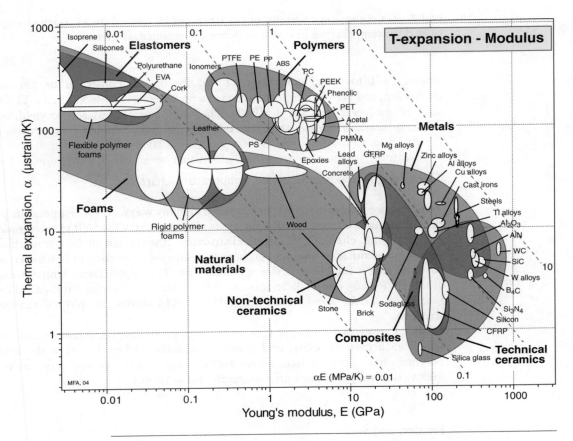

Figure 4.13 The linear expansion coefficient, α, plotted against Young's modulus, E. The contours show the thermal stress created by a temperature change of $1°C$ if the sample is axially constrained. A correction factor C is applied for biaxial or triaxial constraint (see text).

the expansion coefficient varies inversely with the melting point, or (equivalently stated) for all solids the thermal strain, just before they melt, depends only on γ_G, and this is roughly a constant. Equations (4.18) and (4.19) are examples of property correlations, useful for estimating and checking material properties (Chapter 15).

Whenever the thermal expansion or contraction of a body is prevented, thermal stresses appear; if large enough, they cause yielding, fracture, or elastic collapse (buckling). It is common to distinguish between thermal stress caused by external constraint (e.g. a rod, rigidly clamped at both ends) and that which appears without external constraint because of temperature gradients in the body. All scale as the quantity αE, shown as a set of diagonal contours in Figure 4.13. More precisely: the stress $\Delta\sigma$ produced by a temperature change of 1°C in a constrained system, or the stress per °C caused by a sudden change of surface temperature in one that is not constrained, is given by

$$C\Delta\sigma = \alpha E \tag{4.21}$$

where $C = 1$ for axial constraint, $(1 - v)$ for biaxial constraint or normal quenching, and $(1 - 2v)$ for triaxial constraint, where v is Poisson's ratio. These stresses are large: typically 1 MPa/K. They can cause a material to yield, or crack, or spall, or buckle when it is suddenly heated or cooled.

The strength–maximum service temperature chart

Temperature affects material performance in many ways. As the temperature is raised the material may creep, limiting its ability to carry loads. It may degrade or decompose, changing its chemical structure in ways that make it unusable. And it may oxidize or interact in other ways with the environment in which it is used, leaving it unable to perform its function. The approximate temperature at which, for any one of these reasons, it is unsafe to use a material is called its *maximum service temperature* T_{max}. Figure 4.14 shows this plotted against strength.

The chart gives a birds-eye view of the regimes of stress and temperature in which each material class, and material, is usable. Note that even the best polymers have little strength above 200°C; most metals become very soft by 800°C; and only ceramics offer strength above 1500°C.

Friction and wear

God, it is said, created solids, but it was the devil that made surfaces — they are the source of many problems. When surfaces touch and slide, there is friction; and where there is friction, there is wear. Tribologists — the collective noun for those who study friction and wear — are fond of citing the enormous cost, through lost energy and worn equipment, for which these two phenomena are responsible. It is certainly true that, if friction could be eliminated, the efficiency of engines, gear boxes, drive trains and the like would increase; and if wear could be eradicated, they would also last longer. But before accepting this negative image, one should remember that, without wear, pencils would not write on paper or chalk on blackboards; and without friction, one would slither off the slightest incline.

Tribological properties are not attributes of one material alone, but of one material sliding on another with — almost always — a third in between. The number of combinations is far too great to allow choice in a simple, systematic way. The selection of materials for bearings, drives, and sliding seals relies heavily on experience. This experience is captured in reference sources

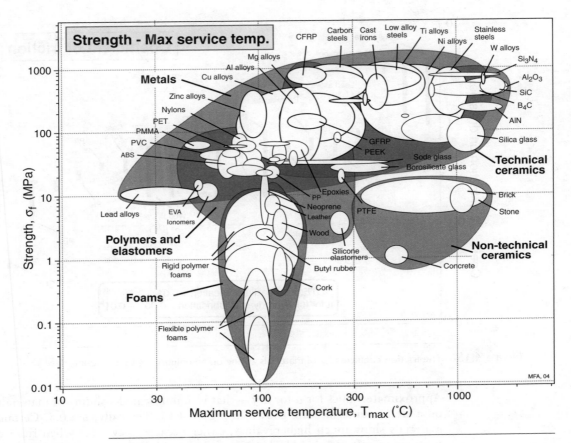

Figure 4.14 Strength plotted against maximum service temperature.

(for which see Chapter 15); in the end it is these that must be consulted. But it does help to have a feel for the magnitude of friction coefficients and wear rates, and an idea of how these relate to material class.

When two surfaces are placed in contact under a normal load F_n and one is made to slide over the other, a force F_s opposes the motion. This force is proportional to F_n but does not depend on the area of the surface — and this is the single most significant result of studies of friction, since it implies that surfaces do not contact completely, but only touch over small patches, the area of which is independent of the apparent, nominal area of contact A_n. The *coefficient friction* μ is defined by

$$\mu = \frac{F_s}{F_n} \tag{4.22}$$

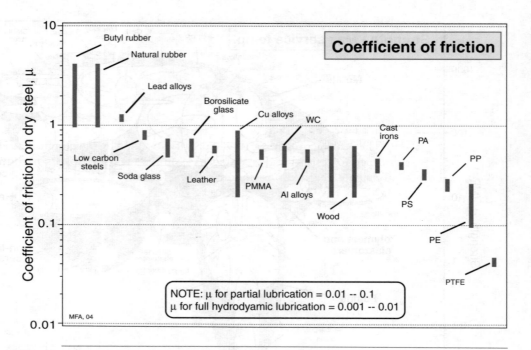

Figure 4.15 The friction coefficient μ of materials sliding on an unlubricated steel counterface.

Approximate values for μ for dry — that is, unlubricated—sliding of materials on a steel counterface are shown in Figure 4.15. Typically, $\mu \approx 0.5$. Certain materials show much higher values, either because they seize when rubbed together (a soft metal rubbed on itself with no lubrication, for instance) or because one surface has a sufficiently low modulus that it conforms to the other (rubber on rough concrete). At the other extreme are sliding combinations with exceptionally low coefficients of friction, such as PTFE, or bronze bearings loaded graphite, sliding on polished steel. Here the coefficient of friction falls as low as 0.04, though this is still high compared with friction for lubricated surfaces, as noted at the bottom of the diagram.

When surfaces slide, they wear. Material is lost from both surfaces, even when one is much harder than the other. The *wear-rate*, W, is conventionally defined as

$$W = \frac{\text{Volume of material removed from contact surface}}{\text{Distance slid}} \quad (4.23)$$

and thus has units of m^2. A more useful quantity, for our purposes, is the specific wear-rate

$$\Omega = \frac{W}{A_n} \quad (4.24)$$

which is dimensionless. It increases with bearing pressure P (the normal force F_n divided by the nominal area A_n), such that the ratio

$$k_a = \frac{W}{F_n} = \frac{\Omega}{P} \qquad (4.25)$$

is roughly constant. The quantity k_a (with units of $(MPa)^{-1}$) is a measure of the propensity of a sliding couple for wear: high k_a means rapid wear at a given bearing pressure.

The bearing pressure P is the quantity specified by the design. The ability of a surface to resist a static contact pressure is measured by its hardness, so we anticipate that the maximum bearing pressure P_{max} should scale with the hardness H of the softer surface:

$$P_{max} = CH$$

where C is a constant. Thus the wear-rate of a bearing surface can be written:

$$\Omega = k_a P = C \left(\frac{P}{P_{max}} \right) k_a H \qquad (4.26)$$

Two material properties appear in this equation: the wear constant k_a and the hardness, H. They are plotted in Figure 4.16. The dimensionless quantity

$$K = k_a H \qquad (4.27)$$

is shown as a set of diagonal contours. Note, first, that materials of a given class (for instance, metals) tend to lie along a downward sloping diagonal across the figure, reflecting the fact that low wear rate is associated with high hardness. The best materials for bearings for a given bearing pressure P are those with the lowest value of k_a, that is, those nearest the bottom of the diagram. On the other hand, an efficient bearing, in terms of size or weight, will be loaded to a safe fraction of its maximum bearing pressure, that is, to a constant value of P/P_{max}, and for these, materials with the lowest values of the product $k_a H$ are best.

Cost bar charts

Properties like modulus, strength or conductivity do not change with time. Cost is bothersome because it does change with time. Supply, scarcity, speculation and inflation contribute to the considerable fluctuations in the cost-per-kg of a commodity like copper or silver. Data for cost-per-kg are tabulated for some materials in daily papers and trade journals; those for others are harder to come by. Approximate values for the cost of materials per kg, and their cost per m^3, are plotted in Figure 4.17(a) and (b). Most commodity materials (glass, steel, aluminum, and the common polymers) cost between $0.5 and $2/kg. Because

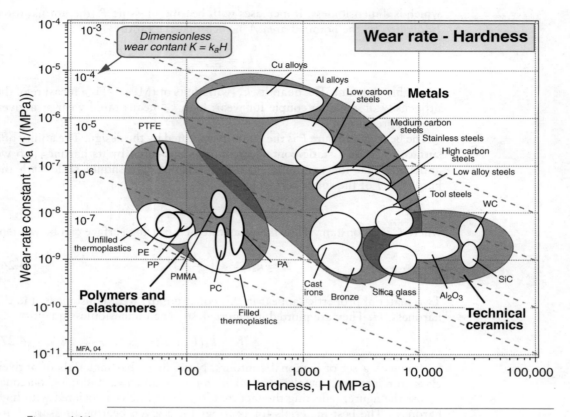

Figure 4.16   The normalized wear rate, k_a, plotted against hardness, H, here expressed in MPa rather than Vickers (H in MPa = $10\,H_v$). The chart gives an overview of the way in which common engineering materials behave.

they have low densities, the cost/m^3 of commodity polymers is less than that of metals.

The modulus–relative cost chart

In design for minimum cost, material selection is guided by indices that involve modulus, strength, and cost per unit volume. To make some correction for the influence of inflation and the units of currency in which cost is measured, we define a *relative cost per unit volume* $C_{v,R}$

$$C_{v,R} = \frac{\text{Cost/kg} \times \text{Density of material}}{\text{Cost/kg} \times \text{Density of mild steel rod}} \tag{4.28}$$

At the time of writing, steel reinforcing rod costs about US\$0.3/kg.

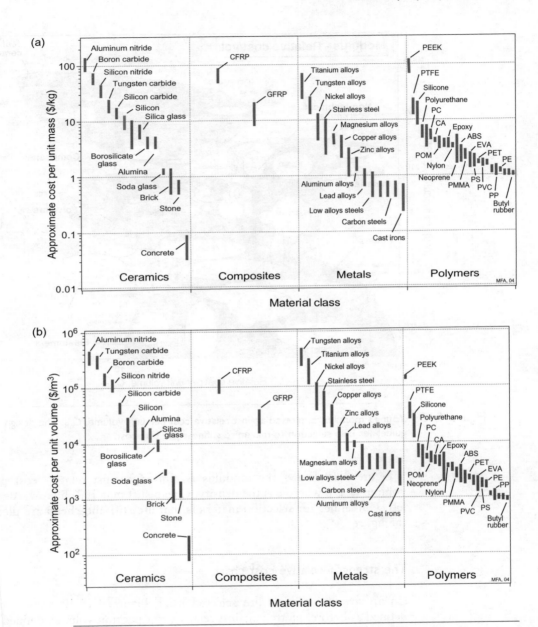

Figure 4.17 (a) The approximate cost/kg of materials. Commodity materials cost about $1/kg special materials cost much more. (b) The approximate cost/m³ of materials. Polymers, because they have low densities, cost less per unit volume than most other materials.

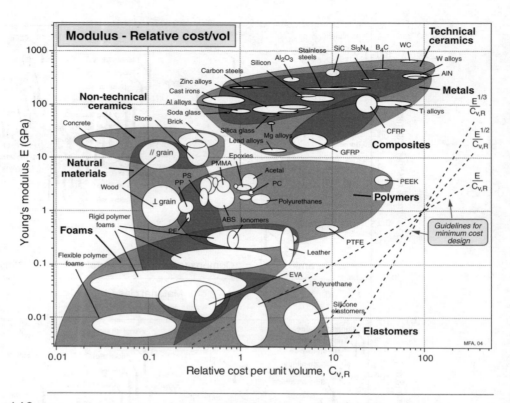

Figure 4.18   Young's modulus, E, plotted against relative cost per unit volume, $C_{v,R}$. The design guidelines help selection to maximize stiffness per unit cost.

Figure 4.18 shows the modulus E plotted against relative cost per unit volume $C_{v,R}$ ρ where ρ is the density. Cheap stiff materials lie towards the top left. Guidelines for selection materials that are stiff and cheap are plotted on the figure.

The strength–relative cost chart

Cheap strong materials are selected using Figure 4.19. It shows strength, defined as before, plotted against relative cost per unit volume, defined above. The qualifications on the definition of strength, given earlier, apply here also.

It must be emphasized that the data plotted here and on the chart of Figure 4.18 are less reliable than those of other charts, and subject to unpredictable change. Despite this dire warning, the two charts are genuinely useful. They allow selection of materials, using the criterion of "function per unit cost". An example is given in Chapter 6, Section 6.5.

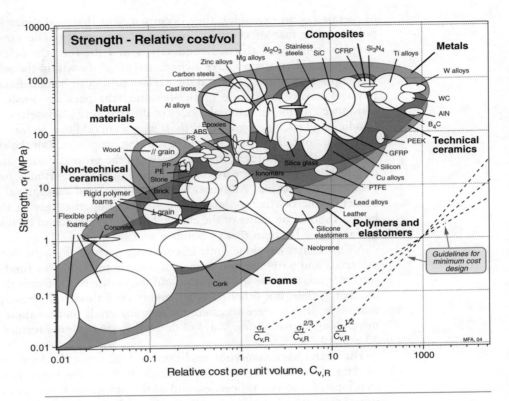

Figure 4.19 Strength, σ_f, plotted against relative cost per unit volume, $C_{v,R}$. The design guidelines help selection to maximize strength per unit cost.

4.4 Summary and conclusions

The engineering properties of materials are usefully displayed as material selection charts. The charts summarize the information in a compact, easily accessible way, they show the range of any given property accessible to the designer and they identify the material class associated with segments of that range. By choosing the axes in a sensible way, more information can be displayed: a chart of modulus E against density ρ reveals the longitudinal wave velocity $(E/\rho)^{1/2}$; a plot of fracture toughness K_{1C} against modulus E shows the toughness G_{1C}; a diagram of thermal conductivity λ against diffusivity, a, also gives the volume specific heat ρC_v; strength, σ_f, against modulus, E, shows the energy-storing capacity σ_f^2/E, and there are many more.

The most striking feature of the charts is the way in which members of a material class cluster together. Despite the wide range of modulus and density of metals (as an example), they occupy a field that is distinct from that of polymers, or that of ceramics, or that of composites. The same is true of strength, toughness, thermal conductivity and the rest: the fields sometimes overlap, but they always have a characteristic place within the whole picture.

The position of the fields and their relationship can be understood in simple physical terms: the nature of the bonding, the packing density, the lattice resistance and the vibrational modes of the structure (themselves a function of bonding and packing), and so forth. It may seem odd that so little mention has been made of micro-structure in determining properties. But the charts clearly show that the first-order difference between the properties of materials has its origins in the mass of the atoms, the nature of the inter-atomic forces and the geometry of packing. Alloying, heat treatment, and mechanical working all influence micro-structure, and through this, properties, giving the elongated bubbles shown on many of the charts; but the magnitude of their effect is less, by factors of 10, than that of bonding and structure.

All the charts have one thing in common: parts of them are populated with materials and parts are not. Some parts are inaccessible for fundamental reasons that relate to the size of atoms and the nature of the forces that bind their atoms together. But other parts are empty even though, in principle, they are accessible. If they were accessed, the new materials that lay there could allow novel design possibilities. Ways of doing this are explored further in Chapters 13 and 14.

The charts have numerous applications. One is the checking and validation of data (Chapter 15); here use is made both of the range covered by the envelope of material properties, and of the numerous relations between them (like $E\Omega = 100\,kT_{m}$), described in Section 4.3. Another concerns the development of, and identification of uses for, new materials; materials that fill gaps in one or more of the charts generally offer some improved design potential. But most important of all, the charts form the basis for a procedure for materials selection. That is developed in the following chapters.

4.5 Further reading

The best general book on the physical origins of the mechanical properties of materials remains that by Cottrell (1964). Values for the material properties that appear on the Charts derive from sources documented in Chapter 13.

Cottrell, A.H. (1964) *Mechanical Properties of Matter*, Wiley, New York Library of Congress Number 65-14262. (*An inspirational book, clear, full of insights and of simple derivations of the basic equations describing the mechanical behavior of solids, liquids and gasses.*)

Tabor, D. (1978) *Properties of Matter*, Penguin Books, London, UK. (*This text, like that of Cottrell, is notable for its clarity and physical insight.*)

Chapter 5

Materials selection — the basics

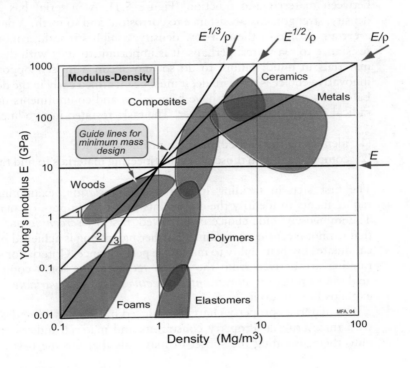

5.1 Introduction and synopsis

This chapter sets out the basic procedure for selection, establishing the link between material and function (Figure 5.1). A material has *attributes*: its density, strength, cost, resistance to corrosion, and so forth. A design demands a certain profile of these: a low density, a high strength, a modest cost and resistance to sea water, perhaps. It is important to start with the full menu of materials in mind; failure to do so may mean a missed opportunity. If an innovative choice is to be made, it must be identified early in the design process. Later, too many decisions have been taken and commitments made to allow radical change: it is now or never. The task, restated in two lines, is that of

(1) identifying the desired attribute profile and then
(2) comparing it with those of real engineering materials to find the best match.

The first step in tackling it is that of *translation*, examining the design requirements to identify the constraints that they impose on material choice. The immensely wide choice is narrowed, first, by *screening-out* the materials that cannot meet the constraints. Further narrowing is achieved by *ranking* the candidates by their ability to maximize performance. Criteria for screening and ranking are derived from the design requirements for a component by an analysis of *function, constraints, objectives*, and *free variables*. This chapter explains how to do it.

The materials property charts introduced in Chapter 4 are designed for use with these criteria. Property constraints and material indices can be plotted onto them, isolating the subset of materials that are the best choice for the

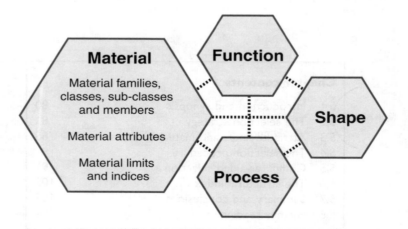

Figure 5.1   Material selection is determined by function. Shape sometimes influences the selection. This chapter and the next deal with materials selection when this is independent of shape.

design. The whole procedure can be implemented in software as a design tool, allowing computer-aided selection. The procedure is fast, and makes for lateral thinking. Examples of the method are given in Chapter 6.

5.2 The selection strategy

Material attributes

Figure 5.2 illustrates how the kingdom of materials is divided into families, classes, sub-classes, and members. Each member is characterized by a set of *attributes*: its properties. As an example, the materials kingdom contains the family "metals", which in turn contains the class "aluminum alloys", the sub-class "6000 series" and finally the particular member "Alloy 6061". It, and every other member of the kingdom, is characterized by a set of attributes that include its mechanical, thermal, electrical, optical, and chemical properties, its processing characteristics, its cost and availability, and the environmental consequences of its use. We call this its *property-profile*. Selection involves seeking the best match between the property-profiles of the materials in the kingdom and that required by the design.

There are four main steps, which we here call *translation, screening, ranking*, and *supporting information* (Figure 5.3). The steps can be likened to those in selecting a candidate for a job. The job is first analyzed and advertised, identifying essential skills and experience required of the candidate ("translation"). Some of these are simple go/no go criteria like the requirement that the applicant "must have a valid driving license", or "a degree in computer science",

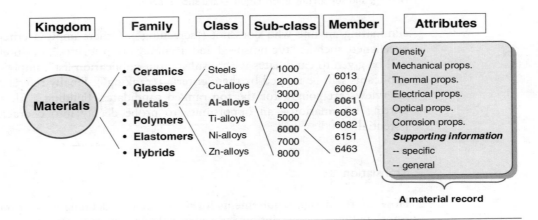

Figure 5.2   The taxonomy of the kingdom of materials and their attributes. Computer-based selection software stores data in a hierarchical structure like this.

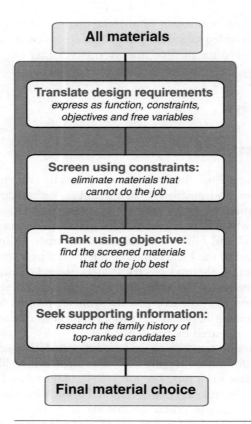

Figure 5.3 The strategy for materials selection. The four main steps — translation, screening, ranking, and supporting information — are shown here.

eliminating anyone who does not ("screening"). Others imply a criterion of excellence, such as "typing speed and accuracy are priorities", or "preference will be given to candidates with a substantial publication list", implying that applicants will be ranked by these criteria ("ranking"). Finally references and interviews are sought for the top ranked candidates, building a file of supporting information — an opportunity to probe deeply into character and potential.

Translation

How are the design requirements for a component (defining what it must do) translated into a prescription for a material? Any engineering component has one or more *functions*: to support a load, to contain a pressure, to transmit heat, and so forth. This must be achieved subject to *constraints*: that certain

Table 5.1   Function, constraints, objectives and free variables

Function	What does component do?
Constraints[*]	What non-negotiable conditions must be met?
	What negotiable but desirable conditions...?
Objective	What is to be maximized or minimized?
Free variables	What parameters of the problem is the designer free to change?

[*]It is sometimes useful to distinguish between "hard" and "soft" constraints. Stiffness and strength might be absolute requirements (hard constraints); cost might be negotiable (a soft constraint).

dimensions are fixed, that the component must carry the design loads or pressures without failure, that it insulates or conducts, that it can function in a certain range of temperature and in a given environment, and many more. In designing the component, the designer has an *objective*: to make it as cheap as possible, perhaps, or as light, or as safe, or perhaps some combination of these. Certain parameters can be adjusted in order to optimize the objective — the designer is free to vary dimensions that have not been constrained by design requirements and, most importantly, free to choose the material for the component. We refer to these as *free variables*. Function and constraints, objective and free variables (Table 5.1) define the boundary conditions for selecting a material and — in the case of load-bearing components — a shape for its cross-section. The first step in relating design requirements to material properties is a clear statement of function, constraints, objective, and free variables.

Screening: attribute limits

Unbiased selection requires that all materials are considered to be candidates until shown to be otherwise, using the steps in the boxes below "translate" in Figure 5.3. The first of these, *screening*, eliminates candidates that cannot do the job at all because one or more of their attributes lies outside the limits set by the constraints. As examples, the requirement that "the component must function in boiling water", or that "the component must be transparent" imposes obvious limits on the attributes of *maximum service temperature* and *optical transparency* that successful candidates must meet. We refer to these as *attribute limits*.

Ranking: material indices

Attribute limits do not, however, help with ordering the candidates that remain. To do this we need optimization criteria. They are found in the material indices, developed below, which measure how well a candidate that has passed the screening step can do the job. Performance is sometimes limited by a single property, sometimes by a combination of them. Thus the best materials for buoyancy are those with the lowest density, ρ; those best for thermal insulation the ones with the smallest values of the thermal conductivity, λ. Here maximizing or minimizing a single property maximizes performance. But — as we shall see – the best materials for a light stiff tie-rod are those with the greatest value of the specific stiffness, E/ρ, where E is Young's modulus. The best materials for a spring are those with the greatest value of σ_f^2/E where σ_f is the failure stress. The property or property-group that maximizes performance for a given design is called its *material index*. There are many such indices, each associated with maximizing some aspect of performance.[1] They provide *criteria of excellence* that allow ranking of materials by their ability to perform well in the given application.

To summarize: screening isolate candidates that are capable of doing the job; ranking identifies those among them that can do the job best.

Supporting information

The outcome of the steps so far is a ranked short-list of candidates that meet the constraints and that maximize or minimize the criterion of excellence, whichever is required. You could just choose the top-ranked candidate, but what bad secrets might it hide? What are its strengths and weaknesses? Does it have a good reputation? What, in a word, is its credit-rating? To proceed further we seek a detailed profile of each: its *supporting information* (Figure 5.3, bottom).

Supporting information differs greatly from the structured property data used for screening. Typically, it is descriptive, graphical or pictorial: case studies of previous uses of the material, details of its corrosion behavior in particular environments, information of availability and pricing, experience of its environmental impact. Such information is found in handbooks, suppliers' data sheets, CD-based data sources and the world-wide web. Supporting information helps narrow the short-list to a final choice, allowing a definitive match to be made between design requirements and material attributes.

Why are all these steps necessary? Without screening and ranking, the candidate-pool is enormous and the volume of supporting information overwhelming. Dipping into it, hoping to stumble on a good material, gets you nowhere. But once a small number of potential candidates have been identified by the screening–ranking steps, detailed supporting information can be sought for these few alone, and the task becomes viable.

---

[1] Maximizing performance often means minimizing something: cost is the obvious example; mass, in transport systems, is another. A low-cost or light component, here, improves performance.

Local conditions

The final choice between competing candidates will, often, depend on local conditions: on in-house expertise or equipment, on the availability of local suppliers, and so forth. A systematic procedure cannot help here — the decision must instead be based on local knowledge. This does not mean that the result of the systematic procedure is irrelevant. It is always important to know which material is best, even if, for local reasons, you decide not to use it.

We will explore supporting information more fully in Chapter 15. Here we focus on the derivation of property limits and indices.

5.3 Attribute limits and material indices

Constraints set property limits. Objectives define material indices, for which we seek extreme values. When the objective in not coupled to a constraint, the material index is a simple material property. When, instead, they are coupled, the index becomes a group of properties like those cited above. Both are explained below. We start with two simple examples of the first — uncoupled objectives.

Heat sinks for hot microchips. A microchip may only consume milliwatts, but the power is dissipated in a tiny volume. The power is low but the *power-density* is high. As chips shrink and clock-speeds grow, heating becomes a problem. The Pentium chip of today's PCs already reaches 85°C, requiring forced cooling. Multiple-chip modules (MCMs) pack as many as 130 chips on to a single substrate. Heating is kept under control by attaching the chip to a heat sink (Figure 5.4), taking pains to ensure good thermal contact between the chip and the sink. The heat sink now becomes a critical component, limiting further development of the electronics. How can its performance be maximized?

To prevent electrical coupling and stray capacitance between chip and heat sink, the heat sink must be a good electrical insulator, meaning a resistivity,

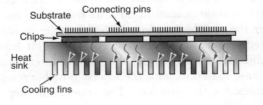

Figure 5.4   A heat sink for power micro-electronics. The material must insulate electrically, but conduct heat as well as possible.

Table 5.2   Function, constraints, objective, and free variables for the heat sink

Function	Heat sink
Constraints	• Material must be "good insulator", or $\rho_e > 10^{19}\,\mu\Omega.\text{cm}$
	• All dimensions are specified
Objective	Maximize thermal conductivity, λ
Free variables	Choice of material

$\rho_e > 10^{19}\,\mu\Omega.\text{cm}$. But to drain heat away from the chip as fast as possible, it must also have the highest possible thermal conductivity, λ. The translation step is summarized in Table 5.2, where we assume that all dimensions are constrained by other aspects of the design.

To explain: resistivity is treated as a *constraint*, a go/no go criterion. Materials that fail to qualify as "good insulator", or have a resistivity greater than the value listed in the table, are screened out. The thermal conductivity is treated as an *objective*: of the materials that meet the constraint, we seek those with the largest values of λ and rank them by this — it becomes the material index for the design. If we assume that all dimensions are fixed by the design, there remains only one *free variable* in seeking to maximize heat-flow: the choice of material. The procedure, then, is to *screen* on resistivity, then *rank* on conductivity.

The steps can be implemented using the $\lambda - \rho_e$ chart of Figure 4.10, reproduced as Figure 5.5. Draw a vertical line at $\rho_e = 10^{19}\mu\Omega.\text{cm}$, then pick off the materials that lie above this line, and have the highest λ. The result: aluminum nitride, AlN, or alumina, Al_2O_3. The final step is to seek supporting information for these two materials. A web-search on "aluminum nitride" leads immediately to detailed data-sheets with the information we seek.

Materials for overhead transmission lines. Electrical power, today, is generated centrally and distributed by overhead or underground cables. Buried lines are costly so cheaper overhead transmission (Figure 5.6) is widely used. A large span is desirable because the towers are expensive, but so too is a low electrical resistance to minimize power losses. The span of cable between two towers must support the tension needed to limit its sag and to tolerate wind and ice loads. Consider the simple case in which the tower spacing L is fixed at a distance that requires a cable with a strength σ_f of at least 80 MPa (a constraint). The objective then becomes that of minimizing resistive losses, and that means seeking materials with the lowest possible resistivity, ρ_e, defining the material index for the problem. The translation step is summarized in Table 5.3.

The prescription, then, is to *screen* on strength and *rank* on resistivity. There is no $\sigma_f - \rho_e$ chart in Chapter 4 (though it is easy to make one using the

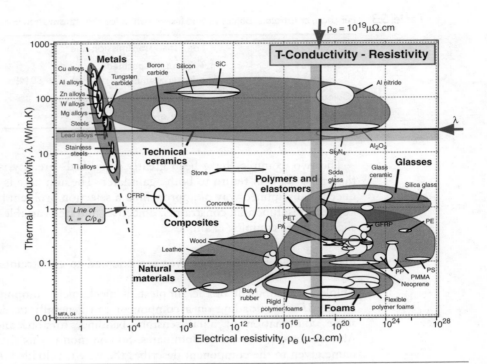

Figure 5.5 The $\lambda - \rho_e$ chart of Figure 4.10 with the attribute limit $\rho_e > 10^{19}$ $\mu\Omega$.cm and the index λ plotted on it. The selection is refined by raising the position of the λ selection line.

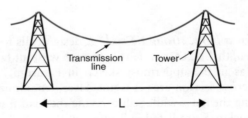

Figure 5.6 A transmission line. The cable must be strong enough to carry its supporting tension, together with wind and ice loads. But it must also conduct electricity as well as possible.

software described in Section 5.5). Instead we use the $\lambda - \rho_e$ chart of Figure 4.10 to identify materials with the lowest resistivity (Cu and Al alloys) and then check, using the $\sigma_f - \rho$ chart of Figure 4.4 that the strength meets the constraint listed in the table. Both do (try it!).

Table 5.3 Function, constraints, objective, and free variables for the transmission line

Function	Long span transmission line
Constraints	• Span L is specified
	• Material must be strength $\sigma_f > 80$ MPa
Objective	Minimize electrical resistivity ρ_e
Free variables	Choice of material

The two examples have been greatly simplified — reality is more complex than this. We will return to both again later. The aim here is simply to introduce the disciplined way of approaching a selection problem by identifying its key features: function, constraints, objective, and free variables. Now for some slightly more complex examples.

Material indices when objectives are coupled to constraints

Think for a moment of the simplest of mechanical components, helped by Figure 5.7. The loading on a component can generally be decomposed into some combination of axial tension, bending, torsion, and compression. Almost always, one mode dominates. So common is this that the functional name given to the component describes the way it is loaded: *ties* carry tensile loads; *beams* carry bending moments; *shafts* carry torques; and *columns* carry compressive axial loads. The words "tie", "beam", "shaft", and "column" each imply a function. Many simple engineering functions can be described by single words or short phrases, saving the need to explain the function in detail. Here we explore property limits and material indices for some of these.

Material index for a light, strong tie-rod. A design calls for a cylindrical tie-rod of specified length L to carry a tensile force F without failure; it is to be of minimum mass, as in the uppermost sketch in Figure 5.7. The length L is specified but the cross-section area A is not. Here, "maximizing performance" means "minimizing the mass while still carrying the load F safely". The design requirements, translated, are listed in Table 5.4.

We first seek an equation describing the quantity to be maximized or minimized. Here it is the mass m of the tie, and it is a minimum that we seek. This equation, called *the objective function*, is

$$m = AL\rho \tag{5.1}$$

where A is the area of the cross-section and ρ is the density of the material of which it is made. The length L and force F are specified and are therefore fixed; the cross-section A, is free. We can reduce the mass by reducing the cross-section, but there is a constraint: the section-area A must be sufficient to

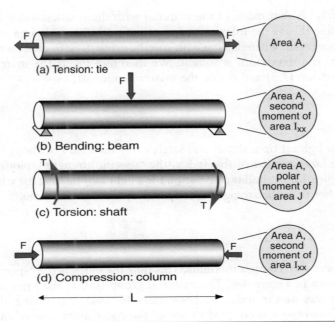

(a) Tension: tie

Area A,

(b) Bending: beam

Area A, second moment of area I_{xx}

(c) Torsion: shaft

Area A, polar moment of area J

(d) Compression: column

Area A, second moment of area I_{xx}

L

Figure 5.7   A cylindrical tie-rod loaded (a) in tension, (b) in bending, (c) in torsion and (d) axially, as a column. The best choice of materials depends on the mode of loading and on the design goal; it is found by deriving the appropriate material index.

Table 5.4   Design requirements for the light tie

Function	Tie rod
Constraints	• Length L is specified • Tie must support axial tensile load F without failing
Objective	Minimize the mass m of the tie
Free variables	• Cross-section area, A • Choice of material

carry the tensile load F, requiring that

$$\frac{F}{A} \le \sigma_f \tag{5.2}$$

where σ_f is the failure strength. Eliminating A between these two equations give

$$m \ge (F)(L)\left(\frac{\rho}{\sigma_f}\right) \tag{5.3}$$

Note the form of this result. The first bracket contains the specified load F. The second bracket contains the specified geometry (the length L of the tie). The last bracket contains the material properties. The lightest tie that will carry F

safely[2] is that made of the material with the smallest value of ρ/σ_f. We could define this as the material index of the problem, seeking a minimum, but it is more usual, when dealing with specific properties, to express them in a form for which a maximum is sought. We therefore invert the material properties in equation (5.3) and define the material index M, as

$$M = \frac{\sigma_f}{\rho} \tag{5.4}$$

The lightest tie-rod that will safely carry the load F without failing is that with the largest value of this index, the "specific strength", plotted in the chart of Figure 4.6. A similar calculation for a light *stiff* tie (one for which the stiffness S rather than the strength σ_f is specified) leads to the index

$$M = \frac{E}{\rho} \tag{5.5}$$

where E is Young's modulus. This time the index is the "specific stiffness", also shown in Figure 4.6. The material group (rather than just a single property) appears as the index in both cases because minimizing the mass m — the objective — was coupled to one of the constraints, that of carrying the load F without failing or deflecting too much.

That was easy. Now for a slightly more difficult (and important) one.

Material index for a light, stiff beam. The mode of loading that most commonly dominates in engineering is not tension, but bending — think of floor joists, of wing spars, of golf-club shafts. Consider, then, a light beam of square section $b \times b$ and length L loaded in bending. It must meet a constraint on its stiffness S, meaning that it must not deflect more than δ under a load F (Figure 5.8). Table 5.5 translates the design requirements.

Appendix A of this book catalogues useful solutions to a range of standard problems. The stiffness of beams is one of these. Turning to Section A3 we find an equation for the stiffness S of an elastic beam. The constraint requires that $S = F/\delta$ be greater than this:

$$S = \frac{F}{\delta} \geq \frac{C_1 EI}{L^3} \tag{5.6}$$

where E is Young's modulus, C_1 is a constant that depends on the distribution of load and I is the second moment of the area of the section, which, for a beam of square section ("Useful Solutions", Appendix A, Section A.2), is

$$I = \frac{b^4}{12} = \frac{A^2}{12} \tag{5.7}$$

---

[2] In reality a safety factor, S_f, is always included in such a calculation, such that equation (5.2) becomes $F/A = \sigma_f/S_f$. If the same safety factor is applied to each material, its value does not influence the choice. We omit it here for simplicity.

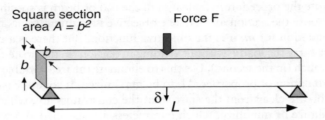

Square section
area $A = b^2$

Force F

δ

L

Figure 5.8   A beam of square section, loaded in bending. Its stiffness is $S = F/\delta$ where F is the load and δ is the deflection.

Table 5.5   Design requirements for the light stiff beam

Function	Beam
Constraints	• Length L is specified
	• Beam must support a bending load F without deflecting too much, meaning that the bending stiffness S is specified
Objective	Minimize the mass of the beam
Free variables	• Cross-section area, A
	• Choice of material

The stiffness S and the length L are specified; the section area A is free. We can reduce the mass of the beam by reducing A, but only so far that the stiffness constraint is still met. Using these two equations to eliminate A in equation (5.1) for the mass gives

$$m \geq \left(\frac{12S}{C_1 L}\right)^{1/2} (L^3)\left(\frac{\rho}{E^{1/2}}\right) \tag{5.8}$$

The brackets are ordered as before: functional requirement, geometry and material. The best materials for a light, stiff beam are those with the smallest values of $\rho/E^{1/2}$. As before, we will invert this, seeking instead large values of the material index

$$M = \frac{E^{1/2}}{\rho} \tag{5.9}$$

In deriving the index, we have assumed that the section of the beam remained square so that both edges changed in length when A changed. If one of the two dimensions is held fixed, the index changes. A panel is a flat plate with a given length L and width W; the only free variable (apart from material) is the thickness t. For this the index becomes (via an identical derivation)

$$M = \frac{E^{1/3}}{\rho} \tag{5.10}$$

Note the procedure. The length of the rod or beam is specified but we are free to choose the section area A. The objective is to minimize its mass, m. We write an equation for m: it is the objective function. But there is a constraint: the rod must carry the load F without yielding in tension (in the first example) or bending too much (in the second). Use this to eliminate the free variable A and read off the combination of properties, M, to be maximized. It sounds easy, and it is so long as you are clear from the start what the constraints are, what you are trying to maximize or minimize, which parameters are specified and which are free.

Deriving indices — how to do it

This is a good moment to describe the method in more general terms. *Structural elements* are components that perform a physical function: they carry loads, transmit heat, store energy, and so on: in short, they satisfy *functional requirements*. The functional requirements are specified by the design: a tie must carry a specified tensile load; a spring must provide a given restoring force or store a given energy, a heat exchanger must transmit heat a given heat flux, and so on.

The performance of a structural element is determined by three things: the functional requirements, the geometry and the properties of the material of which it is made.[3] The performance P of the element is described by an equation of the form

$$P = \left[\left(\begin{array}{c} \text{Functional} \\ \text{requirements, } F \end{array} \right), \left(\begin{array}{c} \text{Geometric} \\ \text{parameters, } G \end{array} \right), \left(\begin{array}{c} \text{Material} \\ \text{properties, } M \end{array} \right) \right]$$

or

$$P = f(F, G, M) \tag{5.11}$$

where P, the *performance metric*, describes some aspect of the performance of the component: its mass, or volume, or cost, or life for example; and "f" means "*a function of*". *Optimum design* is the selection of the material and geometry that maximize or minimize P, according to its desirability or otherwise.

The three groups of parameters in equation (5.11) are said to be *separable* when the equation can be written

$$P = f_1(F) \cdot f_2(G) \cdot f_3(M) \tag{5.12}$$

where f_1, f_2, and f_3 are separate functions that are simply multiplied together. When the groups are separable, as they frequently are, the optimum choice of material becomes independent of the details of the design; it is the same for all geometries, G, and for all values of the function requirement, F. Then the optimum subset of materials can be identified without solving the complete design problem, or even knowing all the details of F and G. This enables

---

[3] In Chapter 11, we introduce a fourth: that of section shape.

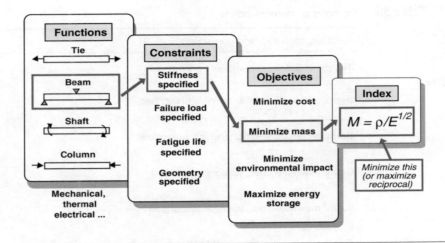

Figure 5.9   The specification of function, objective, and constraint leads to a materials index. The combination in the highlighted boxes leads to the index $E^{1/2}/\rho$.

enormous simplification: the performance for *all F* and *G* is maximized by maximizing $f_3(M)$, which is called the material efficiency coefficient, or *material index* for short. The remaining bit, $f_1(F) \cdot f_2(G)$, is related to *the structural efficiency coefficient*, or *structural index*. We do not need it now, but will examine it briefly in Section 5.7.

Each combination of function, objective and constraint leads to a material index (Figure 5.9); the index is characteristic of the combination, and thus of the function the component performs. The method is general, and, in later chapters, is applied to a wide range of problems. Table 5.6 gives examples of indices and the design problems that they characterize. A fuller catalogue of indices is given in Appendix B. New problems throw up new indices, as the case studies of the next chapter will show.

5.4 The selection procedure

We can now assemble the four steps into a systematic procedure.

Translation

Table 5.7 says it all. Simplified: identify the material attributes that are constrained by the design, decide what you will use as a criterion of excellence (to be minimized or maximized), substitute for any free variables using one of the constraints, and read off the combination of material properties that optimize the criterion of excellence.

Table 5.6   Examples of material-indices

Function, objective, and constraints	Index
Tie, minimum weight, stiffness prescribed	$\dfrac{E}{\rho}$
Beam, minimum weight, stiffness prescribed	$\dfrac{E^{1/2}}{\rho}$
Beam, minimum weight, strength prescribed	$\dfrac{\sigma_y^{2/3}}{\rho}$
Beam, minimum cost, stiffness prescribed	$\dfrac{E^{1/2}}{C_m\rho}$
Beam, minimum cost, strength prescribed	$\dfrac{\sigma_y^{2/3}}{C_m\rho}$
Column, minimum cost, buckling load prescribed	$\dfrac{E^{1/2}}{C_m\rho}$
Spring, minimum weight for given energy storage	$\dfrac{\sigma_y^2}{E\rho}$
Thermal Insulation, minimum cost, heat flux prescribed	$\dfrac{1}{\lambda C_p\rho}$
Electromagnet, maximum field, temperature rise prescribed	$\dfrac{C_p\rho}{\rho_e}$

ρ = density; E = Young's modulus; σ_y = elastic limit; C_m = cost/kg λ = thermal conductivity; ρ_e = electrical resistivity; C_p = specific heat.

Screening: applying attribute limits

Any design imposes certain non-negotiable demands ("constraints") on the material of which it is made. We have explained how these are translated into attribute limits. Attribute limits plot as horizontal or vertical lines on material selection charts, illustrated in Figure 5.10. It shows a schematic $E - \rho$ chart, in the manner of Chapter 4. We suppose that the design imposes limits on these of $E > 10\,\text{GPa}$ and $\rho < 3\,\text{Mg/m}^3$, shown on the figure. The optimizing search is restricted to the window boxed by the limits, labeled "Search region". Less quantifiable properties such as corrosion resistance, wear resistance or formability can all appear as primary limits, which take the form

$$A > A^*$$

or

$$A < A^* \tag{5.13}$$

Table 5.7 Translation

Step	Action
1	Define the design requirements: (a) Function: what does the component do? (b) Constraints: essential requirements that must be met: stiffness, strength, corrosion resistance, forming characteristics, . . . (c) Objective: what is to be maximized or minimized? (d) Free variables: what are the unconstrained variables of the problem?
2	List the constraints (no yield; no fracture; no buckling, etc.) and develop an equation for them if necessary
3	Develop an equation for the objective in terms of the functional requirements, the geometry and the material properties (the objective function)
4	Identify the free (unspecified) variables
5	Substitute for the free variables from the constraint equations into the objective function
6	Group the variables into three groups: functional requirements, F, geometry, G, and material properties, M, thus Performance metric $P \leq f_1(F) \cdot f_2(G) \cdot f_3(M)$ or Performance metric $P \geq f_1(F) \cdot f_2(G) \cdot f_3(M)$
7	Read off the material index, expressed as a quantity M, that optimizes the performance metric P. M is the criterion of excellence.

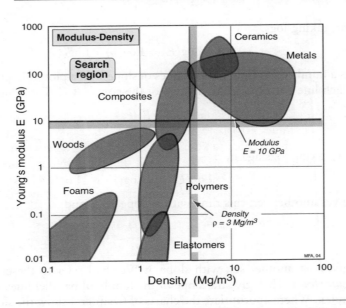

Figure 5.10 A schematic $E - \rho$ chart showing a lower limit for E and an upper one for ρ.

where A is an attribute (service temperature, for instance) and A^* is a critical value of that attribute, set by the design, that must be exceeded, or (in the case of corrosion rate) must *not* be exceeded.

One should not be too hasty in applying attribute limits; it may be possible to engineer a route around them. A component that gets too hot can be cooled; one that corrodes can be coated with a protective film. Many designers apply attribute limits for fracture toughness, K_{1C} and ductility ε_f insisting on materials with, as rules of thumb, $K_{1C} > 15$ MPa.m$^{1/2}$ and $\varepsilon_f > 2\%$ in order to guarantee adequate tolerance to stress concentrations. By doing this they eliminate materials that the more innovative designer is able to use to good purpose (the limits just cited for K_{1C} and ε_f eliminate all polymers and all ceramics, a rash step too early in the design). At this stage, keep as many options open as possible.

Ranking: indices on charts

The next step is to seek, from the subset of materials that meet the property limits, those that maximize the performance of the component. We will use the design of light, stiff components as an example; the other material indices are used in a similar way.

Figure 5.11 shows, as before, modulus E, plotted against density ρ, on log scales. The material indices E/ρ, $E^{1/2}/\rho$, and $E^{1/3}/\rho$ can be plotted onto the figure. The condition

$$\frac{E}{\rho} = C$$

or, taking logs,

$$\mathrm{Log}(E) = \mathrm{Log}(\rho) + \mathrm{Log}(C) \tag{5.14}$$

is a family of straight parallel lines of slope 1 on a plot of $\mathrm{Log}(E)$ against $\mathrm{Log}(\rho)$ each line corresponds to a value of the constant C. The condition

$$\frac{E^{1/2}}{\rho} = C \tag{5.15}$$

or, taking logs again,

$$\mathrm{Log}(E) = 2\,\mathrm{Log}(\rho) + 2\,\mathrm{Log}(C) \tag{5.16}$$

gives another set, this time with a slope of 2; and

$$\frac{E^{1/3}}{\rho} = C \tag{5.17}$$

gives yet another set, with slope 3. We shall refer to these lines as *selection guidelines*. They give the slope of the family of parallel lines belonging to that index. Where appropriate the charts of Chapter 4 show the slopes of guidelines like these.

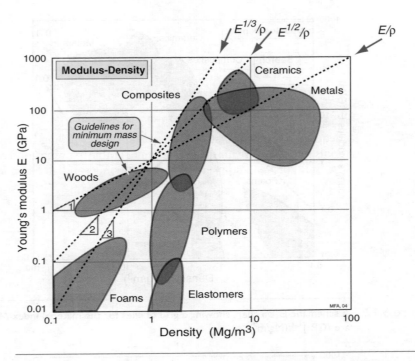

Figure 5.11   A schematic $E - \rho$ chart showing guidelines for the three material indices for stiff, lightweight design.

It is now easy to read off the subset materials that optimally maximize performance for each loading geometry. All the materials that lie on a line of constant $E^{1/2}/\rho$ perform equally well as a light, stiff beam; those above the line are better, those below, worse. Figure 5.12 shows a grid of lines corresponding to values of $E^{1/2}/\rho$ from 0.1 to 3 in units of $GPa^{1/2}/(Mg/m^3)$. A material with $M = 1$ in these units gives a beam that has one tenth the weight of one with $M = 0.1$. The subset of materials with particularly good values of the index is identified by picking a line that isolates a search area containing a reasonably small number of candidates, as shown schematically in Figure 5.13 as a diagonal selection line. Attribute limits can be added, narrowing the search window: that corresponding to $E > 50\,GPa$ is shown as a horizontal line. The short-list of candidate materials is expanded or contracted by moving the index line.

Supporting information

We now have a ranked short-list of potential candidate materials. The last step is to explore their character in depth. The list of constraints usually contains

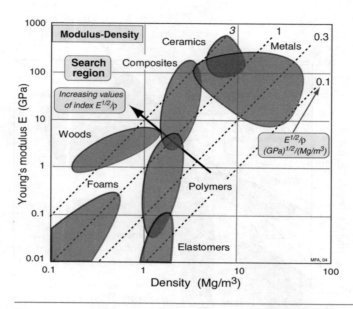

Figure 5.12   A schematic $E - \rho$ chart showing a grid of lines for the material index $M = E^{1/2}/\rho$. The units are $(GPa)^{1/2}/(Mg/m^3)$.

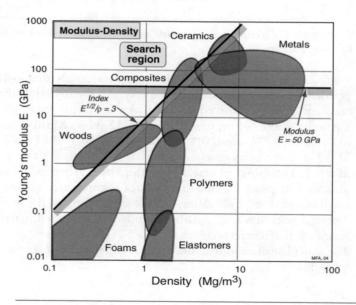

Figure 5.13   A selection based on the index $M = E^{1/2}/\rho$ together with the property limit $E > 50\,GPa$. The shaded band with slope 2 has been positioned to isolate a subset of materials with high $E^{1/2}/\rho$; the horizontal one lies at $E = 50\,GPa$. The materials contained in the search region become the candidates for the next stage of the selection process.

some that cannot be expressed as simple attribute limits. Many of these relate to the behavior of the material in a given environment, or when in contact with another material, or to aspects of the ways in which the material can be shaped, joined, or finished. Such information can be found in handbooks, manufacturers data-sheets, or on the internet. And then — it is to be anticipated — there are the constraints that have been overlooked simply because they were not seen as such. Confidence is built by seeking design guidelines, case studies or failure analyses that document each candidate, building a dossier of its strengths, its weaknesses, and ways in which these can be overcome. All of these come under the heading of supporting information. Finding it is the subject of Chapter 15.

The selection procedure is extended in Chapters 9 and 11 to deal with multiple constraints and objectives and to include section shape. Before moving on to these, it is a good idea to consolidate the ideas so far by applying them to a number of case studies. They follow in Chapter 6.

5.5 Computer-aided selection

The charts of Chapter 4 give an overview, but the number of materials that can be shown on any one of them is obviously limited. Selection using them is practical when there are very few constraints, as the examples of Section 5.3 showed, but when there are many — as there usually are — checking that a given material meets them all is cumbersome. Both problems are overcome by computer implementation of the method.

The CES material and process selection software[4] is an example of such an implementation. A database contains records for materials, organized in the hierarchical manner shown in Figure 5.2. Each record contains structured property-data for a material, each stored as a range spanning the typical (or, often, the permitted) range of values of that property. It also contains limited unstructured data in the form of text, images, and references to sources of information about the material. The data are interrogated by a search engine that offers search interfaces shown schematically in Figure 5.14. On the left is a simple query interface for screening on single properties. The desired upper or lower limits for constrained attributes are entered; the search engine rejects all materials with attributes that lie outside the limits. In the center is shown a second way of interrogating the data: a bar chart like that shown earlier as Figure 4.1. It and the bubble chart shown on the right are the ways both of applying constraints and of ranking. Used for ranking, a selection line or box is super-imposed on the charts with edges that lie at the constrained values of the property (bar chart) or properties (bubble chart), eliminating the material in the shaded areas, and leaving the materials that

---

[4] Granta Design Ltd., Cambridge, UK (www.grantadesign.com).

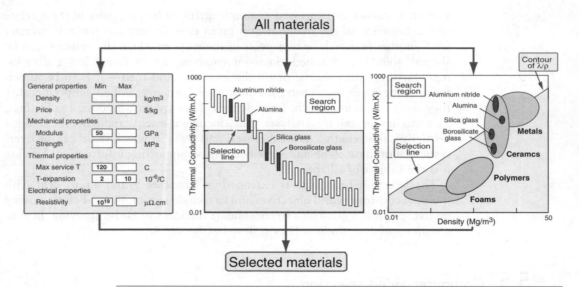

Figure 5.14 Computer-aided selection using the CES software. The schematic shows the three types of selection window. They can be used in any order and any combination. The selection engine isolates the subset of material that pass all the selection stages.

meet all the constraints. If instead, ranking is sought (having already applied all necessary constraints) the line or box is positioned so that a few — say, three — materials are left in the selected area; these are the top ranked candidates.

The figure illustrates an elaboration of the heat sink example given earlier, in which we now add more constraints (Table 5.8). The requirements are as before, plus the requirement that the modulus be greater than 50 GPa, that the expansion coefficient α, lies between 2 and $10 \times 10^{-6}/$°C and that the maximum service temperature exceeds 120°C. All are applied as property limits on the left-hand window, implementing a screening stage.

Ranking on thermal conductivity is shown in the central window. Materials that fail the screening stage on the left are grayed-out; those that pass remain colored. The selection line has been positioned so that two classes of material lie in the search region. The top-ranked candidate is aluminum nitride, the second is alumina. If, for some reason, the mass of the heat sink was also important, it might instead be desired to rank using material index λ/ρ, where ρ is the density. Then the window on the right, showing a $\lambda - \rho$ chart, allows selection by λ/ρ, plotted as diagonal contour on the schematic. The materials furthest above the line are the best choice. Once again, AlN wins.

Table 5.8 Function, expanded constraints, objective, and free variable for the heat sink

Function	Heat sink
Constraints	• Material must be "good insulator", or $\rho_e > 10^{19}\,\mu\Omega.cm$
	• Modulus $E > 50$ GPa
	• Maximum service temperature $T_{max} > 120°C$
	• Expansion coefficient $2 \times 10^{-6} < \alpha < 10 \times 10^{-6}/°C$
	• All dimensions are specified
Objective	Maximize thermal conductivity, λ or conductivity per unit mass λ/ρ
Free variables	Choice of material

Table 5.9 The selection

Material
Diamond
Beryllia (Grade 99)
Beryllia (Grade B995)
Beryllia (Grade BZ)
Aluminum nitride (fully dense)
Aluminum nitride (97 percent dense)

The software contains not one, but two databases. The first of these contains the 68 material classes shown in the charts of Chapter 4 — indeed all these charts were made using the software. They are chosen because they are those most widely used; between them they account for 98 percent of material usage. This database allows a first look at a problem, but it is inadequate for a fuller exploration. The second database is much larger — it contains data for over 3000 materials. By changing the database, the selection criteria already entered are applied instead to the much larger population. Doing this (and ranking on λ as in the central window) gives the top rank candidates listed in Table 5.9, listed in order of decreasing λ. Diamond is outstanding but is probably impracticable for reasons of cost; and compounds of beryllium (beryllia is beryllium oxide) are toxic and for this reason perhaps undesirable. That leaves us with aluminum nitride, our earlier choice. Part of a record for one grade of aluminum nitride is shown in Table 5.10. The upper part lists structured data (there is more, but it's not relevant in this example). The lower part gives the limited unstructured data provided by the record itself, and references to sources that are linked to the record in which more supporting information can be found. The search engine has a further feature, represented by the button labeled "search web" next to the material name at the top. Activating it sends the material name as a string to a web search engine, delivering supporting information available there.

Examples of the use of the software appear later in the book.

Table 5.10 Part of a record for aluminum nitride, showing structured and unstructured data, references and the web-search facility

Aluminum Nitride			
General properties		*Thermal properties*	
Density	3.26–3.33 Mg/m^3	Thermal conductivity	80–200 W/m.K
Price	*70–95 $/kg	Thermal expansion	4.9–6.2 μstrain/K
Mechanical properties		Max. service temperature	*1027–1727 °C
Young's M modulus	302–348 GPa		
Hardness — Vickers	990–1260 HV	*Electrical properties*	
Compressive strength	1970–2700 MPa	Resistivity	1e18–1e21 μΩ.cm
Fracture toughness	2.5–3.4 MPa.m$^{1/2}$	Dielectric constant	8.3–9.3

Supporting information

Design guidelines. Aluminum nitride (AlN) has an unusual combination of properties: it is an electrical insulator, but an excellent conductor of heat. This is just what is wanted for substrates for high-powered electronics; the substrate must insulate yet conduct the heat out of the microchips. This, and its high strength, chemical stability, and low expansion give it a special role as a heat sinks for power electronics. Aluminum nitride starts as a powder, is pressed (with a polymer binder) to the desired shape, then fired at a high temperature, burning off the binder and causing the powder to sinter.

Technical notes. Aluminum nitride is particularly unusual for its high thermal conductivity combined with a high electrical resistance, low dielectric constant, good corrosion, and thermal shock resistance.

Typical uses. Substrates for microcircuits, chip carriers, heat sinks, electronic components; windows, heaters, chucks, clamp rings, gas distribution plates.

References

Handbook of Ceramics, Glasses and Diamonds, (2001) Harper, C.A. editor, McGraw-Hill, New York, NY, USA. ISBN 0-07-026712-X. *(A comprehensive compilation of data and design guidelines.)*

Handbook of structural ceramics, editor: M.M. Schwartz, McGraw-Hill, New York, USA (1992)

Morrell, R. *Handbook of properties of technical & engineering ceramics,* Parts I and II, National Physical Laboratory, Her Majesty's Stationery Office, London, UK (1985)

5.6 The structural index

Books on optimal design of structures (e.g. Shanley, 1960) make the point that the efficiency of material usage in mechanically loaded components depends on the product of three factors: the material index, as defined here; a factor describing section shape, the subject of our Chapter 11; and a *structural index,*[5] which contains elements of the G and F of equation (5.12).

---

[5] Also called the "structural loading coefficient", the "strain number" or the "strain index".

The subjects of this book—material and process selection—focuses on the material index and on shape; but we should examine the structural index briefly, partly to make the connection with the classical theory of optimal design, and partly because it becomes useful (even to us) when structures are scaled in size.

In design for minimum mass (equations (5.3) and (5.8)), a measure of the efficiency of the design is given by the quantity m/L^3. Equation (5.3), for instance, can be written

$$\frac{m}{L^3} \geq \left(\frac{F}{L^2}\right)\left(\frac{\rho}{\sigma_f}\right) \qquad (5.18)$$

and equation (5.8) becomes

$$\frac{m}{L^3} \geq \left(\frac{12}{C_1}\right)^{1/2}\left(\frac{S}{L}\right)^{1/2}\left(\frac{\rho}{E^{1/2}}\right) \qquad (5.19)$$

This m/L^3 has the dimensions of density; the lower this pseudo-density the lighter is the structure for a given scale, and thus the greater is the structural efficiency. The first bracketed term on the right of the equation is merely a constant. The last is the material index. The middle one, F/L^2 for strength-limited design and S/L for stiffness limited design, is called the *structural index*. It has the dimensions of stress; it is a measure of the intensity of loading. Design proportions that are optimal, minimizing material usage, are optimal for structures of any size provided they all have the same structural index. The performance equation (5.8), was written in a way that isolated the structural index, a convention we shall follow in the case studies of Chapter 6.

The structural index for a component of minimum cost is the same as that for one of minimum mass; it is F/L^2 again for strength limited design, S/L when it is stiffness. For beams or columns of minimum mass, cost, or energy content, they is the same. For panels (dimensions $L \times W$) loaded in bending or such that they buckle it is FW/L^3 and SW^2/L^3 where L and W are the (fixed) dimensions of the panel.

5.7 Summary and conclusions

Material selection is tacked in four steps.

- *Translation*—reinterpreting the design requirements in terms of function, constraints, objectives, and free variables.
- *Screening*—deriving attribute limits from the constraints and applying these to isolate a subset of viable materials.
- *Ranking*—ordering the viable candidates by the value of a material index, the criterion of excellence that maximizes or minimizes some measure of performance.

- Seeking supporting information for the top-ranked candidates, exploring aspects of their past history, their established uses, their behavior in relevant environments, their availability and more until a sufficiently detailed picture is built up that a final choice can be made.

Hard-copy material charts allow a first go at the task, and have the merit of maintaining breadth of vision: all material classes are in the frame, so to speak. But materials have many properties, and the number of combinations of these appearing in indices is very much larger. It is impractical to print charts for all of them. Even if you did, their resolution is limited. Both problems are overcome by computer implementation, allowing freedom to explore the whole kingdom of materials and also providing detail when required.

5.8 Further reading

The books listed below discuss optimization methods and their application in materials engineering. None contain the approach developed here.

Dieter, G.E. (1991) *Engineering Design, a Materials and Processing Approach*, 2nd edition, McGraw-Hill, New York, USA. ISBN 0-07-100829-2. *(A well-balanced and respected text focusing on the place of materials and processing in technical design.)*

Gordon, J.E. (1976) *The New Science of Strong Materials, or why you don't Fall Through the Floor*, 2nd edition, Penguin Books, Harmondsworth, UK. ISBN 0-1402-0920-7. *(This very readable book presents ideas about plasticity and fracture, and ways of designing materials to prevent them.)*

Gordon, J.E. (1978) *Structures, or why Things don't Fall Down*, Penguin Books, Harmondsworth, UK. ISBN 0-1402-1961-7. *(A companion to the other book by Gordon (above), this time introducing structural design.)*

Shanley, F.R. (1960) *Weight-Strength Analysis of Aircraft Structures*, 2nd edition, Dover Publications, Inc. New York, USA. Library of Congress Number 60-50107. *(A remarkable text, no longer in print, on the design of light-weight structures.)*

Arora, J.S. (1989) *Introduction to Optimum Design*, McGraw-Hill, New York, USA. ISBN 0-07-002460-X. *(An introduction to the terminology and methods of optimization theory.)*

Chapter 6

Materials selection — case studies

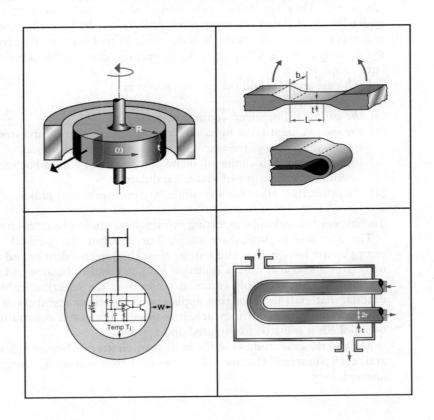

6.1 Introduction and synopsis

Here we have a collection of case studies illustrating the screening methods of Chapter 5. They are deliberately simplified to avoid obscuring the method under layers of detail. In most cases little is lost by this: the best choice of material for the simple example is the same as that for the more complex, for the reasons given in Chapter 5. More realistic case studies are developed in later chapters.

Each case study is laid out in the same way:

(a) *the problem statement*, setting the scene,
(b) *the model*, identifying function, constraints, objectives, and free variables, from which emerge the attribute limits and material indices,
(c) *the selection* in which the full menu of materials is reduced by screening and ranking to a short-list of viable candidates,
(d) *the postscript*, allowing a commentary on results and philosophy.

Techniques for seeking supporting information are left to later chapters.

The first few examples are simple but illustrate the method well. Later examples are less obvious and require clear thinking to identify and distinguish objectives and constraints. Confusion here can lead to bizarre and misleading conclusions. Always apply common sense: does the selection include the traditional materials used for that application? Are some members of the subset obviously unsuitable? If they are, it is usually because a constraint has been overlooked: it must be formulated and applied.

Most of the case studies use the hard-copy charts of Chapter 4; Sections 6.17 and 6.18 illustrate the use of computer-based selection, using the same methodology.

6.2 Materials for oars

Credit for inventing the rowed boat seems to belong to the Egyptians. Boats with oars appear in carved relief on monuments built in Egypt between 3300 and 3000 BC. Boats, before steam power, could be propelled by poling, by sail, or by oar. Oars gave more control than the other two, the military potential of which was well understood by the Romans, the Vikings and the Venetians.

Records of rowing races on the Thames in London extend Back to 1716. Originally the competitors were watermen, rowing the ferries used to carry people and goods across the river. Gradually gentlemen became involved (notably the young gentlemen of Oxford and Cambridge), sophisticating both the rules and the equipment. The real stimulus for development of boat and oar came in 1900 with the establishment of rowing as an Olympic sport. Since then both have drawn to the full on the craftsmanship and materials of their day. Consider, as an example, the oar.

The model. Mechanically speaking, an oar is a beam, loaded in bending. It must be strong enough to carry, without breaking, the bending moment exerted by the oarsman, it must have a stiffness to match the rower's own characteristics and give the right "feel", and — very important — it must be as light as possible. Meeting the strength constraint is easy. Oars are designed on *stiffness*, that is, to give a specified elastic deflection under a given load.

The upper part of Figure 6.1 shows an oar: a blade or "spoon" is bonded to a shaft or "loom" that carries a sleeve and collar to give positive location in the rowlock. The lower part of the figure shows how the oar stiffness is measured: a 10-kg weight is hung on the oar 2.05 m from the collar and the deflection δ at this point is measured. A soft oar will deflect nearly 50 mm; a hard one only 30. A rower, ordering an oar, will specify how hard it should be.

The oar must also be light; extra weight increases the wetted area of the hull and the drag that goes with it. So there we have it: an oar is a beam of specified stiffness and minimum weight. The material index we want was derived in Chapter 5 as equation (5.9). It is that for a light, stiff beam:

$$M = \frac{E^{1/2}}{\rho} \tag{6.1}$$

where E is Young's modulus and ρ is the density. There are other obvious constraints. Oars are dropped, and blades sometimes clash. The material must be tough enough to survive this, so brittle materials (those with a toughness G_{1C}

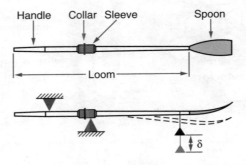

Figure 6.1   An oar. Oars are designed on stiffness, measured in the way shown in the lower figure, and they must be light.

less than 1 kJ/m^2) are unacceptable. Given these requirements, summarized in Table 6.1, what materials would you choose to make oars?

The selection. Figure 6.2 shows the appropriate chart: that in which Young's modulus, E, is plotted against density, ρ. The selection line for the index M has a slope of 2, as explained in Section 5.4; it is positioned so that a small group of materials is left above it. They are the materials with the largest values of M, and it is these that are the best choice, provided they satisfy the other constraint

Table 6.1   Design requirements for the oar

Function	Oar — meaning light, stiff beam
Constraints	• Length L specified
	• Bending stiffness S specified
	• Toughness $G_{1C} > 1$ kJ/m^2
Objective	Minimize the mass
Free variables	• Shaft diameter
	• Choice of material

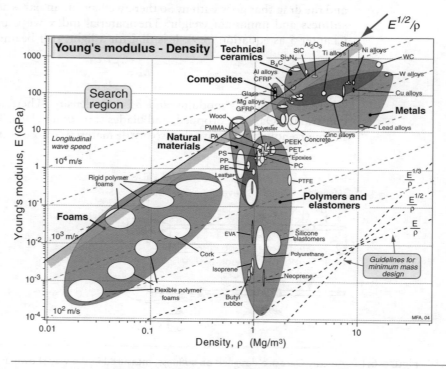

Figure 6.2   Materials for oars. CFRP is better than wood because the structure can be controlled.

(a simple attribute-limits on toughness). They contain three classes of material: woods, carbon reinforced polymers, and certain ceramics (Table 6.2). Ceramics are brittle; the toughness-modulus chart of Figure 4.7 shows that all fail to meet that required by the design. The recommendation is clear. Make your oars out of wood or — better — out of CFRP.

Postscript. Now we know what oars should be made of. What, in reality, is used? Racing oars and sculls are made either of wood or of a high performance composite: carbon-fiber reinforced epoxy.

Wooden oars are made today, as they were 100 years ago, by craftsmen working largely by hand. The shaft and blade are of Sitka spruce from the northern US or Canada, the further north the better because the short growing season gives a finer grain. The wood is cut into strips, four of which are laminated together to average the stiffness and the blade is glued to the shaft. The rough oar is then shelved for some weeks to settle down, and finished by hand cutting and polishing. The final spruce oar weighs between 4 and 4.3 kg, and costs (in 2004) about $250.

Composite blades are a little lighter than wood for the same stiffness. The component parts are fabricated from a mixture of carbon and glass fibers in an epoxy matrix, assembled and glued. The advantage of composites lies partly in the saving of weight (typical weight: 3.9 kg) and partly in the greater control of performance: the shaft is molded to give the stiffness specified by the purchaser. Until recently a CFRP oar cost more than a wooden one, but the price of carbon fibers has fallen sufficiently that the two cost about the same.

Could we do better? The chart shows that wood and CFRP offer the lightest oars, at least when normal construction methods are used. Novel composites, not at present shown on the chart, might permit further weight saving; and functional-grading (a thin, very stiff outer shell with a low density core) might do it. But both appear, at present, unlikely.

Further reading Redgrave, S. (1992) *Complete Book of Rowing*, Partridge Press, London.

Related case studies

6.3	Mirrors for large telescopes
6.4	Table legs
12.2	Spars for man-powered planes
12.4	Forks for a racing bicycle

Table 6.2 Material for oars

Material	Index M $(GPa)^{1/2}/(Mg/m^3)$	Comment
Woods	3.4–6.3	Cheap, traditional, but with natural variability
CFRP	5.3–7.9	As good as wood, more control of properties
Ceramics	4–8.9	Good M but toughness low and cost high

6.3 Mirrors for large telescopes

There are some very large optical telescopes in the world. The newer ones employ complex and cunning tricks to maintain their precision as they track across the sky — more on that in the postscript. But if you want a simple telescope, you make the reflector as a single rigid mirror. The largest such telescope is sited on Mount Semivodrike, near Zelenchukskaya in the Caucasus Mountains of Russia. The mirror is 6 m (236 in.) in diameter. To be sufficiently rigid, the mirror, which is made of glass, is about 1 m thick and weighs 70 tonnes.

The total cost of a large (236 in.) telescope is, like the telescope itself, astronomical — about US$280 m. The mirror itself accounts for only about 5 percent of this cost; the rest is that of the mechanism that holds, positions, and moves it as it tracks across the sky. This mechanism must be stiff enough to position the mirror relative to the collecting system with a precision about equal to that of the wavelength of light. It might seem, at first sight, that doubling the mass m of the mirror would require that the sections of the support-structure be doubled too, so as to keep the stresses (and hence the strains and displacements) the same; but the heavier structure then deflects under its own weight. In practice, the sections have to increase as m^2, and so does the cost.

Before the turn of the century, mirrors were made of speculum metal (density: about $8 \, \text{Mg/m}^3$). Since then, they have been made of glass (density: $2.3 \, \text{Mg/m}^3$), silvered on the front surface, so none of the optical properties of the glass are used. Glass is chosen for its mechanical properties only; the 70 tonnes of glass is just a very elaborate support for 100 nm (about 30 g) of silver. Could one, by taking a radically new look at materials for mirrors, suggest possible routes to the construction of lighter, cheaper telescopes?

The model. At its simplest, the mirror is a circular disk, of diameter $2R$ and mean thickness t, simply supported at its periphery (Figure 6.3). When horizontal, it will deflect under its own weight m; when vertical it will not deflect significantly. This distortion (which changes the focal length and introduces aberrations) must be small enough that it does not interfere with performance; in practice, this means that the deflection δ of the midpoint of the mirror must be less than the wavelength of light. Additional requirements are: high-dimensional stability (no creep), and low thermal expansion (Table 6.3).

The mass of the mirror (the property we wish to minimize) is

$$m = \pi R^2 t \rho \tag{6.2}$$

where ρ is the density of the material of the disk. The elastic deflection, δ, of the center of a horizontal disk due to its own weight is given, for a material with Poisson's ratio of 0.3 (Appendix A), by

$$\delta = \frac{3}{4\pi} \frac{mgR^2}{Et^3} \tag{6.3}$$

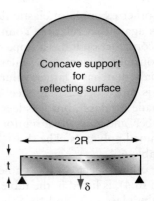

Figure 6.3   The mirror of a large optical telescope is modeled as a disk, simply supported at its periphery. It must not sag by more than a wavelength of light at its center.

Table 6.3   Design requirements for the telescope mirror

Function	Precision mirror
Constraints	• Radius R specified • Must not distort more than δ under self-weight • High dimensional stability: no creep, low thermal expansion
Objective	Minimize the mass, m
Free variables	• Thickness of mirror, t • Choice of material

The quantity g in this equation is the acceleration due to gravity: 9.81 m/s²; E, as before, is Young's modulus. We require that this deflection be less than (say) 10 μm. The diameter $2R$ of the disk is specified by the telescope design, but the thickness t is a free variable. Solving for t and substituting this into the first equation gives

$$m = \left(\frac{3g}{4\delta}\right)^{1/2} \pi R^4 \left[\frac{\rho}{E^{1/3}}\right]^{3/2} \tag{6.4}$$

The lightest mirror is the one with the greatest value of the material index

$$M = \frac{E^{1/3}}{\rho} \tag{6.5}$$

We treat the remaining constraints as attribute limits, requiring a melting point greater than 500°C to avoid creep, zero moisture take up, and a low thermal expansion coefficient ($\alpha < 20 \times 10^{-6}$/K).

The selection. Here we have another example of elastic design for minimum weight. The appropriate chart is again that relating Young's modulus E and density ρ — but the line we now construct on it has a slope of 3, corresponding to the condition $M = E^{1/3}/\rho = $ constant (Figure 6.4). Glass lies at the value $M = 1.7\,(GPa)^{1/3}.m^3/Mg$. Materials that have larger values of M are better, those with lower, worse. Glass is much better than steel or speculum metal (that is why most mirrors are made of glass), but it is less good than magnesium, several ceramics, carbon–fiber, and glass–fiber reinforced polymers, or — an unexpected finding — stiff foamed polymers. The short-list before applying the attribute limits is given in Table 6.4.

One must, of course, examine other aspects of this choice. The mass of the mirror, calculated from equation (6.4), is listed in the table. The CFRP mirror is less than half the weight of the glass one, and that the support-structure could thus be as much as 4 times less expensive. The possible saving by using foam is even greater. But could they be made?

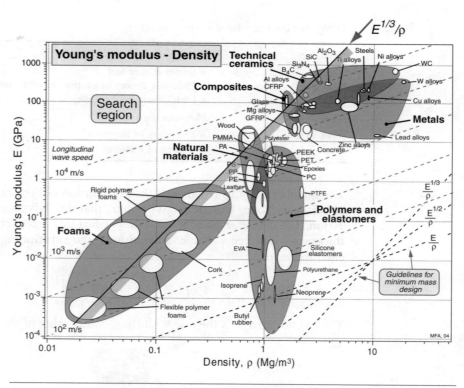

Figure 6.4   Materials for telescope mirrors. Glass is better than most metals, among which magnesium is a good choice. Carbon-fiber reinforced polymers give, potentially, the lowest weight of all, but may lack adequate dimensional stability. Foamed glass is a possible candidate.

Table 6.4   Mirror backing for 200-in. (5.1 m) telescope

Material	$M = E^{1/3}/\rho$ $(GPa)^{1/3}.m^3/Mg$	m (tonne) $2R = 5.1$ m (from equation (6.4))	Comment
Steel (or Speculum)	0.74	73.6	Very heavy. The original choice
GFRP	1.5	25.5	Not dimensionally stable enough — use for radio telescope
Al-alloys	1.6	23.1	Heavier than glass, and with high thermal expansion
Glass	1.7	21.6	The present choice
Mg-alloys	1.9	17.9	Lighter than glass but high thermal expansion
CFRP	3.0	9	Very light, but not dimensionally stable; use for radio telescopes
Foamed polystyrene	4.5	5	Very light, but dimensionally unstable. Foamed glass?

Some of the choices — polystyrene foam or CFRP — may at first seem impractical. But the potential cost-saving (the factor of 16) is so vast that they are worth examining. There are ways of casting a thin film of silicone rubber or of epoxy onto the surface of the mirror-backing (the polystyrene or the CFRP) to give an optically smooth surface that could be silvered. The most obvious obstacle is the lack of stability of polymers — they change dimensions with age, humidity, temperature, and so on. But glass itself can be reinforced with carbon fibers; and it can also be foamed to give a material that is denser than polystyrene foam but much lighter than solid glass. Both foamed and carbon-reinforced glass have the same chemical and environmental stability as solid glass. They could provide a route to large cheap mirrors.

Postscript. There are, of course, other things you can do. The stringent design criterion ($\delta < 10\,\mu m$) can be partially overcome by engineering design without reference to the material used. The 8.2 m Japanese telescope on Mauna Kea, Hawaii and the very large telescope (VLT) at Cerro Paranal Silla in Chile each have a thin glass reflector supported by an array of hydraulic or piezo-electric jacks that exert distributed forces over its back surface, controlled to vary with the attitude of the mirror. The Keck telescope, also on Mauna Kea, is segmented, each segment independently positioned to give optical focus. But the limitations of this sort of mechanical system still require that the mirror meet a stiffness target. While stiffness at minimum weight is the design requirement, the material-selection criteria remain unchanged.

Radio telescopes do not have to be quite as precisely dimensioned as optical ones because they detect radiation with a longer wavelength. But they are much bigger (60 m rather than 6 m) and they suffer from similar distortional problems. Microwaves have wavelengths in the mm band, requiring precision over the mirror face of 0.25 mm. A recent 45 m radio telescope built for the University of Tokyo achieves this, using CFRP. Its parabolic surface is made of 6000 CFRP panels, each servo controlled to compensate for macro-distortion. Recent telescopes have been made from CFRP, for exactly the reasons we deduced.

Related case studies   6.16   Materials to minimize thermal distortion in precision devices

6.4 Materials for table legs

Luigi Tavolino, furniture designer, conceives of a light-weight table of daring simplicity: a flat sheet of toughened glass supported on slender, un-braced, cylindrical legs (Figure 6.5). The legs must be solid (to make them thin) and as light as possible (to make the table easier to move). They must support the table top and whatever is placed upon it without buckling (Table 6.5). What materials could one recommend?

The model. This is a problem with two objectives[2]: weight is to be minimized, and slenderness maximized. There is one constraint: resistance to buckling. Consider minimizing weight first.

The leg is a slender column of material of density ρ and modulus E. Its length, L, and the maximum load, F, it must carry are determined by the design: they are fixed. The radius r of a leg is a free variable. We wish to minimize the mass m of the leg, given by the objective function

$$m = \pi r^2 L \rho \tag{6.6}$$

subject to the constraint that it supports a load P without buckling. The elastic buckling load F_{crit} of a column of length L and radius r (see Appendix A) is

$$F_{\text{crit}} = \frac{\pi^2 EI}{L^2} = \frac{\pi^3 E r^4}{4L^2} \tag{6.7}$$

using $I = \pi r^4/4$ where I is the second moment of the area of the column. The load F must not exceed F_{crit}. Solving for the free variable, r, and substituting it into the equation for m gives

$$m \geq \left(\frac{4F}{\pi}\right)^{1/2} (L)^2 \left[\frac{\rho}{E^{1/2}}\right] \tag{6.8}$$

---

[2] Formal methods for dealing with multiple objectives are developed in Chapter 9.

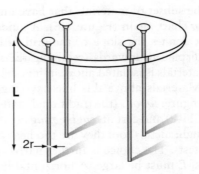

Figure 6.5 A light-weight table with slender cylindrical legs. Lightness and slenderness are independent design goals, both constrained by the requirement that the legs must not buckle when the table is loaded. The best choice is a material with high values of both $E^{1/2}/\rho$ and E.

Table 6.5   Design requirements for table legs

Function	Column (supporting compressive loads)
Constraints	• Length L specified • Must not buckle under design loads • Must not fracture if accidentally struck
Objective	• Minimize the mass, m • Maximize slenderness
Free variables	• Diameter of legs, $2r$ • Choice of material

The material properties are grouped together in the last pair of brackets. The weight is minimized by selecting the subset of materials with the greatest value of the material index

$$M_1 = \frac{E^{1/2}}{\rho}$$

(a result we could have taken directly from Appendix B).

Now slenderness. Inverting equation (6.7) with F_{crit} set equal to F gives an equation for the thinnest leg that will not buckle:

$$r \geq \left(\frac{4F}{\pi^3}\right)^{1/4} (L)^{1/2} \left[\frac{1}{E}\right]^{1/4} \tag{6.9}$$

The thinnest leg is that made of the material with the largest value of the material index

$$M_2 = E$$

The selection. We seek the subset of materials that have high values of $E^{1/2}/\rho$ and E. We need the $E - \rho$ chart again (Figure 6.6). A guideline of slope 2 is drawn on the diagram; it defines the slope of the grid of lines for values of $E^{1/2}/\rho$. The guideline is displaced upwards (retaining the slope) until a reasonably small subset of materials is isolated above it; it is shown at the position $M_1 = 5\ \text{GPa}^{1/2}/(\text{Mg/m}^3)$. Materials above this line have higher values of M_1. They are identified on the figure: *woods* (the traditional material for table legs), *composites* (particularly CFRP) and certain *engineering ceramics*. Polymers are out: they are not stiff enough; metals too: they are too heavy (even magnesium alloys, which are the lightest). The choice is further narrowed by the requirement that, for slenderness, E must be large. A horizontal line on the diagram links materials with equal values of E; those above are stiffer. Figure 6.6 shows that placing this line at $M_1 = 100\ \text{GPa}$ eliminates woods and GFRP. If the legs must be really thin, then the short-list is reduced to CFRP and ceramics: they give legs that weigh the same as the wooden ones but are barely half as thick. Ceramics, we know, are brittle: they have low values of fracture toughness.

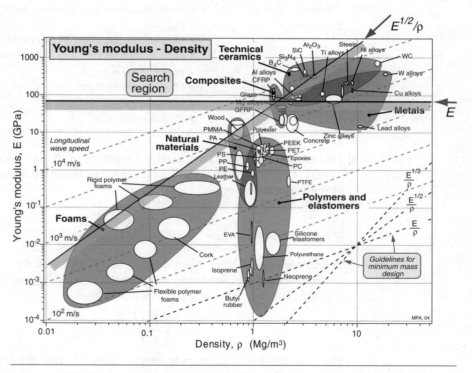

Figure 6.6   Materials for light, slender legs. Wood is a good choice; so is a composite such as CFRP, which, having a higher modulus than wood, gives a column that is both light and slender. Ceramics meet the stated design goals, but are brittle.

Table legs are exposed to abuse — they get knocked and kicked; common sense suggest that an additional constraint is needed, that of adequate toughness. This can be done using Figure 4.7; it eliminates ceramics, leaving CFRP. The cost of CFRP (Figure 4.17) may cause Snr. Tavolino to reconsider his design, but that is another matter: he did not mention cost in his original specification.

It is a good idea to lay out the results as a table, showing not only the materials that are best, but those that are second-best — they may, when other considerations are involved, become the best choice. Table 6.6 shows the way to do it.

Postscript. Tubular legs, the reader will say, must be lighter than solid ones. True; but they will also be fatter. So it depends on the relative importance Snr. Tavolino attaches to his two objectives — lightness and slenderness — and only he can decide that. If he can be persuaded to live with fat legs, tubing can be considered — and the material choice may be different. Materials selection when section-shape is a variable comes in Chapter 11.

Ceramic legs were eliminated because of low toughness. If (improbably) the goal was to design a light, slender-legged table for use at high temperatures, ceramics should be reconsidered. The brittleness problem can be by-passed by protecting the legs from abuse, or by pre-stressing them in compression.

Related case studies

6.2	Materials for oars	
6.3	Mirrors for large telescopes	
12.2	Spars for man-powered planes	
12.4	Forks for a racing bicycle	
12.7	Table legs again: thin or light?	

6.5 Cost: structural materials for buildings

The most expensive thing that most people buy is the house they live in. Roughly half the cost of a house is the cost of the materials of which it is made, and they are used in large quantities (family house: around 200 tonnes; large apartment block: around 20,000 tonnes). The materials are used in three ways:

Table 6.6   Materials for table legs

Material	Typical M_1 (GPa$^{1/2}$.m^3/Mg)	Typical M_2 GPa	Comment
GFRP	2.5	20	Cheaper than CFRP, but lower M_1 and M_2
Woods	4.5	10	Outstanding M_1; poor M_2 Cheap, traditional, reliable
Ceramics	6.3	300	Outstanding M_1 and M_2. Eliminated by brittleness
CFRP	6.6	100	Outstanding M_1 and M_2, but expensive

structurally to hold the building up; as cladding, to keep the weather out; and as "internals", to insulate against heat, sound, and so forth.

Consider the selection of materials for the structure (Figure 6.7). They must be stiff, strong, and cheap. Stiff, so that the building does not flex too much under wind loads or internal loading. Strong, so that there is no risk of it collapsing. And cheap, because such a lot of material is used. The structural frame of a building is rarely exposed to the environment, and is not, in general, visible, so criteria of corrosion resistance or appearance are not important here. The design goal is simple: strength and stiffness at minimum cost. To be more specific: consider the selection of material for floor beams. Table 6.7 summarizes the requirements.

The model. The material index for a stiff beam of minimum mass, m, was developed in Chapter 5 (equations (5.6)–(5.9)). The cost C of the beam is just its mass, m, times the cost per kg, C_m, of the material of which it is made:

$$C = mC_m = AL\rho C_m \qquad (6.10)$$

which becomes the objective function of the problem. Proceeding as in Chapter 5, we find the index for a stiff beam of minimum cost to be:

$$M_1 = \frac{E^{1/2}}{\rho C_m}$$

The index when strength rather than stiffness is the constraint was not derived earlier. Here it is. The objective function is still equation (6.10), but the

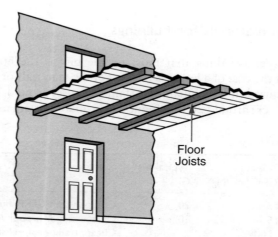

Floor
Joists

Figure 6.7   The materials of a building perform three broad roles. The frame gives mechanical support; the cladding excludes the environment; and the internal surfacing controls heat, light and sound. The selection criteria depend on the function.

Table 6.7   Design requirements for floor beams

Function	Floor beam
Constraints	• Length L specified
	• Stiffness: must not deflect too much under design loads
	• Strength: must not fail under design loads
Objective	• Minimize the cost, C
Free variables	• Cross-section area of beam, A
	• Choice of material

constraint is now that of strength: the beam must support F without failing. The failure load of a beam (Appendix A, Section A.4) is:

$$F_f = C_2 \frac{I \sigma_f}{y_m L} \tag{6.11}$$

where C_2 is a constant, σ_f is the failure strength of the material of the beam and y_m is the distance between the neutral axis of the beam and its outer filament for a rectangular beam of depth d and width b). We assume the proportions of the beam are fixed so that $d = \alpha b$ where α is the aspect ratio, typically 2. Using this and $I = bd^3/12$ to eliminate A in equation (6.10) gives the cost of the beam that will just support the load F_f:

$$C = \left(\frac{6\sqrt{\alpha}}{C^2} \frac{F_f}{L^2} \right)^{2/3} (L^3) \left[\frac{\rho C_m}{\sigma_f^{2/3}} \right] \tag{6.12}$$

The mass is minimized by selecting materials with the largest values of the index

$$M_2 = \frac{\sigma_f^{2/3}}{\rho C_m}$$

The selection. Stiffness first. Figure 6.8(a) shows the relevant chart: modulus E against relative cost per unit volume, $C_m \, \rho$ (the chart uses a relative cost C_R, defined in Chapter 4, in place of C_m but this makes no difference to the selection). The shaded band has the appropriate slope for M_1; it isolates concrete, stone, brick, woods, cast irons, and carbon steels. Figure 6.8(b) shows strength against relative cost. The shaded band — M_2 this time — gives almost the same selection. They are listed, with values, in Table 6.8. They are exactly the materials with which buildings have been, and are, made.

Postscript. Concrete, stone, and brick have strength only in compression; the form of the building must use them in this way (columns, arches). Wood, steel, and reinforced concrete have strength both in tension and compression, and steel, additionally, can be given efficient shapes (I-sections, box sections, tubes,

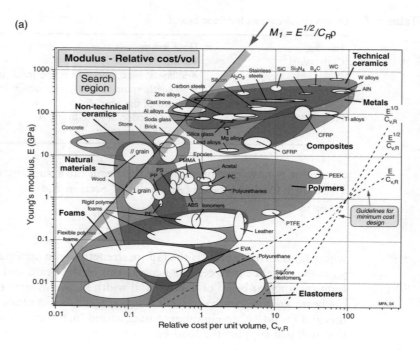

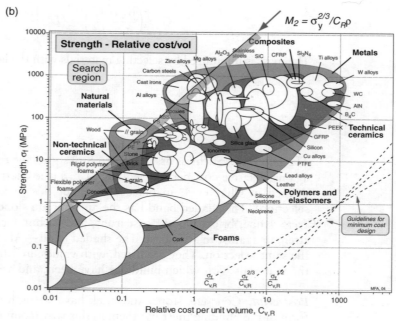

Figure 6.8   The selection of cheap (a) stiff and (b) strong materials for the structural frames of buildings.

Table 6.8   Structural materials for buildings

Material	M_1 $(GPa^{1/2}/(kg/m^3))$	M_2 $(MPa^{2/3} (kg/m^3))$	Comment
Concrete	160	14	
Brick	12	12	Use in compression only
Stone	9.3	12	
Woods	21	90	Tension and compression, with freedom of section shape
Cast Iron	17	90	
Steel	14	45	

discussed in Chapter 11); the form of the building made from these has much greater freedom.

It is sometimes suggested that architects live in the past; that in the late 20th century they should be building with fiberglass (GFRP), aluminum alloys and stainless steel. Occasionally they do, but the last two figures give an idea of the penalty involved: the cost of achieving the same stiffness and strength is between 5 and 20 times greater. Civil construction (buildings, bridges, roads, and the like) is materials-intensive: the cost of the material dominates the product cost, and the quantity used is enormous. Then only the cheapest of materials qualify, and the design must be adapted to use them.

Further reading   Cowan, H.J. and Smith, P.R. (1988) *The Science and Technology of Building Materials*, Van Nostrand-Reinhold, New York.
Doran, D.K. (1992) *The Construction Reference Book*, Butterworth-Heinemann, Oxford, UK.

Related case
studies
6.2   Materials for oars
6.4   Materials for table legs
12.5   Floor joists: wood, bamboo or steel?

6.6 Materials for flywheels

Flywheels store energy. Small ones — the sort found in children's toys — are made of lead. Old steam engines have flywheels; they are made of cast iron. Cars have them too (though you cannot see them) to smooth power-transmission. More recently flywheels have been proposed for power storage and regenerative braking systems for vehicles; a few have been built, some of high-strength steel, some of composites. Lead, cast iron, steel, composites — there is a strange diversity here. What *is* the best choice of material for a flywheel?

An efficient flywheel stores as much *energy per unit weight* as possible. As the flywheel is spun up, increasing its angular velocity, ω, it stores more energy. The limit is set by failure caused by centrifugal loading: if the centrifugal stress exceeds the tensile strength (or fatigue strength), the flywheel flies apart. One constraint, clearly, is that this should not occur.

The flywheel of a child's toy is not efficient in this sense. Its velocity is limited by the pulling-power of the child, and never remotely approaches the burst velocity. In this case, and for the flywheel of an automobile engine — we wish to maximize the *energy stored per unit volume* at a constant (specified) *angular velocity*. There is also a constraint on the outer radius, R, of the flywheel so that it will fit into a confined space.

The answer therefore depends on the application. The strategy for optimizing flywheels for efficient energy-storing systems differs from that for children's toys. The two alternative sets of design requirements are listed in Table 6.9(a) and (b).

The model. An efficient flywheel of the first type stores as much energy per unit weight as possible, without failing. Think of it as a solid disk of radius R and thickness t, rotating with angular velocity ω (Figure 6.9). The energy U stored in the flywheel is (Appendix A)

$$U = \frac{1}{2}J\omega^2 \tag{6.13}$$

Here $J = (\pi/2)\rho R^4 t$ is the polar moment of inertia of the disk and ρ the density of the material of which it is made, giving

$$U = \frac{\pi}{4}\rho R^4 t\omega^2 \tag{6.14}$$

Table 6.9   Design requirements for maximum-energy flywheel and fixed velocity

(a) For maximum-energy flywheel	
Function	Flywheel for energy storage
Constraints	• Outer radius, *R*, fixed
	• Must not burst
	• Adequate toughness to give crack-tolerance
Objective	Maximize kinetic energy per unit mass
Free variables	Choice of material

(b) For fixed velocity	
Function	Flywheel for child's toy
Constraints	Outer radius, *R*, fixed
Objective	Maximize kinetic energy per unit volume at fixed angular velocity
Free variables	Choice of material

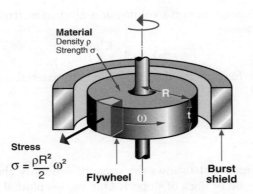

Figure 6.9   A flywheel. The maximum kinetic energy it can store is limited by its strength.

The mass of the disk is

$$m = \pi R^4 t \rho \tag{6.15}$$

The quantity to be maximized is the kinetic energy per unit mass, which is the ratio of the last two equations:

$$\frac{U}{m} = \frac{1}{4} R^2 \omega^2 \tag{6.16}$$

As the flywheel is spun up, the energy stored in it increases, but so does the centrifugal stress. The maximum principal stress in a spinning disk of uniform thickness (Appendix A) is

$$\sigma_{max} = \left(\frac{3+\nu}{8}\right) \rho R^2 \omega^2 \approx \frac{1}{2} \rho R^2 \omega^2 \tag{6.17}$$

where ν is Poisson's ratio ($\nu \approx 1/3$). This stress must not exceed the failure stress σ_f (with an appropriate factor of safety, here omitted). This sets an upper limit to the angular velocity, ω, and disk radius, R (the free variables). Eliminating $R\omega$ between the last two equations gives

$$\frac{U}{m} = \frac{1}{2}\left(\frac{\sigma_f}{\rho}\right) \tag{6.18}$$

The best materials for high-performance flywheels are those with high values of the material index

$$M = \frac{\sigma_f}{\rho} \tag{6.19}$$

It has units of kJ/kg.

And now the other sort of flywheel — that of the child's toy. Here we seek the material that stores the most energy per unit volume V at constant velocity, ω. The energy per unit volume at a given ω is (from equation (6.2)):

$$\frac{U}{V} = \frac{1}{4}\rho R^2 \omega^2$$

Both R and ω are fixed by the design, so the best material is now that with the greatest value of

$$M_2 = \rho \tag{6.20}$$

The selection. Figure 6.10 shows the strength—density chart. Values of M_1 correspond to a grid of lines of slope 1. One such is plotted as a diagonal line at the value $M_1 = 200\,\text{kJ/kg}$. Candidate materials with high values of M_1 lie in the search region towards the top left. The best choices are unexpected ones: composites, particularly CFRP, high strength titanium alloys and some ceramics, but these are ruled out by their low toughness.

But what of the lead flywheels of children's toys? There could hardly be two more different materials than CFRP and lead: the one, strong and light,

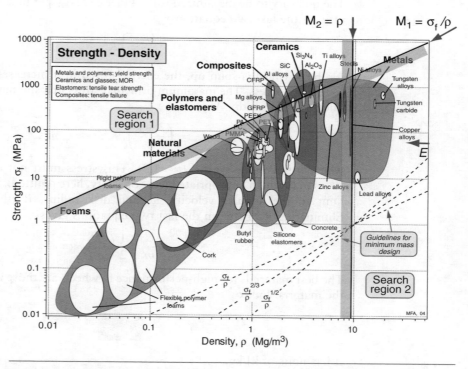

Figure 6.10   Materials for flywheels. Composites are the best choices. Lead and cast iron, traditional for flywheels, are good when performance is limited by rotational velocity, not strength.

the other, soft and heavy. Why lead? It is because, in the child's toy, the constraint is different. Even a super-child cannot spin the flywheel of his toy up to its burst velocity. The angular velocity ω is limited instead by the drive mechanism (pull-string, friction drive). Then as we have seen, the best material is that with the largest density. The second selection line on Figure 6.10 shows the index M_2 at the value 10 Mg/m^3. We seek materials in Search Area 2 to the right of this line. Lead is good. Cast iron is less good, but cheaper. Gold, platinum, and uranium (not shown on the chart) are better, but may be thought unsuitable for other reasons.

Postscript. A CFRP rotor is able to store around 400 kJ/kg. A lead flywheel, by contrast, can store only 1 kJ/kg before disintegration; a cast-iron flywheel, about 30. All these are small compared with the energy density in gasoline: roughly 20,000 kJ/kg. Even so, the energy density in the flywheel is considerable; its sudden release in a failure could be catastrophic. The disk must be surrounded by a burst-shield and precise quality control in manufacture is essential to avoid out-of-balance forces. This has been achieved in a number of composite energy-storage flywheels intended for use in trucks and buses, and as an energy reservoir for smoothing wind-power generation.

And now a digression: the electric car. Hybrid petrol-electric cars are already on the roads, using advanced lead-acid battery technology to store energy. But batteries have their problems: the energy density they can contain is low (see Table 6.10); their weight limits both the range and the performance of the car. It is practical to build flywheels with an energy density of roughly equal to that of the best batteries. Serious consideration is now being given to a flywheel for electric cars. A pair of counter-rotating CFRP disks are housed in a steel burst-shield. Magnets embedded in the disks pass near coils in the housing, inducing a current and allowing power to be drawn to the electric motor that drives the

Table 6.10   Energy density of power sources

Source	Energy density (kJ/kg)	Comment
Gasoline	20,000	Oxidation of hydrocarbon — mass of oxygen not included
Rocket fuel	5000	Less than hydrocarbons because oxidizing agent forms part of fuel
Flywheels	Up to 400	Attractive, but not yet proven
Lithium-ion battery	Up to 350	Attractive but expensive, and with limited life
Nickel-cadmium battery	170–200	
Lead-acid battery	50–80	Large weight for acceptable range
Springs rubber bands	Up to 5	Much less efficient method of energy storage than flywheel

wheels. Such a flywheel could, it is estimated, give an electric car an adequate range, at a cost competitive with the gasoline engine and with none of the local pollution.

Further reading   Christensen, R.M. (1979) *Mechanics of Composite Materials*, Wiley Interscience, New York, p. 213 et seq.

Lewis, G. (1990) *Selection of Engineering Materials*, Part 1, Prentice Hall, NJ, p. 1.

Medlicott, P.A.C. and Potter, K.D. (1986) The development of a composite flywheel for vehicle applications, in Brunsch, K., Golden, H-D., and Horkert, C-M. (eds) *High Tech — the Way into the Nineties*, Elsevier, Amsterdam, p. 29.

6.7 Materials for springs

Springs come in many shapes (Figure 6.11 and Table 6.11) and have many purposes: think of axial springs (e.g. a rubber band), leaf springs, helical springs, spiral springs, torsion bars. Regardless of their shape or use, the best material for a spring of minimum volume is that with the greatest

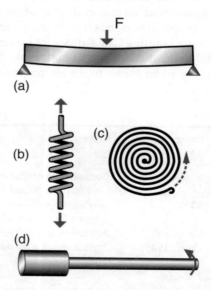

Figure 6.11   Springs store energy. The best material for any spring, regardless of its shape or the way in which it is loaded, is that with the highest value of σ_f^2/E, or, if weight is important, $\sigma_f^2/E\rho$.

Table 6.11   Design requirements for springs

Function	Elastic spring
Constraints	No failure, meaning $\sigma < \sigma_f$ throughout the spring
Objective	• Maximum stored elastic energy per unit volume, or
	• Maximum stored elastic energy per unit weight
Free variables	Choice of material

value of σ_f^2/E, and for minimum weight it is that with the greatest value of $\sigma_f^2/\rho E$ (derived below). We use them as a way of introducing two of the most useful of the charts: Young's modulus E plotted against strength σ_f, and specific modulus E/ρ plotted against specific strength σ_f/ρ (Figures 4.5 and 4.6).

The model. The primary function of a spring is to store elastic energy and — when required — release it again. The elastic energy stored per unit volume in a block of material stressed uniformly to a stress σ is

$$W_v = \frac{1}{2}\frac{\sigma^2}{E} \qquad (6.21)$$

where E is Young's modulus. We wish to maximize W_v. The spring will be damaged if the stress σ exceeds the yield stress or failure stress σ_f; the constraint is $\sigma < \sigma_f$. Thus the maximum energy density is

$$W_v = \frac{1}{2}\frac{\sigma_f^2}{E} \qquad (6.22)$$

Torsion bars and leaf springs are less efficient than axial springs because much of the material is not fully loaded: the material at the neutral axis, for instance, is not loaded at all. For leaf springs

$$W_v = \frac{1}{4}\frac{\sigma_f^2}{E}$$

and for torsion bars

$$W_v = \frac{1}{3}\frac{\sigma_f^2}{E}$$

But — as these results show — this has no influence on the choice of material. The best stuff for a spring regardless of its shape is that with the biggest value of

$$M_1 = \frac{\sigma_f^2}{E} \qquad (6.23)$$

If weight, rather than volume, matters, we must divide this by the density ρ (giving energy stored per unit weight), and seek materials with high values of

$$M_2 = \frac{\sigma_f^2}{\rho E} \qquad (6.24)$$

The selection. The choice of materials for springs of minimum volume is shown in Figure 6.12(a). A family lines of slope 2 link materials with equal values of $M_1 = \sigma_f^2/E$; those with the highest values of M_1 lie towards the bottom right. The heavy line is one of the family; it is positioned so that a subset of materials is left exposed. The best choices are a *high-strength steel* lying near the top end of the line. Other materials are suggested too: CFRP (now used for truck springs), *titanium alloys* (good but expensive), and *nylon* (children's toys often have nylon springs), and, of course, *elastomers*. Note how the procedure has identified a candidate from almost every class of materials: metals, polymers, elastomers and composites. They are listed, with commentary, in Table 6.12(a).

Materials selection for light springs is shown in Figure 6.12(b). A family of lines of slope 2 link materials with equal values of

$$M_2 = \left(\frac{\sigma_f}{\rho}\right)^2 \bigg/ \left(\frac{E}{\rho}\right) = \frac{\sigma_f^2}{E\rho} \tag{6.25}$$

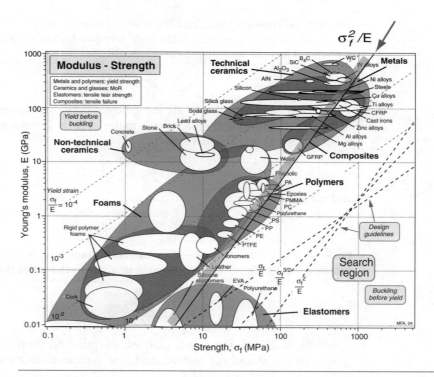

Figure 6.12(a)   Materials for small springs. high strength ("spring") steel is good. Glass, CFRP and GFRP all, under the right circumstances, make good springs. Elastomers are excellent. Ceramics are eliminated by their low tensile strength.

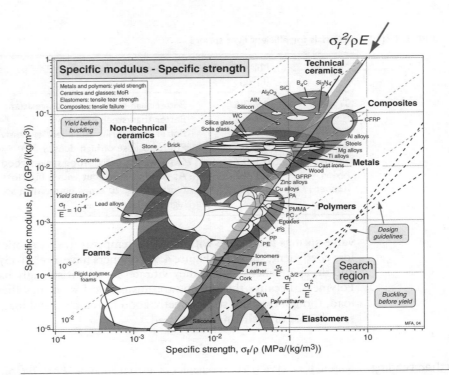

Figure 6.12(b) Materials for light springs. Metals are disadvantaged by their high densities. Composites are good; so is wood. Elastomers are excellent.

Table 6.12(a)   Materials for efficient small springs

Material	$M_1 = \sigma_f^2/E$ (MJ/m³)	Comment
Ti alloys	4–12	Expensive, corrosion-resistant
CFRP	6–10	Comparable in performance with steel; expensive
Spring steel	3–7	The traditional choice: easily formed and heat treated
Nylon	1.5–2.5	Cheap and easily shaped, but high loss factor
Rubber	20–50	Better than spring steel; but high loss factor

One is shown at the value $M_2 = 2\,\text{kJ/kg}$. Metals, because of their high density, are less good than composites, and much less good than elastomers. (You can store roughly eight times more elastic energy, per unit weight, in a rubber band than in the best spring steel.) Candidates are listed in Table 6.12(b). Wood—the traditional material for archery bows, now appears.

Postscript. Many additional considerations enter the choice of a material for a spring. Springs for vehicle suspensions must resist fatigue and corrosion; engine

Table 6.12(b)   Materials for efficient light springs

Material	$M_1 = \sigma_f^2/\rho E$ (kJ/kg)	Comment
Ti alloys	0.9–2.6	Better than steel; corrosion-resistant; expensive
CFRP	3.9–6.5	Better than steel; expensive
GFRP	1.0–1.8	Better than spring steel; less expensive than CFRP
Spring steel	0.4–0.9	Poor, because of high density
Wood	0.3–0.7	On a weight basis, wood makes good springs
Nylon	1.3–2.1	As good as steel, but with a high loss factor
Rubber	18–45	Outstanding; 20 times better than spring steel; but with high loss factor

valve-springs must cope with elevated temperatures. A subtler property is the loss coefficient, shown in Figure 4.9. Polymers have a relatively high loss factor and dissipate energy when they vibrate; metals, if strongly hardened, do not. Polymers, because they creep, are unsuitable for springs that carry a steady load, though they are still perfectly good for catches and locating springs that spend most of their time unstressed.

Further reading
Boiton, R.G. (1963) The mechanics of instrumentation, *Proc. Int. Mech. Eng.* 177(10), 269–288.
Hayes, M. (1990) Materials update 2: springs, *Engineering*, May, p. 42.

Related case studies
6.8   Elastic hinges and couplings
12.3   Ultra-efficient springs
12.8   Shapes that flex: leaf and strand structures
14.4   Connectors that do not relax their grip

6.8 Elastic hinges and couplings

Nature makes much use of elastic hinges: skin, muscle, cartilage all allow large, recoverable deflections. Man, too, design with *flexure* and *torsion hinges*: ligaments that connect or transmit load between components while allowing limited relative movement between them by deflecting elastically (Figure 6.13 and Table 6.13). Which materials make good hinges?

The model. Consider the hinge for the lid of a box. The box, lid and hinge are to be molded in one operation. The hinge is a thin ligament of material that flexes elastically as the box is closed, as in the figure, but it carries no significant axial loads. Then the best material is the one that (for given ligament

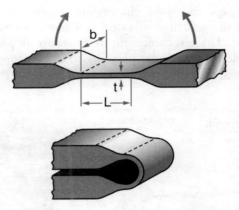

Figure 6.13 Elastic or "natural" hinges. The ligaments must bend repeatedly without failing. The cap of a shampoo bottle is an example; elastic hinges are used in high performance applications too, and are found widely in nature.

Table 6.13 Design requirements for elastic hinges

Function	Elastic hinge
Constraints	No failure, meaning $\sigma < \sigma_f$ throughout the hinge
Objective	Maximize elastic flexure
Free variables	Choice of material

dimensions) bends to the smallest radius without yielding or failing. When a ligament of thickness t is bent elastically to a radius R, the surface strain is

$$\varepsilon = \frac{t}{2R} \tag{6.26}$$

and — since the hinge is elastic — the maximum stress is

$$\sigma = E\frac{t}{2R} \tag{6.27}$$

This must not exceed the yield or failure strength σ_f. Thus the minimum radius to which the ligament can be bent without damage is

$$R \geq \frac{1}{2}\left[\frac{E}{\sigma_f}\right] \tag{6.28}$$

The best material is the one that can be bent to the smallest radius, that is, the one with the greatest value of the index

$$M = \frac{\sigma_f}{E} \tag{6.29}$$

The selection. We need the $\sigma_f - E$ chart again (Figure 6.14). Candidates are identified by using the guideline of slope 1; a line is shown at the position $M = \sigma_f/E = 3 \times 10^{-2}$. The best choices for the hinge are all polymeric materials. The short-list (Table 6.14) includes polyethylene, polypropylene, nylon, and, best of all, elastomers, though these may be too flexible for the body of the box

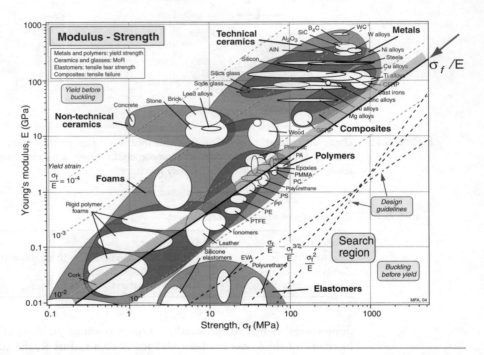

Figure 6.14   Materials for elastic hinges. Elastomers are best, but may not be rigid enough to meet other design needs. Then polymers such as nylon, PTFE and PE are better. Spring steel is less good, but much stronger.

Table 6.14   Materials for elastic hinges

Material	M $(\times 10^{-3})$	Comment
Polyethylene	32	Widely used for cheap hinged bottle caps, etc.
Polypropylene	30	Stiffer than polyethylene. Easily molded
Nylon	30	Stiffer than polyethylene. Easily molded
PTFE	35	Very durable; more expensive than PE, PP, etc.
Elastomers	100–1000	Outstanding, but low modulus
High strength copper alloys	4	M less good than polymers. Use when high tensile stiffness is required
Spring steel	6	

itself. Cheap products with this sort of elastic hinge are generally molded from polyethylene, polypropylene, or nylon. Spring steel and other metallic spring materials (like phosphor bronze) are possibilities: they combine usable σ_f/E with high E, giving flexibility with good positional stability (as in the suspensions of relays). Table 6.14 gives further details.

Postscript. Polymers give more design-freedom than metals. The elastic hinge is one example of this, reducing the box, hinge and lid (3 components plus the fasteners needed to join them) to a single box-hinge-lid, molded in one operation. Their spring-like properties allow snap-together, easily-joined parts. Another is the elastomeric coupling—a flexible universal joint, allowing high angular, parallel, and axial flexibility with good shock absorption characteristics. Elastomeric hinges offer many opportunities, to be exploited in engineering design.

Related case studies

6.7 Materials for springs
6.9 Materials for seals
6.10 Deflection-limited design with brittle polymers
12.8 Shapes that flex: leaf and strand structures

6.9 Materials for seals

A reusable elastic seal consists of a cylinder of material compressed between two flat surfaces (Figure 6.15). The seal must form the largest possible contact width, b, while keeping the contact stress, σ, sufficiently low that it does not damage the flat surfaces; and the seal itself must remain elastic so that it can be

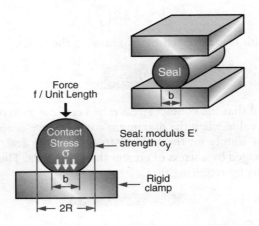

Figure 6.15 An elastic seal. A good seal gives a large conforming contact-area without imposing damaging loads on itself or on the surfaces with which it mates.

Table 6.15   Design requirements for elastic seals

Function	Elastic seal
Constraints	• Limit on contact pressure • Low cost
Objective	Maximum conformability to surface
Free variables	Choice of material

reused many times. What materials make good seals? Elastomers — everyone know that. But let us do the job properly; there may be more to be learnt. We build the selection around the requirements of Table 6.15.

The model. A cylinder of diameter $2R$ and modulus E, pressed on to a rigid flat surface by a force f per unit length, forms an elastic contact of width b (Appendix A) where

$$b \approx 2\left(\frac{fR}{E}\right)^{1/3} \tag{6.30}$$

This is the quantity to be maximized: the objective function. The contact stress, both in the seal and in the surface, is adequately approximated (Appendix A) by

$$\sigma = 0.6\left(\frac{fE}{R}\right)^{1/3} \tag{6.31}$$

The constraint: the seal must remain elastic, that is, σ must be less than the yield or failure strength, σ_f, of the material of which it is made. Combining the last two equations with this condition gives

$$b \leq 3.3R\left(\frac{\sigma_f}{E}\right) \tag{6.32}$$

The contact width is maximized by maximizing the index

$$M_1 = \frac{\sigma_f}{E}$$

It is also required that the contact stress σ be kept low to avoid damage to the flat surfaces. Its value when the maximum contact force is applied (to give the biggest width) is simply σ_f, the failure strength of the seal. Suppose the flat surfaces are damaged by a stress of greater than 100 MPa. The contact pressure is kept below this by requiring that

$$M_2 = \sigma_f \leq 100\,\text{MPa}$$

The selection. The two indices are plotted on the $\sigma_f - E$ chart in Figure 6.16 isolating elastomers, foams and cork. The candidates are listed

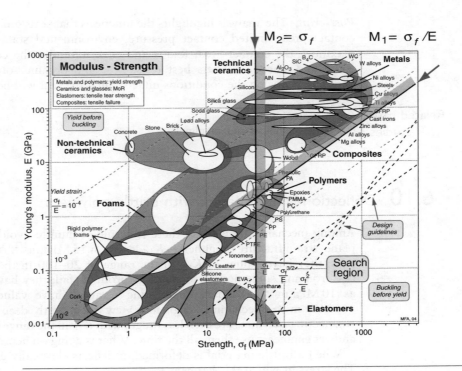

Figure 6.16   Materials for elastic seals. Elastomers, compliant polymers and foams make good seals.

Table 6.16   Materials for reusable seals

Material	$M_1 = \dfrac{\sigma_f}{E}$	Comment
Elastomeric EVA	0.7–1	The natural choice; poor resistance to heat and to some solvents
Polyurethanes	2–5	Widely used for seals
Silicone rubbers	0.2–0.5	Higher temperature capability than carbon-chain elastomers, chemically inert
PTFE	0.05–0.1	Expensive but chemically stable and with high temperature capability
Polyethylenes	0.02–0.05	Cheap but liable to take a permanent set
Polypropylenes	0.2–0.04	Cheap but liable to take a permanent set
Nylons	0.02–0.03	Near upper limit on contact pressure
Cork	0.03–0.06	Low contact stress, chemically stable
Polymer foams	up to 0.03	Very low contact pressure; delicate seals

in Table 6.16 with commentary. The value of $M_2 = 100\,\text{MPa}$ admits all elastomers as candidates. If M_2 were reduced to 10 MPa, all but the most compliant elastomers are eliminated, and foamed polymers become the best bet.

Postscript. The analysis highlights the functions that seals must perform: large contact area, limited contact pressure, environmental stability. Elastomers maximize the contact area; foams and cork minimize the contact pressure; PTFE and silicone rubbers best resist heat and organic solvents. The final choice depends on the conditions under which the seal will be used.

<table>
<tr><td>Related case studies</td><td>6.7</td><td>Materials for springs</td></tr>
<tr><td></td><td>6.8</td><td>Elastic hinges and couplings</td></tr>
</table>

6.10 Deflection-limited design with brittle polymers

Among mechanical engineers there is a rule-of-thumb: avoid materials with plane–strain fracture toughnesses K_{1C} less than 15 MPa.m$^{1/2}$. Almost all metals pass: they have values of K_{1C} in the range of 20–100 in these units. White cast iron and some powder-metallurgy products fail; they have values as low as 10 MPa.m$^{1/2}$. Ordinary engineering ceramics have values in the range 1–6 MPa.m$^{1/2}$; mechanical engineers view them with deep suspicion. But engineering polymers are even less tough, with K_{1C} in the range 0.5–3 MPa.m$^{1/2}$ and yet engineers use them all the time. What is going on here?

When a brittle material is deformed, it deflects elastically until it fractures. The stress at which this happens is

$$\sigma_f = \frac{CK_c}{\sqrt{\pi a_c}} \tag{6.33}$$

where K_c is an appropriate fracture toughness, a_c is the length of the largest crack contained in the material and C is a constant that depends on geometry, but is usually about 1. In a *load-limited* design a tension member of a bridge, say — the part will fail in a brittle way if the stress exceeds that given by equation (6.33). Here, obviously, we want materials with high values of K_c.

But not all designs are load-limited; some are *energy-limited*, others are *deflection limited*. Then the criterion for selection changes. Consider, then, the three scenarios created by the three alternative constraints of Table 6.17.

Table 6.17 Design requirements for deflection limited structures

Function	Resist brittle fracture
Constraints	• Design load specified or
	• Design energy specified or
	• Design deflection specified
Objective	Minimize volume (mass, cost)
Free variables	Choice of material

The model. In load-limited design the component must carry a specified load or pressure without fracturing. It is usual to identify K_c, with the plane-strain fracture toughness, K_{1C}, corresponding to the most highly constrained cracking conditions, because this is conservative. Then, as equation (6.33) shows, the best choice of materials for minimum volume design are those with high values of

$$M_1 = K_{1C} \tag{6.34}$$

For load-limited design using thin sheet, a plane-stress fracture toughness may be more appropriate; and for multi-layer materials, it may be an interface fracture toughness that matters. The point, though, is clear enough: the best materials for load-limited design are those with large values of the appropriate K_c.

But, as we have said, not all design is load-limited. Springs, and containment systems for turbines and flywheels are *energy-limited*. Take the spring (Figure 6.11) as an example. The elastic energy per unit volume stored in it is the integral over the volume of

$$U_e = \frac{1}{2}\sigma\varepsilon = \frac{1}{2}\frac{\sigma^2}{E}$$

The stress is limited by the fracture stress of equation (6.33) so that — if "failure" means "fracture" — the maximum energy the spring can store is

$$U_e^{max} = \frac{C^2}{2\pi a_c}\left(\frac{K_{1C}^2}{E}\right)$$

For a given initial flaw size, energy is maximized by choosing materials with large values of

$$M_2 = \frac{K_{1C}^2}{E} \approx J_c \tag{6.35}$$

where J_c is the toughness (usual units: kJ/m^2).

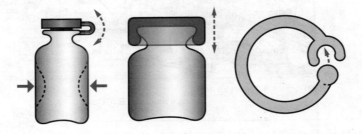

Figure 6.17   Load and deflection-limited design. Polymers, having low moduli, frequently require deflection-limited design methods.

There is a third scenario: that of *displacement-limited* design (Figure 6.17). Snap-on bottle tops, snap together fasteners, and such like are displacement-limited: they must allow sufficient elastic displacement to permit the snap-action without failure, requiring a large failure strain ε_f. The strain is related to the stress by Hooke's law $\varepsilon = \sigma/E$ and the stress is limited by the fracture equation (6.33). Thus the failure strain is

$$\varepsilon_f = \frac{C}{\sqrt{\pi a_c}} \frac{K_{1C}}{E} \tag{6.36}$$

The best materials for displacement-limited design are those with large values of

$$M_3 = \frac{K_{1C}}{E} \tag{6.37}$$

The selection. Figure 6.18 shows a chart of fracture toughness, K_{1C}, plotted against modulus E. It allows materials to be compared by values of fracture

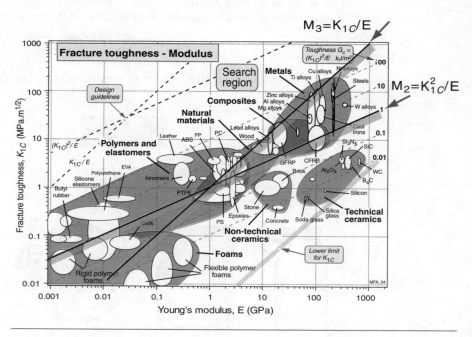

Figure 6.18   The selection of materials for load, deflection, and energy-limited design. In deflection-limited design, polymers are as good as metals, despite having very low values of the fracture toughness.

Table 6.18   Materials fracture-limited design

Design type and rule-of-thumb	Material
Load-limited design $K_{1C} > 15\,MPa.m^{1/2}$	Metals, polymer-matrix composites
Energy-limited design $J_c > 1\,kJ/m^2$	Metals, composites and some polymers
Displacement-limited design $K_{1C}/E > 10^{-3}\,m^{1/2}$	Polymers, elastomers and the toughest metals

toughness, M_1, by toughness, M_2, and by values of the deflection-limited index M_3. As the engineer's rule-of-thumb demands, almost all metals have values of K_{1C} that lie above the $15\,MPa.m^{1/2}$ acceptance level for load-limited design, shown and a horizontal selection line in Figure 6.18. Polymers and ceramics do not.

The line showing M_2 on Figure 6.18 is placed at the value $1\,kJ/m^2$. Materials with values of M_2 greater than this have a degree of shock-resistance with which engineers feel comfortable (another rule-of-thumb). Metals, composites, and some polymers qualify; ceramics do not. When we come to deflection-limited design, the picture changes again. The line shows the index $M_3 = K_{1C}/E$ at the value $10^{-3}\,m^{1/2}$. It illustrates why polymers find such wide application: when the design is deflection-limited, polymers—particularly nylons, polycarbonates and polystyrene—are better than the best metals (Table 6.18).

Postscript. The figure gives further insights. The mechanical engineers' love of metals (and, more recently, of composites) is inspired not merely by the appeal of their K_{1C} values. They are good by all three criteria (K_{1C}, K_{1C}^2/E and K_{1C}/E). Polymers have good values of K_{1C}/E and are acceptable by K_{1C}^2/E. Ceramics are poor by all three criteria. Herein lie the deeper roots of the engineers' distrust of ceramics.

Further reading   Background in fracture mechanics and safety criteria can be found in:

Brock, D. (1984) *Elementary Engineering Fracture Mechanics*, Martinus Nijoff, Boston.
Hellan, K. (1985) *Introduction to Fracture Mechanics*, McGraw-Hill.
Hertzberg, R.W. (1989) *Deformation and Fracture Mechanics of Engineering Materials*, Wiley, New York.

Related case studies   6.7   Materials for springs
6.8   Elastic hinges and couplings
6.11   Safe pressure vessels

6.11 Safe pressure vessels

Pressure vessels, from the simplest aerosol-can to the biggest boiler, are designed, for safety, to yield or leak before they break. The details of this design method vary. Small pressure vessels are usually designed to allow general yield at a pressure still too low to cause any crack the vessel may contain to propagate ("yield before break"); the distortion caused by yielding is easy to detect and the pressure can be released safely. With large pressure vessels this may not be possible. Instead, safe design is achieved by ensuring that the smallest crack that will propagate unstably has a length greater than the thickness of the vessel wall ("leak before break"); the leak is easily detected, and it releases pressure gradually and thus safely (Table 6.19). The two criteria lead to different material indices. What are they?

The model. The stress in the wall of a thin-walled spherical pressure vessel of radius R (Figure 6.19) is

$$\sigma = \frac{pR}{2t} \tag{6.38}$$

In pressure vessel design, the wall thickness, t, is chosen so that, at the working pressure p, this stress is less than the yield strength σ_f of the wall. A small

Table 6.19 Design requirements for safe pressure vessels

Function	Pressure vessel (contain pressure p safely)
Constraints	Radius R specified
Objective	• Maximize safety using yield-before-break criterion, or • Maximize safety using leak-before-break criterion
Free variables	*Choice of material*

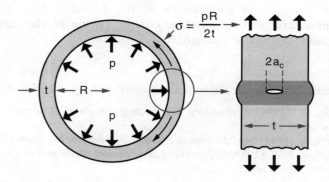

Figure 6.19 A pressure vessel containing a flaw. Safe design of small pressure vessels requires that they yield before they break; that of large pressure vessels may require, instead, that they leak before they break.

pressure vessel can be examined ultrasonically, or by X-ray methods, or proof tested, to establish that it contains no crack or flaw of diameter greater than $2a_c^*$; then the stress required to make the crack propagate[3] is

$$\sigma = \frac{CK_{1C}}{\sqrt{\pi a_c^*}}$$

where C is a constant near unity and K_{1C} is the plane-strain fracture toughness. Safety can be achieved by ensuring that the working stress is less than this, giving

$$p \leq \frac{2t}{R} \frac{K_{1C}}{\sqrt{\pi a_c^*}}$$

The largest pressure (for a given R, t and a_c^*) is carried by the material with the greatest value of

$$M_1 = K_{1C} \tag{6.39}$$

But this design is not fail-safe. If the inspection is faulty, or if, for some other reason a crack of length greater than a_c^* appears, catastrophe follows. Greater security is obtained by requiring that the crack will not propagate even if the stress reaches the general yield stress — for then the vessel will deform stably in a way that can be detected. This condition is expressed by setting σ equal to the yield stress σ_f giving

$$\pi a_c \leq C^2 \left[\frac{K_{1C}}{\sigma_f}\right]^2$$

The tolerable crack size, and thus the integrity of the vessel, is maximized by choosing a material with the largest value of

$$M_2 = \frac{K_{1C}}{\sigma_f} \tag{6.40}$$

Large pressure vessels cannot always be X-rayed or sonically tested; and proof testing them may be impractical. Further, cracks can grow slowly because of corrosion or cyclic loading, so that a single examination at the beginning of service life is not sufficient. Then safety can be ensured by arranging that a crack just large enough to penetrate both the inner and the outer surface of the vessel is still stable, because the leak caused by the crack can be detected. This is achieved if the stress is always less than or equal to

$$\sigma = \frac{CK_{1C}}{\sqrt{\pi t/2}} \tag{6.41}$$

---

[3] If the wall is sufficiently thin, and close to general yield, it will fail in a plane-stress mode. Then the relevant fracture toughness is that for plane stress, not the smaller value for plane strain.

The wall thickness t of the pressure vessel was, of course, designed to contain the pressure p without yielding. From equation (6.38), this means that

$$t \geq \frac{pR}{2\sigma_f} \tag{6.42}$$

Substituting this into the previous equation (with $\sigma = \sigma_f$) gives

$$p \leq \frac{4C^2}{\pi R}\left(\frac{K_{1C}^2}{\sigma_f}\right) \tag{6.43}$$

The maximum pressure is carried most safely by the material with the greatest value of

$$M_3 = \frac{K_{1C}^2}{\sigma_f} \tag{6.44}$$

Both M_1 and M_2 could be made large by making the yield strength of the wall, σ_f, very small: lead, for instance, has high values of both, but you would not choose it for a pressure vessel. That is because the vessel wall must also be as thin as possible, both for economy of material, and to keep it light. The thinnest wall, from equation (6.42), is that with the largest yield strength, σ_f. Thus we wish also to maximize

$$M_4 = \sigma_f$$

narrowing further the choice of material.

The selection. These selection criteria are explored by using the chart shown in Figure 6.20: the fracture toughness, K_{1C}, plotted against elastic limit σ_f. The indices M_1, M_2, M_3 and M_4 appear as lines of slope 0, 1, 1/2 and as lines that are vertical. Take "yield before break" as an example. A diagonal line corresponding to a constant value of $M_1 = K_{1C}/\sigma_f$ links materials with equal performance; those above the line are better. The line shown in the figure at $M_1 = 0.6\,\mathrm{m}^{1/2}$ (corresponding to a process zone of size 100 mm) excludes everything but the toughest steels, copper, aluminum and titanium alloys, though some polymers nearly make it (pressurized lemonade and beer containers are made of these polymers). A second selection line at $M_3 = 50\,\mathrm{MPa}$ eliminates aluminum alloys. Details are given in Table 6.20.

The leak-before-break criterion

$$M_2 = \frac{K_{1C}^2}{\sigma_f} \tag{6.45}$$

favors low alloy steel, stainless, and carbon steels more strongly, but does not greatly change the conclusions.

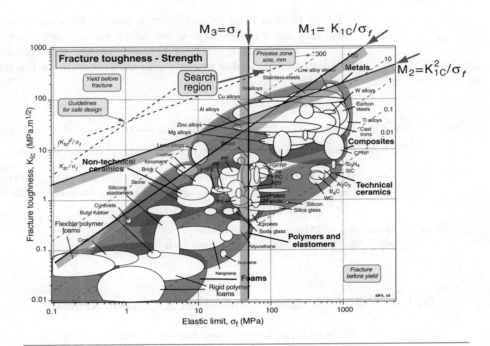

Figure 6.20   Materials for pressure vessels. Steel, copper alloys, and aluminum alloys best satisfy the "yield-before-break" criterion. In addition, a high yield strength allows a high working pressure. The materials in the "search areas" triangle are the best choice. The leak-before-break criterion leads to essentially the same selection.

Table 6.20   Materials for safe pressure vessels

Material	$M_1 = K_{IC}/\sigma_f$ (m$^{1/2}$)	$M_3 = \sigma_f$ (MPa)	Comment
Stainless steels	0.35	300	Nuclear pressure vessels are made of grade 316 stainless steel
Low alloy steels	0.2	800	These are standard in this application
Copper	0.5	200	Hard drawn copper is used for small boilers and pressure vessels
Aluminum alloys	0.15	200	Pressure tanks of rockets are aluminum
Titanium alloys	0.13	800	Good for light pressure vessels, but expensive

Postscript. Large pressure vessels are always made of steel. Those for models — a model steam engine, for instance — are made of copper. It is chosen, even though it is more expensive, because of its greater resistance to corrosion. Corrosion rates do not scale with size. The loss of 0.1 mm through corrosion is not serious in a pressure vessel that is 10 mm thick; but if it is only 1 mm thick it becomes a concern.

Boiler failures used to be common place — there are even songs about it. Now they are rare, though when safety margins are pared to a minimum (rockets, new aircraft designs) pressure vessels still occasionally fail. This (relative) success is one of the major contributions of fracture mechanics to engineering practice.

Further reading   Background in fracture mechanics and safety criteria can be found in:

Brock, D. (1984) *Elementary Engineering Fracture Mechanics*, Martinus Nijoff, Boston.
Hellan, K. (1985) *Introduction to Fracture Mechanics*, McGraw-Hill.
Hertzberg, R.W. (1989) *Deformation and Fracture Mechanics of Engineering Materials*, Wiley, New York.

Related case   6.6   Materials for flywheels
studies   6.10   Deflection-limited design with brittle polymers

6.12 Stiff, high damping materials for shaker tables

Shakers, if you live in Pennsylvania, are the members of an obscure and declining religious sect, noted for their austere wooden furniture. To those who live elsewhere they are devices for vibration-testing (Figure 6.21). This second sort of shaker consists of an electromagnetic actuator driving a table, at frequencies up to 1000 Hz, to which the test-object (a space probe, an automobile, an aircraft component, or the like) is clamped. The shaker applies a spectrum of vibration frequencies, f, and amplitudes, A, to the test-object to explore its response.

A big table operating at high frequency dissipates a great deal of power. The primary objective is to minimize this, but subject to a number of constraints itemized in Table 6.21. What materials make good shaker tables?

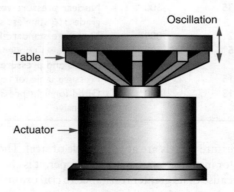

Figure 6.21   A shaker table. It is required to be stiff, but have high intrinsic "damping" or loss coefficient.

Table 6.21   Design requirements for shaker tables

Function	Table for vibration tester ("shaker table")
Constraints	• Radius, R, specified • Must be stiff enough to avoid distortion by clamping forces • Natural frequencies above maximum operating frequency (to avoid resonance) • High damping to minimize stray vibrations • Tough enough to withstand mishandling and shock
Objective	Minimize power consumption
Free variables	• Choice of material • Table thickness, t

The model. The power p (Watts) consumed by a dissipative vibrating system with a sinusoidal input is

$$p = C_1 m A^2 \omega^3 \tag{6.46}$$

where m is the mass of the table, A is the amplitude of vibration, ω is the frequency (rad s^{-1}) and C_1 is a constant. Provided the operating frequency ω is significantly less than the resonant frequency of the table, then $C_1 \approx 1$. The amplitude A and the frequency ω are prescribed. To minimize the power lost in shaking the table itself, we must minimize its mass m. We idealize the table as a disk of given radius, R. Its thickness, t, is a free variable. Its mass is

$$m = \pi R^2 t \rho \tag{6.47}$$

where ρ is the density of the material of which it is made. The thickness influences the bending-stiffness of the table—and this is important both to prevent the table flexing too much under clamping loads, and because it determines its lowest natural vibration frequency. The bending stiffness, S, is

$$S = \frac{C_2 E I}{R^3}$$

where C_2 is a constant. The second moment of the section, I, is proportional to $t^3 R$. Thus, for a given stiffness S and radius R,

$$t = C_3 \left(\frac{S R^2}{E} \right)^{1/3}$$

where C_3 is another constant. Inserting this into equation (6.47) we obtain

$$m = C_3 \pi R^{8/3} S^{1/3} \left(\frac{\rho}{E^{1/3}} \right) \tag{6.48}$$

The mass of the table, for a given stiffness and minimum vibration frequency, is therefore minimized by selecting materials with high values of

$$\boxed{M_1 = \frac{E^{1/3}}{\rho}} \tag{6.49}$$

There are three further requirements. The first is that of high mechanical damping, measured by the loss coefficient, η. The second that the fracture toughness K_{1C} of the table be sufficient to withstand mishandling and clamping forces. And the third is that the material should not cost too much.

The selection. Figure 6.22 shows the chart of loss coefficient η plotted against modulus E. The vertical line shows the constraint $E \geq 30$ GPa, the horizontal one, the constraint $\eta > 0.001$. The search region contains CFRP and a number of metals: magnesium, titanium, cast irons and steels. All are possible candidates. Table 6.22 compares their properties.

Postscript. Stiffness, high natural frequencies and damping are qualities often sought in engineering design. The shaker table found its solution (in real life as well as this case study) in the choice of a cast magnesium alloy.

Sometimes a solution is possible by combining materials (more on this in Chapter 13). The loss coefficient chart shows that polymers and elastomers have high damping. Sheet steel panels, prone to lightly-damped vibration, can be damped by coating one surface with a polymer, a technique

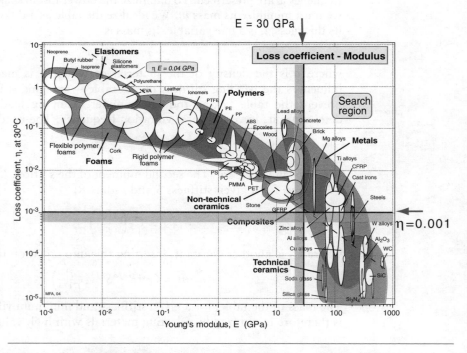

Figure 6.22   Selection of materials for the shaker table. Magnesium alloys, cast irons, GFRP, concrete and the special high-damping Mn–Cu alloys are candidates.

Table 6.22   Materials for shaker tables

Material	Loss coeff, η	$M_1 = E^{1/3}/\rho$ $GPa^{1/3}/(Mg/m^3)$	ρ $Mg/(m)^3$	Comment
Mg-alloys	Up to 2×10^{-2}	1.9	1.75	The best combination of properties
Titanium alloys	Up to 5×10^{-3}	1.0	4.6	Good damping but expensive
CFRP	Up to 4×10^{-3}	3.0	1.8	Less damping than Mg-alloys, but possible
Cast irons	Up to 4×10^{-3}	0.7	7.8	Good damping but heavy
Zinc alloys	Up to 7×10^{-3}	0.7	5.5	Less damping than Mg-alloys, but possible for a small table

exploited in automobiles, typewriters and machine tools. Aluminum structures can be stiffened (raising natural frequencies) by bonding carbon fiber to them: an approach sometimes use in aircraft design. And structures loaded in bending or torsion can be made lighter, for the same stiffness (again increasing natural frequencies), by shaping them efficiently: by attaching ribs to their underside, for instance. Shaker tables — even the austere wooden tables of the Pennsylvania Shakers — exploit shape in this way.

Further reading

Tustin, W. and Mercado, R. (1984) *Random Vibrations in Perspective*, Tustin Institute of Technology Inc, Santa Barbara, CA, USA.

Cebon, D. and Ashby, M.F. (1994) Materials selection for precision instruments, *Meas. Sci. Technol.* 5, 296–306.

Related case studies

6.4   Materials for table legs
6.7   Materials for springs
6.16   Materials to minimize thermal distortion in precision devices

6.13 Insulation for short-term isothermal containers

Each member of the crew of a military aircraft carries, for emergencies, a radio beacon. If forced to eject, the crew member could find himself in trying circumstances — in water at 4°C, for example (much of the earth's surface is ocean with a mean temperature of roughly this). The beacon guides friendly rescue services, minimizing exposure time.

But microelectronic metabolisms (like those of humans) are upset by low temperatures. In the case of the beacon, it is its transmission frequency that starts to drift. The design specification for the egg-shaped package containing

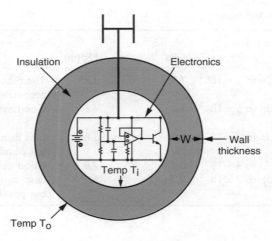

Figure 6.23 An isothermal container. It is designed to maximize the time before the inside temperature changes after the outside temperature has suddenly changed.

Table 6.23 Design requirements for short-term insulation

Function	Short term thermal insulation
Constraints	Wall thickness must not exceed w
Objective	Maximize time t before internal temperature changes when external temperature suddenly drops
Free variables	Choice of material

the electronics (Figure 6.23) requires that, when the temperature of the outer surface is changed by 30°C, the temperature of the inner surface should not change significantly for an hour. To keep the device small, the wall thickness is limited to a thickness w of 20 mm. What is the best material for the package? A dewar system is out — it is too fragile.

A foam of some sort, you might think. But here is a case in which intuition leads you astray. So let us formulate the design requirements (Table 6.23) and do the job properly.

The model. We model the container as a wall of thickness w, thermal conductivity λ. The heat flux q through the wall, once a steady-state has been established, is given by Fick's first law:

$$q = -\lambda \frac{\mathrm{d}T}{\mathrm{d}x} = \lambda \frac{(T_i - T_o)}{w} \tag{6.50}$$

where T_o is the temperature of the outer surface, T_i is that of the inner one and $\mathrm{d}T/\mathrm{d}x$ is the temperature gradient (Figure 6.23). The only free variable here is the thermal conductivity, λ. The flux is minimized by choosing a wall material

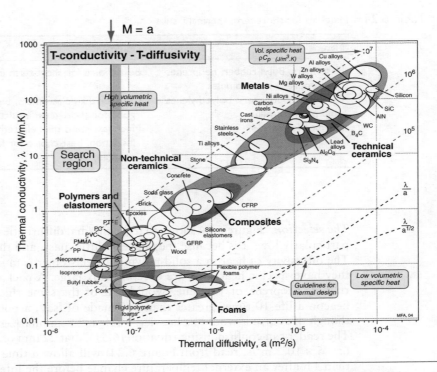

Figure 6.24   Materials for short-term isothermal containers. Elastomers are good; foams are not.

with the lowest possible value of λ. The $\lambda - \alpha$ chart (Figure 6.24) shows that this is, indeed, a foam.

But we have answered the wrong question. The design brief was not to minimize the *heat flux* through the wall, but the *time* before the temperature of the inner wall changed appreciably. When the surface temperature of a body is suddenly changed, a temperature wave, so to speak, propagates inwards. The distance x it penetrates in time t is approximately $\sqrt{2at}$. Here a is the thermal diffusivity, defined by

$$a = \frac{\lambda}{\rho C_p} \tag{6.51}$$

where ρ is the density and C_p is the specific heat (Appendix A). Equating this to the wall thickness w gives

$$t \approx \frac{w^2}{2a} \tag{6.52}$$

The time is maximized by choosing the smallest value of the thermal diffusivity, a, not the conductivity λ.

Table 6.24   Materials for short-term thermal insulation

Material	Comment
Elastomers: Butyl rubber, neoprene and isoprene are examples	Best choice for short-term insulation
Commodity polymers: polyethylenes and polypropylenes	Cheaper than elastomers, but somewhat less good for short-term insulation
Polymer foams	Much less good than elastomers for short-term insulation; best choice for long-term insulation at steady state

The selection. Figure 6.24 shows that the thermal diffusivities of foams are not particularly low; it is because they have so little mass, and thus heat capacity. The diffusivity of heat in a solid polymer or elastomer is much lower because they have specific heats that are exceptionally large. A package made of solid rubber, neoprene or isoprene, would — if of the same thickness — give the beacon a life 10 times greater than one made of (say) a polystyrene foam — though of course it would be heavier. Table 6.24 summarizes the conclusions. The reader can confirm, using equation (6.51), that 22 mm of a solid elastomer ($a = 5 \times 10^{-8} \, m^2/s$, read from Figure 6.24) will allow a time interval of more than 1 h after an external temperature change before the internal temperature shifts much.

Postscript. One can do better than this. The trick is to exploit other ways of absorbing heat. If a liquid — a low-melting wax, for instance — can be found that solidifies at a temperature equal to the minimum desired operating temperature for the transmitter (T_i), it can be used as a "latent-heat sink". Channels in the package are filled with the liquid; the inner temperature can only fall below the desired operating temperature when all the liquid has solidified. The latent heat of solidification must be supplied to do this, giving the package a large (apparent) specific heat, and thus an exceptionally low diffusivity for heat at the temperature T_i. The same idea is used, in reverse, in "freezer packs" that solidify when placed in the freezer compartment of a refrigerator and remain cold (by melting, at 4°C) when packed around warm beer cans in a portable cooler.

Further reading   Holman, J.P. (1981) *Heat Transfer*, 5th edition, McGraw-Hill, New York, USA.

Related case studies   6.14   Energy-efficient kiln walls
6.15   Materials for passive solar heating

6.14 Energy-efficient kiln walls

The energy cost of one firing cycle of a large pottery kiln (Figure 6.25) is considerable. Part is the cost of the energy that is lost by conduction through the kiln walls; it is reduced by choosing a wall material with a low conductivity, and by making the wall thick. The rest is the cost of the energy used to raise the kiln to its operating temperature; it is reduced by choosing a wall material with a low heat capacity, and by making the wall thin. Is there a material index that captures these apparently conflicting design goals? And if so, what is a good choice of material for kiln walls? The choice is based on the requirements of Table 6.25.

The model. When a kiln is fired, the internal temperature rises quickly from ambient, T_o, to the operating temperature, T_i, where it is held for the firing time t. The energy consumed in the firing time has, as we have said, two contributions. The first is the heat conducted out: at steady state the heat loss by

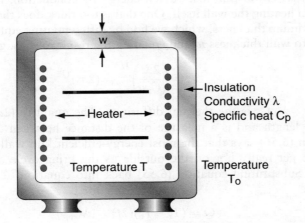

| Figure 6.25 | A kiln. On firing, the kiln wall is first heated to the operating temperature, then held at this temperature. A linear gradient is then expected through the kiln wall. |

Table 6.25   Design requirements for kiln walls

Function	Thermal insulation for kiln (cyclic heating and cooling)
Constraints	• Maximum operating temperature 1000°C • Possible limit on kiln-wall thickness for space reasons
Objective	Minimize energy consumed in firing cycle
Free variables	• Kiln wall thickness, w • Choice of material

conduction, Q_1, per unit area, is given by the first law of heat flow. If held for time t it is

$$Q_1 = -\lambda \frac{dT}{dx} t = \lambda \frac{(T_i - T_o)}{w} t \qquad (6.53)$$

Here λ is the thermal conductivity, dT/dx is the temperature gradient and w is the insulation wall-thickness. The second contribution is the heat absorbed by the kiln wall in raising it to T_i, and this can be considerable. Per unit area, it is

$$Q_2 = C_p \rho w \left(\frac{T_i - T_o}{2} \right) \qquad (6.54)$$

where C_p is the specific heat of the wall material and ρ is its density. The total energy consumed per unit area is the sum of these two:

$$Q = Q_1 + Q_2 = \frac{\lambda(T_i + T_o)t}{w} + \frac{C_p \rho w (T_i - T_o)}{2} \qquad (6.55)$$

A wall that is too thin loses much energy by conduction, but absorbs little energy in heating the wall itself. One that is too thick does the opposite. There is an optimum thickness, which we find by differentiating equation (6.54) with respect to wall thickness w and equating the result to zero, giving:

$$w = \left(\frac{2\lambda t}{C_p \rho} \right)^{1/2} = (2at)^{1/2} \qquad (6.56)$$

where $a = \lambda/\rho C_p$ is the thermal diffusivity. The quantity $(2at)^{1/2}$ has dimensions of length and is a measure of the distance heat can diffuse in time t. Equation (6.56) says that the most energy-efficient kiln wall is one that only starts to get really hot on the outside as the firing cycle approaches completion. Substituting equation (6.55) back into equation (6.55) to eliminate w gives:

$$Q = (T_i - T_o)(2t)^{1/2}(\lambda C_p \rho)^{1/2}$$

Q is minimized by choosing a material with a low value of the quantity $(\lambda C_p \rho)^{1/2}$, that is, by maximizing

$$M = (\lambda C_p \rho)^{-1/2} = \frac{a^{1/2}}{\lambda} \qquad (6.57)$$

By eliminating the wall thickness w we have lost track of it. It could, for some materials, be excessively large. Before accepting a candidate material we must check, by evaluating equation (6.56) how thick the wall made from it will be.

The selection. Figure 6.26 shows the $\lambda - a$ chart with a selection line corresponding to $M = a^{1/2}/\lambda$ plotted on it. Polymer foams, cork and solid polymers are good, but only if the internal temperature is less than 150°C. Real kilns operate near 1000°C requiring materials with a maximum service temperature above this value. The figure suggests brick (Table 6.26), but

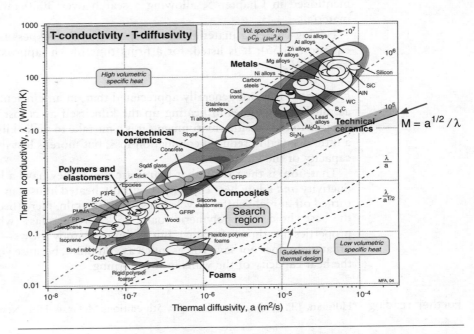

Figure 6.26   Materials for kiln walls. Low density, porous or foam-like ceramics are the best choice.

Table 6.26   Materials for energy-efficient kilns

Material	$M = a^{1/2}/\lambda$ $(m^2K/W.s^{1/2})$	Thickness w (mm)	Comment
Brick	10^{-3}	90	The obvious choice: the lower the density, the better the performance. Special refractory bricks have values of M as high as 3×10^{-3}
Concrete	5×10^{-4}	110	High-temperature concrete can withstand temperatures up to 1000°C
Woods	2×10^{-3}	60	The boiler of Stevenson's "Rocket" steam engine was insulated with wood
Solid elastomers and solid polymers	2×10^{-3}–3×10^{-3} 2×10^{-3}	50	Good values of material index. Useful if the wall must be very thin. Limited to temperatures below 150°C
Polymer foam, cork	$3 \times 10^{-3} = 3 \times 10^{-2}$	50–100	The highest value of M—hence their use in house insulation. Limited to temperatures below 150°C

here the limitation of the hard-copy charts becomes apparent: there is not enough room to show specialized materials such as refractory bricks and concretes. The limitation is overcome by the computer-based methods mentioned in Chapter 5, allowing a search over 3000 rather than just 68 materials.

Having chosen a material, the acceptable wall thickness is calculated from equation (6.55). It is listed, for a firing time of 3 h (approximately 10^4 s) in Table 6.26.

Postscript. It is not generally appreciated that, in an efficiently-designed kiln, as much energy goes in heating up the kiln itself as is lost by thermal conduction to the outside environment. It is a mistake to make kiln walls too thick; a little is saved in reduced conduction-loss, but more is lost in the greater heat capacity of the kiln itself.

That, too, is the reason that foams are good: they have a low thermal conductivity *and* a low heat capacity. Centrally heated houses in which the heat is turned off at night suffer a cycle like that of the kiln. Here (because T_i is lower) the best choice is a polymeric foam, cork, or fiberglass (which has thermal properties like those of foams). But as this case study shows — turning the heat off at night does not save you as much as you think, because you have to supply the heat capacity of the walls in the morning.

Further reading   Holman, J.P. (1981) *Heat Transfer*, 5th edition, McGraw-Hill, New York, USA.

Related case
studies
6.13   Insulation for short-term isothermal containers
6.15   Materials for passive solar heating

6.15 Materials for passive solar heating

There are a number of schemes for capturing solar energy for home heating: solar cells, liquid filled heat exchangers, and solid heat reservoirs. The simplest of these is the heat-storing wall: a thick wall, the outer surface of which is heated by exposure to direct sunshine during the day, and from which heat is extracted at night by blowing air over its inner surface (Figure 6.27). An essential of such a scheme is that the time-constant for heat flow through the wall be about 12 h; then the wall first warms on the inner surface roughly 12 h after the sun first warms the outer one, giving out at night what it took in during the day. We will suppose that, for architectural reasons, the wall must not be more than $\frac{1}{2}$ m thick. What materials maximize the thermal energy captured by the wall while retaining a heat-diffusion time of up to 12 h? Table 6.27 summarizes the requirements.

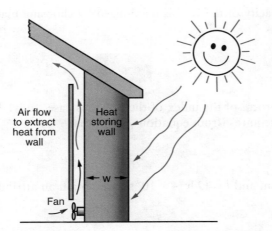

Figure 6.27 A heat-storing wall. The sun shines on the outside during the day; heat is extracted from the inside at night. The heat diffusion-time through the wall must be about 12 hours.

Table 6.27 Design requirements for passive solar heating

Function	Heat storing medium
Constraints	• Heat diffusion time through wall $t \approx 12\,\text{h}$
	• Wall thickness $\leq 0.5\,\text{m}$
	• Adequate working temperature $T_{\text{max}} > 100°\text{C}$
Objective	Maximize thermal energy stored per unit material cost
Free variables	• Wall thickness, w
	• Choice of material

The model. The heat content, Q, per unit area of wall, when heated through a temperature interval ΔT gives the objective function

$$Q = w\rho C_p \Delta T \qquad (6.58)$$

where w is the wall thickness, and ρC_p is the volumetric specific heat (the density ρ times the specific heat C_p). The 12-h time constant is a constraint. It is adequately estimated by the approximation used earlier for the heat-diffusion distance in time t (see Appendix A):

$$w = \sqrt{2at} \qquad (6.59)$$

where a is the thermal diffusivity. Eliminating the free variable w gives

$$Q = \sqrt{2t}\,\Delta T a^{1/2} \rho C_p \qquad (6.60)$$

or, using the fact that $a = \lambda/\rho C_p$ where λ is the thermal conductivity,

$$Q = \sqrt{2t}\Delta T\left(\frac{\lambda}{a^{1/2}}\right)$$

The heat capacity of the wall is maximized by choosing material with a high value of

$$M = \frac{\lambda}{a^{1/2}} \tag{6.61}$$

it is the reciprocal of the index of the previous case study. The restriction on thickness w requires (from equation (6.59)) that

$$a \leq \frac{w^2}{2t}$$

with $w \leq 0.5\,\mathrm{m}$ and $t = 12\,\mathrm{h}$ ($4 \times 10^4\,\mathrm{s}$), we obtain an attribute limit

$$a \leq 3 \times 10^{-6}\,\mathrm{m}^2/\mathrm{s} \tag{6.62}$$

The selection. Figure 6.28 shows thermal conductivity λ plotted against thermal diffusivity a with M and the limit on a plotted on it. It identifies the group of materials, listed in Table 6.28: they maximize M_1 while meeting the

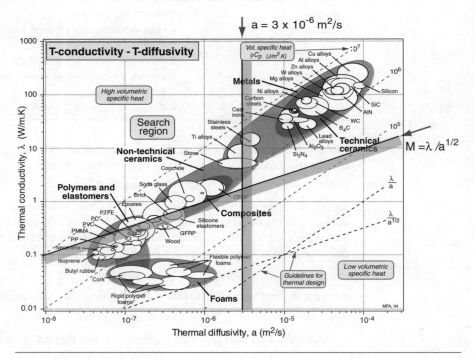

Figure 6.28   Materials for heat-storing walls. Cement, concrete and stone are practical choices; brick is less good.

Table 6.28   Materials for passive solar heat-storage

Material	$M_1 = \lambda/a^{1/2}$ (W.s$^{1/2}$/m^2.K)	Approx. cost $/m^3	Comment
Concrete	2.2×10^3	200	The best choice—good performance at minimum cost
Stone	3.5×10^3	1400	Better performance than concrete because specific heat is greater, but more expensive
Brick	10^3	1400	Less good than concrete
Glass	1.6×10^3	10,000	Useful—part of the wall could be glass
Titanium	4.6×10^3	200,000	An unexpected, but valid, selection. Expensive

constraint on wall thickness. Solids are good; porous materials and foams (often used in walls) are not.

Postscript. All this is fine, but what of cost? If this scheme is to be used for housing, cost is an important consideration. The approximate costs per unit volume, read from Figure 4.17(b), are listed in the table—it points to the selection of concrete, with stone and brick as alternatives.

Related case studies   6.13   Insulation for short-term isothermal containers
6.14   Energy-efficient kiln walls

6.16 Materials to minimize thermal distortion in precision devices

The precision of a measuring device, like a sub-micrometer displacement gauge, is limited by its stiffness and by the dimensional change caused by temperature gradients. Compensation for elastic deflection can be arranged; and corrections to cope with thermal expansion are possible too—provided the device is at a uniform temperature. *Thermal gradients* are the real problem: they cause a change of shape—that is, a distortion of the device—for which compensation is not possible. Sensitivity to vibration is also a problem: natural excitation introduces noise and thus imprecision into the measurement. So it is permissible to allow expansion in precision instrument design, provided distortion does not occur (Chetwynd, 1987). Elastic deflection is allowed, provided natural vibration frequencies are high.

What, then, are good materials for precision devices? Table 6.29 lists the requirements.

The model. Figure 6.29 shows, schematically, such a device: it consists for a force loop, an actuator and a sensor. We aim to choose a material for the force loop. It will, in general, support heat sources: the fingers of the operator of the

Table 6.29 Design requirements for precision devices

Function	Force loop (frame) for precision device
Constraints	• Must tolerate heat flux • Must tolerate vibration
Objective	Maximize positional accuracy (minimize distortion)
Free variables	Choice of material

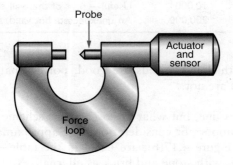

Figure 6.29 A schematic of a precision measuring device. Super-accurate dimension-sensing devices include the atomic-force microscope and the scanning tunneling microscope.

device in the figure, or, more usually, electrical components that generate heat. The relevant material index is found by considering the simple case of one-dimensional heat flow through a rod insulated except at its ends, one of which is at ambient and the other connected to the heat source. In the steady state, Fourier's law is

$$q = -\lambda \frac{dT}{dx} \tag{6.63}$$

where q is heat input per unit area, λ is the thermal conductivity and dT/dx is the resulting temperature gradient. The strain is related to temperature by

$$\varepsilon = \alpha(T_o - T) \tag{6.64}$$

where α is the thermal conductivity and T_o is ambient temperature. The distortion is proportional to the gradient of the strain:

$$\frac{d\varepsilon}{dx} = \frac{\alpha \, dT}{dx} = \left(\frac{\alpha}{\lambda}\right)q \tag{6.65}$$

Thus for a given geometry and heat flow, the distortion $d\varepsilon/dx$ is minimized by selecting materials with large values of the index

$$M_1 = \frac{\lambda}{\alpha}$$

The other problem is vibration. The sensitivity to external excitation is minimized by making the natural frequencies of the device as high as possible. The flexural vibrations have the lowest frequencies; they are proportional to

$$M_2 = \frac{E^{1/2}}{\rho}$$

A high value of this index will minimize the problem. Finally, of course, the device must not cost too much.

The selection. Figure 6.30 shows the expansion coefficient, α, plotted against the thermal conductivity, λ. Contours show constant values of the quantity λ/α. A search region is isolated by the line $\lambda/\alpha = 10^7$ W/m, giving the short list of Table 6.30. Values of $M_2 = E^{1/2}/\rho$ read from the $E - \rho$ chart of Figure 4.3 are included in the table. Among metals, copper, tungsten and the special nickel alloy Invar have the best values of M_1 but are disadvantaged by having high densities and thus poor values of M_2. The best choice is silicon, available in large sections, with high purity. Silicon carbide is an alternative.

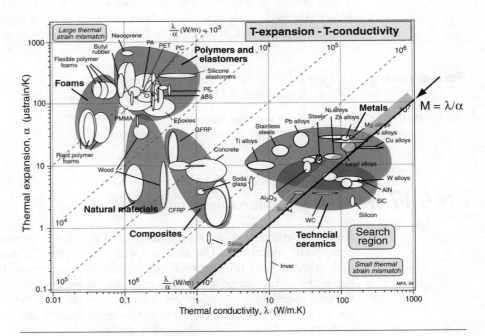

Figure 6.30   Materials for precision measuring devices. Metals are less good than ceramics because they have lower vibration frequencies. Silicon may be the best choice.

Table 6.30 Materials to minimize thermal distortion

Material	$M_1 = \lambda/\alpha$ (W/m)	$M_2 = E^{1/2}/\rho$ (GPa$^{1/2}$/(Mg/m^3))	Comment
Silicon	6×10^7	5.2	Excellent M_1 and M_2
Silicon carbide	3×10^7	6.4	Excellent M_1 and M_2 but more difficult to shape than silicon
Copper	2×10^7	1.3	High density gives poor value of M_2
Tungsten	3×10^7	1.1	Better than copper, silver or gold, but less good than silicon or SiC.
Aluminum alloys	10^7	3.3	The cheapest and most easily shaped choice

Postscript. Nano-scale measuring and imaging systems present the problem analyzed here. The atomic-force microscope and the scanning-tunneling microscope both rely on a probe, supported on a force loop, typically with a piezo-electric actuator and electronics to sense the proximity of the probe to the test surface. Closer to home, the mechanism of a video recorder and that of a hard disk drive qualify as precision instruments; both have a sensor (the read head) attached, with associated electronics, to a force loop. The materials identified in this case study are the best choice for force loop.

Further reading   Chetwynd, D.G. (1987) *Precision Engineering*, **9** (1), 3.
Cebon, D. and Ashby, M.F. (1994) *Meas. Sci. Technol.*, 5, 296.

Related case   6.3   Mirrors for large telescopes
studies   6.13   Insulation for short-term isothermal containers

6.17 Nylon bearings for ships' rudders

Rudder bearings of ships (Figure 6.31 and Table 6.31) operate under the most unpleasant conditions. The sliding speed is low, but the bearing pressure is high and adequate lubrication is often difficult to maintain. The rudder lies in the wake of the propeller, which generates severe vibration and consequent fretting. Sand and wear debris tend to get trapped between the bearing surfaces. Add to this the environment — aerated salt water — and you can see that bearing design is something of a challenge.

Ship bearings are traditionally made of bronze. The wear resistance of bronzes is good, and the maximum bearing pressure (important here) is high.

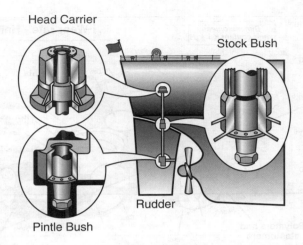

Figure 6.31   A ship's rudder and its bearings.

Table 6.31   Design requirements for rudder bearings

Function	Sliding bearing
Constraints	• Wear resistant with water lubrication
	• Resist corrosion in sea water
	• High damping desirable
Objective	Maximize life, meaning minimize wear rate
Free variables	• Choice of material
	• Bearing diameter and length

But, in sea water, galvanic cells are set up between the bronze and any other metal to which it is attached by a conducting path (no matter how remote), and in a ship such connections are inevitable. So galvanic corrosion, as well as abrasion by sand, is a problem. Is there a better choice than bronze?

The model. We assume (reasonably) that the bearing *force, F,* is fixed by the design of the ship. The bearing *pressure, P,* can be controlled by changing the area A of the bearing surface:

$$P \propto \frac{F}{A}$$

This means that we are free to choose a material with a lower maximum bearing pressure provided the length of the bearing itself is increased to compensate. With this thought in mind, we seek a bearing material that will not corrode in salt water and can function without full lubrication.

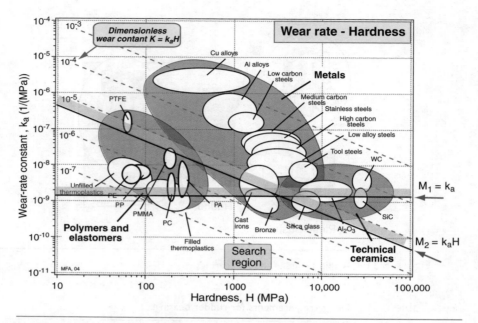

Figure 6.32 Materials for rudder bearings. Wear is very complex, so the chart gives qualitative guidance only. It suggests that polymers like nylon or filled or reinforced polymers might be an alternative to bronze provided the bearing area is increased appropriately.

The selection. Figure 6.32 shows the chart of wear-rate constant, k_a, and hardness, H. The wear-rate, W, is given by equation (4.26), which, repeated, is

$$\Omega = k_a P = C\left(\frac{P}{P_{max}}\right)k_a H$$

where C is a constant, P is the bearing pressure, P_{max} the maximum allowable bearing pressure for the material, and H is its hardness. If the bearing is not re-sized when a new material is used, the bearing pressure P is unchanged and the material with the lowest wear-rate is simply that with the smallest value of the quantity

$$M_1 = k_a$$

Bronze performs well, but filled thermoplastics are nearly as good and have superior corrosion resistance in salt water. If, on the other hand, the bearing is re-sized so that it operates at a set fraction of P_{max} (0.5, say), the material with the lowest wear-rate is that with the smallest value of

$$M_2 = k_a H$$

Here polymers are clearly superior. Table 6.32 summarizes the conclusions.

Table 6.32   Materials for rudder bearings

Material	Comment
PTFE, polyethylenes polypropylenes, nylon	Low friction and good wear resistance at low bearing pressures
Glass-reinforced PTFE, filled polyethylenes and polypropylenes	Excellent wear and corrosion resistance in sea water. A viable alternative to bronze if bearing pressures are not too large
Silicon carbide SiC, alumina Al_2O_3, tungsten carbide WC	Good wear and corrosion resistance but poor impact properties and very low damping

Postscript. Recently, at least one manufacturer of marine bearings has started to supply cast Nylon-6 bearings for large ship rudders. The makers claim just the advantages we would expect from this case study:

(a) wear and abrasion resistance with water lubrication is improved;
(b) deliberate lubrication is unnecessary;
(c) corrosion resistance is excellent;
(d) the elastic and damping properties of Nylon-6 protect the rudder from shocks (see the damping/modulus chart);
(e) there is no fretting;
(f) the material is easy to handle and install, and is inexpensive to machine.

Figure 6.32 suggests that a filled polymer or composite might be even better. Carbon–fiber filled nylon has better wear resistance than unfilled nylon, but it is less tough and flexible, and it does not damp vibration as effectively. As in all such problems, the best material is the one that comes closest to meeting *all* the demands made on it, not just the primary design criterion (in this case, wear resistance). The suggestion of the chart is a useful one, worth a try. It would take sea-tests to tell whether it should be adopted.

6.18 Materials for heat exchangers

This and the next case study illustrate the output of the CES software described in Sections 5.5.

Heat exchangers take heat from one fluid and pass it to a second (Figure 6.33). The fire-tube array of a steam engine is a heat exchanger, taking heat from the hot combustion gases of the firebox and transmitting it to the water in the boiler. The network of finned tubes in an air conditioner is a heat exchanger, taking heat from the air of the room and dumping it into the working fluid of the conditioner. A key element in all heat exchangers is the tube wall or membrane that separates the two fluids. It is required to transmit heat, and there is frequently a pressure difference across it, which can be large.

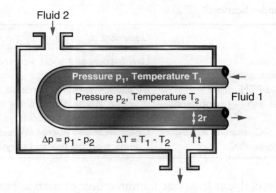

Figure 6.33 A heat exchanger. There is a pressure difference Δp and a temperature difference ΔT across the tube wall that also must resist attack by chloride ions.

Table 6.33 Design requirements for a heat exchanger

Function	Heat exchanger
Constraints	• Support pressure difference, Δp • Withstand chloride ions • Operating temperature up to 150°C • Modest cost
Objective	• Maximize heat flow per unit area (minimum volume exchanger) or • Maximize heat flow per unit mass (minimum mass exchanger)
Free variables	• Tube-wall thickness, t • Choice of material

What are the best materials for making heat exchangers? Or, to be specific, what are the best materials for a conduction-limited exchanger with substantial pressure difference between the two fluids, one of them containing chloride ions (sea water). Table 6.33 summarizes these requirements.

The model. First, a little background on heat flow. Heat transfer from one fluid, through a membrane to a second fluid, involves *convective* transfer from fluid 1 into the tube wall, *conduction* through the wall, and *convection* again to transfer it into fluid 2. The heat flux into the tube wall by convection (W/m^2) is described by the heat transfer equation:

$$q = h_1 \Delta T_1 \tag{6.66}$$

in which h_1 is the heat transfer coefficient and ΔT_1 is the temperature drop across the surface from fluid 1 into the wall. Conduction is described by the conduction (or Fourier) equation, which, for one-dimensional heat-flow takes the form:

$$q = \lambda \frac{\Delta T}{t} \tag{6.67}$$

where λ is the thermal conductivity of the wall (thickness t) and ΔT is the temperature difference across it. It is helpful to think of the *thermal resistance* at surface 1 as $1/h_1$; that of surface 2 is $1/h_2$; and that of the wall itself is t/λ. Then continuity of heat flux requires that the total resistance $1/U$ is

$$\frac{1}{U} = \frac{1}{h_1} + \frac{t}{\lambda} + \frac{1}{h_2} \tag{6.68}$$

where U is called the "total heat transfer coefficient". The heat flux from fluid 1 to fluid 2 is then given by

$$q = U(T_1 - T_2) \tag{6.69}$$

where $(T_1 - T_2)$ is the difference in temperature between the two working fluids.

When one of the fluids is a gas — as in an air conditioner — convective heat transfer at the tube surfaces contributes most of the resistance; then fins are used to increase the surface area across which heat can be transferred. But when both working fluids are liquid, convective heat transfer is rapid and conduction through the wall dominates the thermal resistance; $1/h_1$ and $1/h_2$ are negligible compared with t/λ. In this case, simple tube or plate elements are used, making their wall as thin as possible to minimize t/λ. We will consider the second case: conduction-limited heat transfer, where the heat flow is adequately described by equation (6.63).

Consider, then, a heat exchanger with n tubes of length L, each of radius r and wall thickness t. Our aim is to select a material to maximize the total heat flow:

$$Q = qA = \frac{A\lambda}{t} \Delta T \tag{6.70}$$

where $A = 2\pi r L n$ is the total surface are of tubing.

This is the objective function. The constraint is that the wall thickness must be sufficient to support the pressure Δp between the inside and outside, as in Figure 6.33. This requires that the stress in the wall remain below the elastic limit, σ_y, of the material of which the tube is made (multiplied by a safety factor — which we can leave out):

$$\sigma = \frac{\Delta p r}{t} < \sigma_y \tag{6.71}$$

This constrains the minimum value of t. Eliminating t between equations (6.70) and (6.71) gives

$$Q = \frac{A \Delta T}{r \Delta p} (\lambda \sigma_y) \tag{6.72}$$

The heat flow per unit area of tube wall, Q/A, is maximized by maximizing

$$M_1 = \lambda \sigma_y \tag{6.73}$$

Four further considerations enter the selection. It is essential to choose a material that can withstand corrosion in the working fluids, which we take to be water containing chloride ions (sea water). Cost, too, will be of concern. The maximum operating temperature must be adequate and the materials must have sufficient ductility to be drawn to tube or rolled to sheet. Cost, too, will be of concern.

The selection. A preliminary search (not shown) for materials with large values of M_1, using the CES Level 1/2 database, suggests *copper alloys* as one possibility. We therefore turn to the Level 3 database for more help. The first selection stage applies limits of 150°C on maximum service temperature, 30 percent on elongation, a material cost of less than \$4/kg and requires a rating of "very good" resistance to sea water. The second stage (Figure 6.34) is a chart of σ_y versus λ enabling $M_1 = \sigma_y \lambda$ to be maximized. The materials with large M_1 are listed in Table 6.34.

Postscript. Conduction may limit heat flow in theory, but unspeakable things go on inside heat exchangers. Sea water — often one of the working fluids — seethes with bio-fouling organisms that attach themselves to tube walls and thrive there, like barnacles on a boat, creating a layer of high thermal resistance

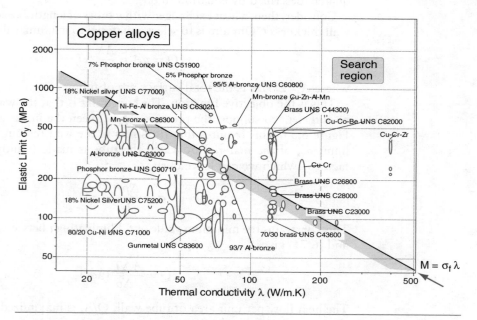

Figure 6.34   A chart of yield strength (elastic limit) σ_y against thermal conductivity, λ, showing the index M_1, using the Level 3 CES database.

Table 6.34   Materials for heat exchangers

Material	Comment
Brasses	Liable to dezincification
Phosphor bronzes	Cheap, but not as corrosion resistant as aluminum-bronzes
Aluminum-bronzes, wrought	An economical and practical choice
Nickel-iron-aluminum-bronzes	More corrosion resistant, but more expensive

impeding fluid flow. A search for supporting information reveals that some materials are more resistant to biofouling than others; copper-nickel alloys are particularly good, probably because the organisms dislike copper salts, even in very low concentrations. Otherwise the problem must be tackled by adding chemical inhibitors to the fluids, or by scraping—the traditional winter pastime of boat owners.

It is sometimes important to minimize the weight of heat exchangers. Repeating the calculation to seek materials the maximum value of Q/m (where m is the mass of the tubes) gives, instead of M_1, the index

$$M_2 = \frac{\lambda \sigma_y^{\,2}}{\rho} \tag{6.74}$$

where ρ is the density of the material of which the tubes are made. (The strength σ_y is now raised to the power of 2 because the weight depends on wall thickness as well as density, and wall thickness varies as $1/\sigma_y$ (equation 6.71).) Similarly, the cheapest heat exchangers are those made of the material with the greatest value of

$$M_3 = \frac{\lambda \sigma_y^{\,2}}{C_m \rho} \tag{6.75}$$

where C_m is the cost per kg of the material. In both cases aluminum alloys score highly because they are both light and cheap. The selections are not shown but can readily be explored using the CES system.

Further reading   Holman, J.P. (1981) *Heat Transfer*, 5th edition, McGraw-Hill, New York, USA.

Related case   6.11   Safe pressure vessels
studies   6.16   Materials to minimize thermal distortion in precision devices

6.19 Materials for radomes

This and the previous case study illustrate the output of the CES software described in Sections 5.6.

When the BBC[4] want to catch you watching television without a license, they park outside your house an unmarked van equipped to detect high-frequency radiation. The vehicle looks normal enough, but it differs from the norm alone in one important respect: the body-skin is not made of pressed steel, but of a material transparent to microwaves. The requirements of the body are much the same as those for the protective dome enclosing the delicate detectors that pick up high frequency signals from space; or those that protect the radar equipment in ships, aircraft, and spacecraft. What are the best materials to make them?

The function of a radome is to shield a microwave antenna from the adverse effects of the environment, while having as little effect as possible on the electrical performance. When trying to detect incoming signals that are weak to begin with, even a small attenuation of the signal as it passes through the radome decreases the sensitivity of the system. Yet the radome must withstand structural loads, loads caused by pressure difference between the inside and outside of the dome, and — in the case of supersonic flight — high temperatures. Table 6.35 summarizes the design requirements.

The model. Figure 6.35 shows an idealized radome. It is a hemispherical skin of microwave-transparent material of radius R and thickness t, supporting a pressure-difference Δp between its inner and outer surfaces. The two critical material properties in determining radome performance are the dielectric constant, ε, and the electric loss tangent $\tan \delta$. Losses are of two types: *reflection* and *absorption*. The fraction of the signal that is reflected is related to the dielectric constant ε and the higher the frequency, the higher the reflected fraction. Air has a dielectric constant of 1; a radome with the same dielectric constant, if it were possible, would not reflect any radiation ("stealth" technology seeks to achieve this).

The second, and often more important loss is that due to absorption as the signal passes through the skin of the radome. When an electro-magnetic wave of frequency f (cycles/s) passes through a dielectric with loss tangent $\tan \delta$, the fractional *power loss* in passing through a thickness dt is

$$\left| \frac{du}{u} \right| = \frac{fA^2 \varepsilon_0}{2} \left(\varepsilon \tan \delta \right) dt \tag{6.76}$$

---

[4] The British Broadcasting Corporation derives its income from the license fee paid by owners of television receivers. Failure to pay the fee deprives the BBC of its income – hence the sophisticated detection scheme.

Table 6.35   Design requirements for a radome

Function	Radome
Constraints	• Support pressure difference Δp • Tolerate temperature up to T_{max}
Objective	Minimize dielectric loss in transmission of microwaves
Free variables	• Thickness of skin, t • Choice of material

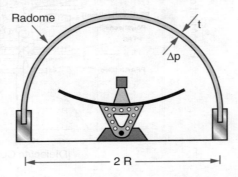

Figure 6.35   A radome. It must be transparent to microwaves yet support wind loads and, in many application, a pressure difference.

where A is the electric amplitude of the wave and ε_0 the permitivity of vacuum. For a thin shell (thickness t) the loss per unit area is thus

$$\left|\frac{\Delta U}{U}\right| = \frac{fA^2\varepsilon_0 t}{2}(\varepsilon \tan \delta) \qquad (6.77)$$

This is the quantity we wish to minimize—the objective function—and this is achieved by making the skin as thin as possible. But the need to support a pressure difference Δp imposes a constraint. The pressure difference creates a stress

$$\sigma = \frac{\Delta p R}{2t} \qquad (6.78)$$

in the skin. If it is to support Δp, this stress must be less than the failure stress σ_f of the material of which it is made, imposing a constraint on the thickness:

$$t \geq \frac{\Delta p R}{2\sigma_f}$$

Substituting this into the equation (6.77) gives

$$\left|\frac{\Delta U}{U}\right| = \frac{fA^2\varepsilon_0 \Delta p R}{4}\left(\frac{\varepsilon \tan \delta}{\sigma_f}\right) \qquad (6.79)$$

(a)

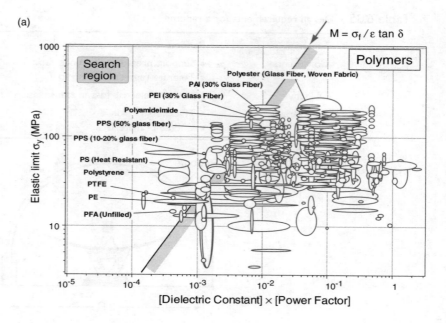

(b)

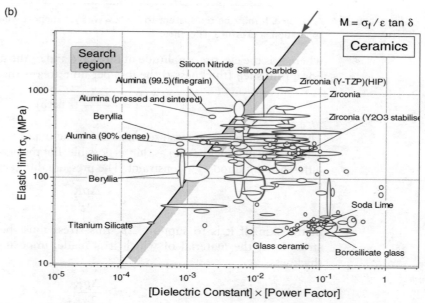

Figure 6.36 (a) The elastic limit, σ_f, plotted against the power factor, $\varepsilon \tan \delta$, using the Level 3 CES database. Here the selection is limited to polymers and polymer–matrix composites. (b) The same chart as Figure 6.36(a), imposing the requirement that $T_{max} > 300°C$. Only ceramics have low dielectric loss and the ability to carry load at high temperature.

The power loss is minimized by maximizing the index

$$M_1 = \frac{\sigma_f}{\varepsilon \tan \delta} \qquad (6.80)$$

There are further constraints. Resistance to abrasion (impact of small particles) scales with hardness, which in turn scales with yield or fracture strength, σ_f, so — when abrasion is important — one might seek also to maximize

$$M_2 = \sigma_f$$

Toughness may also be a consideration. In supersonic flight heating becomes important; then a constraint on maximum service temperature applies also.

The selection. A preliminary survey using the Level 1/2 database shows that polymers have attractive values of M_1, but have poor values of M_2 and can only be used at and near ambient temperature (Figure 3.36). Certain ceramics, too, are good when measured by M_1 and are stable to high temperatures. For help we turn to the Level 3 database. Appropriate charts are shown in Figure 6.36(a) and (b). The axes are σ_f and $\varepsilon \tan \delta$. Both have a selection line of slope 1, corresponding to M_1. The first uses data for polymers and polymer composites. In the second a constraint of maximum service temperature $>300°C$ has been imposed; only ceramics survive. The selection is summarized in Table 6.36. The materials of the first row — teflon, (PTFE) polyethylene, and polypropylene — maximize M_1. If greater strength or impact resistance is required, the fiber-reinforced polymers of the second row are the best choice. When, additionally, high temperatures are involved, the ceramics listed in the third row become candidates.

Postscript. What are real radomes made of? Among polymers, PTFE and polycarbonate are the commonest. Both are very flexible. Where structural rigidity is required (as in the BBC van) GFRP (epoxy or polyester reinforced with woven glass cloth) are used, though with some loss of performance.

Table 6.36   Materials for radomes

Material	Comment
PTFE, polyethylenes, polypropylenes, polystyrene and polyphenylene sulfide (PPS)	Minimum dielectric loss, but limited to near room temperature
Glass-reinforced polyester, PTFE, polyethylenes and polypropylenes, polyamideimide	Slightly greater loss, but greater strength and temperature resistance
Silica, alumina, beryllia, silicon carbide	The choice for re-entry vehicles and rockets where heating is great

When performance is at a premium, glass-reinforced PTFE is used instead. For skin-heating up to 300°C, polymides meet the requirements; beyond that temperature it has to be ceramics. Silica (SiO_2), alumina (Al_2O_3), beryllia (BeO) and silicon nitride (Si_3N_4) are all employed. The choices we have identified are all there.

Further reading Huddleston, G.K. and Bassett, H.L. (1993) in Johnson, R.C. and Jasik, H. (eds), *Antenna Engineering Handbook*, 2nd edition. McGraw-Hill, New York, Chapter 44.

Lewis, C.F. (1988) Materials keep a low profile, *Mech. Eng.*, June, 37–41.

Related case studies   6.11   Safe pressure vessels

6.20 Summary and conclusions

The case studies of this chapter illustrate how the choice of material is narrowed from the initial, broad, menu to a small subset that can be tried, tested, and examined further. Most designs make certain non-negotiable demands on a material: it must withstand a temperature greater than T, it must resist corrosive fluid F, and so forth. These constraints narrow the choice to a few broad classes of material. The choice is narrowed further by seeking the combination of properties that maximize performance (combinations like $E^{1/2}/\rho$) or maximize safety (combinations like K_{1C}/σ_f) or conduction or insulation (like $a^{1/2}/\lambda$) These, plus economics, isolate a small subset of materials for further consideration.

The final choice between these will depend on more detailed information on their properties, considerations of manufacture, economics and aesthetics. These are discussed in the chapters that follow.

6.21 Further reading

The texts listed below give detailed case studies of materials selection. They generally assume that a short-list of candidates is already known and argue their relative merits, rather than starting with a clean slate, as we do here.

Callister, W.D. (2003) *Materials Science and Engineering, An Introduction*, 6th edition, John Wiley, New York, USA. ISBN 0-471-13576-3.

Charles, J.A., Crane, F.A.A. and Furness, J.A.G. (1997) *Selection and Use of Engineering Materials*, 3rd edition, Butterworth-Heinemann, Oxford, UK. ISBN 0-7506-3277-1. (*A Materials-Science approach to the selection of materials.*)

Dieter, G.E. (1991) *Engineering Design, a Materials and Processing Approach*, 2nd edition, McGraw-Hill, New York, USA. ISBN 0-07-100829-2. (*A well-balanced and respected text focusing on the place of materials and processing in technical design.*)

Farag, M.M. (1989) *Selection of Materials and Manufacturing Processes for Engineering Design*, Prentice-Hall, Englewood Cliffs, NJ, USA. ISBN 0-13-575192-6. (*A materials-science approach to the selection of materials.*)

Lewis, G. (1990) *Selection of Engineering Materials*, Prentice-Hall, Englewood Cliffs, NJ, USA. ISBN 0-13-802190-2 (*A text on material selection for technical design, based largely on case studies.*)

Chapter 7

Processes and process selection

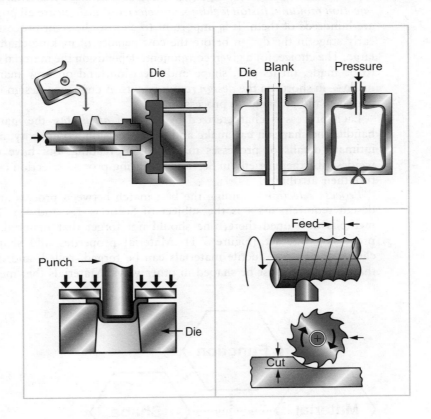

Chapter Contents

7.1   Introduction and synopsis

A *process* is a method of shaping, joining, or finishing a material. *Sand casting, injection molding, fusion welding,* and *electro-polishing* are all processes; there are hundreds of them. It is important to choose the right process-route at an early stage in the design before the cost-penalty of making changes becomes large. The choice, for a given component, depends on the material of which it is to be made, on its size, shape and precision, and on how many are to be made — in short, on the *design requirements.* A change in design requirements may demand a change in process route.

Each process is characterized by a set of *attributes:* the materials it can handle, the shapes it can make and their precision, complexity, and size. The intimate details of processes make tedious reading, but have to be faced: we describe them briefly in Section 7.3, using process selection charts to capture their attributes.

Process selection — finding the best match between process attributes and design requirements — is the subject of Sections 7.4 and 7.5. In using the methods developed there, one should not forget that material, shape, and processing interact (Figure 7.1). Material properties and shape limit the choice of process: ductile materials can be forged, rolled, and drawn; those that are brittle must be shaped in other ways. Materials that melt at modest

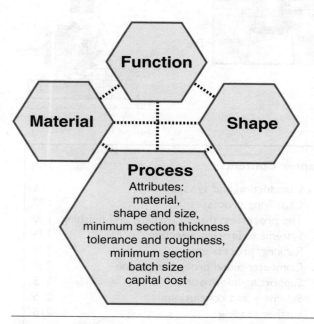

Figure 7.1   Processing selection depends on material and shape. The "process attributes" are used as criteria for selection.

temperatures to low-viscosity liquids can be cast; those that do not have to be processed by other routes. Shape, too, can influence the choice of process. Slender shapes can be made easily by rolling or drawing but not by casting. Hollow shapes cannot be made by forging, but they can by casting or molding. Conversely, processing affects properties. Rolling and forging change the hardness and texture of metals, and align the inclusions they contain, enhancing strength, and ductility. Composites only acquire their properties during processing; before, they are just a soup of polymer and a sheaf of fibers.

Like the other aspects of design, process selection is an iterative procedure. The first iteration gives one or more possible processes-routes. The design must then be re-thought to adapt it, as far as possible, to ease of manufacture by the most promising route. The final choice is based on a comparison of *process-cost,* requiring the use of cost models developed in Section 7.6, and on *supporting information*: case histories, documented experience and examples of process-routes used for related products (Section 7.7). Supporting information helps in another way: that of dealing with the coupling between process and material properties. Processes influence properties, sometimes in a desired way (e.g. heat treatment) sometimes not (uncontrolled casting defects, for instance). This coupling cannot be described by simple processes attributes, but requires empirical characterization or process modeling.

The chapter ends, as always, with a summary and annotated recommendations for further reading.

7.2 Classifying processes

Manufacturing processes can be classified under the headings shown in Figure 7.2. *Primary processes* create *shapes*. The first row lists seven primary forming processes: casting, molding, deformation, powder methods, methods for forming composites, special methods, and rapid prototyping. *Secondary processes* modify shapes or properties; here they are shown as "machining", which adds features to an already shaped body, and "heat treatment, which enhances surface or bulk properties. Below these comes *joining,* and, finally, *finishing.*

The merit of Figure 7.2 is as a flow chart: a progression through a manufacturing route. It should not be treated too literally: the order of the steps can be varied to suit the needs of the design. The point it makes is that there are three broad process families: those of shaping, joining, and finishing. The attributes of one family differ so greatly from those of another that, in assembling and structuring data for them, they must be treated separately.

To structure processes attributes for screening, we need a hierarchical classification of the processes families themselves, in the spirit of that used in Chapter 5 for materials. This gives each process a place, enabling the development of a computer-based tool. Figure 7.3 shows part of it. The process

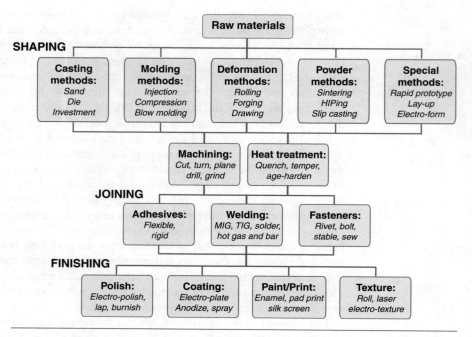

Figure 7.2   The classes of process. The first row contains the primary shaping processes; below lie
the secondary processes of machining and heat treatment, followed by the families of
joining and finishing processes.

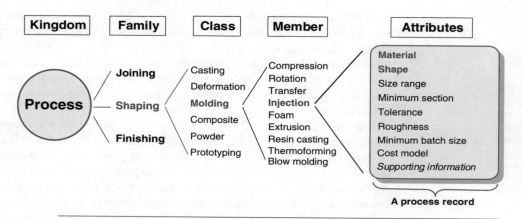

Figure 7.3   The taxonomy of the kingdom of process with part of the *shaping* family expanded. Each
member is characterized by a set of attributes. Process selection involves matching
these to the requirements of the design.

kingdom has three families: shaping, joining, and finishing. In this figure, the shaping family is expanded to show classes: casting, deformation, molding, etc. One of these — molding — is again expanded to show its members: rotation molding, blow molding, injection molding, and so forth. Each of these have certain attributes: the materials it can handle, the shapes it can make, their size, precision, and an optimum batch size (the number of units that it can make economically). This is the information that you would find in a record for a shaping-process in a selection database.

The other two families are partly expanded in Figure 7.4. There are three broad joining classes: adhesives, welding, and fasteners. In this figure one of them — welding — is expanded to show its members. As before each member has attributes. The first is the material or materials that the process can join. After that the attribute-list differs from that for shaping. Here the geometry of the joint and the way it will be loaded are important, as are requirements that the joint can, or cannot, be disassembled, be watertight, be electrically conducting and the like.

The lower part of the figure expands the family of finishing. Some of the classes it contains are shown; one — coating — is expanded to show some of its members. As with joining, the material to be coated is an important attribute but the others again differ. Most important is the purpose of the treatment, followed by properties of the coating itself.

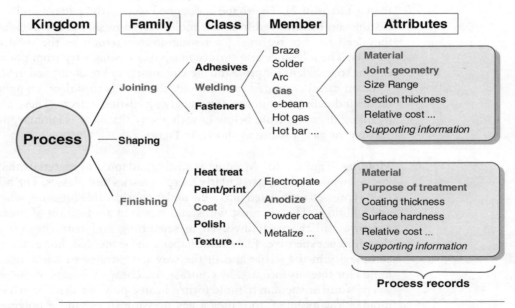

Figure 7.4   The taxonomy of the process kingdom again, with the families of *joining* and *finishing* partly expanded.

With this background we can embark on our lightning tour of processes. It will be kept as concise as possible; details can be found in the numerous books listed in Section 7.9.

7.3 The processes: shaping, joining, and finishing

Shaping processes

In *casting* (Figure 7.5), a liquid is poured or forced into a mold where it solidifies by cooling. Casting is distinguished from molding, which comes next, by the low viscosity of the liquid: it fills the mold by flow under its own weight (as in gravity sand and investment casting) or under a modest pressure (as in die casting and pressure sand casting). Sand molds for one-off castings are cheap; metal dies for die-casting large batches can be expensive. Between these extremes lie a number of other casting methods: shell, investment, plaster-mold and so forth.

Cast shapes must be designed for easy flow of liquid to all parts of the mold, and for progressive solidification that does not trap pockets of liquid in a solid shell, giving shrinkage cavities. Whenever possible, section thicknesses are made uniform (the thickness of adjoining sections should not differ by more than a factor of 2). The shape is designed so that the pattern and the finished casting can be removed from the mold. Keyed-in shapes are avoided because they lead to "hot tearing" (a tensile creep-fracture) as the solid cools and shrinks. The tolerance and surface finish of a casting vary from poor for sand-casting to excellent for precision die-castings; they are quantified in Section 7.5.

When metal is poured into a mold, the flow is turbulent, trapping surface oxide and debris within the casting, giving casting defects. These are avoided by filling the mold from below in such a way that flow is laminar, driven by a vacuum or gas pressure as shown in Figure 7.4.

Molding (Figure 7.6). Molding is casting, adapted to materials that are very viscous when molten, particularly thermoplastics and glasses. The hot, viscous fluid is pressed or injected into a die under considerable pressure, where it cools and solidifies. The die must withstand repeated application of pressure, temperature and the wear involved in separating and removing the part, and therefore is expensive. Elaborate shapes can be molded, but at the penalty of complexity in die shape and in the way it separates to allow removal. The molds for thermo-forming, by contrast, are cheap. Variants of the process use gas pressure or vacuum to mold form a heated polymer sheet onto a single-part mold. Blow-molding, too, uses a gas pressure to expand a polymer or glass blank into a split outer-die. It is a rapid, low-cost process well suited for mass-production of cheap parts like milk bottles. Polymers, like metals, can be

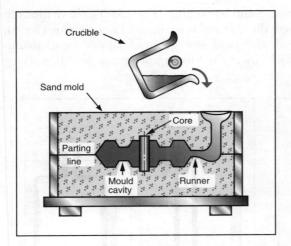

Sand casting

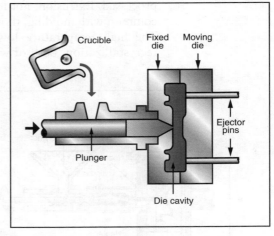

Die casting

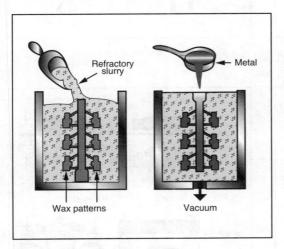

Investment casting

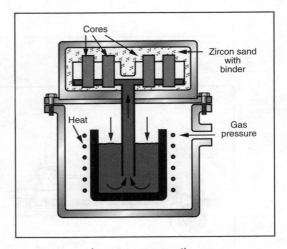

Low pressure casting

Figure 7.5 Casting processes. In *sand casting*, liquid metal is poured into a split sand mold. In *die casting*, liquid is forced under pressure into a metal mold. In *investment casting*, a wax pattern in embedded in refractory, melted out, and the cavity filled with metal. In *pressure casting*, a die is filled from below, giving control of atmosphere and of the flow of metal into the die.

extruded; virtually all rods, tubes and other prismatic sections are made in this way.

Deformation processing (Figure 7.7). This process can be hot, warm or cold—cold, that is, relative to the melting point of the T_m material being processed. Extrusion, hot forging and hot rolling ($T > 0.55\ T_m$) have much in common with molding, though the material is a true solid not a viscous liquid. The high temperature lowers the yield strength and allows simultaneous recrystallization, both of which lower the forming pressures. Warm working

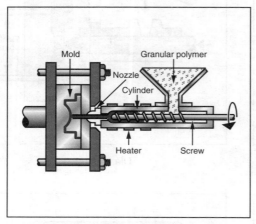

Injection-molding

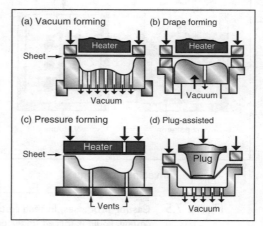

Blow-molding

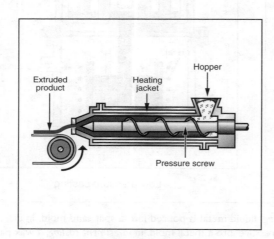

Polymer extrusion

Thermo-forming

Figure 7.6   Molding processes. In *injection-molding*, a granular polymer (or filled polymer) is heated, compressed and sheared by a screw feeder, forcing it into the mold cavity. In *blow-molding*, a tubular blank of hot polymer or glass is expanded by gas pressure against the inner wall of a split die. In *polymer extrusion*, shaped sections are formed by extrusion through a shaped die. In *thermo-forming*, a sheet of thermoplastic is heated and deformed into a female die by vacuum or gas pressure.

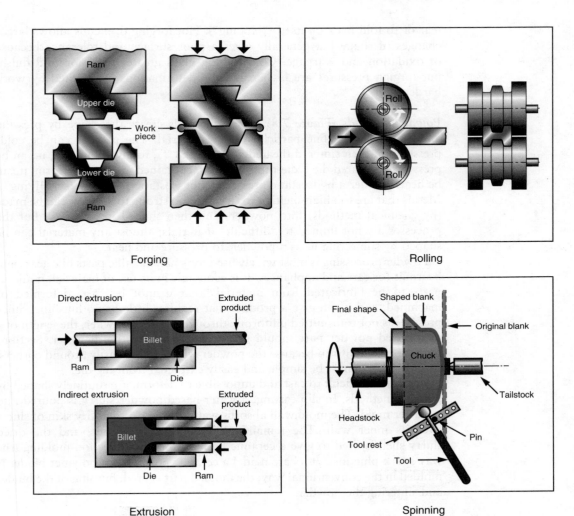

Forging

Rolling

Extrusion

Spinning

Figure 7.7 Deformation processes. In *forging*, a slug of metal is shaped between two dies held in the jaws of a press. In *rolling*, a billet or bar is reduced in section by compressive deformation between the rolls. In *extrusion*, metal is forced to flow through a die aperture to give a continuous prismatic shape. All three process can be hot ($T > 0.85T_m$), warm ($0.55T_m < T < 0.85T_m$) or cold ($T < 0.35T_m$). In *spinning*, a spinning disk of ductile metal is shaped over a wooden pattern by repeated sweeps of the smooth, rounded, tool.

($0.35T_m < T < 0.55T_m$) allows recovery but not recrystallization. Cold forging, rolling, and drawing ($T < 0.35T_m$) exploit work hardening to increase the strength of the final product, but at the penalty of higher forming pressures.

Forged parts are designed to avoid rapid changes in thickness and sharp radii of curvature since both require large local strains that can cause the material to tear or to fold back on itself ("lapping"). Hot forging of metals allows larger changes of shape but generally gives a poor surface and tolerance because of oxidation and warpage. Cold forging gives greater precision and finish, but forging pressures are higher and the deformations are limited by work hardening.

Powder methods (Figure 7.8). These methods create the shape by pressing and then sintering fine particles of the material. The powder can be cold-pressed and then sintered (heated at up to 0.8 T_m to give bonding); it can be pressed in a heated die ("die-pressing"); or, contained in a thin preform, it can be heated under a hydrostatic pressure ("hot isostatic pressing" or "HIPing"). Metals that are too high-melting to cast and too strong to deform, can be made (by chemical methods) into powders and then shaped in this way. But the processes are not limited to "difficult" materials; almost any material can be shaped by subjecting it, as a powder, to pressure and heat.

Powder processing is most widely used for small metallic parts like gears and bearings for cars and appliances. It is economic in its use of material, it allows parts to be fabricated from materials that cannot be cast, deformed or machined, and it can give a product that requires little or no finishing. Since pressure is not transmitted uniformly through a bed of powder, the length of a die-pressed powder part should not exceed 2.5 times its diameter. Sections must be near-uniform because the powder will not flow easily around corners. And the shape must be simple and easily extracted from the die.

Ceramics, difficult to cast and impossible to deform, are routinely shaped by powder methods. In slip casting, a water-based powder slurry is poured into a plaster mold. The mold wall absorbs water, leaving a semi-dry skin of slurry over its inner wall. The remaining liquid is drained out, and the dried slurry shell is fired to give a ceramic body. In powder injection molding (the way spark-plug insulators are made) a ceramic powder in a polymer binder is molded in the conventional way; the molded part is fired, burning of the binder and sintering the powder.

Composite fabrication methods (Figure 7.9). These make polymer–matrix composites reinforced with continuous or chopped fibers. Large components are fabricated by filament winding or by laying-up pre-impregnated mats of carbon, glass or Kevlar fiber ("pre-preg") to the required thickness, pressing and curing. Parts of the process can be automated, but it remains a slow manufacturing route; and, if the component is a critical one, extensive ultrasonic testing may be necessary to confirm its integrity. Higher integrity is given by vacuum- or pressure-bag molding, which squeezes bubbles out of the matrix before it polymerizes. Lay-up methods are best suited to a small number of high-performance, tailor-made, components. More routine components (car bumpers, tennis racquets) are made from chopped-fiber composites by pressing and heating a "dough" of resin containing the fibers, known as bulk molding compound (BMC) or sheet molding compound

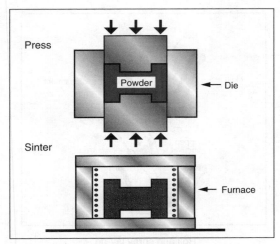

Die-pressing and sintering

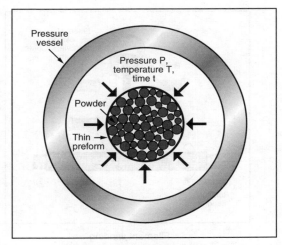

Hot isostatic pressing

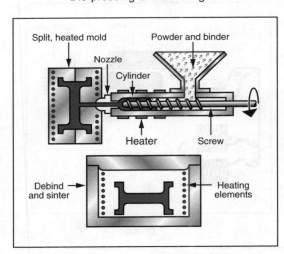

Powder injection molding

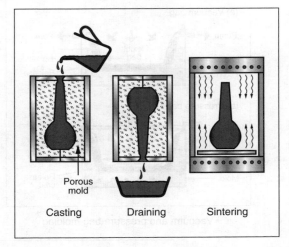

Slip casting

Figure 7.8   Powder processing. In *die-pressing and sintering* the powder is compacted in a die, often with a binder, and the green compact is then fired to give a more or less dense product. In *hot isostatic pressing*, powder in a thin, shaped, shell or pre-form is heated and compressed by an external gas pressure. In *powder injection molding*, powder and binder are forced into a die to give a green blank that is then fired. In *slip casting*, a water-based powder slurry is poured into a porous plaster mold that absorbs the water, leaving a powder shell that is subsequently fired.

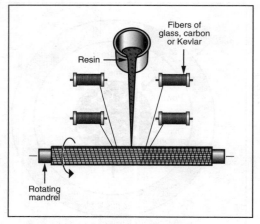

Filament winding

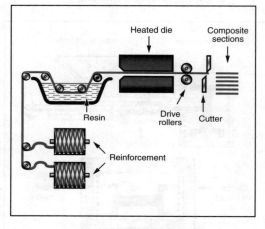

Roll and spray lay-up

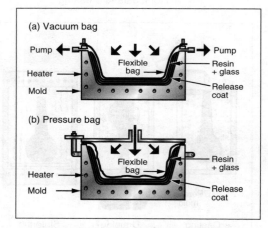

Vacuum- and pressure-bag molding

Pultrusion

Figure 7.9 Composite forming methods. In *filament winding*, fibers of glass, Kevlar or carbon are wound onto a former and impregnated with a resin-hardener mix. In *roll* and *spray lay-up*, fiber reinforcement is laid up in a mold onto which the resin-hardener mix is rolled or sprayed. In *vacuum-* and *pressure-bag molding*, laid-up fiber reinforcement, impregnated with resin-hardener mix, is compressed and heated to cause polymerization. In *pultrusion*, fibers are fed through a resin bath into a heated die to form continuous prismatic sections.

(SMC), in a mold, or by injection molding a rather more fluid mixture into a die. The flow pattern is critical in aligning the fibers, so that the designer must work closely with the manufacturer to exploit the composite properties fully.

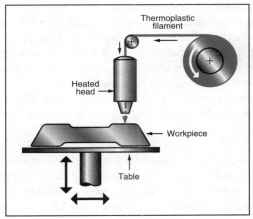

Deposition modeling

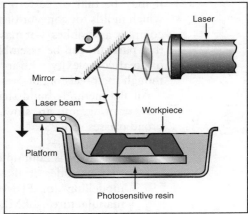

Stereo-lithography, SLA

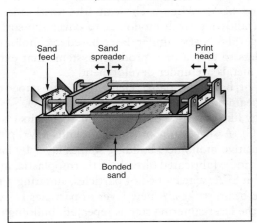

Direct mold modeling

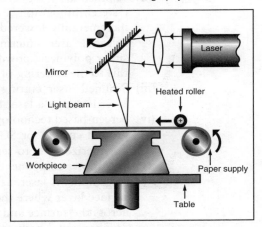

Laminated object manufacture, LOM

Figure 7.10   Rapid prototyping. In *deposition modeling* and *ballistic particle manufacture* (BPM), a solid
body is created by the layer-by-layer deposition of polymer droplets. In *stereo-
lithography* (SLA), a solid shape is created layer-by-layer by laser-induced polymerization
of a resin. In *direct mold modeling*, a sand mold is built up layer-by-layer by selective
spraying of a binder from a scanning print-head. In *laminated object manufacture* (LOM),
a solid body is created from layers of paper, cut by a scanning laser beam and bonded
with a heat-sensitive polymer.

Rapid prototyping systems (RPS — Figure 7.10). The RPS allow single
examples of complex shapes to be made from numerical data generated by
CAD solid-modeling software. The motive may be that of visualization: the
aesthetics of an object may be evident only when viewed as a prototype.

It may be that of pattern-making: the prototype becomes the master from which molds for conventional processing, such as casting, can be made or — in complex assemblies — it may be that of validating intricate geometry, ensuring that parts fit, can be assembled, and are accessible. All RPS can create shapes of great complexity with internal cavities, overhangs and transverse features, though the precision, at present, is limited to ±0.3 mm at best.

All RP methods build shapes layer-by-layer, rather like three-dimensional (3D) printing, and are slow (typically 4–40 h per unit). There are at least six broad classes of RPS:

  (i) The shape is built up from a thermoplastic fed to a single scanning head that extrudes it like a thin layer of toothpaste ("fused deposition modelling" or FDM), exudes it as tiny droplets ("ballistic particle manufacture", BPM), or ejects it in a patterned array like a bubble-jet printer ("3D printing").

 (ii) Scanned-laser induced polymerization of a photo-sensitive monomer ("stereo-lithography" or SLA). After each scan, the work piece is incrementally lowered, allowing fresh monomer to cover the surface. Selected laser sintering (SLS) uses similar laser-based technology to sinter polymeric powders to give a final product. Systems that extend this to the sintering of metals are under development.

(iii) Scanned laser cutting of bondable paper elements. Each paper-thin layer is cut by a laser beam and heat bonded to the one below.

 (iv) Screen-based technology like that used to produce microcircuits ("solid ground curing" or SGC). A succession of screens admits UV light to polymerize a photo-sensitive monomer, building shapes layer by layer.

  (v) SLS allows components to be fabricated directly in thermoplastic, metal or ceramic. A laser, as in SLA, scans a bed of particles, sintering a thin surface layer where the beam strikes. A new layer of particles is swept across the surface and the laser-sintering step is repeated, building up a 3-dimensional body.

 (vi) Bonded sand molding offers the ability to make large complex metal parts easily. Here a multi-jet print-head squirts a binder onto a bed of loose casting sand, building up the mold shape much as selected laser sintering does, but more quickly. When complete the mold is lifted from the remaining loose sand and used in a conventional casting process.

To be useful, the prototypes made by RPS are used as masters for silicone molding, allowing a number of replicas to be cast using high-temperature resins or metals.

Machining (Figure 7.11). Almost all engineering components, whether made of metal, polymer, or ceramic, are subjected to some kind of machining during manufacture. To make this possible they should be designed to make gripping and jigging easy, and to keep the symmetry high: symmetric shapes need fewer

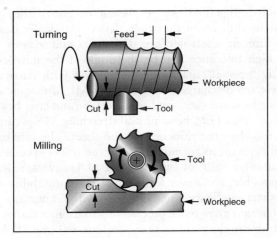

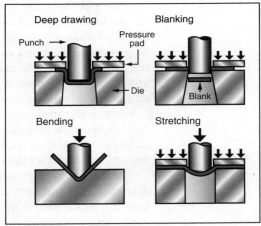

Turning and milling

Drawing, blanking, bending and stretching

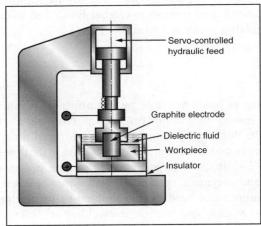

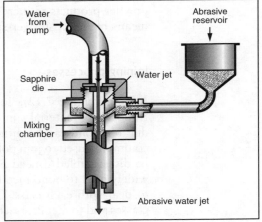

Electro-discharge machining

Water-jet cutting

Figure 7.11 Machining operations. In *turning* and *milling*, the sharp, hardened tip of a tool cuts a chip from the workpiece surface. In *drawing, blanking, bending* and *stretching*, a sheet is shaped and cut to give flat and dished shapes. In *electro-discharge machining*, electric discharge between a graphite electrode and the workpiece, submerged in a dielectric such as paraffin, erodes the workpiece to the desired shape. In water-jet cutting, an abrasive entrained in a high speed water-jet erodes the material in its path.

operations. Metals differ greatly in their *machinability*, a measure of the ease of chip formation, the ability to give a smooth surface, and the ability to give economical tool life (evaluated in a standard test). Poor machinability means higher cost.

Most polymers are molded to a final shape. When necessary they can be machined but their low moduli mean that they deflect elastically during the machining operation, limiting the tolerance. Ceramics and glasses can be ground and lapped to high tolerance and finish (think of the mirrors of telescopes). There are many "special" machining techniques with particular applications; they include electro-discharge machining (EDM), ultrasonic cutting, chemical milling, cutting by water-jets, sand-jets, electron and laser beams.

Sheet metal forming involves punching, bending and stretching. Holes cannot be punched to a diameter less than the thickness of the sheet. The minimum radius to which a sheet can be bent, its *formability*, is sometimes expressed in multiples of the sheet thickness t: a value of 1 is good; one of 4 is average. Radii are best made as large as possible, and never less than t. The formability also determines the amount the sheet can be stretched or drawn without necking and failing. The *forming limit diagram* gives more precise information: it shows the combination of principal strains in the plane of the sheet that will cause failure. The part is designed so that the strains do not exceed this limit.

Machining is often a secondary operations applied to castings, moldings or powder products to increase finish and tolerance. Higher finish and tolerance means higher cost; over-specifying either is a mistake.

Joining processes

Joining (Figure 7.12). Joining is made possible by a number of techniques. Almost any material can be joined with adhesives, though ensuring a sound, durable bond can be difficult. Bolting, riveting, stapling, and snap fitting are commonly used to join polymers and metals, and have the feature that they can be disassembled if need be. Welding, the largest class of joining processes, is widely used to bond metals and polymers; specialized techniques have evolved to deal with each class. Ceramics can be diffusion-bonded to themselves, to glasses and to metals. Friction welding and friction-stir welding rely on the heat and deformation generated by friction to create a bond.

If components are to be welded, the material of which they are made must be characterized by a high *weldability*. Like *machinability*, it measures a combination of basic properties. A low thermal conductivity allows welding with a low rate of heat input but can lead to greater distortion on cooling. Low thermal expansion gives small thermal strains with less risk of distortion. A solid solution is better than an age-hardened alloy because, in the heat-affected zone on either side of the weld, over-ageing and softening can occur.

Welding always leaves internal stresses that are roughly equal to the yield strength of the parent material. They can be relaxed by heat treatment but this is expensive, so it is better to minimize their effect by good design. To achieve this, parts to be welded are made of equal thickness whenever possible, the welds are located where stress or deflection is least critical, and the total number of welds is minimized.

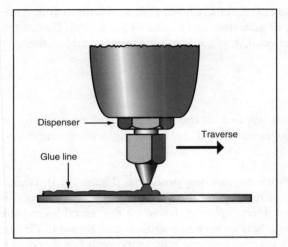

Adhesives

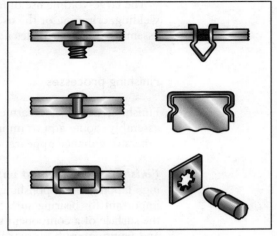

Fasteners

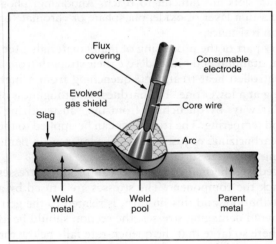

Welding: manual metal arc

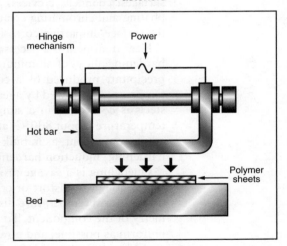

Welding: hot bar polymer welding

Figure 7.12 Joining operations. In *adhesive bonding*, a film of adhesive is applied to one surface, which is then pressed onto the mating one. *Fastening* is achieved by bolting, riveting, stapling, push-through snap fastener, push-on snap fastener or rod-to-sheet snap fastener. In *metal fusion-welding*, metal is melted, and more added from a filler rod, to give a bond or coating. In *thermoplastic polymer welding*, heat is applied to the polymer components, which are simultaneously pressed together to form a bond.

The large-volume use of fasteners is costly because it is difficult to automate; welding, crimping or the use of adhesives can be more economical. Design for assembly (DFA) provides a check-list to guide minimizing assembly time.

Finishing processes

Finishing describes treatments applied to the surface of the component or assembly. Some aim to improve mechanical and other engineering properties, others to enhance appearance.

Finishing treatments to improve engineering properties (Figure 7.13). Grinding, lapping, and polishing increase precision and smoothness, particularly important for bearing surfaces. Electro-plating deposits a thin metal layer onto the surface of a component to give resistance to corrosion and abrasion. Plating and painting are both made easier by a simple part shape with largely convex surfaces: channels, crevices, and slots are difficult to reach. Anodizing, phosphating and chromating create a thin layer of oxide, phosphate or chromate on the surface, imparting corrosion resistance.

Heat treatment is a necessary part of the processing of many materials. Age-hardening alloys of aluminum, titanium and nickel derive their strength from a precipitate produced by a controlled heat treatment: quenching from a high temperature followed by ageing at a lower one. The hardness and toughness of steels is controlled in a similar way: by quenching from the "austenitizing" temperature (about 800°C) and tempering. The treatment can be applied to the entire component, as in bulk carburizing, or just to a surface layer, as in flame hardening, induction hardening and laser surface hardening.

Quenching is a savage procedure; thermal contraction can produce stresses large enough to distort or crack the component. The stresses are caused by a non-uniform temperature distribution, and this, in turn, is related to the geometry of the component. To avoid damaging stresses, the section should be as uniform as possible, and nowhere so large that the quench-rate falls below the critical value required for successful heat treatment. Stress concentrations should be avoided: they are the source of quench cracks. Materials that have been molded or deformed may contain internal stresses that can be removed, at least partially, by stress-relief anneals another — sort of heat treatment.

Finishing treatments that enhance aesthetics (Figure 7.14). The processes just described can be used to enhance the visual and tactile attributes of a material: electroplating and anodizing are examples. There are many more, of which painting is the most widely used. Organic-solvent based paints give durable coatings with high finish, but the solvent poses environmental problems. Water-based paints overcome these, but dry more slowly and the resulting paint film is less perfect. In polymer powder-coating and polymer powder-spraying a film of thermoplastic — nylon, polypropylene or polyethylene — is deposited on

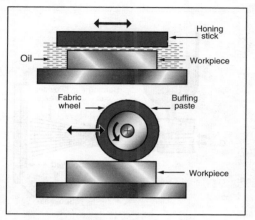

Mechanial polishing

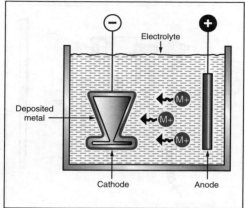

Electro-plating

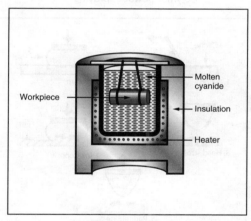

Heat treatment: carburizing

Anodizing

Figure 7.13   Finishing processes to protect and enhance properties. In *mechanical polishing*, the roughness of a surface is reduced, and its precision increased, by material removal using finely ground abrasives. In *electro-plating*, metal is plated onto a conducting workpiece by electro-deposition in a plating bath. In *heat treatment*, a surface layer of the workpiece is hardened, and made more corrosion resistant, by the inward diffusion of carbon, nitrogen, phosphorous or aluminum from a powder bed or molten bath. In *anodizing*, a surface oxide layer is built up on the workpiece (which must be aluminum, magnesium, titanium, or zinc) by a potential gradient in an oxidizing bath.

the surface, giving a protective layer that can be brightly coloured. In screen printing an oil-based ink is squeegeed through a mesh on which a blocking-film holds back the ink where it is not wanted; full colour printing requires the successive use of up to four screens. Screen printing is widely used to print

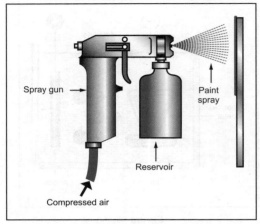

Paint spraying

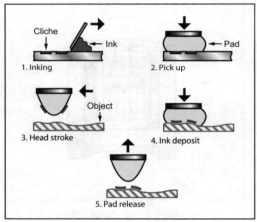

Polymer powder spraying

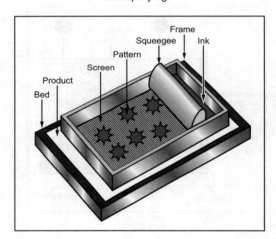

Silk-screen printing

Pad printing

Figure 7.14 Finishing processes to enhance appearance. In *paint spraying*, a pigment in an organic or water-based solvent is sprayed onto the surface to be decorated. In *polymer powder-coating* a layer of thermoplastic is deposited on the surface by direct spraying in a gas flame, or by immersing the hot workpiece in a bed of powder. In *silk-screen printing*, ink is wiped onto the surface through a screen onto which a blocking-pattern has been deposited, allowing ink to pass in selected areas only. In *pad printing*, an inked pattern is picked up on a rubber pad and applied to the surface, which can be curved or irregular.

designs onto flat surfaces. Curved surfaces require the use of pad printing, in which a pattern, etched onto a metal "cliche", is inked and picked up on a soft rubber pad. The pad is pressed onto the product, depositing the pattern on its surface; the compliant rubber conforms to the curvature of the surface.

Enough of the processes themselves; for more detail the reader will have to consult Further reading, Section 7.9.

7.4 Systematic process selection

The strategy

The strategy for selecting processes parallels that for materials. The starting point is that all processes are considered as possible candidates until shown to be otherwise. Figure 7.15 shows the now-familiar steps: *Translation, screening, ranking* and search for *supporting information*. In translation, the design requirements are expressed as constraints on material, shape, size, tolerance, roughness, and other process-related parameters. It is helpful to list these in the

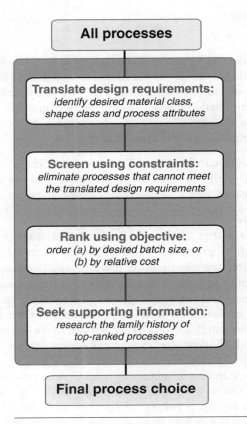

Figure 7.15   A flow chart of the procedure for process selection. It parallels that for material selection.

Table 7.1   Translation of process requirements

Function	What must the process do? (Shape? Join? Finish?)
Constraints	What material, shape, size, precision, etc. must it provide?
Objective	What is to be maximized or minimized? (Cost? Time? Quality?)
Free variables	• Choice of process
	• Process chain options

way suggested by Table 7.1. The constraints are used to screen out processes that are incapable of meeting them, using process selection diagrams (or their computer-based equivalents) shown in a moment. The surviving processes are then ranked by economic measures, also detailed below. Finally, the top-ranked candidates are explored in depth, seeking as much supporting information as possible to enable a final choice. Chapter 8 gives examples.

Selection charts

As already explained, each process is characterized by a set of attributes. These are conveniently displayed as simple matrices and bar charts. They provide the selection tools we need for screening. The hard-copy versions, shown here, are necessarily simplified, showing only a limited number of processes and attributes. Computer implementation, the subject of Section 7.6, allows exploration of a much larger number of both.

The process-material matrix (Figure 7.16). A given process can shape, or join, or finish some materials but not others. The matrix shows the links between material and process classes. Processes that cannot shape the material of choice are non-starters.

The process-shape matrix (Figure 7.17). Shape is the most difficult attribute to characterize. Many processes involve rotation or translation of a tool or of the material, directing our thinking towards axial symmetry, translational symmetry, uniformity of section and such like. Turning creates axisymmetric (or circular) shapes; extrusion, drawing and rolling make prismatic shapes, both circular and non-circular. Sheet-forming processes make flat shapes (stamping) or dished shapes (drawing). Certain processes can make 3D shapes, and among these some can make hollow shapes whereas others cannot. Figure 7.18 illustrates this classification scheme.

   The process-shape matrix displays the links between the two. If the process cannot make the shape we want, it may be possible to combine it with a secondary process to give a process-chain that adds the additional features: casting followed by machining is an obvious example. But remember: every additional process step adds cost.

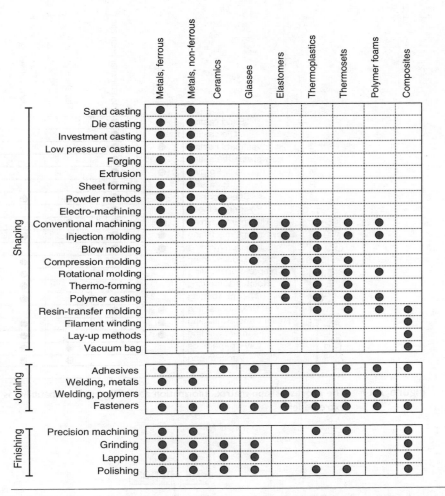

Figure 7.16   The process–material matrix. A red dot indicates that the pair are compatible.

The mass bar-chart (Figure 7.19). The bar-chart — laid on its side to make labeling easier — shows the typical mass-range of components that each processes can make. It is one of four, allowing application of constraints on size (measured by mass), section thickness, tolerance and surface roughness. Large components can be built up by joining smaller ones. For this reason the ranges associated with joining are shown in the lower part of the figure. In applying a constraint on mass, we seek single shaping-processes or shaping-joining combinations capable of making it, rejecting those that cannot.

The processes themselves are grouped by the material classes they can treat, allowing discrimination by both material and shape.

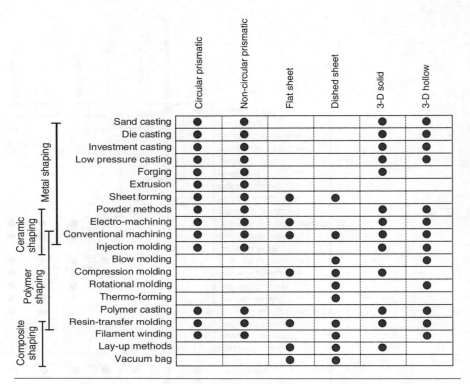

Figure 7.17   The process–shape matrix. Information about material compatibility is included at the extreme left.

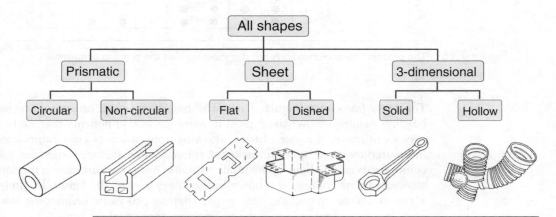

Figure 7.18   The shape classification. More complex schemes are possible, but none are wholly satisfactory.

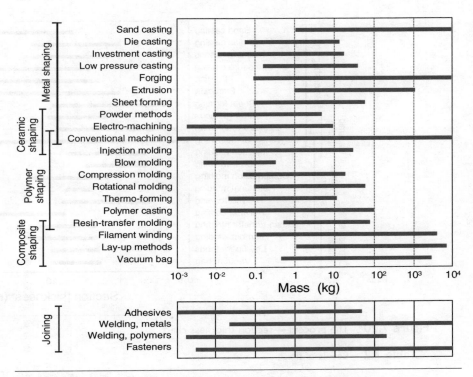

Figure 7.19   The process—mass-range chart. The inclusion of joining allows simple process chains to be explored.

The section thickness bar-chart (Figure 7.20). Surface tension and heat-flow limit the minimum section and slenderness of gravity cast shapes. The range can be extended by applying a pressure as in centrifugal casting and pressured die casting, or by pre-heating the mold. But there remain definite lower limits for the section thickness achievable by casting. Deformation processes—cold, warm, and hot—cover a wider range. Limits on forging-pressures set a lower limit on thickness and slenderness, but it is not nearly as severe as in casting. Machining creates slender shapes by removing unwanted material. Powder-forming methods are more limited in the section thicknesses they can create, but they can be used for ceramics and very hard metals that cannot be shaped in other ways. Polymer-forming methods—injection molding, pressing, blow-molding, etc.—share this regime. Special techniques, which include electro-forming, plasma-spraying and various vapor-deposition methods, allow very slender shapes.

The bar-chart of Figure 7.20 allows selection to meet constraints on section thickness.

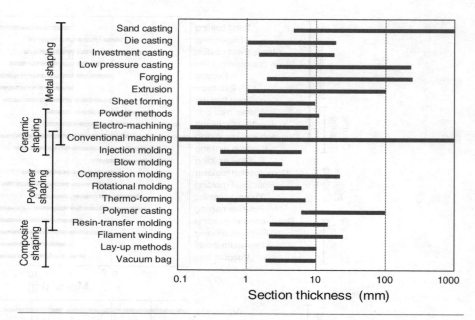

Figure 7.20 The process — section thickness chart.

Table 7.2 Levels of finish

Finish (μm)	Process	Typical application
$R = 0.01$	Lapping	Mirrors
$R = 0.1$	Precision grind or lap	High quality bearings
$R = 0.2-0.5$	Precision grinding	Cylinders, pistons, cams, bearings
$R = 0.5-2$	Precision machining	Gears, ordinary machine parts
$R = 2-10$	Machining	Light-loaded bearings. Non-critical components
$R = 3-100$	Unfinished castings	Non-bearing surfaces

The tolerance and surface-roughness bar-charts (Figures 7.21, 7.22 and Table 7.2). No process can shape a part *exactly* to a specified dimension. Some deviation Δx from a desired dimension x is permitted; it is referred to as the *tolerance, T,* and is specified as $x = 100 \pm 0.1$ mm, or as $x = 50^{+0.01}_{-0.001}$ mm. Closely related to this is the *surface roughness, R,* measured by the root-mean-square amplitude of the irregularities on the surface. It is specified as $R < 100 \, \mu$m (the rough surface of a sand casting) or $R < 100 \, \mu$m (a highly polished surface).

Manufacturing processes vary in the levels of tolerance and roughness they can achieve economically. Achievable tolerances and roughness are shown in Figures 7.21 and 7.22. The tolerance T is obviously greater than $2R$; indeed, since R is the root-mean-square roughness, the peak roughness is more like $5R$. Real processes give tolerances that range from about $10R$ to $1000R$. Sand

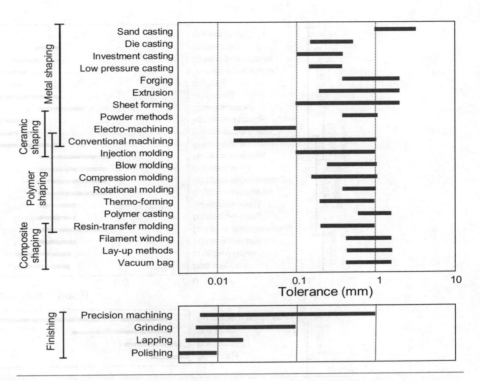

Figure 7.21   The process—tolerance chart. The inclusion of finishing processes allows simple process chains to be explored.

casting gives rough surfaces; casting into metal dies gives a better finish. Molded polymers inherit the finish of the molds and thus can be very smooth, but tolerances better than ±0.2 mm are seldom possible because internal stresses left by molding cause distortion and because polymers creep in service. Machining, capable of high-dimensional accuracy and surface finish, is commonly used after casting or deformation processing to bring the tolerance or finish up to the desired level. Metals and ceramics can be surface-ground and lapped to a high tolerance and smoothness: a large telescope reflector has a tolerance approaching 5 μm over a dimension of a meter or more, and a roughness of about 1/100 of this. But such precision and finish are expensive: processing costs increase almost exponentially as the requirements for tolerance and surface finish are made more severe. It is an expensive mistake to over-specify precision.

Use of hard-copy process selection charts. The charts presented here provide an overview: an initial at-a-glance graphical comparison of the capabilities of various process classes. In a given selection exercise they are not all equally useful: sometimes one is discriminating, another not — it depends on the design

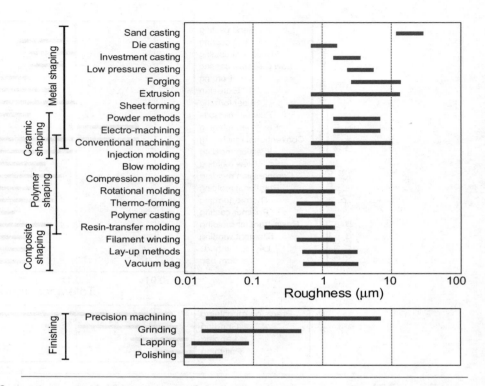

Figure 7.22 The process — surface roughness chart. The inclusion of finishing processes allows simple process chains to be explored.

requirements. They should not be used blindly, but used to give guidance in selection and engender a feel for the capabilities and limitations of various process types, remembering that some attributes (precision, for instance) can be added later by using secondary processes. That is as far as one can go with hard-copy charts. The number of processes that can be presented on hard-copy charts is obviously limited and the resolution is poor. But the procedure lends itself well to computer implementation, overcoming these deficiencies. It is described in Section 7.6.

The next step is to rank the survivors by economic criteria. To do this we need to examine process cost.

7.5 Ranking: process cost

Part of the cost of a component is that of the material of which it is made. The rest is the cost of manufacture, that is, of forming it to a shape, and of joining and finishing. Before turning to details, there are four common-sense rules for minimizing cost that the designer should bear in mind. They are these.

Keep things standard. If someone already makes the part you want, it will almost certainly be cheaper to buy it than to make it. If nobody does, then it is cheaper to design it to be made from standard stock (sheet, rod, tube) than from non-standard shapes or from special castings or forgings. Try to use standard materials, and as few of them as possible: it reduces inventory costs and the range of tooling the manufacturer needs, and it can help with recycling.

Keep things simple. If a part has to be machined, it will have to be clamped; the cost increases with the number of times it will have to be re-jigged or re-oriented, specially if special tools are necessary. If a part is to be welded or brazed, the welder must be able to reach it with his torch and still see what he is doing. If it is to be cast or molded or forged, it should be remembered that high (and expensive) pressures are required to make fluids flow into narrow channels, and that re-entrant shapes greatly complicate mold and die design. Think of making the part yourself: will it be awkward? Could slight re-design make it less awkward?

Make the parts easy to assemble. Assembly takes time, and time is money. If the overhead rate is a mere $60 per hour, every minute of assembly time adds another $1 to the cost. Design for assembly (DFA) addresses this problem with a set of common-sense criteria and rules. Briefly, there are three:

- minimize part count,
- design parts to be self-aligning on assembly,
- use joining methods that are fast. Snap-fits and spot welds are faster than threaded fasteners or, usually, adhesives.

Do not specify more performance than is needed. Performance must be paid for. High strength metals are more heavily alloyed with expensive additions; high performance polymers are chemically more complex; high performance ceramics require greater quality control in their manufacture. All of these increase material costs. In addition, high strength materials are hard to fabricate. The forming pressures (whether for a metal or a polymer) are higher; tool wear is greater; ductility is usually less so that deformation processing can be difficult or impossible. This can mean that new processing routes must be used: investment casting or powder forming instead of conventional casting and mechanical working; more expensive molding equipment operating at higher temperatures and pressures, and so on. The better performance of the high strength material must be paid for, not only in greater material cost but also in the higher cost of processing. Finally, there are the questions of tolerance and roughness. Cost rises exponentially with precision and surface finish. It is an expensive mistake to specify tighter tolerance or smoother surfaces than are necessary. The message is clear. Performance costs money. Do not over-specify it.

To make further progress, we must examine the contributions to process costs, and their origins.

Economic criteria for selection

If you have to sharpen a pencil, you can do it with a knife. If, instead, you had to sharpen a thousand pencils, it would pay to buy an electric sharpener. And if you had to sharpen a million, you might wish to equip yourself with an automatic feeding, gripping, and sharpening system. To cope with pencils of different length and diameter, you could go further and devise a micro-processor-controlled system with sensors to measure pencil dimensions, sharpening pressure and so on—an "intelligent" system that can recognize and adapt to pencil size. The choice of process, then, depends on the number of pencils you wish to sharpen, that is, on the *batch size*. The best choice is that one that costs least per pencil sharpened.

Figure 7.23 is a schematic of how the cost of sharpening a pencil might vary with batch size. A knife does not cost much but it is slow, so the labor cost is high. The other processes involve progressively greater capital investment but do the job more quickly, reducing labor costs. The balance between capital cost and rate gives the shape of the curves. In this figure the best choice is the lowest curve—a knife for up to 100 pencils; an electric sharpener for 10^2 to 10^4, an automatic system for 10^4 to 10^6, and so on.

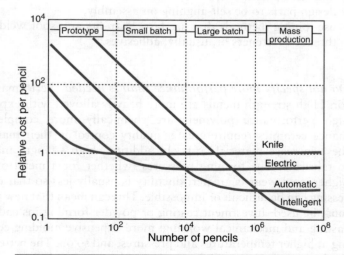

Figure 7.23   The cost of sharpening a pencil plotted against batch size for four processes. The curves all have the form of equation (7.5).

Economic batch size. Process cost depends on a large number of independent variables, not all within the control of the modeler. Cost modeling is described in the next section, but — given the disheartening implications of the last sentence — it is comforting to have an alternative, if approximate, way out. The influence of many of the inputs to the cost of a process are captured by a single attribute: the *economic batch size*; those for the processes described in this chapter are shown in Figure 7.24. A process with an economic batch size with the range B_1–B_2 is one that is found by experience to be competitive in cost when the output lies in that range, just as the electric sharpener was economic in the range 10^2 to 10^4. The economic batch size is commonly cited for processes. The easy way to introduce economy into the selection is to rank candidate processes by economic batch size and retain those that are economic in the range you want. But do not harbor false illusions: many variables cannot be rolled into one without loss of discrimination. A cost model gives deeper insight.

Cost modeling

The manufacture of a component consumes resources (Figure 7.25), each of which has an associated cost. The final cost is the sum of those of the resources it consumes. They are detailed in Table 7.3. Thus the cost of producing a

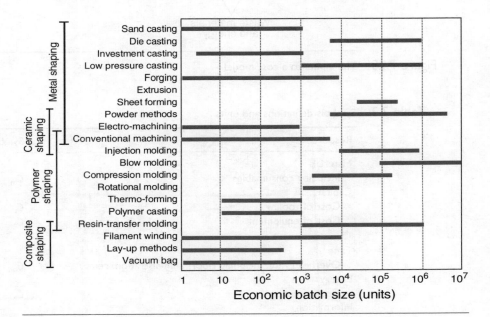

Figure 7.24   The economic batch-size chart.

component of mass m entails the cost C_m ($/kg) of the materials and feed-stocks from which it is made. It involves the cost of dedicated tooling, C_t ($), and that of the capital equipment, C_c ($), in which the tooling will be used. It requires time, chargeable at an overhead rate $\dot{C}_{oh}$ (thus with units of $/h), in which we include the cost of labor, administration and general plant costs. It requires energy, which is sometimes charged against a process-step if it is very energy intensive but more usually is treated as part of the overhead and lumped into $\dot{C}_{oh}$, as we shall do here. Finally there is the cost of information, meaning that of research and development, royalty or licence fees; this, too, we view as a cost per unit time and lump it into the overhead.

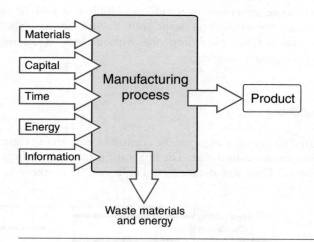

Figure 7.25   The inputs to a cost model.

Table 7.3   Symbols definitions and units

Resource	Symbol	Unit
Materials		
Including consumables	C_m	$/kg
Capital		
Cost of tooling	C_t	$
Cost of equipment	C_c	$
Time		
Overhead rate, including labor, administration, rent, ...	$\dot{C}_{oh}$	$/h
Energy		
Cost of energy	C_e	$/h
Information		
R & D or royalty payments	C_i	$/year

Think now of the manufacture of a component (the "unit of output") weighing m kg, made of a material costing C_m \$/kg. The first contribution to the unit cost is that of the material mC_m magnified by the factor $1/(1-f)$ where f is the scrap fraction — the fraction of the starting material that ends up as sprues, risers, turnings, rejects or waste:

$$C_1 = \frac{mC_m}{(1-f)} \tag{7.1}$$

The cost C_t of a set of tooling — dies, molds, fixtures, and jigs — is what is called a *dedicated cost*: one that must be wholly assigned to the production run of this single component. It is written off against the numerical size n of the production run. Tooling wears out. If the run is a long one, replacement will be necessary. Thus tooling cost per unit takes the form

$$C_2 = \frac{C_t}{n}\left\{ \text{Int}\left(\frac{n}{n_t} + 0.51\right)\right\} \tag{7.2}$$

where n_t is the number of units that a set of tooling can make before it has to be replaced, and 'Int' is the integer function. The term in curly brackets simply increments the tooling cost by that of one tool-set every time n exceeds n_t.

The capital cost of equipment, C_c, by contrast, is rarely dedicated. A given piece of equipment — for example a powder press — can be used to make many different components by installing different die-sets or tooling. It is usual to convert the capital cost of *non-dedicated* equipment, and the cost of borrowing the capital itself, into an overhead by dividing it by a capital write-off time, t_{wo} (5 years, say), over which it is to be recovered. The quantity C_c/t_{wo} is then a cost per hour — provided the equipment is used continuously. That is rarely the case, so the term is modified by dividing it by a load factor, L — the fraction of time for which the equipment is productive. The cost per unit is then this hourly cost divided by the rate $\dot{n}$ at which units are produced:

$$C_3 = \frac{1}{\dot{n}}\left(\frac{C_c}{Lt_{wo}}\right) \tag{7.3}$$

Finally there is the overhead rate $\dot{C}_{oh}$. It becomes a cost per unit when divided by the production rate $\dot{n}$ units per hour:

$$C_4 = \frac{\dot{C}_{oh}}{\dot{n}} \tag{7.4}$$

The total shaping cost per part, C_s, is the sum of these four terms, taking the form:

$$C_S = \frac{mC_m}{(1-f)} + \frac{C_t}{n}\left\{\text{Int}\left(\frac{n}{n_t} + 0.51\right)\right\} + \frac{1}{\dot{n}}\left(\frac{C_c}{Lt_{wo}} + \dot{C}_{oh}\right) \tag{7.5}$$

The equation says: the cost has three essential contributions—a material cost per unit of production that is independent of batch size and rate, a dedicated cost per unit of production that varies as the reciprocal of the production volume ($1/n$), and a gross overhead per unit of production that varies as the reciprocal of the production rate ($1/\dot{n}$). The equation describes a set of curves, one for each process. Each has the shape of the pencil-sharpening curves of Figures 7.23.

Figure 7.26 illustrates a second example: the manufacture of an aluminum con-rod by three alternative processes: sand casting, die casting and low pressure casting. At small batch sizes the unit cost is dominated by the "fixed" costs of tooling (the second term on the right of equation (7.5)). As the batch size n increases, the contribution of this to the unit cost falls (provided, of course, that the tooling has a life that is greater than n) until it flattens out at a value that is dominated by the "variable" costs of material, labour and other overheads. Competing processes usually differ in tooling cost C_t and production rate $\dot{n}$, causing their $C - n$ curves to intersect, as shown in the schematic. Sand casting equipment is cheap but the process is slow. Die casting equipment costs much more but it is also much faster. Mold costs for low pressure casting are greater than for sand casting, and the process is a little slower. Data for the terms in equation (7.5) are listed in Table 7.4. They show that the capital cost of the die-casting equipment is much greater than that for sand casting, but it is also much faster. The material cost, the labour cost per hour and the capital write-off time are, of course, the same for all. Figure 7.26 is a plot of equation (7.5), using these data, all of which are normalized to the material cost. The curves for sand and die casting intersect at a batch size of 3000:

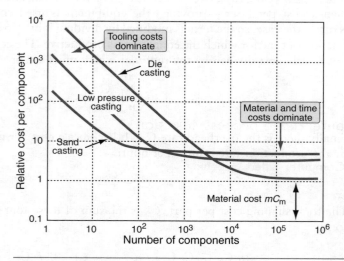

Figure 7.26 The relative cost of casting as a function of the number of components to be cast.

Table 7.4   Data for the cost equation

Relative cost*	Sand casting	Die casting	Low pressure casting	Comment
Material, $mC_m/(1-f)$	1	1	1	
Basic overhead, $\dot{C}_{oh}$(per hour)	10	10	10	Process independent parameters
Capital write-off time, t_{wo} (yrs)	5	5	5	
Load factor	0.5	0.5	0.5	
Dedicated tool cost, C_t	210	16,000	2000	
Capital cost, C_c	800	30,000	8000	Process dependent parameters
Batch rate, $\dot{n}$ (per hour)	3	100	10	
Tool life, n_t (number of units)	200,000	1,000,000	500,000	

* All costs normalized to the material cost.

below this, sand casting is the most economical process; above, it is die casting. Low pressure casting is more expensive for all batch sizes, but the higher quality casting it allows may be worth the extra. Note that, for small batches, the component cost is dominated by that of the tooling—the material cost hardly matters. But as the batch size grows, the contribution of the second term in the cost equation diminishes; and if the process is fast, the cost falls until it is typically about twice that of the material of which the component is made.

Technical cost modeling. Equation (7.5) is the first step in modeling cost. Greater predictive power is possible with technical cost models that exploit understanding of the way in which the design, the process and cost interact. The capital cost of equipment depends on size and degree of automation. Tooling cost and production rate depend on complexity. These and many other dependencies can be captured in theoretical or empirical formulae or look-up tables that can be built into the cost model, giving more resolution in ranking competing processes. For more advanced analyses the reader is referred to the literature listed at the end of this chapter.

7.6 Computer-aided process selection

Computer-aided screening

Shaping. If process attributes are stored in a database with an appropriate user-interface, selection charts can be created and selection boxes manipulated with much greater freedom. The CES platform, mentioned earlier, is an example of such a system. The way it works is described here; examples of its use are given in Chapter 8. The database contains records, each describing the attributes of a single process. Table 7.5 shows part of a typical record: it is that

Table 7.5 Part of a record for a process

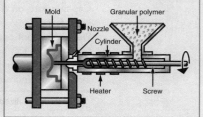

Injection molding

The process. No other process has changed product design more than injection molding. Injection molded products appear in every sector of product design: consumer products, business, industrial, computers, communication, medical and research products, toys, cosmetic packaging, and sports equipment. The most common equipment for molding thermoplastics is the reciprocating screw machine, shown schematically in the figure. Polymer granules are fed into a spiral press where they mix and soften to a dough-like consistency that can be forced through one or more channels ("sprues") into the die. The polymer solidifies under pressure and the component is then ejected.

Thermoplastics, thermosets, and elastomers can all be injection molded. Co-injection allows molding of components with different materials, colors and features. Injection foam molding allows economical production of large molded components by using inert gas or chemical blowing agents to make components that have a solid skin and a cellular inner structure.

Physical attributes

Mass range	0.01–25 kg
Range of section thickness	0.4–6.3 mm
Tolerance	0.2–1 mm
Roughness	0.2–1.6 μm
Surface roughness (A = v. smooth)	A

Economic attributes

Economic batch size (units)	1E4–1E6
Relative tooling cost	very high
Relative equipment cost	high
Labor intensity	low

Cost modeling

Relative cost index (per unit)	15.65–47.02
Capital cost	* 3.28E4–7.38E5 USD
Material utilization fraction	* 0.6–0.9
Production rate (units)	* 60–1000/h
Tooling cost	* 3280–3.28E4 USD
Tool life (units)	* 1E4–1E6

Shape

Circular prismatic	True
Non-circular prismatic	True
Solid 3D	True
Hollow 3D	True

Table 7.5   *(Continued)*

Supporting Information

Design guidelines. Injection molding is the best way to mass-produce small, precise, polymer components with complex shapes. The surface finish is good; texture and pattern can be easily altered in the tool, and fine detail reproduces well. Decorative labels can be molded onto the surface of the component (see in-mold decoration). The only finishing operation is the removal of the sprue.

Technical notes. Most thermoplastics can be injection molded, although those with high melting temperatures (e.g. PTFE) are difficult. Thermoplastic based composites (short fiber and particulate filled) can be processed providing the filler-loading is not too large. Large changes in section area are not recommended. Small re-entrant angles and complex shapes are possible, though some features (e.g. undercuts, screw threads, inserts) may result in increased tooling costs.

Typical uses. Extremely varied. Housings, containers, covers, knobs, tool handles, plumbing fittings, lenses, etc.

---

for injection molding. A schematic indicates how the process works; it is supported by a short description. This is followed by a listing of attributes: the shapes it can make, the attributes relating to shape and physical characteristics, and those that describe economic parameters; a brief description of typical uses, references and notes. The numeric attributes are stored as ranges, indicating the range of capability of the process. Each record is linked to records for the materials with which it is compatible, allowing choice of material to be used as a screening criterion, like the material matrix of Figure 7.16 but with greater resolution.

The starting point for selection is the idea that all processes are potential candidates until screened out because they fail to meet design requirements. A short list of candidates is extracted in two steps: screening to eliminate processes that cannot meet the design specification, and ranking to order the survivors by economic criteria. A typical three stage selection takes the form shown in Figure 7.27. It shows, on the left, a screening on material, implemented by selecting the material of choice from the hierarchy described in Chapter 5 (Figure 5.2). To this is added a limit stage in which desired limits on numeric attributes are entered in a dialog box, and the required shape class is selected. Alternatively, bar-charts like those of Figures 7.19–7.22 can be created, selecting the desired range of the attribute with a selection box like that on the right of Figure 7.27. The bar-chart shown here is for economic batch size, allowing approximate economic ranking. The system lists the processes that pass all the stages. It overcomes the obvious limitation of the hard-copy charts in allowing selection over a much larger number of processes and attributes.

The cost model is implemented in the CES system. The records contain approximate data for the ranges of capital and tooling costs (C_c and C_t) and for the rate of production ($\dot{n}$). Equation (7.5) contains other parameters not listed in the record because they are not attributes of the process itself but depend on the design, or the material, or the economics (and thus the location) of the plant in which the processing will be done. The user of the cost model must provide this information, conveniently entered through a dialog box like that of Figure 7.28.

One type of output is shown in Figure 7.29. The user-defined parameters are listed on the figure. The shaded band brackets a range of costs. The lower edge of the band uses the lower limits of the ranges for the input parameters—it

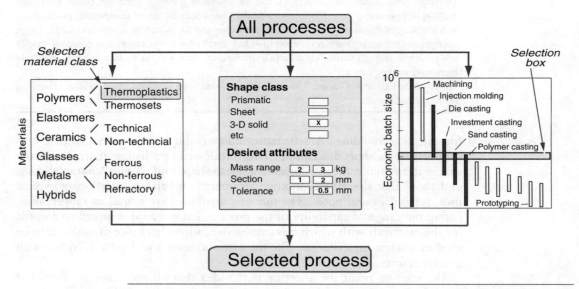

Figure 7.27   The steps in computer-based process selection. The first imposes the constraint of material, the second of shape and numeric attributes, and the third allows ranking by economic batch size.

Mass of component	3.5	kg
Overhead rate	60	$/hr
Capital write-off time	5	Yrs
Load factor	0.5	
Desired batch size	1000	

Figure 7.28   A dialog box that allows the user-defined conditions to be entered into the cost model.

characterizes simple parts requiring only a small machine and an inexpensive mold. The upper edge uses the upper limits of the ranges; it describes large complex parts requiring both a larger machine and a more complex mold. Plots of this sort allow two processes to be compared and highlight cost drivers, but they do not easily allow a ranking of a large population of competing processes. This is achieved by plotting unit cost for each member of the population for a chosen batch size. The output is shown as a bar chart in Figure 7.30. The software evaluates equation (7.5) for each member of the population and orders them by the mean value of the cost suggesting those that are the most economic. As explained earlier, the ranking is based on very approximate data; but since the most expensive processes in the figure are over 100 times more costly than the cheapest, an error of a factor of 2 in the inputs changes the ranking only slightly.

Joining. The dominant constraints in selecting a joining process are usually set by the material or materials to be joined and by the geometry of the joint itself (butt, sleeve, lap, etc.). When these and secondary constraints (e.g. requirements that the joint be water-tight, or demountable, or electrical conducting) are met, the relative cost becomes the discriminator.

Detailed studies of assembly costs — particularly those involving fasteners — establish that it is a significant cost driver. The key to low-cost assembly is to make it fast. Design for assembly (DFA) addresses this issue. The method has three steps. The first is an examination of the design of the product, questioning whether each individual part is necessary. The second is to estimate the

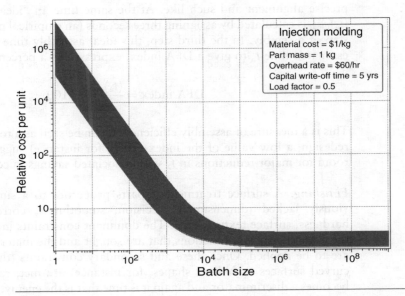

Figure 7.29 The output of a computer-based cost model for a single process, here injection molding.

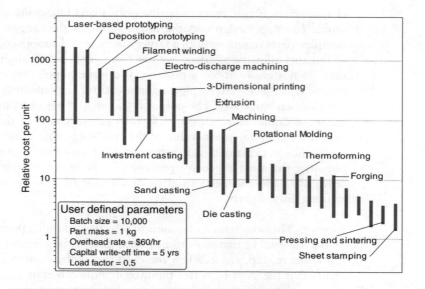

Figure 7.30   The output of the same computer-based model allowing comparison between competing processes to shape a given component.

assembly time t_a by summing the times required for each step of assembly, using historical data that relate the time for a single step to the features of the joint — the nature of the fastener, the ease of access, the requirement for precise alignment and such like. At the same time an "ideal" assembly time $(t_a)_{ideal}$ is calculated by assigning three seconds (an empirical minimum) to each step of assembly. In the third step, this ideal assembly time is divided by the estimated time t_a to give a DFA index, expressed as a percentage:

$$\text{DFA index} = \frac{(t_a)_{ideal}}{t_a} \times 100 \qquad (7.6)$$

This is a measure of assembly efficiency. It can be seen as a tool for motivating redesign: a low value of the index (10%, for instance) suggests that there is room for major reductions in t_a with associated savings in cost.

Finishing. A surface treatment imparts properties to a surface that it previously lacked (dimensional precision, smoothness, corrosion resistance, hardness, surface texture, etc.). The dominant constraints in selecting a treatment are the surface functions that are sought and the material to which they are to be applied. Once these and secondary constraints (the ability to treat curved surfaces or hollow shapes, for instance) are met, relative cost again becomes a discriminator and again it is time that is the enemy. Organic solvent-based paints dry more quickly than those that are water-based, electro-less plating is faster that electro-plating, and laser surface-hardening can be quicker than more conventional heat-treatments. Economics recommend the selection of the faster process. But minimizing time is not the only discriminator. Increasingly environmental concerns restrict the use of certain finishing processes: the air pollution associated with organic solvent-based paints, for instance, is sufficiently serious to prompt moves to phase out their use. We return to environmental issues in Chapter 16.

7.7 Supporting information

Systematic screening and ranking based on attributes common to all processes in a given family yields a short list of candidates. We now need supporting information — details, case studies, experience, warnings, anything that helps form a final judgement. The CES database contains some. But where do you go from there?

Start with texts and handbooks — they don't help with systematic selection, but they *are* good on supporting information. They and other sources are listed in Section 7.9 and in Chapter 15. The handbook of Bralla and the ASM Handbook series are particularly helpful, though increasingly dated. Next look at the data sheets and design manuals available from the makers and suppliers of process equipment, and, often, from material suppliers. Leading suppliers exhibit at major conferences and exhibitions and an increasing number have helpful web sites. Specialized software allowing greater discrimination within a narrow process domain is still largely a research topic, but will in time become available.

7.8 Summary and conclusions

A wide range of shaping, joining, and finishing processes is available to the design engineer. Each has certain characteristics, which, taken together, suit it to the processing of certain materials in certain shapes, but disqualify it for others. Faced with the choice, the designer has, in the past, relied on locally-available expertise or on common practice. Neither of these lead to innovation, nor are they well matched to current design methods. The structured, systematic approach of this chapter provides a way forward. It ensures that potentially interesting processes are not overlooked, and guides the user quickly to processes capable of achieving the desired requirements.

The method parallels that for selection of material, using process selection matrices and charts to implement the procedure. A component design dictates a certain, known, combination of process attributes. These design requirements are plotted onto the charts, identifying a subset of possible processes. The method lends itself to computer implementation, allowing selection from a large portfolio of processes by screening on attributes and ranking by economic criteria.

There is, of course, much more to process selection than this. It is to be seen, rather, as a first systematic step, replacing a total reliance on local experience and past practice. The narrowing of choice helps considerably: it is now much easier to identify the right source for more expert knowledge and to ask of it the right questions. But the final choice still depends on local economic and organizational factors that can only be decided on a case-by-case basis.

7.9 Further reading

ASM Handbook Series (1971–2004), "Volume 4: Heat treatment; Volume 5: Surface engineering; Volume 6: Welding, brazing and soldering; Volume 7: Powder metal technologies; Volume 14: Forming and forging; Volume 15: Casting; Volume 16: Machining", ASM International, Metals Park, OH USA. (*A comprehensive set of handbooks on processing, occasionally updated, and now available on-line at www.asminternational.org/hbk/index.jsp*)

Bralla, J.G. (1998) *Design for Manufacturability Handbook*, 2nd edition, McGraw-Hill, New York, USA. ISBN 0-07-007139-X. (*Turgid reading, but a rich mine of information about manufacturing processes.*)

Bréchet, Y., Bassetti, D., Landru, D. and Salvo, L (2001) Challenges in materials and process selection, *Prog. Mater. Sci.* **46**, 407–428. (*An exploration of knowledge-based methods for capturing material and process attributes.*)

Campbell, J. (1991) *Casting*, Butterworth-Heinemann, Oxford, UK. ISBN 0-7506-1696-2. (*The fundamental science and technology of casting processes.*)

Clark, J.P. and Field, F.R. III (1997) Techno-economic issues in materials selection, in *ASM Metals Handbook*, Vol 20. American Society for Metals, Metals Park, OH USA. (*A paper outlining the principles of technical cost-modeling and its use in the automobile industry.*)

Dieter, G.E. (1991) *Engineering Design, a Materials and Processing Approach*, 2nd edition, McGraw-Hill, New York, USA. ISBN 0-07-100829-2. (*A well-balanced and respected text focusing on the place of materials and processing in technical design.*)

Esawi, A. and Ashby, M.F. (1998) Computer-based selection of manufacturing processes: methods, software and case studies, *Proc. Inst. Mech. Eng.* **212**, 595–610. (*A paper describing the development and use of the CES database for process selection.*)

Esawi, A. and Ashby, M.F. (2003) Cost estimates to guide pre-selection of processes, *Mater. Design*, **24**, 605–616. (*A paper developing the cost-estimation methods used in this chapter.*)

Grainger, S. and Blunt, J. (1998) *Engineering Coatings, Design and Application*, Abington Publishing, Abington, Cambridge UK. ISBN 1-85573-369-2. (*A handbook of surface treatment processes to improve surface durability — generally meaning surface hardness.*)

Houldcroft, P. (1990) *Which Process?*, Abington Publishing, Abington, Cambridge UK. ISBN 1-85573-008-1. (*The title of this useful book is misleading — it deals only with a subset of joining process: the welding of steels. But here it is good, matching the process to the design requirements.*)

Kalpakjian, S. and Schmid, S.R. (2003) *Manufacturing Processes for Engineering Materials*, 4th edition, Prentice Hall, Pearson Education, Inc, New Jersey, USA. ISBN 0-13-040871-9. (*A comprehensive and widely used text on material processing.*)

Lascoe, O.D. (1988) *Handbook of Fabrication Processes*, ASM International, Metals Park, Columbus, OH, USA. ISBN 0-87170-302-5. (*A reference source for fabrication processes.*)

Shercliff, H.R. and Lovatt, A.M. (2001) Selection of manufacturing processes in design and the role of process modelling *Prog. Mater. Sci.* **46**, 429–459. (*An introduction to ways of dealing with the coupling between process and material attributes.*)

Swift, K.G. and Booker, J.D. (1997) *Process Selection, from Design to Manufacture*, Arnold, London, UK. ISBN 0-340-69249-9. (*Details of 48 processes in a standard format, structured to guide process selection.*)

Wise, R.J. (1999) *Thermal Welding of Polymers*, Abington Publishing, Abington, Cambridge UK. ISBN 1-85573-495-8. (*An introduction to the thermal welding of thermoplastics.*)

Chapter 8

Process selection case studies

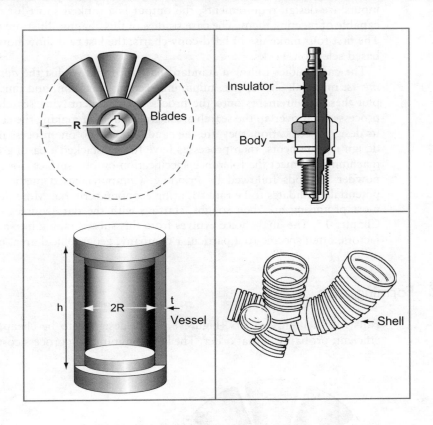

8.1 Introduction and synopsis

The last chapter described a systematic procedure for process selection. The inputs are design requirements; the output is a ranked short-list of processes capable of meeting them. The case studies of this chapter illustrate the method. The first four make use of hard-copy charts; the last two show how computer-based selection works.

The case studies follow a standard pattern. First, we list the *design requirements*: material, shape, size, minimum section, precision, and finish. Then we plot these requirements onto the process charts, identifying search areas. The processes that overlap the search areas are capable of making the component to its design specification: they are the candidates. If no one process meets all the design requirements, then processes have to be "stacked": casting followed by machining (to meet the tolerance specification on one surface, for instance); or powder methods followed by grinding. Computer-based methods allow the potential candidates to be ranked, using economic criteria. More details for the most promising are then sought, starting with the data sources described in Chapter 15. The final choice evolves from this subset, taking into account local factors, often specific to a particular company, geographical area, or country.

8.2 Forming a fan

Fans for vacuum cleaners (Figure 8.1) are designed to be cheap, quiet, and efficient, probably in that order. The key to minimizing process costs is to form

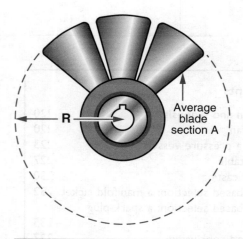

Figure 8.1   A fan. The blades are to be made of nylon, require low roughness and a certain precision, and are to be produced in large numbers.

the fan to its final shape in a single operation — that is, to achieve net-shape forming — leaving only the central hub to be machined to fit the shaft with which it mates. This means the selection of a single process that can meet the specifications on precision and tolerance, avoiding the need for machining or finishing of the disk or blades.

The design requirements. The material choice for the fan is nylon. The pumping rate of a fan is determined by its radius and rate of revolution: it is this that determines its size. The designer calculates the need for a fan of radius 60 mm, with 20 profiled blades of average thickness 4 mm. The volume of material in the fan is, roughly, its surface area times its thickness — about $10^{-4} m^3$ — giving (when multiplied by the density of nylon, $1100 kg/m^3$) a weight in the range 0.1 to 0.2 kg. The fan has a fairly complex shape, though its high symmetry simplifies it somewhat. We classify it as 3D solid. In the designer's view, balance and surface smoothness are what really matter. They (and the geometry) determine the pumping efficiency of the fan and influence the noise it makes. He specifies a tolerance of ± 0.5 mm a surface roughness of $\leq 1 \mu m$. A production run of 10,000 fans is envisaged. The design requirements are summarized in Table 8.1.

What processes can meet them?

The selection. We turn first to the material- and shape-process matrices (Figures 8.2 and 8.3) on which selection boxes have been drawn. The intersection of the two leaves five classes of shaping process — those boxed in the second figure. Screening on mass and section thickness (Figures 8.4 and 8.5) eliminates polymer casting and RTM, leaving the other three. The constraints on tolerance and roughness are upper limits only (Figures 8.6 and 8.7); all three process classes survive. The planned batch size of 10,000 plotted on the economic batch-size chart (Figure 8.8), eliminates machining from solid.

The surviving processes are listed in Table 8.2.

Table 8.1   Process requirements for the fan

Function	Fan
Constraints	• Material: nylon
	• Shape: 3D solid
	• Mass: 0.1–0.2 kg
	• Minimum section: 4 mm
	• Precision: ± 0.5 mm
	• Roughness: $< 1 \mu m$
	• Batch size: 10,000
Objective	Minimize cost
Free variables	Choice of process

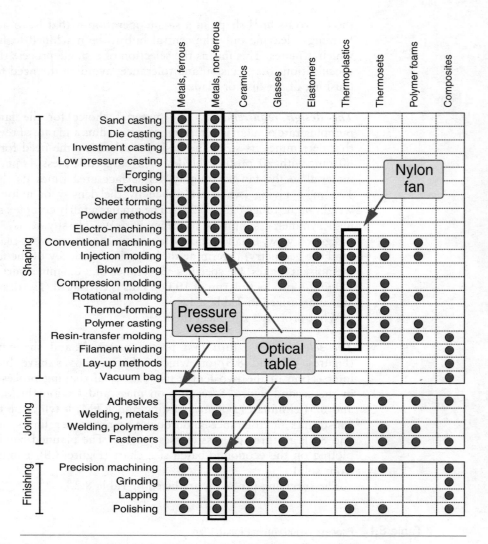

Figure 8.2   The process—material compatibility matrix, showing the requirements of the three case studies. By including the joining and finishing processes it becomes possible to check that the more restrictive requirements can be met by combining processes.

Table 8.2   Processes for forming the fan

Process	Comment
Injection molding	Meets all design constraints
Compression molding	May need further finishing operations

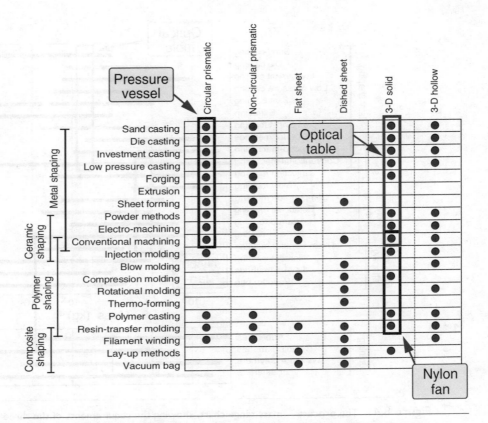

Figure 8.3   The process—shape compatibility matrix, showing the requirements of the three case studies. A summary of the material compatibility appears at the left. The intersection of this selection stage and the last narrows the choice.

Postscript. There are (as always) other considerations. There are the questions of capital investment, local skills, overhead rate and so forth. The charts cannot answer these. But the procedure has been helpful in narrowing the choice, suggesting alternatives, and providing a background against which a final selection can be made.

Related case studies   6.6   Materials for flywheels

8.3 Fabricating a pressure vessel

A pressure vessel is required for a hot-isostatic press or HIP. Materials for pressure vessels were the subject of an earlier case study (Section 6.11); tough steels are the best choice.

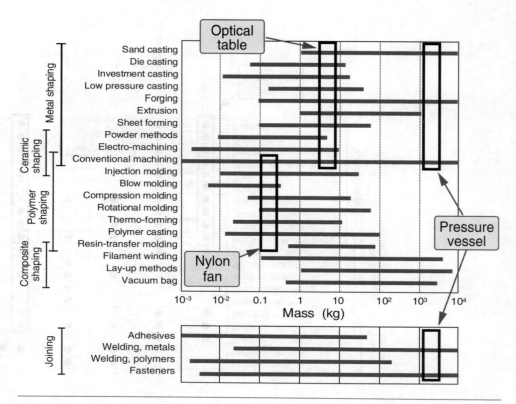

Figure 8.4 The process—mass range chart, showing the requirements of the three case studies. The inclusion of joining processes allows the possibility of fabrication of large structures to be explored.

The design requirements. The design asks for a cylindrical pressure vessel with an inside radius R of 0.4 m and a height h of 2 m, with removable end-caps (Figure 8.9). It must safely contain a pressure p of 50 MPa. A steel with a yield strength σ_y of 500 MPa has been selected. The necessary wall thickness t is given approximately by equating the hoop stress in the wall, roughly pR/t, to the yield strength of the material of which it is made, σ_y, divided by a safety factor S_f that we will take to be 2. Solving for t gives:

$$t = \frac{2pR}{\sigma_y} = 80 \text{ mm} \tag{8.1}$$

The outer radius R_0 is, therefore, 0.38 m. The volume V of steel in the cylinder is approximately 0.34 m^3. Lest that sounds small, consider the weight. The density of steel is just under 8000 kg/m^3. The cylinder weighs 2.7 tonnes.

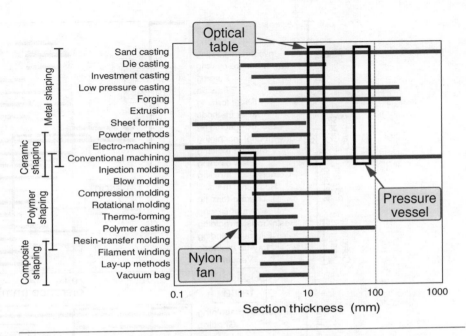

Figure 8.5   The process—section thickness chart, showing the requirements of the three case studies.

A range of presses is envisaged, centered on this one, but with inner radii and pressures that range by a factor of 2 on either side. (A constant pressure implies a constant "aspect ratio", R/t.) Neither the precision nor the surface roughness of the vessel are important in selecting the primary forming operation because the end faces and internal threads will be machined, regardless of how it is made. The order is for 10 cylinders. What processes are available to shape them?

The selection. Table 8.3 summarizes the requirements. They are plotted on the charts, Figures 8.2–8.5 and 8.8. The discriminating requirements, this time, are mass and batch size. They single out the four possibilities listed in Table 8.4: the vessel can be machined from the solid, forged, cast, or fabricated (by welding plates together, for instance).

Material constraints are worth checking (Figure 8.2), but they do not add any further restrictions. Tolerance and roughness do not matter except on the end faces and threads (where the end-caps must mate) and any ports in the sides—these require a precision of ±0.1 mm. The answer here (Figure 8.6, lower part) is to add a second process: an additional machining or grinding step can achieve it.

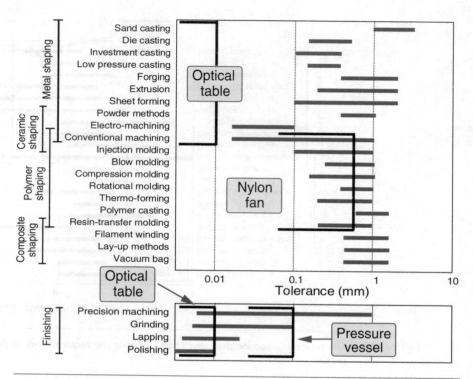

Figure 8.6   The process — tolerance chart, showing the requirements of the three case studies. The inclusion of joining and finishing processes allows the possibility of fabrication of large structures to be explored. Tolerance and surface roughness are specified as an upper limit only, so the selection boxes are open-ended to the left.

Postscript. A "systematic" procedure is one that allows a conclusion to be reached without prior specialised knowledge. This case study is an example. We can get so far (Table 8.4) systematically, and it is a considerable help. But we can get no further without adding some expertise.

A cast pressure-vessel is not impossible, but it would be viewed with suspicion by an expert because of the risk of casting defects; safety might then require proof testing or elaborate non-destructive inspection. The only way to make very large pressure vessels is to weld them, and here we encounter the same problem: welds are defect-prone and can only be accepted after elaborate inspection. Forging, or machining from a previously-forged billet are the best because the large compressive deformation heals defects and aligns oxides and other muck in a harmless, strung-out way.

That is only the start of the expertise. You will have to go to an expert for the rest.

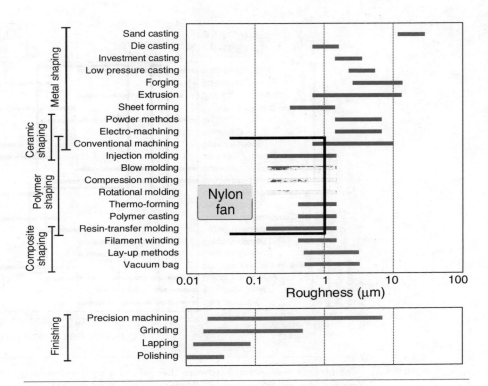

Figure 8.7 The process—surface roughness chart. Only one case study—that of the fan—imposed restrictions on this.

Related case studies   6.11   Safe pressure vessels

8.4 An optical table

An optical table is a flat plate mounted in a way that minimizes vibration pick-up, on which optical systems can be positioned with an accuracy comparable with the wavelength of light. They used to be made of polished granite or cast iron; more recently stainless steel or aluminum is used. The key feature of the table is its flatness: a good one is flat to within 0.01 mm over its entire surface.

The design requirements. A design for such a table is sketched in Figure 8.10. The table is a plate 350 mm square and 16 mm thick, weighing 10 kg, with edge grooves and threaded fixing points, but these features can be added later by simple machining and need not concern us here. It is to be

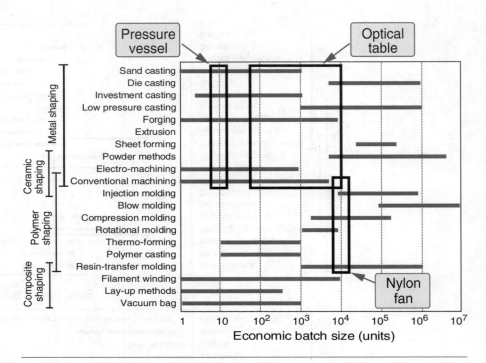

Figure 8.8 The process — economic batch-size chart, showing the requirements of the three case studies. The box for the optical table spans the range of possible production volumes listed in the requirements.

made of an aluminum–silicon alloy, Alloy 356, chosen for its ability to resist thermal distortion (see Section 6.16). What is important is the flatness of ±0.01 mm. How is the plate to be made? Table 8.5 summarizes the requirements.

The selection. Not surprisingly, many primary manufacturing processes can shape aluminum alloys to a plate (Figures 8.2 and 8.3). Similarly, the constraints on mass (Figure 8.4) and on section thickness (Figure 8.5) eliminate only powder forming and electro-machining, leaving many others. The tolerance constraint of ±0.01 mm is the critical one (Figure 8.6): none of the primary shaping processes can achieve it. We must add a finishing stage, shown in the lower part of the figure: precision machining, grinding, lapping and polishing are all capable of adding the necessary refinement.

The conclusion: use a casting process to make the blanks for the table, choosing the cheapest. Then add a precision machining step to create the flat working surface.

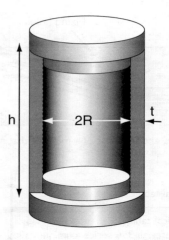

Figure 8.9   A pressure vessel. The task is to form the cylinder; adding ports, threads, and connections will be done later by machining.

Table 8.3   Process requirements for the pressure vessel

Function	Pressure vessel
Constraints	• Material: steel • Shape: circular prismatic • Mass: 2720 kg • Minimum section: 80 mm • Batch size: 10
Objective	Minimize cost
Free variables	• Choice of process • Process chain options

Table 8.4   Processes for forming the pressure vessels

Process	Comment
Machining	Machine from solid (rolled or forged) billet Much material discarded, but a reliable product Might select for one-off
Hot forging	Steel forged to thick-walled tube, and finished by machining end faces, ports, etc. Preferred route for economy of material use
Casting	Cast cylindrical tube, finished by machining end-faces and ports. Casting-defects a problem
Fabrication	Weld previously-shaped plates. Not suitable for the HIP; use for very large vessels (e.g. nuclear pressure vessels)

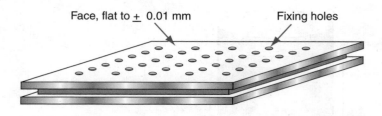

Face, flat to ± 0.01 mm   Fixing holes

Figure 8.10   An optical table. It must be very flat, and be made of a material that minimizes distortion caused by temperature gradients.

Table 8.5   Process requirements for the optical table

Function	Optical table
Constraints	• Material: 356 series aluminum alloy
	• Shape: 3D solid
	• Mass: 10 kg
	• Section thickness 16 mm
	• Tolerance ±0.01 mm
Objective	Minimize cost
Free variables	• Choice of process
	• Process chain options

Postscript. The second last sentence above contained the throw-away phrase "... choosing the cheapest". But how? With a cost model, of course. The next case study does so.

Related case studies   6.16   Materials to minimize thermal distortion in precision devices
8.5   Economical casting

8.5 Economical casting

The last case study described optical tables and their manufactue. Here we explore aspects of cost.

The design requirements. The designer, uncertain of the market for the tables, asks for advice on the best way, first, to cast 10 prototype tables, followed by a preliminary run of 200 tables, and (if these succeed) enough tables to satisfy a potential high-school market of about 10,000. The last case study showed that the high precision demanded by the design can only be met by machining the working surfaces of the table, so the tolerance and roughness of the casting itself do not matter. The best choice of casting method is the cheapest.

Process data for four casting methods for aluminum alloy 356 (cost C_m/kg) are listed in Table 8.6. All four have passed the screening stage and thus are potential candidates. The costs are given in units of the material cost, $mC_m/(1-f)$, of one table of mass m (that is, $mC_m/(1-f) = 1$). In these units, the overhead rate, $\dot{C}_{oh}$, is 10 per hour. The tooling cost C_t, the capital cost C_c of equipment and the batch rate for each process, $\dot{n}$ per hour, are listed in the table, all in units of mC_m. Which is the best choice?

The selection. Provided the many components of cost have been properly distributed between mC_m, $\dot{C}_{oh}$, C_t and C_c, the manufacturing cost per table (equation (7.5), simplified by assuming that the tool life exceeds the desired batch size) is

$$C = \frac{mC_m}{(1-f)} + \frac{C_t}{n} + \frac{1}{\dot{n}}\left(\frac{C_c}{Lt_{wo}} + \dot{C}_{oh}\right) \qquad (8.2)$$

Analytical solutions for the cheapest process are possible, but the most helpful way to solve the problem is by plotting the equation for each of the four casting methods, using the data in Table 8.6. The result is shown in Figure 8.11.

The selection can now be read off: for 10 tables, sand casting is the cheapest. But since a production run of 200 is certain, for which permanent mold casting is cheaper, it probably makes sense to use this for the prototype as well. If the product is adopted by schools, creating a demand for 10,000 tables, die casting becomes competitive, though probably not worth implementing unless an even larger batch is envisaged.

Postscript. All this is deceptively easy. The difficult part is that of assembling the data of Table 8.6, partitioning costs between the three heads of material, labor and capital. In practice this requires a detailed, in-house, study of costs and involves information not just for the optical table but for the entire

Table 8.6   Process costs for four casting methods

Relative cost*	Sand casting	Die casting	Permanent mold	Low pressure casting	Comment
Material, $mC_m/(1-f)$	1	1	1	1	Process independent parameters
Basic overhead $\dot{C}_{oh}$ (per hour)	10	10	10	10	
Capital write-off time t_{wo} (years)	5	5	5	5	
Load factor	0.5	0.5	0.5	0.5	
Dedicated tool cost, C_t	210	16,000	1000	2000	Process dependent parameters
Capital cost, C_c	800	30,000	2000	8000	
Batch rate, $\dot{n}$ (per hour)	3	100	7	10	

* All costs normalized to the material cost mC_m.

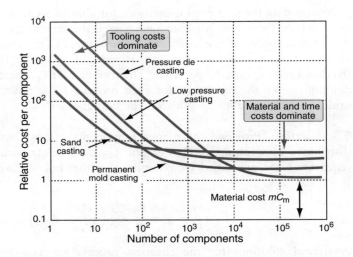

Figure 8.11 The relative cost of the four alternative casting processes.

product line of the company. But when — for a given company — the data for competing processes are known, selecting the cheapest route for a new design can be guided by the method.

Related case studies

6.16 Materials to minimize thermal distortion in precision devices

8.4 An optical table

8.6 Computer-based selection: a manifold jacket

The difficulties of using hard-copy charts for process selection will, by now, be obvious: the charts have limited resolution, and are cumbersome to use. They give a helpful overview but they are not the way to get a definitive selection. Computer-based methods overcome both problems.

The CES system, which builds on the methods of Chapter 7, was described earlier. The best way to use it is to employ a limit stage to impose constraints on material, shape, mass, section, tolerance, etc. The output is the subset of processes that meet all the limits. The economics are then examined by plotting the desired batch size on a bar-chart of economic batch size, or by implementing the cost model that is built into the software. If the requirements are very demanding, no single process can meet them all. Then the procedure is to relax the most demanding of them (often tolerance and surface roughness) seeking processes that can meet the others. A second process is then sought to add the desired refinement.

The next two case studies show how the method works.

The design requirements. The manifold jacket shown in Figure 8.12 is part of the propulsion system of a space vehicle. It is to be made of nickel. It is large, weighing about 7 kg, and complex, having an unsymmetrical 3D-hollow shape. The minimum section thickness is between 2 and 5 mm. The requirement on precision and surface finish are strict (tolerance $< \pm 0.1$ mm, roughness < 20 μm). Because of its limited application, only 10 units are to be made. Table 8.7 lists the requirements.

The selection. The constraints are applied by entering them in a limit-selection dialog box like that of Figure 8.13. The results are summarized in the economic batch-size bar-chart of Figure 8.14. Each bar represents a process that can shape nickel. The 10 that meet the design requirements are identified and shown in red; the others are grey. The figure shows that, for a batch size of 10, four are better than the rest: electro-forming, manual investment casting, ceramic mold prototyping, and machining from the solid. They are compared in Table 8.8.

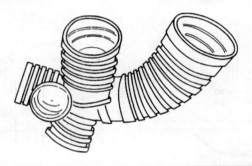

Figure 8.12   A manifold jacket (redrawn from: Bralla, J.G., 1986, *Handbook of Product Design for Manufacturing*, McGraw-Hill, New York, USA).

Table 8.7   Process requirements for the manifold jacket

Function	Manifold jacket
Constraints	• Material: nickel
	• Shape: 3D hollow
	• Mass: 7 kg
	• Minimum section: 2–5 mm
	• Tolerance: $< \pm 0.1$ mm
	• Surface roughness: < 20 μm
	• Batch size: 10
Objective	Minimize cost
Free variables	• Choice of process
	• Process chain options

Class attributes

Material class	Non-ferrous metal
Process type	Primary
Shape class	3-D hollow

Numeric attributes

	Min	Max	
Mass range	6	7	kg
Section	2	5	mm
Tolerance		0.1	mm
Roughness		20	μm

Figure 8.13   The design requirements for the manifold jacket entered in the dialog box.

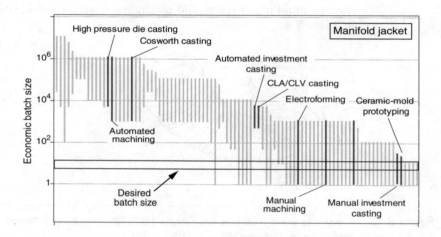

Figure 8.14   The ranking by economic batch size of the processes that pass all the selection stages. The box captures the desired batch size.

Table 8.8   Processes capable of making the manifold jacket

Process	Comment
Electro-forming	A practical choice for pure nickel
Investment casting (manual)	A possible choice, but requiring elaborate gating
Direct mold prototyping (followed by casting)	A prototyping method in which a mold is made by printing binder onto a sand bed—see Chapter 7, Figure 7.10
Machining from solid	Numerically controlled 5-axis head required

Postscript. Electro-forming and *investment casting* emerged as the most promising candidates for making the manifold jacket. A search for further information in the sources listed in Chapter 15 reveals that electro-forming of nickel is established practice and that components as large as 20 kg are routinely made by this process. It looks like the best choice.

Related case studies   8.7   Computer-based selection: a spark plug insulator

8.7   Computer-based selection: a spark plug insulator

This is the second of two case studies illustrating the use of computer-based selection methods.

The design requirements. The anatomy of a spark plug is shown schematically in Figure 8.15. It is an assembly of components, one of which is the insulator. This is to be made of a ceramic, *alumina*, with the shape shown in the figure: an axisymmetric-hollow-stepped shape. It weighs about 0.05 kg and has a minimum section of 1.2 mm. Precision is important, since the insulator is part of an assembly; the design specifies a precision of ±0.3 mm and a surface finish of better than 10 μm. The insulators are to be made in large numbers: the projected batch size is 100,000. Cost should be as low as possible. Table 8.9 summarizes the requirements.

The selection. As in the previous case study, the constraints are applied using a limit-stage dialog box (Figure 8.16). Only two processes survive. They are listed in Table 8.10: die pressing of powder followed by sintering, and powder

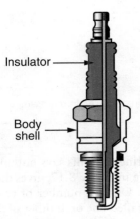

Figure 8.15   A spark plug. We seek a process to make the insulator.

Table 8.9   Process requirements for the spark plug insulator

Function	Insulator
Constraints	• Material: alumina
	• Shape: 3D hollow
	• Mass: 0.04–0.06 kg
	• Minimum section: 1.2 mm
	• Tolerance: $<\pm0.3$ mm
	• Surface roughness: $<10\,\mu$m
	• Batch size: 100,000
Objective	Minimize cost
Free variables	Choice of process

Table 8.10   Processes capable of making the spark plug insulator

Process	Comment
Die pressing and sintering	Practical choice
PIM	Practical choice

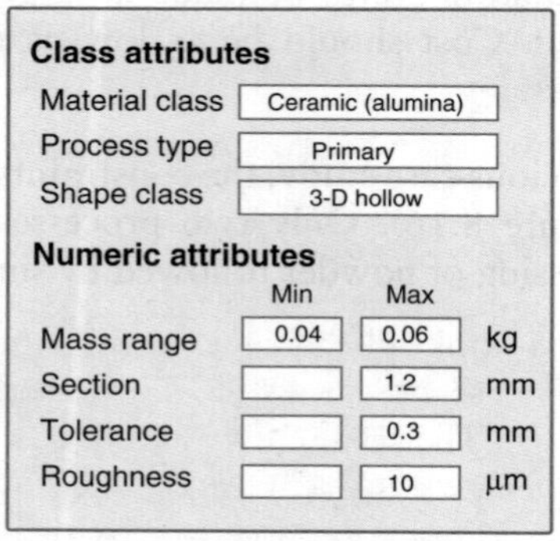

Class attributes

Material class	Ceramic (alumina)
Process type	Primary
Shape class	3-D hollow

Numeric attributes

	Min	Max	
Mass range	0.04	0.06	kg
Section		1.2	mm
Tolerance		0.3	mm
Roughness		10	μm

Figure 8.16   The design requirements for the spark plug, entered in the dialog box.

injection molding (PIM) with sintering. But this says nothing of the economics of manufacture. A final stage, shown in Figure 8.17, gives the output of the cost model for a batch size of 100,000. On it, a number of ceramic-forming processes are identified. Most of these failed one or another of the constraints. The two survivors, shown in red, are cheaper than any of the others. The model is not precise enough to distinguish between them.

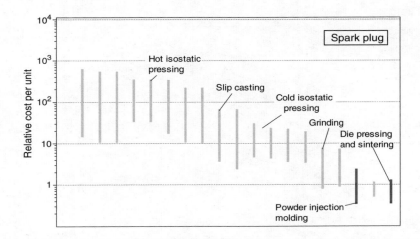

Figure 8.17   The ranking, this time by relative cost, evaluated at a batch size of 10^5, of the processes that pass all the selection stages. Only two survive; they are of comparable cost.

> *Postscript.* Insulators are made commercially by PIM. More detailed cost analysis would be required before a final decision is made. Spark plugs have a very competitive market and, therefore, the cost of manufacturing should be kept low by choosing the cheapest route.

Related case studies   8.6   Computer-based selection: a manifold jacket

8.8 Summary and conclusions

Process selection, at first sight, looks like a black art: the initiated know; the rest of the world cannot even guess how they do it. But this — as the chapter demonstrates — is not really so. The systematic approach, developed in Chapter 7 and illustrated here, identifies a sub-set of viable processes using design information only: size, shape, complexity, precision, roughness, and material — itself chosen by the systematic methods of Chapter 5. It does not identify the single, best, choice; that depends on too many case-specific considerations. But, by identifying candidates, it directs the user to data sources (starting with those listed in Chapters 15) that provide the details needed to make a final selection.

The case studies, deliberately, span an exceptional range of size, shape, and material. In each, the systematic method leads to helpful conclusions.

Chapter 9

Multiple constraints and objectives

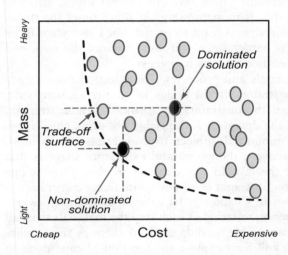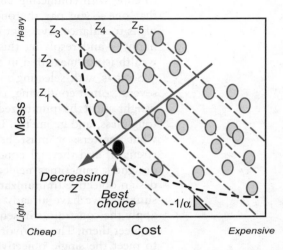

9.1 Introduction and synopsis

Most decisions you make in life involve trade-offs. Sometimes the trade-off is to cope with conflicting constraints: I must pay this bill but I must also pay that one — you pay the one that is most pressing. At other times the trade-off is to balance divergent objectives: I want to be rich but I also want to be happy — and resolving this is harder since you must balance the two, and wealth is not measured in the same units as happiness.

So it is with selecting materials and processes. The selection must satisfy several, often conflicting, constraints. In the design of an aircraft wing-spar, weight must be minimized, with constraints on stiffness, fatigue strength, toughness, and geometry. In the design of a disposable hot-drink cup, cost is what matters; it must be minimized subject to constraints on stiffness, strength, and thermal conductivity, though painful experience suggests that designers sometimes neglect the last. In this class of problem there is one design objective (minimization of weight or of cost) with many constraints, a situation we have already met in Chapter 5. Its solution is straightforward: apply the constraints in sequence, rejecting at each step the materials that fail to meet them. The survivors are viable candidates. Rank them by their ability to meet the single objective, and then explore the top-ranked candidates in depth. Usually this does the job, but sometimes there is an extra twist. It is described in Section 9.2.

A second class of problem involves more than one objective, and here the conflict is more severe. Nature being what it is, the choice of materials that best meets one objective will not usually be that which best meets the others. The designer charged with selecting a material for a wing-spar that must be both as light *and* as cheap as possible faces an obvious difficulty: the lightest material will certainly not be the cheapest, and vice versa. To make any progress, the designer needs a way of trading weight against cost, and this is a problem we have not encountered till now. Strategies for dealing with both classes of problem are summarized in Figure 9.1 on which we now expand.

There are a number of quick though subjective ways of dealing with conflicting constraints and objectives: the *method of weight-factors* and methods employing *fuzzy logic* — they are discussed in the Appendix at the end of this chapter. They are a good way of getting into the problem, so to speak, but their limitations must be recognized. Subjectivity is eliminated by employing the *active constraint method* to resolve conflicting constraints, and by combining conflicting objectives into a single *penalty function*. These are standard tools of multi-criteria optimization. To use them, we must adopt a convention that all objectives are expressed as quantities to be *minimized*; without it the penalty-function method does not work.

So now the important stuff. Figure 9.1 is the road map. We start on the top path and work down.

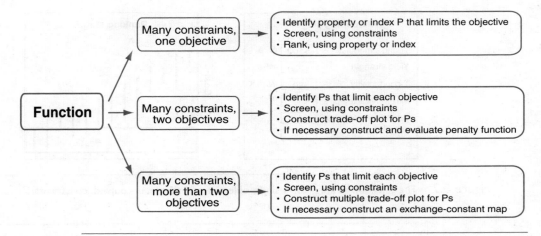

Figure 9.1 Strategies for tackling selection with multiple constraints and conflicting objectives.

9.2 Selection with multiple constraints

Nearly all material-selection problems are over-constrained, meaning that there are more constraints than free variables. We have already seen multiple constraints in Chapters 5 and 6. Recapitulating, they are tackled by identifying the constraints and the objective imposed by the design requirements, and applying the following steps:

- *Screen*, using each constraint in turn.
- *Rank*, using the performance metric describing the objective (often mass, volume, or cost) or simply by the value of the material property or index that enters the equation for the metric.
- *Seek supporting information* for the top-ranked candidates and use this to make the final choice.

Steps 1 and 2 are illustrated in Figure 9.2, which we think of as the central methodology. The icon on the left represents screening by imposing constraints on properties, on requirements such as corrosion resistance or on the ability to be processed in a certain way. That on the right — here a bar chart for cost for the surviving candidates — indicates how they are ranked. All very simple.

But not so fast. There is one little twist. It concerns the special case of a single objective that can be limited by more than one constraint. As an example, the requirements for a tie-rod of minimum mass might specify both stiffness *and* strength, leading to two independent equations for the mass. Following exactly the steps of Chapter 5, equation (5.3), the situation is summarized by the chain of reasoning shown in the box below.

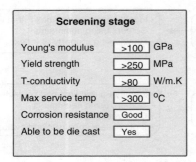

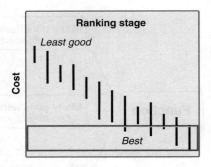

Figure 9.2 The selection strategy for a single objective and simple, uncoupled, constraints.

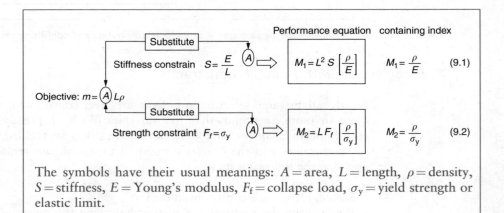

The symbols have their usual meanings: A = area, L = length, ρ = density, S = stiffness, E = Young's modulus, F_f = collapse load, σ_y = yield strength or elastic limit.

If stiffness is the dominant constraint the mass of the rod is m_1; if it is strength the mass is m_2. If the tie is to meet the requirements on both its mass has to be the greater of m_1 and m_2. Writing

$$\tilde{m} = (\max\ m_1,\ m_2) \tag{9.3}$$

we search for the material that offers the smallest value of $\tilde{m}$. This is an example of a "min–max" problem, not uncommon in the world of optimization. We seek the smallest value (min) of a metric that is the larger (max) of two or more alternatives.

The analytical method. Powerful methods exist for solving min–max problems when the metric (here, mass) is a continuous function of control variables (the things on the right of the performance equations (9.1) and (9.2)). But here one of the control variables is the material, and we are dealing with a population of materials each of which has its own unique values of material properties. The problem is discreet, not continuous.

One way to tackle it is to evaluate both m_1 and m_2 for each member of the population and assign the larger of the two to each member, and then rank the members by the assigned value, seeking a minimum. Table 9.1 illustrates the use of the method to select a material for a light, stiff, strong tie of length L, stiffness S, and collapse load F_f with the values

$$L = 1\,\text{m}, \quad S = 3 \times 10^7\,\text{N/m}, \quad F_f = 10^5\,\text{N}$$

Substituting these values and the material properties shown in the table into equations (9.1) and (9.2) gives the values for m_1 and m_2 shown in the table. The last column shows $\tilde{m}$ calculated from equation (9.3). For these design requirements Ti-6-4 is emphatically the best choice: it allows the lightest tie that satisfies both constraints.

When there are 3000 rather than three materials from which to choose, simple computer codes can be used to sort and rank them. But this numerical approach lacks visual immediacy and the stimulus for creative thinking that a more graphical method allows. We describe this next.

The graphical method. Suppose, for a population of materials, that we plot m_1 against m_2 as suggested by Figure 9.3(a). Each bubble represents a material. (All the variables in both equations are specified except the material, so the only difference between one bubble and another is the material.) We wish to minimize mass, so the best choices lie somewhere near the bottom left. But where, exactly? The choice if stiffness is paramount and strength is unimportant must surely differ from that if the opposite were true. The line $m_1 = m_2$ separates the chart into two regions. In one, $m_1 > m_2$ and constraint 1 is dominant. In the other, $m_2 > m_1$ and constraint 2 dominates. In region 1 our objective is to minimize m_1, since it is the larger of the two; in region 2, the opposite. This defines a box-shaped selection envelope with its corner on the $m_1 = m_2$ line. The nearer the box is pulled to the bottom left, the smaller is $\tilde{m}$. The best choice is the last material left in the box.

This explains the idea, but there is a better way to implement it. Figure 9.3(a), with m_1 and m_2 as axes, is specific to single values of S, L, and F_f; if these change we need a new chart. Suppose, instead, that we plot the material indices $M_1 = \rho/E$ and $M_2 = \rho/\sigma_y$ that are contained in the performance equations, as shown in Figure 9.3(b). Now each bubble depends only on material

Table 9.1   Selection of a material for a light, stiff, strong tie

Material	ρ (kg/m³)	E (GPa)	σ_y (MPa)	m_1 (kg)	m_2 (kg)	$\tilde{m}$ (kg)
1020 Steel	7850	205	320	1.15	2.45	2.45
6061 Al	2700	70	120	1.16	2.25	2.25
Ti-6-4	4400	115	950	1.15	0.46	**1.15**

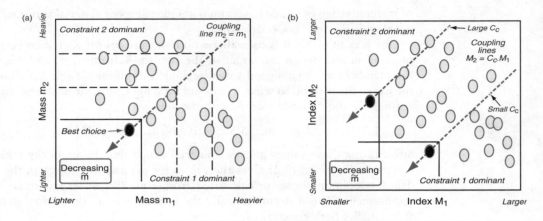

Figure 9.3 The graphical approach to min–max problems. (a): coupled selection using performance metrics (here, mass). (b): a more general approach: coupled selection using material indices and a coupling constant C_c.

properties; its position does not depend on the values of S, L, or F_f. The condition $m_1 = m_2$, substituting from equations (9.1) and (9.2), yields the relationship

$$M_2 = \left(\frac{LS}{F_f}\right) M_1 \tag{9.4a}$$

or, on logarithmic scales

$$\text{Log}(M_2) = \text{Log}(M_1) + \log\left(\frac{LS}{F_f}\right) \tag{9.4b}$$

This describes a line of slope 1, in a position that depends on the value of LS/F_f. We refer to this line as the *coupling line*, and to LS/F_f as the *coupling constant*, symbol C_c. The selection strategy remains the same: a box, with its corner on the coupling line, is pulled down towards the bottom left. But the chart is now more general, covering all values of S and F_f. Changing either one of these, or the geometry of the component (here described by L) moves the coupling line and changes the selections.

Stated more formally, the steps are these:

(a) Express the objective as an equation.
(b) Eliminate the free variable using each constraint in turn, giving sets of performance equations with the form:

$$P_1 = f_1(F) \cdot g_1(G) \cdot M_1 \tag{9.5a}$$
$$P_2 = f_2(F) \cdot g_2(G) \cdot M_2 \tag{9.5b}$$
$$P_3 = f_3(F) \cdots \text{etc.}$$

where the f_s and g_s are expressions containing the functional requirements F and geometry G, and M_1 and M_2 are material indices. We seek the material that gives the minimum value of the quantity

$$\tilde{P} = \max(P_1,\, P_2,\, P_3, \ldots)$$

(c) If the first constraint is the most restrictive (i.e. it is the *active* constraint), the performance is given by equation (9.5a), and this is optimized by seeking materials with the best (here meaning the smallest) values of M_1. When the second constraint is the active one, the performance equation is given by equation (9.5b) and the best values of M_2 must be sought. And so on.

A chart with axes of M_1 and M_2 can be divided into two domains in each of which one constraint is active, the other inactive, separated by the line along which the equations (9.5a) and (9.5b) are equal. Equating them and rearranging gives:

$$M_2 = \left[\frac{f_1(F) \cdot g_1(G)}{f_2(F) \cdot g_2(G)} \right] \cdot M_1 \tag{9.6a}$$

or

$$M_2 = [C_c] \cdot M_1 \tag{9.6b}$$

This equation couples the two indices M_1 and M_2; hence the name "coupling equation". The quantity in square brackets — the coupling constant, C_c — is fixed by the specification of the design. Change in the value of the functional requirements F or the geometry G changes the coupling constant, shifts the line, moves the box, and changes the selection.

Worked examples are given in Chapter 10.

9.3 Conflicting objectives, penalty-functions, and exchange constants

Real-life materials selection almost always requires that a compromise be reached between conflicting objectives. Three crop up all the time. They are these.

- Minimizing *mass* — a common target in designing things that move or have to be moved, or oscillate, or must response quickly to a limited force (think of aerospace and ground transport systems).
- Minimizing *volume* — because less material is used, and because space is increasingly precious (think of the drive for ever thinner, smaller mobile phones, portable computers, MP3 players . . . and the need to pack more and more functionality into a fixed volume).
- Minimizing *cost* — profitability depends on the difference between cost and value (more on this in a moment); the most obvious way to increase this difference is to reduce cost.

To this we must now add a fourth objective;

- Minimizing *environmental impact* — the damage to our surroundings caused by product manufacture and use.

There are, of course, other objectives, specific to particular applications. Some are just one of the four above expressed in different words. The objective of maximizing *power-to-weight ratio* translates into minimizing mass for a given power output. Maximizing *energy storage* in a spring, battery or flywheel means minimizing the volume for a given stored energy. Some can be quantified in engineering terms, such as maximizing reliability (although this can translate into achieving a given reliability at a minimum cost) and others cannot, such as maximizing consumer appeal, an amalgam of performance, styling, image, and marketing.

So we have four common objectives, each characterized by a performance metric P_i. At least two are involved in the design of almost any product. Conflict arises because the choice that optimizes one objective will not, in general, do the same for the others; then the best choice is a compromise, optimizing none, but pushing all as close to their optima as their interdependence allows. And this highlights the central problem: how is mass to be compared with cost, or volume with environmental impact? Unlike the Ps of the last section, each is measured in different units; they are incommensurate. We need strategies to deal with this. They come in a moment. First, some definitions.

Trade-off strategies. Consider the choice of material to minimize both mass (performance metric P_1) and cost (performance metric P_2) while also meeting a set of constraints such as a required maximum service temperature, or corrosion resistance in a certain environment. Following the standard terminology of optimizations theory, we define a *solution* as a viable choice of material, meeting all the constraints but not necessarily optimal by either of the objectives. Figure 9.4 is a plot of P_1 against P_2 for alternative solutions, each bubble describing a solution. The solutions that minimize P_1 do not minimize P_2, and *vice versa*. Some solutions, such as that at **A**, are far from optimal — all the solutions in the box attached to have lower values of both P_1 and P_2. Solutions like **A** are said to be *dominated* by others. Solutions like those at **B** have the characteristic that no other solutions exists with lower values of both P_1 and P_2. These are said to be *non-dominated* solutions. The line or surface on which they lie is called the non-dominated or optimal *trade-off surface*. The values of P_1 and P_2 corresponding to the non-dominated set of solutions are called the *Pareto set*.

There are three strategies for progressing further. The solutions on or near the trade-off surface offer the best compromise; the rest can be rejected. Often, this is enough to identify a short-list, using intuition to rank them, for which supporting information can now be sought (Strategy 1). Alternatively (Strategy 2)

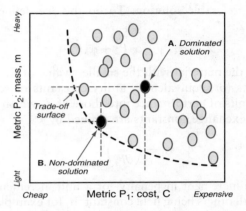

Figure 9.4   Multiple objectives: we seek the material that co-minimizes mass and cost. Mass and cost for a component made from alternative material choices. Each bubble is a *solution*—a material choice that meets all constraints. The trade-off surface links non-dominated solutions.

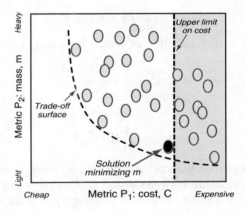

Figure 9.5   The trade-off plot with a simple constraint imposed on cost. The solution with the lowest mass can now be read off, but it is not a true optimization.

one objective can be reformulated as a constraint, as illustrated in Figure 9.5. Here an upper-limit is set on cost; the solution that minimizes the other constraint can then be read off. But this is cheating: it is not a true optimization. To achieve that, we need Strategy 3: that of *penalty functions*.

Penalty functions. The trade-off surface identifies the subset of solutions that offer the best compromises between the objectives. Ultimately, though, we want a single solution. One way to do this is to aggregate the various objectives into a single objective function, formulated such that its minimum defines the most preferable solution. To do this we define a locally linear penalty function[1] Z:

$$Z = \alpha_1 P_1 + \alpha_2 P_2 + \alpha_3 P_3 + \cdots \tag{9.7}$$

The best choice is the material with the smallest value of Z. The αs are called exchange constants (or, equivalently, utility constants or scaling constants); they convert the units of performance into those of Z, which is usually that of currency (\$). The exchange constants are defined by

$$\alpha_i = \left(\frac{\partial Z}{\partial P_i} \right)_{P_j, j \neq i} \tag{9.8}$$

They measure the increment in penalty for a unit increment in a given performance metric, all others being held constant. If, for example, the metric P_2 is mass m, then α_2 is the change in Z associated with unit increase in m.

Frequently one of the objectives to be minimized is cost, C, so that $P_1 = C$. Then it makes sense to measure penalty in units of currency. With this choice, unit change in C gives unit change in Z, with the result that $\alpha_1 = 1$ and equation (9.7) becomes

$$Z = C + \alpha_2 P_2 + \alpha_3 P_3 + \cdots \tag{9.9}$$

Consider now the earlier example in which $P_1 = \text{cost}$, C, and $P_2 = \text{mass}$, m, so that

$$Z = C + \alpha m$$

or

$$m = -\frac{1}{\alpha} C + \frac{1}{\alpha} Z \tag{9.10}$$

Then α is the change in Z associated with unit increase in m. Equation (9.10) defines a linear relationship between m and C. This plots as a family of parallel penalty-lines each for a given value of Z, as shown in Figure 9.6, left. The slope of the lines is the reciprocal of the exchange constant, $-1/\alpha$. The value of Z decreases towards the bottom left: best choices lie there. The optimum solution is the one nearest the point at which a penalty line is tangential to the trade-off surface, since it is the one with the smallest value of Z. Narrowing the choice to just one candidate at this stage is not sensible — we do not yet know what the search for supporting information will reveal. Instead, we choose the subset of solutions that lie closest to the tangent-point.

One little quirk. Almost all the material selection charts use logarithmic scales, for very good reasons (Chapter 4). A linear relationship, plotted on log

---

[1] Also called a *value function* or *utility function*. The method allows a local minimum to be found. When the search space is large, it is necessary to recognize that the values of the exchange constants α_i may themselves depend on the values of the performance metrics P_i.

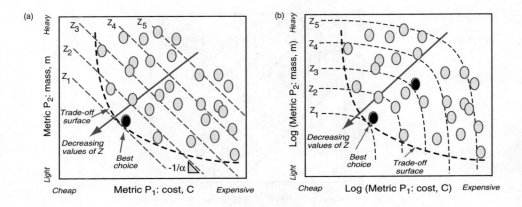

Figure 9.6 (a): the penalty function Z superimposed on the trade-off plot. The contours of Z have a slope of $-1/\alpha$. The contour that is tangent to the trade-off surface identifies the optimum solution. (b): the same, plotted on logarithmic scales; the linear relation now plots as curved lines.

scales, appears as a curve, as shown in Figure 9.6(b). But the procedure remains the same: the best candidates are those nearest the point at which one of these curves just touches the trade-off surface.

Relative penalty functions. When, as is often the case, we seek a better material for an *existing* application, it is more helpful to compare the new material choice with the existing one. To do this we define a *relative* penalty function

$$Z^* = \frac{C}{C_o} + \alpha^* \frac{m}{m_o} \tag{9.11}$$

in which the subscript "o" means properties of the existing material and the asterisk * on Z^* and α^* is a reminder that both are now dimensionless. The relative exchange constant α^* measures the fractional gain in value for a given fractional gain in performance. Thus $\alpha^* = 1$ means that, at constant Z,

$$\frac{\Delta C}{C_o} = -\frac{\Delta m}{m_o}$$

and that halving the mass is perceived to be worth twice the cost.

Figure 9.7 shows the relative trade-off plot, here on linear scales. The axes are C/C_o and m/m_o. The material currently used in the application appears at the co-ordinates $(1,1)$. Solutions is sector **A** are both lighter and cheaper than the existing material, those in sector **B** are cheaper but heavier, those in sector **C** are lighter but more expensive, and those in **D** are uninteresting. Contours of Z^* can be plotted onto the figure. The contour that is tangent to the relative trade-off surface again identifies the optimum search area. As before, when

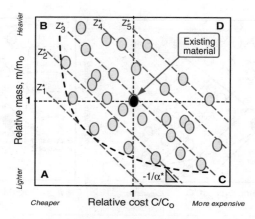

Figure 9.7 A relative trade-off plot, useful when exploring substitution of an existing material with the aim of reducing mass, or cost, or both. The existing material sits at the co-ordinates (1,1). Solutions in sector **A** are both lighter and cheaper.

logarithmic scales are used the contours of Z^* become curves. The case studies of Chapter 10 make use of relative penalty functions.

So if values for the exchange constants are known, a completely systematic selection is possible. But that is a big "if". It is discussed next.

Values for the exchange constants, α_i. An exchange constant is a measure of the penalty of unit increase in a performance metric, or—more easily understood—it is the value or "utility" of a unit decrease in the metric. Its magnitude and sign depend on the application. Thus the utility of weight saving in a family car is small, though significant; in aerospace it is much larger. The utility of heat transfer in house insulation is directly related to the cost of the energy used to heat the house; that in a heat-exchanger for power electronics can be much higher. The utility can be real, meaning that it measures a true saving of cost. But it can, sometimes, be perceived, meaning that the consumer, influenced by scarcity, advertising or fashion, will pay more or less than the true value of these metrics.

In many engineering applications the exchange constants can be derived approximately from technical models for the life-cost of a system. Thus the utility of weight saving in transport systems is derived from the value of the fuel saved or that of the increased payload, evaluated over the life of the system. Table 9.2 gives approximate values for α. The most striking thing about them is the enormous range. The exchange constant depends in a dramatic way on the application in which the material will be used. It is this that lies behind the difficulty in adopting aluminum alloys for cars despite their universal use in aircraft, the much greater use of titanium alloys in military than in civil aircraft, and the restriction of beryllium to use in space vehicles.

Table 9.2   Exchange constants α for the mass—cost trade-of for transport systems

Sector: transport systems	Basis of estimate	Exchange constant, α (US$/kg)
Family car	Fuel saving	1–2
Truck	Payload	5–20
Civil aircraft	Payload	100–500
Military aircraft	Payload, performance	500–1000
Space vehicle	Payload	3000–10,000

Exchange constants can be estimated approximately in various ways. The cost of launching a payload into space lies in the range $3000 to $10,000/kg; a reduction of 1 kg in the weight of the launch structure would allow a corresponding increase in payload, giving the ranges of α shown in Table 9.2. Similar arguments based on increased payload or decreased fuel consumption give the values shown for civil aircraft, commercial trucks, and automobiles. The values change with time, reflecting changes in fuel costs, legislation to increase fuel economy and such like. Special circumstances can change them dramatically—an aero-engine builder who has guaranteed a certain power/weight ratio for his engine, may be willing to pay more than $1000 to save a kilogram if it is the only way in which the guarantee can be met, or (expressed as a penalty) will face a cost of $1000 kg if it is not.

These values for the exchange constant are based on engineering criteria. More difficult to assess are those based on perceived value. That for the weight/cost trade-off for a bicycle is an example. To the enthusiast, a lighter bike is a better bike. Figure 9.8 shows just how much the cyclist values reduction in weight. It is a trade-off plot of mass and cost of bicycles, using data from a bike magazine. The tangent to the trade-of line at any point gives a measure of the exchange constant: it ranges from $20/kg to $2000/kg, depending on the mass. Does it make sense for the ordinary cyclist to pay $2000 to reduce the mass of the bike by 1 kg when, by dieting, he could reduce the mass of the system (himself plus the bike) by more without spending a penny? Possibly. But mostly it is perceived value. Advertising aims to increase the perceived value of a product, increasing its value without increasing its cost. It influences the exchange constants for family cars and it is the driver for the development of titanium watches, carbon fiber spectacle frames and much sports equipment. Then the value of α is harder to pin down.

There are other circumstances in which establishing the exchange constant can be difficult. An example in that for environmental impact—the damage to the environment caused by manufacture, or use, or disposal of a given product. Minimizing environmental impact has now become an important objective, almost as important as minimizing cost. Ingenious design can reduce the first without driving the second up too much. But how much is unit decrease in impact worth? Until an exchange constant is agreed or imposed, it is difficult for the designer to respond.

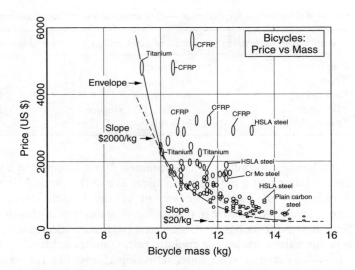

Figure 9.8   A cost–mass trade-off plot for bicycles, using data from a bike magazine. Solutions are labeled with the material of which the frame is made. The tangent to the trade-off surface at any point gives an estimate of the exchange constant. It depends on the application. To a consumer seeking a cheap bike for shopping the value of weight saving is low ($20/kg). To an enthusiast wanting performance, it appears to be high ($2000/kg).

Things, though, are not quite as difficult as they at first appear. Useful engineering decisions can be reached even when exchange constants are imprecisely known, as explained in the next section.

How do exchange constants influence choice? The discreteness of the search space for material selection means that a given solution on the trade-off surface is optimal for a certain range of values of α; outside this range another solution becomes the optimal choice. The range can be large, so any value of exchange constant within the range leads to the same choice of material. This is illustrated in Figure 9.9. For simplicity solutions have been moved so that, in this figure, only three are potentially optimal. For $\alpha < 0.1$, solution **A** is the optimum; for $0.1 < \alpha < 10$, solution **B** is the best choice; and for $\alpha > 10$, it is solution **C**. This information is captured in the bar on the right of Figure 9.9 showing the range of values of α, subdivided at the points at which a change of optimum occurs and labeled with the solution that is optimal in each range.

This suggests a way of extending the visualization to three objectives. It is illustrated in Figures 9.10 and 9.11. The first shows a hypothetical trade-off surface for three performance metrics, one of which is cost. The penalty function takes the form

$$Z = C + \alpha_1 P_1 + \alpha_2 P_2 \tag{9.12}$$

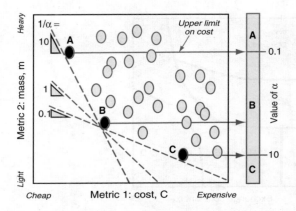

Figure 9.9   It is often the case that a single material (or subset of materials) is optimal over a wide range of values of the exchange constant. Then approximate values for exchange constants are sufficient to reach precise conclusions about the choice of materials.

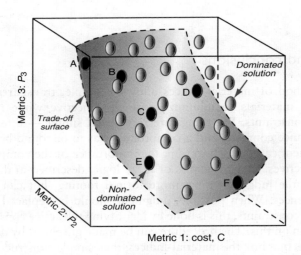

Figure 9.10   A trade-off plot for three objectives: Cost, C, and P_1 and P_2. The black eggs are solutions that lie on the trade-off surface, the lighter eggs are dominated solutions.

in which the two exchange constants α_1 and α_2 relate P_1 and P_2 to cost C. The segments of the bar in Figure 9.9 now become areas in Figure 9.11, each defining the range of α_1 and α_2 for which a given solution on the trade-off surface is optimal.

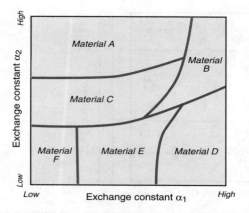

Figure 9.11   An exchange-constant chart showing the optimum material choice from those that lie on the trade-off surface of Figure 9.10, as a function of the two exchange constants. The exchange constants need only be known approximately to identify a segment of the diagram, and thus the choice of material.

9.4 Summary and conclusions

The method of material-indices allows a simple, transparent procedure for selecting materials to minimize a single objective while meeting a set of simple constraints. But things are rarely that simple — different measures of performance compete, and a compromise has to be found between them.

Judgment can be used to rank the importance of the competing constraints and objectives. Weight-factors or fuzzy logic, described in the chapter appendix, put the judgment on a more formal footing, but can also obscure its consequences. When possible, judgment should be replaced by analysis. For multiple constraints, this is done by identifying the active constraint and basing the design on this. The procedure can be made graphical by deriving coupling-equations that link the material-indices; then simple constructions on material-selection charts with the indices as axes identify unambiguously the subset of materials that maximize performance while meeting all the constraints. Compound objectives require the formulation of a penalty function, Z, containing one or more exchange constants; α_i; it allows all objectives to be expressed in the same units (usually cost). Minimizing Z identifies the optimum choice.

When multiple constraints operate, or a compound objective is involved, the best choice of material is far from obvious. It is here that the methods developed here have real power.

9.5 Further reading

Ashby, M.F. (2000) Multi-objective optimization in material design and selection, *Acta Mater.* **48**, pp. 359–369. (*An exploration of the use of trade-off surfaces and utility functions for material selection.*)

Bader, M.G. (1977) Proceedings of ICCM-11, Gold Coast, Australia, Vol. 1: Composites applications and design, ICCM, London. (*An example of trade-off methods applied to the choice of composite systems.*)

Bourell, D.L. (1997) Decision matrices in materials selection, in Dieter G.E. (ed.), *ASM Handbook, Vol. 20: Materials Selection and Design*, ASM International, Materials Park, OH, USA, pp. 291–296. ISBN 0-87170-386-6. (*An introduction to the use of weight-factors and decision matrices.*)

Clark, J.P., Roth, R. and Field, F.R. (1997) Techno-economic issues in material science, *ASM Handbook, Vol. 20: Materials Selection and Design*, ASM International, Materials Park, OH, USA, pp. 255–265. ISBN 0-87170-386-6. (*The authors explore methods of cost and utility analysis, and of environmental issues, in materials selection.*)

Dieter, G.E. (2000) *Engineering Design, a Materials and Processing Approach*, 3rd edition, McGraw-Hill, New York, USA, pp. 150–153 and 255–257. ISBN 0-07-366136-8. (*A well-balanced and respected text, now in its 3rd edition, focusing on the role of materials and processing in technical design.*)

Field, F.R. and de Neufville, R. (1988) Material selection — maximizing overall utility, *Metals Mater.*, **June** 378–382. (*A summary of utility analysis applied to material selection in the automobile industry.*)

Goicoechea, A., Hansen, D.R. and Druckstein, L. (1982) *Multi-Objective Decision Analysis with Engineering and Business Applications*, Wiley, New York, USA. (*A good starting point for the theory of multi-objective decision-making.*)

Keeney, R.L. and Raiffa, H. (1993) *Decisions with Multiple Objectives: Preferences and Value Tradeoffs*, 2nd edition, Cambridge University Press, Cambridge, UK. ISBN 0-521-43883-7. (*A notably readable introduction to methods of decision-making with multiple, competing objectives.*)

Landru, D. (2000) Aides informatisées à la sélectiin des matériaux et des procédés dans la conception des pièces de structure, PhD Thesis, L'Institut National Polytechnique de Grenoble, France. (*A wide ranging exploration of selection methods for materials and processes, outlining novel ways of estimating exchange constants.*)

Papalambros, P.Y. and Wilde, D.J. (2000) *Principles of Optimal Design, Modeling and Computation*, 2nd edition, Cambridge University Press, Cambridge, UK. ISBN 0-521-62727-3. (*An introduction to methods of optimal engineering design.*)

Pareto, V. (1906) *Manuale di Economica Politica*, Societa Editrice Libraria, Milano, Italy, translated into English by Schwier, A.S. (1971) as "*Manual of Political Economics*", Macmillan, New York, USA. (*A book that is much quoted but little read; the origin of the concept of a trade-off surface as an approach to multi-objective optimization.*)

Sawaragi, Y., Nakayama, H. and Tanino, T. (1985) *Theory of Multi-Objective Optimisation*, Academic Press Inc. Orlando, Florida, USA. ISBN 0-12-620370-9.

(Multi-objective optimization in all is gruesome detail. Exhaustive, but not the best place to start.)

Sirisalee, P., Parks, G.T., Clarkson, P.J. and Ashby, M.F. (2003) A new approach to multi-criteria material selection in engineering design. International Conference On Engineering Design, ICED '03 Stockholm, August 19–21, 2003. (*The paper in which the idea of exchange-constant charts like that of Figure 9.11 is proposed and explored.*)

Appendix: Traditional methods of dealing with multiple constraints and objectives

Suppose you want a component with a required stiffness (constraint 1) and strength (constraint 2) and it must be as light as possible (one objective). You could choose materials with high modulus E for stiffness, and then the subset of these that have high elastic limits σ_y for strength, and the subset of those that have low density ρ for light weight. Then again, if you wanted a material with a required stiffness (one constraint) that was simultaneously as light (objective 1) and as cheap (objective 2) as possible, you could apply the constraint and then find the subset of survivors that were light and the subset of *those* survivors that were cheap. Some selection systems work that way, but it is not a good idea because there is no guidance in deciding relative importance of the limits on stiffness, strength, weight and cost. This is not a trivial difficulty: it is exactly this relative importance that makes aluminum the prime structural material for aerospace and steel that for ground-based structures.

These problems of relative importance are old; engineers have sought methods to overcome them for at least a century. The traditional approach is that of assigning *weight-factors* to each constraint and objective and using them to guide choice in ways that are summarized below. The up-side: experienced engineers can be good at assessing relative weights. The down-side: the method relies on judgment. In assessing weights judgments can differ, and there are subtler problems, one of them discussed below. For this reason, this chapter has focused on systematic methods. But one should know about the traditional methods because they are still widely used.

The method of weight-factors. Weight-factors seek to quantify judgment. The method works like this. The key properties or indices are identified and their values M_i are tabulated for promising candidates. Since their absolute values can differ widely and depend on the units in which they are measured, each is first scaled by dividing it by the largest index of its group, $(M_i)_{max}$, so that the largest, after scaling, has the value 1. Each is then multiplied by a weight factor, w_i, with a value between 0 and 1, expressing its relative importance for the performance of the component. This give a weighted index W_i:

$$W_i = w_i \frac{M_i}{(M_i)_{max}} \tag{9.13}$$

For properties that are not readily expressed as numerical values, such as weldability or wear resistance, rankings such as A to E are expressed instead by a numeric rating,

$A = 5$ (very good) to $E = 1$ (very bad), then dividing by the highest rating value as before. For properties that are to be minimized, like corrosion rate, the scaling uses the minimum value $(M_i)_{min}$, expressed in the form

$$W_i = w_i \frac{(M_i)_{min}}{M_i} \qquad (9.14)$$

The weight-factors w_i are chosen such that they add up to 1, that is: $w_i < 1$ and $\Sigma w_i = 1$. There are numerous schemes for assigning their values (see Further reading); all require, in varying degrees, the use of judgment. The most important property is given the largest w, the second most important, the second largest and so on. The W_is are calculated from equations (9.12) and (9.13) and summed. The best choice is the material with the largest value of the sum

$$W = \Sigma_i W_i \qquad (9.15)$$

Sounds simple, but there are problems, some obvious (like that of subjectivity in assigning the weights), some more subtle. Here is an example: the selection of a material to make a light component (low ρ) that must be strong (high σ_y). Table 9.3 gives values for four possible candidates. Weight, in our judgment, is more important than strength, so we assign it the weight factor

$$w_1 = 0.7$$

That for strength is then

$$w_2 = 0.25$$

Normalize the index values (as in equations (9.13) and (9.14)) and sum them (equation (9.15)) to give W. The second last column of the table shows the result: beryllium wins easily; Ti-6-4 comes second, 6061 aluminum third. But observe what happens if beryllium (which is very expensive and can be toxic) is omitted from the selection, leaving only the first three materials. The same procedure now leads to the values of W in the last column: 6061 aluminum wins, Ti-6-4 is second. Removing one, non-viable, material from the selection has reversed the ranking of those that remain. Even if the weight factors could be chosen with accuracy, this dependence of the outcome on the population from which the choice is made is disturbing. The method is inherently unstable, sensitive to irrelevant alternatives.

Table 9.3   Example of use of weight factors

Material	ρ (Mg/m³)	σ_y (MPa)	W (inc. Be)	W (excl. Be)
1020 Steel	7.85	320	0.27	0.34
6061 Al (T4)	2.7	120	0.55	**0.78**
Ti-6-4	4.4	950	0.57	0.71
Beryllium	1.86	170	**0.79**	—

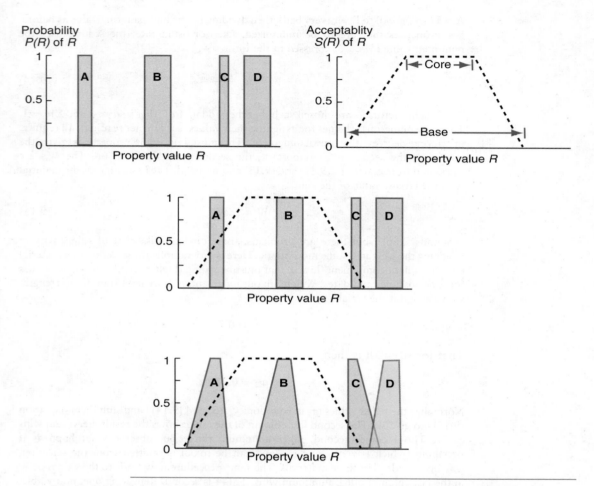

Figure 9.12   Fuzzy selection methods. Sharply-defined properties and a fuzzy selection criterion, shown in the top row, are combined to give weight-factors for each material, center. The properties themselves can be given fuzzy ranges, as shown at the bottom.

Fuzzy logic. Fuzzy logic takes weight-factors one step further. Figure 9.12 at the upper left, shows the probability $M(R)$ of a material having a property with a value R in a given range. Here the property has a well-defined range for each of the four materials **A**, **B**, **C**, and **D** (the values are *crisp* in the terminology of the field). The selection criterion, shown at the top right, identifies the range of R that is sought for the properties, and it is *fuzzy*, that is to say, it has a well-defined *core* defining the ideal range sought for the property, with a wider *base*, extending the range to include boundary-regions in which the value of the property is allowable, but with decreasing acceptability as the edges of the base are approached. This defines the probability $S(R)$ of a choice being a successful one.

The superposition of the two figures, shown at the center of Figure 9.12 illustrates a single selection stage. Desirability is measured by the product $M(R) \cdot S(R)$. Here material B is fully acceptable—it acquires a weight of 1. Material A is acceptable but with a lower weight, here 0.5; C is acceptable with a weight of roughly 0.25, and D is unacceptable—it has a weight of 0. At the end of the first selection stage, each material in the database has one weight-factor associated with it. The procedure is repeated for successive stages, which could include indices derived from other constraints and objectives. The weights for each material are aggregated—by multiplying them together, for instance—to give each a super-weight with a value between 0 (totally unacceptable) and 1 (fully acceptable by all criteria). The method can be refined further by giving fuzzy boundaries to the material properties or indices as well as to the selection criteria, as illustrated in the lower part of Figure 9.12. Techniques exist to choose the positions of the cores and the bases, but despite the sophistication the basic problem remains: the selection of the ranges $S(R)$ is a matter of judgment.

Weight factors and fuzzy methods all have merit when more rigorous analysis is impractical. They can be a good first step. But if you really want to identify the best material for a complex design, you need the methods of Sections 9.2 and 9.3.

Chapter 10

Case studies—multiple constraints and conflicting objectives

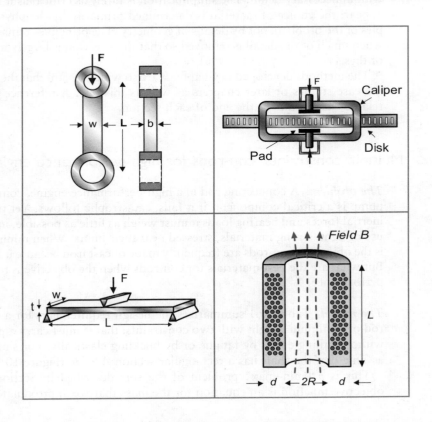

10.1 Introduction and synopsis

These case studies illustrate how the techniques described in the last chapter really work. They are deliberately simplified to avoid clouding the illustration with unnecessary detail. The simplification is rarely as critical as it may at first appear: the choice of material is determined primarily by the physical principles of the problem, not by details of geometry. The principles remain the same when much of the detail is removed so that the selection is largely independent of these.

The methods developed in Chapter 9 are so widely useful that they appear in the case studies of later chapters as well as this one. A reference is made to related case studies at the end of each section.

10.2 Multiple constraints: con-rods for high-performance engines

The problem. A connecting rod in a high-performance engine, compressor, or pump is a critical component: if it fails, catastrophe follows. Yet to minimize inertial forces and bearing loads it must weigh as little as possible, implying the use of light, strong materials, stressed near their limits. When minimizing cost is the objective, con-rods are frequently made of cast iron because it is so cheap. But what are the best materials for con-rods when the objective is to maximize performance?

The model. Table 10.1 summarizes the design requirements for a connecting rod of minimum weight with two constraints: that it must carry a peak load F without failing either by fatigue or by buckling elastically. For simplicity, we assume that the shaft has a rectangular section $A = bw$ (Figure 10.1).

This is a "min–max" problem of the sort described in Section 9.2. The objective function is an equation for the mass that we approximate as

$$m = \beta A L \rho \qquad (10.1)$$

Table 10.1   The design requirements: connecting rods

Function	Connecting rod for reciprocating engine or pump
Constraints	• Must not fail by high-cycle fatigue, or
	• Must not fail by elastic buckling
	• Stroke, and thus con-rod length *L*, specified
Objective	Minimize mass
Free variables	• Cross-section A
	• Choice of material

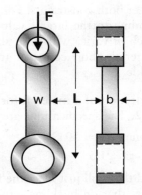

Figure 10.1 A connecting rod. It must not buckle or fail by—example of multiple constraints.

where L is the length of the con-rod, ρ the density of the material of which it is made, A the cross-section of the shaft, and β a constant multiplier to allow for the mass of the bearing housings.

The fatigue constraint requires that

$$\frac{F}{A} \leq \sigma_e \tag{10.2}$$

where σ_e is the endurance limit of the material of which the con-rod is made. (Here, and elsewhere, we omit the safety factor that would normally enter an equation of this sort, since it does not influence the selection.) Using equation (10.2) to eliminate A in (10.1) gives the mass of a con-rod that will just meet the fatigue constraint:

$$m_1 = \beta FL\left(\frac{\rho}{\sigma_e}\right) \tag{10.3}$$

containing the material-index $M_l = \rho/\sigma_e$.

The buckling constraint requires that the peak compressive load F does not exceed the Euler buckling load:

$$F \leq \frac{\pi^2 EI}{L^2} \tag{10.4}$$

with $(I = b^3 w/12)$ (Appendix A). Writing $b = \alpha w$, where α is a dimensionless "shape-constant" characterizing the proportions of the cross-section, and eliminating A from equation (10.1) gives a second equation for the mass

$$m_2 = \beta\left(\frac{12F}{\alpha\pi^2}\right)^{1/2} L^2\left(\frac{\rho}{E^{1/2}}\right) \tag{10.5}$$

containing the second material index $M_2 = \rho/E^{1/2}$.

The con-rod, to be safe, must meet both constraints. For a given length, L, the active constraint is the one leading to the largest value of the mass, m. Figure 10.2 shows the way in which m varies with L (a sketch of equations (10.3) and (10.5)), for a single material. Short con-rods are liable to fatigue failure, long ones are prone to buckle.

The selection. Consider first the selection of a material for the con-rod from among the limited list of Table 10.2. The specifications are

$$L = 200\,\text{mm}, \quad F = 50\,\text{kN}$$
$$\alpha = 0.8, \quad \beta = 1.5$$

The table lists the mass m_1 of a rod that will just meet the fatigue constraint, and the mass m_2 that will just meet that on buckling (equations (10.3) and (10.5)). For three of the materials the active constraint is that of fatigue; for two it is that of buckling. The quantity $\tilde{m}$ in the last column of the table is the larger of m_1 and m_2 for each material; it is the lowest mass that meets both constraints. The material offering the lightest rod of that with the smallest value of $\tilde{m}$. Here it is the titanium alloy Ti 6Al 4V. The metal-matrix composite Duralcan 6061–20% SiC is a close second. Both weigh less than half as much as a cast-iron rod.

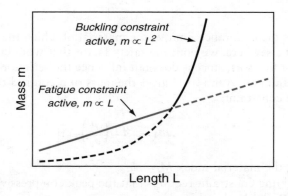

Figure 10.2   The equations for the mass of the con-rod are shown schematically as a function of L.

Table 10.2   Selection of a material for the con-rod

Material	ρ (kg/m^3)	E (GPa)	σ_e (MPa)	m_1 (kg)	m_2 (kg)	$\tilde{m} = max\,(m_1, m_2)$ (kg)
Nodular cast iron	7150	178	250	0.43	0.22	0.43
HSLA steel 4140 (o.q. T-315)	7850	210	590	0.20	0.28	0.28
Al S355.0 casting alloy	2700	70	95	0.39	0.14	0.39
Duralcan Al–SiC(p) composite	2880	110	230	0.18	0.12	0.18
Titanium 6Al 4V	4400	115	530	0.12	0.17	0.17

Well, that is one way to use the method, but it is not the best. First, it assumes some "pre-selection" procedure has been used to obtain the materials listed in the table, but does not explain how this is to be done; and second, the results apply only to the values of F and L listed above — change these, and the selection changes. If we wish to escape these restrictions, the graphical method is the way to do it.

The mass of the rod that will survive both fatigue and buckling is the larger of the two masses m_1 and m_2 (equations (10.3) and (10.5)). Setting them equal gives the equation of the coupling line (defined in Section 9.2):

$$M_2 = \left[\left(\frac{\alpha \pi^2}{12} \cdot \frac{F}{L^2} \right)^{1/2} \right] \cdot M_1 \tag{10.6}$$

The quantity in square brackets is the coupling constant C introduced in Section 9.2. It contains the quantity F/L^2 — the "structural loading coefficient" of Section 5.6.

Materials with the optimum combination of M_1 and M_2 are identified by creating a chart with these indices as axes. Figure 10.3 illustrates this, using a database of light alloys, but including cast iron for comparison. Coupling lines for two values of F/L^2 are plotted on it, taking $\alpha = 0.8$. Two extreme selections are shown, one isolating the best subset when the structural loading

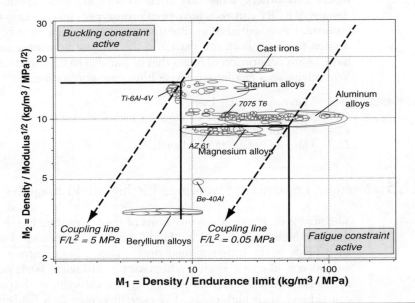

Figure 10.3   Over-constrained design leads to two or more performance indices linked by coupling equations. The diagonal broken lines show the coupling equation for two extreme values of F/L^2. The selection lines intersect on the appropriate coupling line given the box-shaped search areas. (Chart created with CES 4, 2004.)

Table 10.3   Materials for high-performance con-rods

Material	Comment
Magnesium alloys	AZ61 and related alloys offer good all-round performance
Titanium alloys	Ti-6-4 is the best choice for high F/L^2
Beryllium alloys	The ultimate choice, but difficult to process and very expensive
Aluminum alloys	Cheaper than titanium or magnesium, but lower performance

coefficient F/L^2 is high, the other when it is low. Beryllium and its alloys emerge as the best choice for all values of C within this range. Leaving them aside, the best choice when F/L^2 is large ($F/L^2 = 5$ MPa) are titanium alloys such as Ti 6Al 4V. For the low value ($F/L^2 = 0.05$ MPa), magnesium alloys such as AZ61 offer lighter solutions than aluminum or titanium. Table 10.3 lists the conclusions.

Postscript. Con-rods have been made from all the materials in the table: aluminum and magnesium in road cars, titanium and (rarely) beryllium in racing engines. Had we included CFRP in the selection, we would have found that it, too, performs well by the criteria we have used. This conclusion has been reached by others, who have tried to do something about it: at least three designs of CFRP con-rods have been prototyped. It is not easy to design a CFRP con-rod. It is essential to use continuous fibers, which must be wound in such a way as to create both the shaft and the bearing housings; and the shaft must have a high proportion of fibers that lie parallel to the direction in which F acts. You might, as a challenge, devise how you would do it.

Related case studies

10.3 Multiple constraints: windings for high-field magnets

The problem. Physicists, for reasons of their own, like to see what happens to things in high magnetic fields. "High" means 50 T or more. The only way to get such fields is the old-fashioned one: dump a huge current through a wire-wound coil like that shown schematically in Figure 10.4; neither permanent magnets (practical limit: 1.5 T), nor super-conducting coils (present limit: 25 T) can achieve such high fields. The current generates a field-pulse that lasts as long as the current flows. The upper limits on the field and its duration are set by the material of the coil itself: if the field is too high, the coil blows itself apart; if too long, it melts. So choosing the right material for the coil is critical. What should it be? The answer depends on the pulse-length.

Pulsed fields are classified according to their duration and strength as in Table 10.4. The requirements for the survival of the magnet producing them are summarized in Table 10.5. There is one objective—to maximize the field—with two constraints deriving from the requirement of survivability: that the windings are strong enough to withstand the radial force on them caused by the field, and that they do not heat up too much.

The model. Detailed modeling gets a little complicated, so let us start with some intelligent guesses (IGs). First, if the windings must carry load (the first

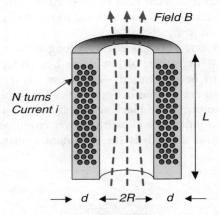

Figure 10.4 Windings for high-powered magnets. There are two constraints: the magnet must not overheat, and it must not fail under the radial magnetic forces.

Table 10.4 Duration and strengths of pulsed fields

Classification	Duration	Field strength (T)
Continuous	1 s–∞	<30
Long	100 ms–1 s	30–60
Standard	10–100 ms	40–70
Short	10–1000 μs	70–80
Ultra-short	0.1–10 μs	>100

Table 10.5 The design requirements: high field magnet windings

Function	Magnet windings
Constraints	• No mechanical failure • Temperature rise <100°C • Radius R and length L of coil specified
Objective	Maximize magnetic field
Free variables	Choice of material for the winding

constraint) they must be strong — the higher the strength, the greater the field they can tolerate. So (IG 1) we want materials with a high elastic limit, σ_y. Second, a current i flowing for a time t_p through a coil of resistance R_e generates $i^2 R_e t_p$ joules of energy, and if this takes place in a volume V, the temperature rise is

$$\Delta T = \frac{i^2 R_e t_p}{V C_p \rho}$$

where C_p is the specific heat of the material and ρ its density. So (IG 2) to maximize the current (and thus the field B) we need materials with low values of $R_e/C_p \rho$ or since resistance R_e is proportional to resistivity ρ_e for a fixed coil geometry, materials with low $\rho_e/C_p \rho$.

Both guesses are correct. This has got us a long way; a simple search for material with high σ_y — or rather, of low $M_1 = 1/\sigma_y$ (since we must express objectives in a form to be minimized) — and low $M_2 = \rho_e/C_p \rho$ will find a sensible subset. The two are plotted in Figure 10.5 for some 1200 metals and alloys (ignore for the moment the black selection boxes). The materials with the best combination of indices lie alone the lower envelope of the populated region. For a short pulse strength is the dominant constraint and we need

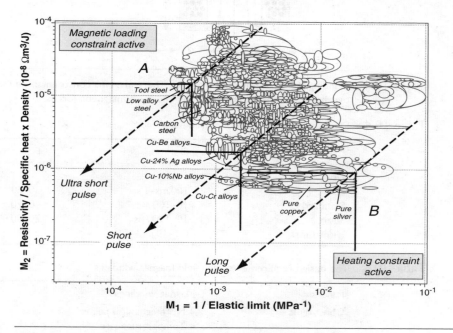

Figure 10.5   The two material groups that determine the choice of material for winding of high powered magnets or electric motors. The axes are the two "guesses" made in the text — the modeling confirms the choice and allows precise positioning of the selection lines for a given pulse duration. (Chart created with CES 4, 2004.)

materials with low M_1; those near **A** are the best choice. For a long pulse heating is the dominant constraint, and materials near **B**, with low M_2, are the answer.

This is progress, and it may be enough. If we want greater resolution, we must abandon guesswork (intelligent though it was) and apply min–max methods, requiring more detailed modeling—they lead to the selection boxes on Figure 10.6. The modeling gets a bit involved—if it looks too grim, skip to the next section, headed "The selection".

Consider first destruction by magnetic loading. The field, B (units: weber/m²), in a long solenoid like that of Figure 10.4 is:

$$B = \frac{\mu_0 Ni}{L} \cdot \lambda_f \cdot F(\alpha, \beta) \tag{10.7}$$

where μ_0 is the permeability of air ($4\pi \times 10^{-7}$ Wb/A m), N is the number of turns, i is the current, L is the length of the coil, λ_f is the fill-factor that accounts for the thickness of insulation (λ_f = cross-section of conductor/cross-section of coil), and $F(\alpha, \beta)$ is a geometric constant (the "shape factor") that depends on the proportions of the magnet (defined on Figure 10.4), the value of which need not concern us.

The field creates a force on the current-carrying coil. It acts radially outwards, rather like the pressure in a pressure vessel, with a magnitude

$$p = \frac{B^2}{2\mu_0 \cdot F(\alpha, \beta)} \tag{10.8}$$

though it is actually a body force, not a surface force. The pressure generates a stress σ in the windings and their casing

$$\sigma = \frac{pR}{d} = \frac{B^2}{2\mu_0 \cdot F(\alpha, \beta)} \cdot \frac{R}{d} \tag{10.9}$$

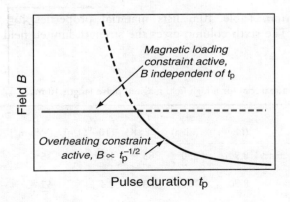

Figure 10.6   The two equations for B are sketched, indicating the active constraint.

This must not exceed the yield strength σ_y of the windings, giving the first limit on B:

$$B_1 \leq \left(\frac{2\mu_0 d\sigma_y \cdot F(\alpha, \beta)}{R}\right)^{1/2} \quad (10.10)$$

The field is maximized by maximizing σ_y, that is, by minimizing $M_1 = 1/\sigma_y$, vindicating IG 1.

Now consider destruction by overheating. High-powered magnets are initially cooled in liquid nitrogen to $-196°C$ in order to reduce the resistant of the windings; if the windings warm above room temperature, the resistance, R_e, in general, becomes too large. The entire energy of the pulse, $\int i^2 R_e \, dt \approx i^2 \overline{R}_e t_p$ is converted into heat (here $\overline{R}_e$ is the average of the resistance over the heating cycle and t_p is the length of the pulse); and since there is insufficient time for the heat to be conducted away, this energy causes the temperature of the coil to rise by ΔT, where

$$\Delta T = \frac{i^2 \overline{R}_e t_p}{C_p \rho V} = \frac{B^2}{\mu_0{}^2} \cdot \frac{\rho_e t_p}{d^2 C_p \rho} \quad (10.11)$$

Here ρ_e is the resistivity of the material of the windings, V its volume, C_p its specific heat (J/kg.K) and ρ its density. If the upper limit for the temperature is 200 K, $\Delta T_{max} \leq 100$ K, giving the second limit on B:

$$B_2 \leq \left(\frac{\mu_0{}^2 d^2 C_p \rho \lambda_f \Delta T_{max}}{t_p \rho_e}\right)^{1/2} F(\alpha, \beta) \quad (10.12)$$

The field is maximized by minimizing $M_2 = \rho_e/C_p\rho$, in accord with IG 2. The two equations for B are sketched as a function of pulse-time t_p in Figure 10.6. For short pulses, the strength constraint is active; for long ones, the heating constraint is dominant.

The selection. Table 10.6 lists material properties for three alternative windings. The sixth column gives the strength-limited field strength, B_1; the

Table 10.6   Selection of a material for a high-field magnet, pulse length 10 ms

Material	ρ (Mg/m³)	σ_y (MPa)	C_p (J/kg.K)	ρ_e ($10^{-8}\Omega.m$)	B_1 (Wb/m²)	B_2 (Wb/m²)	$\tilde{B}$ (Wb/m²)
High-conductivity copper	8.94	250	385	1.7	**35**	113	35
Cu–15% Nb composite	8.90	780	368	2.4	**62**	92	**62**
HSLA steel	7.85	1600	450	25	89	**30**	30

seventh column, the heat-limited field B_2 evaluated for the following values of the design requirements:

$$t_p = 10\,\text{ms}, \quad \lambda_f = 0.5, \quad \Delta T_{max} = 100\,\text{K}$$
$$F(\alpha, \beta) = 1, \quad R = 0.05\,\text{m}, \quad d = 0.1\,\text{m}$$

Strength is the active constraint for the copper-based alloys; heating for the steels. The last column lists the limiting field $\tilde{B}$ for the active constraint. The Cu–Nb composites offer the largest $\tilde{B}$.

So far, so good. But we have the same problem that appeared in the last case study — someone pre-selected the three materials in the table; surely there must be others? And the choice we reached is specific to a magnet with the dimensions listed above and a pulse time t_p of 10 ms. What happens if we change these? We need the graphical method.

The cross-over point in Figure 10.6 is that at which equations (10.10) and (10.12) are equal, giving the coupling the line

$$M_2 = \left[\frac{\mu_0 R d \lambda_f \, F(\alpha, \beta) \, \Delta T_{max}}{2 t_p} \right] \cdot M_1 \qquad (10.13)$$

The quantity in square brackets is the coupling constant C; it depends on the pulse length, t_p.

Now back to Figure 10.5. The axes, as already said, are the two indices M_1 and M_2. Three selections are shown, one for ultra short-pulse magnets, the other two for longer pulses. Each selection box is a contour of constant field, B; its corner lies on the coupling line for the appropriate pulse duration. The best choice, for a given pulse length, is that contained in the box that lies farthest down its coupling line. The results are summarized in Table 10.7.

Table 10.7   Materials for high-field magnet windings

Material	Comment
Continuous and long pulse High conductivity coppers Pure silver	Best choice for low-field, long-pulse magnets (heat-limited)
Short pulse Copper-Al_2O_3 composites (Glidcop) H–C copper cadmium alloys H–C copper zirconium alloys H–C copper chromium alloys Drawn copper-niobium composites	Best choice for high-field, short-pulse magnets (heat- and strength-limited)
Ultra short pulse, ultra high field Copper–beryllium–cobalt alloys High-strength, low-alloy steels	Best choice for high-field, short-pulse magnets (strength-limited)

Postscript. The case study, as developed here, is an oversimplification. Magnet design, today, is very sophisticated, involving nested sets of electro and super-conducting magnets (up to 9 deep), with geometry the most important variable. But the selection scheme for coil materials has validity: when pulses are long, resistivity is the primary consideration; when they are very short, it is strength, and the best choice for each is that developed here. Similar considerations enter the selection of materials for very high-speed motors, for bus-bars and for relays.

Further reading

Herlach, F. (1988) The technology of pulsed high-field magnets, IEEE Trans. Magnet. **24**, 1049.

Wood, J.T., Embury, J.D. and Ashby, M.F. (1995) An approach to material selection for high field magnet design, *Acta Metal. Mater.* **43**, 212.

Related case studies

10.2   Multiple constraints: con-rods for high-performance engines

10.4   Conflicting objectives: casings for a mini-disk player

The problem. The miniaturization of consumer electronics — mobile phones, PDAs, MP3, and mini-disc players — is a major design driver. The ideal is a device that you can slip into a shirt pocket and not even know that it is there. Some are now less than 12 mm thick. The casing has to be stiff and strong enough to protect the electronics — the display, particularly — from damage. Casings, typically, are made of moulded ABS. To be stiff enough they have to be at least 1 mm thick, meaning that the casing occupies 20% of the volume of the device. The problem is particularly acute with mobile phones (because people

Figure 10.7   A mini-disk player.

sit on them) and with portable computers, in which thinness and lightness are very highly valued. But the consequences of making the casing too thin are severe: if it is insufficiently stiff its flexure will damage the screen. The challenge: to identify materials for thin, light, casings that are at least as stiff as the current ABS case. We must recognize that the thinnest may not be the lightest, and vice versa. A trade-off will be needed. Table 10.8 summarizes the requirements.

The model. We idealize one panel of the casing in the way shown in Figure 10.8. External loads cause it to bend. If it bends too much the display will be damaged. The bending stiffness is

$$S = \frac{48EI}{L^3}$$

with

$$I = \frac{Wt^3}{12} \tag{10.14}$$

where E is Young's modulus, I the second moment of the area of the panel, and the dimensions L, W, and t are shown on the figure. The stiffness S must equal or exceed a design requirement S^* if the panel is to perform its function properly. Combining the two equations and solving for the thickness t gives

$$t \geq \left(\frac{S^*L^3}{4EW}\right)^{1/3} \propto \left(\frac{1}{E}\right)^{1/3} \tag{10.15}$$

Table 10.8   The design requirements: casing for mini-disk player

Function	Light, thin (cheap) casing
Constraints	• Bending stiffness, S^*, specified • Dimensions L and W specified
Objective	• Minimize thickness of casing • Minimize mass of casing
Free variables	• Thickness t of casing wall • Choice of material

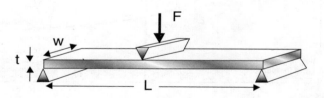

Figure 10.8   The casing can be idealized as a panel of dimensions $L \times W$, and thickness t, loaded in bending.

the thinnest panel is that made from the material with the smallest value of the index $1/E^{1/3}$. The mass of the panel per unit area, m_a, is just ρt, where ρ is its density — the lightest panel is that made from the material with the smallest value of

$$m_a \propto \frac{\rho}{E^{1/3}} \qquad (10.16)$$

We use the existing ABS panel, stiffness S^*, as the standard for comparison. If ABS has a modulus E_0 and a density ρ_0, then a panel made from any other material (modulus E, density ρ) will, according to equations (10.15) and (10.16) have a thickness t relative to that of the ABS panel, t_0, given by

$$\frac{t}{t_0} = \left(\frac{E_0}{E}\right)^{1/3} \qquad (10.17)$$

and a relative mass per unit area

$$\frac{m_a}{m_{a,0}} = \left(\frac{\rho}{E^{1/3}}\right)\left(\frac{E_0^{1/3}}{\rho_0}\right) \qquad (10.18)$$

We wish to explore the trade-off between t/t_0 and $m_a/m_{a,0}$ for possible solutions.

The selection. Figure 10.9 shows the necessary plot, here limited to a few material classes for simplicity. It is divided into four sectors with ABS at the

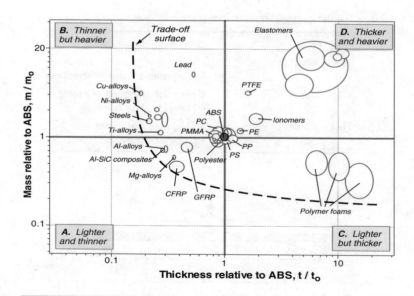

Figure 10.9   The relative thickness and mass of casings made from alternative materials.

center at the co-ordinates (1, 1). The solutions in sector **A** are both thinner and lighter than ABS, some by a factor of 2. Those in sector **B** and **C** are better by one metric but worse by the other. Those in sector **D** are worse by both. To narrow in on an optimal choice we sketch in a trade-off surfaces, shown as the broken line. The solutions nearest to this surface are, in terms of one metric or the other, good choices. Intuition guides us to those in or near Sector **A**.

This is already enough to suggest choices that offer considerable savings in thickness and in weight. If we want to go further we must formulate a relative penalty function. Define Z^*, measured in units of currency, as

$$Z^* = \alpha_t^* \frac{t}{t_0} + \alpha_m^* \frac{m_a}{m_{a,0}} \tag{10.19}$$

The exchange constant α_t^* measures the decrease in penalty — or gain in value — for a fractional decrease in thickness, α_m^* for a fractional decrease in mass. As an example, set $\alpha_t^* = \alpha_m^*$, meaning that we value both equally. Then solutions with equal penalty Z^* are those on the contour

$$\frac{m_a}{m_{a,0}} = \frac{t}{t_0} + \frac{Z^*}{\alpha_m^*} \tag{10.20}$$

This linear relationship plots as a family of curves (not straight lines because of the log scales) with Z^*/α_m^* decreasing towards the bottom left. The absolute value of Z^*/α_m^* does not matter — all we need it for is to identify the point at which a contour is tangent to the trade-off surface as shown in Figure 10.10.

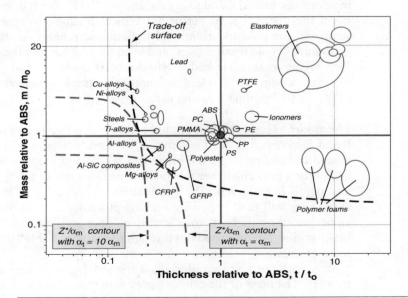

Figure 10.10   The trade-off plot with two values of the relative exchange constants. (Chart created with CES 4, 2004.)

The solutions nearest this point are the optimum choices: CFRP, magnesium alloys, and Al-SiC composites. If, instead, we set $\alpha_t^* = 10\alpha_m^*$, meaning that thinness is much more highly valued than low weight, the contour moves to the second position shown on Figure 10.10. Now titanium, and even steel become attractive candidates.

Postscript. Back in 1997 when extreme thinness and lightness first became major design drivers, the conclusions reached here were new. At that time almost all casings for hand-held electronics were made of ABS, polycarbonate, or, occasionally, steel. Now, 7 or more years later, examples of aluminum, magnesium, titanium and even CFRP casings can be found in currently marketed products. The value of the case study (which dates from 1997) is as an illustration of the way in which systematic methods can be applied to multi-objective selection.

Related case
studies
10.5   Conflicting objectives: materials for a disk-brake caliper

10.5   Conflicting objectives: materials for a disk-brake caliper

The problem. It is unusual — very unusual — to ask whether cost is important in selecting a material and to get the answer "NO". But it does happen, notably when the material is to perform a critical function in space (beryllium for structural components, iridium for radiation screening), in medical procedures (just think of gold tooth fillings) and in equipment for highly competitive sports (one racing motorcycle had a cylinder-head made of solid silver for its high thermal conductivity). Here is another example — material for the brake calipers of a Formula 1 racing car.

The model. The brake caliper can be idealized as two beams of length L, depth b and thickness h, locked together at their ends (Figure 10.11). Each beam is loaded in bending when the brake is applied, and because braking generates heat, it gets hot. The lower schematic represents one of the beams. Its length L and depth b are given. The beam stiffness S is critical: if it is inadequate the caliper will flex, impairing braking efficiency and allowing vibration. Its ability to transmit heat, too, is critical since part of the heat generated in braking must be conducted out through the caliper. Table 10.9 summarizes the requirements.

Start with the first two objectives: minimizing mass and maximizing heat-transfer. The mass of the caliper scales with that of one of the beams. Its mass per unit area is simply

$$m_a = h\rho \quad \text{(units: kg/m}^2) \tag{10.21}$$

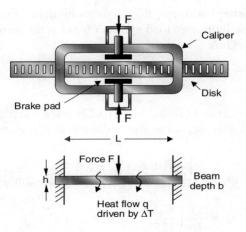

Figure 10.11   A schematic of a brake caliper. The long arms of the caliper are loaded in bending, and must conduct heat well to prevent overheating.

Table 10.9   The design requirements: brake caliper

Function	Brake caliper
Constraints	• Bending stiffness, S^*, specified
	• Dimensions L and b specified
Objective	• Minimize mass of caliper
	• Maximize heat transfer through caliper
	• Minimize material cost
Free variables	• Thickness h of caliper wall
	• Choice of material for the winding

where ρ is the density of the material of which it is made. Heat transfer q depends on the thermal conductivity λ of the material of the beam; the heat flux per unit area is

$$q_a = \lambda \frac{\Delta T}{h} \quad \text{(units: Watts/m}^2\text{)} \tag{10.22}$$

where ΔT is the temperature difference between the surfaces.

The quantities L, b, and ΔT are specified. The only free variable is the thickness h. But there is a constraint: the caliper must be stiff enough to ensure that it does not flex or vibrate excessively. To achieve this we require that

$$S = \frac{C_1 EI}{L^3} = \frac{C_1 E b h^3}{12 L^3} \geq S^* \quad \text{(units: N/m)} \tag{10.23}$$

where S^* is the desired stiffness, E is Young's modulus, C_1 is a constant that depends on the distribution of load and $I = bb^3/12$ is the second moment of the area of the beam. Thus

$$b \geq \left(\frac{12S^*}{C_1 bE}\right)^{1/3} L \tag{10.24}$$

Inserting this in equations (10.21) and (10.22) gives equations for the performance metrics: the mass m_a of the arm and the heat q_a transferred to it, per unit area:

$$m_a \geq \left(\frac{12S^*}{C_1 b}\right)^{1/3} L\left(\frac{\rho}{E^{1/3}}\right) \quad \text{(units kg/m}^2) \tag{10.25}$$

$$q_a = \frac{\Delta T}{L}\left(\frac{C_1 b}{12S^*}\right)^{1/3}(\lambda E^{1/3}) \quad \text{(W/m}^2) \tag{10.26}$$

The first equation contains the material index $M_1 = \rho/E^{1/3}$, the second (expressed such that a minimum is sought), the index $M_2 = 1/\lambda E^{1/3}$.

The standard material for a brake caliper is nodular cast iron — it is cheap and stiff, but it is also heavy and a relatively poor conductor. We use this as a standard for comparison, normalizing equations (10.25) and (10.26) by the values for cast iron (density, ρ_0 modulus E_0 and conductivity λ_0), giving

$$\frac{m_a}{m_{a,0}} = \left(\frac{\rho}{E^{1/2}}\right)\left(\frac{E_0^{1/3}}{\rho_0}\right) \tag{10.27}$$

and

$$\frac{q_{a,0}}{q_a} = \frac{\lambda_0 E_0^{1/3}}{\lambda E^{1/3}} \tag{10.28}$$

The equation for q_a has been inverted so that the best choice of material is that which minimizes both of these. Figure 10.12 shows a chart with these as axes. It is divided into four quadrant, centered on cast iron at the point $(1,1)$. Each bubble describes a material. Those in the lower left are better than cast iron by both objectives; an aluminum caliper, for example, has half the weight and offers twice the heat transfer. The ultimate choice is beryllium or its alloy Be 40%Al.

To go further we formulate the relative penalty function

$$Z^* = \alpha_m^*\left(\frac{m_a}{m_{a,0}}\right) + \alpha_q^*\left(\frac{q_{a,0}}{q_a}\right) \tag{10.29}$$

in which the terms in brackets are given by equations (10.27) and (10.28) and the exchange constants α_m^* and α_q^* measure the relative value of a fractional saving of weight or increase in heat transfer relative to cast iron. The penalty function is plotted in Figure 10.12 for three values of the ratio α_q^*/α_m^* of the

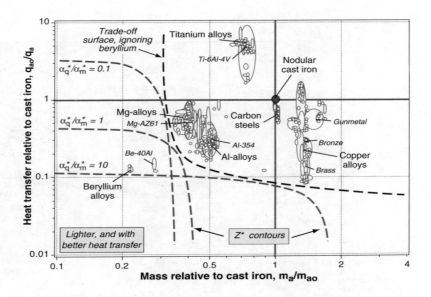

Figure 10.12 A chart with equations (10.27) and (10.28) as axes. Beryllium and its alloys are the preferred choice, minimizing both mass and maximizing heat transfer. But if these exotics are excluded, the choice becomes dependent on the ratio α_q^*/α_m^*. The trade-off surface and penalty contours for three values of α_q^*/α_m^* are shown. (Chart created with CES 4, 2004.)

exchange constants. Each is tangent to a trade-off surface that excludes the "exotic" beryllium alloys, which otherwise dominate the selection for all values. For $\alpha_q^*/\alpha_m^* = 0.1$, meaning mass reduction is of prime importance, magnesium alloys are the best choice. If mass reduction and heat transfer are given equal weight ($\alpha_q^*/\alpha_m^* = 1$) aluminum alloys become a good choice. If heat transfer is the over-riding consideration ($\alpha_q^*/\alpha_m^* = 10$), alloys based on copper win.

So far we have ignored cost (beryllium is exceedingly expensive). To include it we extend the penalty function, introducing the relative cost per unit area, $C_m m_a/C_{m,0} m_{a,0}$, where C_m is the cost per kg of the material and $C_{m,0}$ that of cast iron, giving

$$Z^* = \frac{C_m\, m_a}{C_{m,0}\, m_{a,0}} + \alpha_m^*\left(\frac{m_a}{m_{a,0}}\right) + \alpha_q^*\left(\frac{q_{a,0}}{q_a}\right) \qquad (10.30)$$

A two-dimensional materials chart is no longer adequate because three groups of material properties are involved. The way forward is to create an exchange-constant chart, introduced in Chapter 9, showing the optimum choice as a function of the values of the two exchange constants α_m^* and α_q^*. Figure 10.13 is the result. When α_m^* and α_q^* are small (lower left of the figure) cost dominates the selection, giving cast iron as the choice that minimizes Z^*. With increasing importance of weight reduction and increased heat transfer,

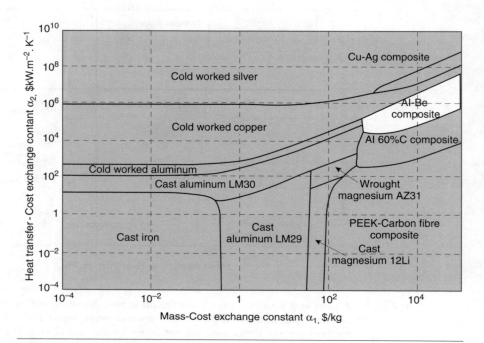

Figure 10.13   An exchange-constant chart for the selection of brake-caliper materials.

Figure 10.14   An Al–Be composite disk brake caliper for a Ferrari Formula 1 racing car (Marder, 1997).

aluminum and magnesium alloys appear. Only when both low weight and high heat flow are exceedingly highly valued do the beryllium alloys appear. Figure 10.14 is a photograph of a brake caliper designed for Ferrari Racing. It is made of precisely this material.

Postscript. Note how the method points to applications for new materials, of which the Al–Be alloy is an example. If it were not in the database, the materials that surround it in Figure 10.13 would occupy its space. Add it to the database and it pops up at the position shown. Its "co-ordinates" in $\alpha_1 - \alpha_2$

space point to applications in which stiffness at low mass and high heat transfer are highly valued and identify the materials that compete with it for this market.

10.6 Summary and conclusions

Most designs are over-constrained: they must simultaneously meet several competing, and often conflicting, requirements. But although they conflict, an optimum selection is still possible. The "active constraint" method, developed in Chapter 9, allows the selection of materials that optimally meet two or more constraints. It is illustrated here by two case studies, one of them mechanical, one electro-mechanical.

Greater challenges arise when the design must meet two or more conflicting objectives (such as minimizing mass, volume, cost, and environmental impact). Here we need a way to express all the objectives in the same units, a "common currency", so to speak. The conversion factors are called the "exchange constants". Establishing the value of the exchange constant is an important step in solving the problem. With it, a penalty function Z is constructed that combines the objectives. Materials that minimize Z meet all the objectives in a properly balanced way. The most obvious common currency is cost itself, requiring an "exchange rate" to be established between cost and the other objectives. This can be done for mass, and—at least in principle—for other objectives too. The method is illustrated by two further case studies.

Chapter 11

Selection of material and shape

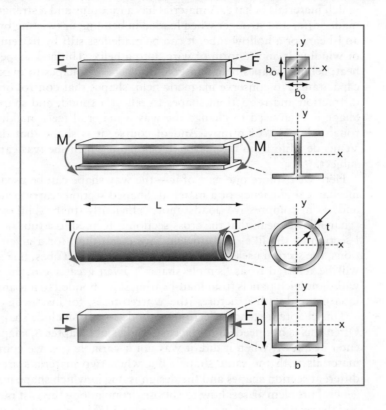

11.1 Introduction and synopsis

Pause now for a moment and reflect on how shape is used to modify the ways in which materials behave. A material has a modulus and a strength, but it can be made stiffer and stronger when loaded in bending or twisting by shaping it into an I-beam or a hollow tube. It can be made less stiff by flattening it into a leaf or winding it, in the form of wire, into a helix. Thinned shapes help dissipate heat; cellular shapes help conserve it. There are shapes to maximize electrical capacitance, to conserve magnetic field, shapes that control optical reflection, diffraction and refraction, shapes to reflect a sound, and shapes to absorb it. Shape is even used to change the way a material feels, making it smooth or rough, slippery, or grippy. And of course, it is shape that distinguishes the Venus de Milo from the marble block from which she was carved. It is a rich subject.

Here we explore one part of it — the way shape can be used to increase the mechanical efficiency of a material. Shaped sections carry bending, torsional, and axial-compressive loads more efficiently than solid sections do. By "shaped" we mean that the cross-section is formed to a tube, a box-section, an I-section or the like. By "efficient" we mean that, for a given loading conditions, the section uses as little material as possible. Tubes, boxes and I-sections will be referred to as "simple shapes". Even greater efficiencies are possible with sandwich panels (thin load-bearing skins bonded to a foam or honeycomb interior) and with structures (the Warren truss, for instance).

This chapter extends selection methods so as to include shape (Figure 11.1). Often this is not needed: in the case studies of Chapter 6, shape either did not enter at all, or, when it did, it was not a variable (i.e. we compared different materials with the same shape). But when two materials are available with different section shapes and the design is one in which shape matters, the more general problem arises: how to choose, from among the vast range of materials and the section shapes in which they are available or could potentially be made, the combinations that maximize performance. Take the example of a bicycle: its forks are loaded in bending. They could, say, be made of steel or of wood — early bikes *were* made of wood. But steel is available as thin-walled tube, whereas the wood is not; wood, usually, has a solid section. A solid wood bicycle is certainly lighter for the same stiffness than a solid steel one, but is it lighter than one made of steel tubing? Might a magnesium I-section be lighter still? How, in short, is one to choose the best combination of material and shape?

A procedure for answering these and related questions is outlined in this chapter. It involves the definition of *shape factors*. A *material* can be thought of as having properties but no shape. A *component* or *structure* is a material made into a shape (Figure 11.2). The shape factor is a measure of the efficiency material usage. Further, they enable the definition of material indices like those of Chapter 5, but that now include shape. When shape is constant, the

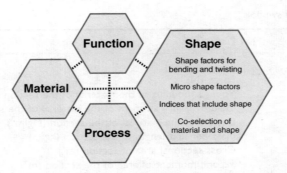

Figure 11.1   Section shape is important for certain modes of loading. When shape is a variable a new term, the shape factor ϕ, appears in some of the material indices: they then allow optimum selection of material and shape.

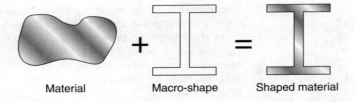

Figure 11.2   Mechanical efficiency is obtained by combining material with macroscopic shape. The shape is characterized by a dimensionless shape factor, ϕ. The schematic is suggested by Parkhouse (1984).

indices reduce exactly to those of Chapter 5; but when shape is a variable, the shape factor appears in the expressions for the indices.

The ideas in this chapter are a little more difficult than those of Chapter 5; their importance lies in the connection they make between materials selection and the design of load-bearing structures. A feel for the method can be had by reading the first and last sections (11.2, 11.7 and 11.8) alone; these, plus the results listed in Tables 11.1 and 11.2, should be enough to allow the case studies of Chapter 12 (which apply the methods) to be understood. The reader who wishes to grasp how the results arise will have to read the whole thing.

The symbols used in the development are listed, for convenience, in Table 11.1.

11.2 Shape factors

The loads on a component can be decomposed into those that are axial, those that exert bending moments and those that exert torques. One of these usually

Table 11.1   Definition of symbols

Symbol	Definition
M	Moment (Nm)
F	Force (N)
E	Young's modulus of the material of the section (GPa)
σ_f	Yield or failure strength of the material of the section (MPa)
ρ	Density of the material of the section (Mg/m^3)
m_l	Mass per unit length of the section (kg/m)
A	Cross-sectional area of section (m^2)
I	Second moment of area of the section (m^4)
I_o	Second moment of area of the square reference section (m^4)
Z	Section modulus of the section (m^3)
Z_o	Section modulus of the square reference section (m^3)
φ_B^e	Macro shape factor for elastic bending deflection ($-$)
φ_B^f	Macro shape factor for onset of plasticity or failure in bending ($-$)
φ_T^e	Macro shape factor for elastic torsional deflection ($-$)
φ_T^f	Macro shape factor for onset of plasticity or failure in torsion ($-$)
ψ_B^e	Micro shape factor for elastic bending deflection ($-$)
ψ_B^f	Micro shape factor for onset of plasticity or failure in bending ($-$)
(EI)	Essential term in bending stiffness (N.m^2)
$(Z\sigma_f)$	Essential term in bending strength (N.m)
t	Web and flange thickness (m)
c	Web height (m)
d	Section height ($2t + c$) of sandwich (m)
b	Section (flange) width (m)
L	Section length (m)

dominates to such an extent that structural elements are specially designed to carry it, and these have common names. Thus *ties* carry tensile loads; *beams* carry bending moments; *shafts* carry torques; *columns* carry compressive axial loads. Figure 11.3 shows these modes of loading applied to shapes that resist them well. The point it makes is that the best material-and-shape combination depends on the mode of loading. In what follows, we separate the modes, dealing with each separately.

In axial tension, the area of the cross-section is important but its shape is not: all sections with the same area will carry the same load. Not so in bending: beams with hollow-box or I-sections are better than solid sections of the same cross-sectional area. Torsion too, has its efficient shapes: circular tubes, for instance, are more efficient than either solid sections or I-sections. To characterize this we need a metric — a way of measuring the structural efficiency of a section shape, independent of the material of which it is made. An obvious one is that given by the ratio of ϕ the stiffness or strength of the shaped section to that of a 'neutral' reference shape, which we take to be that of a solid square

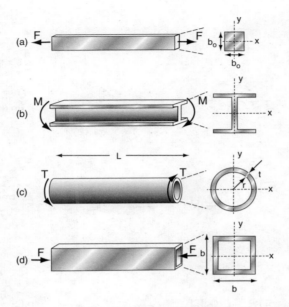

Figure 11.3   Common modes of loading and the section-shapes that are chosen to support them: (a) axial tension (b) bending (c) torsion and (d) axial compression, which can lead to buckling.

section with the same cross-sectional area, and thus the same mass per unit length, as the shaped section.[1]

Elastic bending of beams and twisting of shafts (Figure 11.3b and c)

The bending stiffness S of a beam is proportional to the product EI

$$S \propto EI$$

Here E is Young's modulus and I is the second moment of area of the beam about the axis of bending (the x axis):

$$I = \int_{\text{section}} y^2 \, dA \tag{11.1}$$

where y is measured normal to the bending axis and dA is the differential element of area at y (Figure 11.3). Values of the moment I and of the area A for common sections are listed in the first two columns of Table 11.2. Those for the more complex shapes are approximate, but completely adequate for

---

[1] The definitions of the shape factors ϕ differ from those in earlier editions of this book, which were based on a solid cylinder rather than a solid square section as the standard shape. The change allows simplification, and makes almost no difference to the numerical values of the shape factors.

Table 11.2 Moments of sections, and units

Section shape	Area A (m)	Moment I (m⁴)	Moment K (m⁴)	Moment Z (m⁴)	Moment Q (m⁴)	Moment Z_p (m⁴)
	bh	$\dfrac{bh^3}{12}$	$\dfrac{bh^3}{3}\left(1-0.58\dfrac{b}{h}\right)$ $(h>b)$	$\dfrac{bh^2}{6}$	$\dfrac{b^2h^2}{(3h+1.8b)}$ $(h>b)$	$\dfrac{bh^2}{4}$
	$\dfrac{\sqrt{3}}{4}a^2$	$\dfrac{a^4}{32\sqrt{3}}$	$\dfrac{\sqrt{3}\,a^4}{80}$	$\dfrac{a^3}{32}$	$\dfrac{a^3}{20}$	$\dfrac{3a^3}{64}$
	πr^2	$\dfrac{\pi}{4}r^4$	$\dfrac{\pi}{2}r^4$	$\dfrac{\pi}{4}r^3$	$\dfrac{\pi}{2}r^3$	$\dfrac{\pi}{3}r^3$
	πab	$\dfrac{\pi}{4}a^3b$	$\dfrac{\pi a^3 b^3}{(a^2+b^2)}$	$\dfrac{\pi}{4}a^2b$	$\dfrac{\pi}{2}a^2b$ $(a<b)$	$\dfrac{\pi}{3}a^2b$
	$\pi(r_o^2-r_i^2)$ $\approx 2\pi rt$	$\dfrac{\pi}{4}(r_o^4-r_i^4)$ $\approx \pi r^3 t$	$\dfrac{\pi}{2}(r_o^4-r_i^4)$ $\approx 2\pi r^3 t$	$\dfrac{\pi}{4r_o}(r_o^4-r_i^4)$ $\approx \pi r^2 t$	$\dfrac{\pi}{2r_o}(r_o^4-r_i^4)$ $\approx 2\pi r^2 t$	$\dfrac{\pi}{3}(r_o^3-r_i^3)$ $\approx \pi r^2 t$
	$2t(h+b)$ $(h,b \gg t)$	$\dfrac{1}{6}h^3t\left(1+3\dfrac{b}{h}\right)$	$\dfrac{2tb^2h^2}{(h+b)}\left(1-\dfrac{t}{h}\right)^4$	$\dfrac{1}{3}h^2t\left(1+3\dfrac{b}{h}\right)$	$2tbh\left(1-\dfrac{t}{h}\right)^2$	$bht\left(1+\dfrac{h}{2b}\right)$

Section	$\pi(a+b)t$ $(a,\,b \gg t)$	$\dfrac{\pi}{4}a^3t\left(1+\dfrac{3b}{a}\right)$	$\dfrac{4\pi(ab)^{5/2}t}{(a^2+b^2)}$	$\dfrac{\pi}{4}a^2t\left(1+\dfrac{3b}{a}\right)$	$2\pi(a^3b)^{1/2}$ $(b>a)$	$\pi abt\left(2+\dfrac{a}{b}\right)$
(ellipse, $2a$, $2b$, t)	$\pi(a+b)t$ $(a,\,b \gg t)$	$\dfrac{\pi}{4}a^3t\left(1+\dfrac{3b}{a}\right)$	$\dfrac{4\pi(ab)^{5/2}t}{(a^2+b^2)}$	$\dfrac{\pi}{4}a^2t\left(1+\dfrac{3b}{a}\right)$	$2\pi(a^3b)^{1/2}$ $(b>a)$	$\pi abt\left(2+\dfrac{a}{b}\right)$
(two bars, h_o, h_i, b)	$b(h_o-h_i)$ $\approx 2bt$ $(h,\,b \gg t)$	$\dfrac{b}{12}(h_o^3-h_i^3)$ $\approx \tfrac{1}{2}bth_o^2$	—	$\dfrac{b}{6h_o}(h_o^3-h_i^3)$ $\approx bth_o$	—	$\dfrac{b}{4}(h_o^2-h_i^2)$ $\approx bth_o$
(H, h, b, t)	$2t(h+b)$ $(h,\,b \gg t)$	$\dfrac{1}{6}h^3t\left(1+\dfrac{3b}{h}\right)$	$\dfrac{2}{3}bt^3\left(1+\dfrac{4h}{b}\right)$	$\dfrac{1}{3}h^2t\left(1+\dfrac{3b}{h}\right)$	$\dfrac{2}{3}bt^2\left(1+\dfrac{4h}{b}\right)$	$bht\left(1+\dfrac{h}{2b}\right)$
(T, h, $2t$)	$2t(h+b)$ $(h,\,b \gg t)$	$\dfrac{t}{6}(h^3+4bt^2)$	$\dfrac{t^3}{3}(8b+h)$	$\dfrac{t}{3h}(h^3+4bt^2)$	$\dfrac{t^2}{3}(8b+h)$	$\dfrac{th^2}{2}\left\{1+\dfrac{2t(b-2t)}{h^2}\right\}$
(I, h, $2t$, b, t)	$2t(h+b)$ $h,\,b \gg t$	$\dfrac{t}{6}(h^3+4bt^2)$	$\dfrac{2}{3}ht^3\left(1+\dfrac{4b}{h}\right)$	$\dfrac{t}{3h}(h^3+4bt^2)$	$\dfrac{2}{3}ht^2\left(1+\dfrac{4b}{h}\right)$	$\dfrac{th^2}{2}\left\{1+\dfrac{2t(b-2t)}{h^2}\right\}$

present needs. The second moment of area, I_o, for a reference beam of square section with edge-length b_o and section-area $A = b_o^2$ is simply

$$I_o = \frac{b_o^4}{12} = \frac{A^2}{12} \tag{11.2}$$

(Here and elsewhere the subscript 'o' refers to the solid square section.) The bending stiffness of the shaped section differs from that of a square one with the same area A by the factor ϕ_B^e where

$$\phi_B^e = \frac{S}{S_o} = \frac{EI}{EI_o} = \frac{12\,I}{A^2} \tag{11.3}$$

We call ϕ_B^e the *shape factor for elastic bending*. Note that it is dimensionless — I has dimensions of $(\text{length})^4$ and so does A^2. It depends only on shape, not on scale: big and small beams have the same value of ϕ_B^e if their section shapes are the same.[2] This is shown in Figure 11.4. The three members of each group differ in scale but have the same shape factor — each member is a magnified or shrunken version of its neighbors. Shape-efficiency factors ϕ_e for common shapes, calculated from the expressions for A and I in Table 11.2, are listed in the first column of Table 11.3. Solid equiaxed sections (circles, squares, hexagons, octagons) all have values very close to 1 — for practical purposes they can be set equal to 1. But if the section is elongated, or hollow, or of I-section, things change; a thin-walled tube or a slender I-beam can have a value of ϕ_B^e of 50 or more. A beam with $\phi_B^e = 50$ is 50 times stiffer than a solid beam of the same weight.

Figure 11.5 is a plot of I against A for values of ϕ_B^e (equation (11.3)). The contour for $\phi_B^e = 1$ describes the square-section reference beam. Those for $\phi_B^e = 10$ and $\phi_B^e = 100$ describe more efficient shapes, as suggested by the icons at the bottom left, in each of which the axis of bending is horizontal. But it is not always high stiffness that is wanted. Springs, cradles, suspensions, cables, and other structures that must flex yet have high tensile strength, rely on having a low bending stiffness. Then we want low shape efficiency. It is achieved by spreading the material in a plane containing the axis of bending to form sheets or wires, as suggested by the contours for $\phi_B^e = 0.1$ and 0.01.

Shapes that resist bending well may not be so good when twisted. The stiffness of a shaft — the torque T divided by the angle of twist, θ (Figure 11.3c) — is proportional to GK, where G is its shear modulus and K its torsional moment of area. For circular sections K is identical with the polar moment of area, J:

$$J = \int_{\text{section}} r^2 \, dA \tag{11.4}$$

---

[2] This elastic shape-efficiency factor is related to the radius of gyration, R_g, by $\phi_B^e = 12R_g^2/A$. It is related to the "shape parameter", k_1, of Shanley (1960) by $\phi_B^e = 12k_1$.

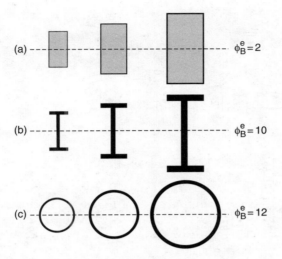

Figure 11.4 (a) A set of rectangular sections with $\phi_B^e = 2$; (b) a set of I-sections with $\phi_B^e = 10$; and (c) a set of tubes with $\phi_B^e = 12$. Members of a set differ in size but not in shape.

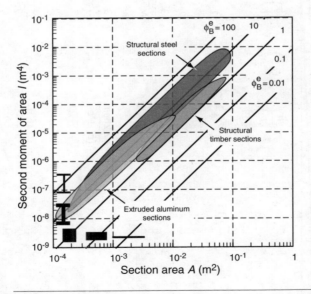

Figure 11.5 The second moment of area, I, plotted against section area A. Efficient structures have high values of the ratio Z/A^2; inefficient structures (ones that bend easily) have low values. Real structural sections have values of I and A that lie in the shaded zones. Note that there are limits on A and on the maximum shape efficiency ϕ_B^e that depend on material.

Table 11.3 Shape efficiency factors

Section shape	Bending factor, φ_B^e	Torsional factor, φ_T^e	Bending factor, φ_B^f	Torsional factor, φ_T^f	Bending factor, φ_B^{pl}
	$\dfrac{h}{b}$	$2.38\dfrac{h}{b}\left(1-0.58\dfrac{b}{h}\right)$ $(h>b)$	$\left(\dfrac{h}{b}\right)^{0.5}$	$1.6\sqrt{\dfrac{b}{h}}\dfrac{1}{(1+0.6b/h)}$ $(h>b)$	$\left(\dfrac{h}{b}\right)^{0.5}$
	$\dfrac{2}{\sqrt{3}}=1.15$	0.832	$\dfrac{3^{1/4}}{2}=0.658$	0.83	$\dfrac{3^{1/4}}{2}=0.658$
	$\dfrac{3}{\pi}=0.955$	1.14	$\dfrac{3}{2\sqrt{\pi}}=0.846$	1.35	$\dfrac{4}{3\sqrt{\pi}}=0.752$
	$\dfrac{3}{\pi}\dfrac{a}{b}$	$\dfrac{2.28ab}{(a^2+b^2)}$	$\dfrac{3}{2\sqrt{\pi}}\sqrt{\dfrac{a}{b}}$	$1.35\sqrt{\dfrac{a}{b}}$ $(a<b)$	$\dfrac{4}{3\sqrt{\pi}}\sqrt{\dfrac{a}{b}}=0.752\sqrt{\dfrac{a}{b}}$
	$\dfrac{3}{\pi}\left(\dfrac{r}{t}\right)$ $(r\gg t)$	$1.14\left(\dfrac{r}{t}\right)$	$\dfrac{3}{\sqrt{2\pi}}\sqrt{\dfrac{r}{t}}$	$1.91\sqrt{\dfrac{r}{t}}$	$\sqrt{\dfrac{2}{\pi}}\sqrt{\dfrac{r}{t}}$
	$\dfrac{1}{2}\dfrac{h}{t}\dfrac{(1+3b/h)}{(1+b/h)^2}$ $(h,b\gg t)$	$\dfrac{3.57b^2(1-t/h)^4}{th(1+b/h)^3}$	$\dfrac{1}{\sqrt{2}}\sqrt{\dfrac{h}{t}}\dfrac{(1+3b/h)}{(1+b/h)^{3/2}}$	$3.39\sqrt{\dfrac{h^2}{bt}}\dfrac{1}{(1+h/b)^{3/2}}$	$\sqrt{2}\sqrt{\dfrac{h^2}{bt}}\dfrac{(1+h/2b)}{(1+h/b)^{3/2}}$

Cross-section					
Hollow ellipse (2a, 2b, wall t)	$\dfrac{3a}{\pi}\dfrac{(1+3b/a)}{t(1+b/a)^2}$ (a,b ≫ t)	$\dfrac{9.12(ab)^{5/2}}{t(a^2+b^2)(a+b)^2}$	$\dfrac{3}{2\sqrt{\pi}}\sqrt{\dfrac{a}{t}}\dfrac{(1+3b/a)}{(1+b/a)^{3/2}}$	$5.41\sqrt{\dfrac{a}{t}}\dfrac{1}{(1+a/b)^{3/2}}$	$\dfrac{4}{\sqrt{\pi}}\sqrt{\dfrac{a^2}{bt}}\dfrac{(2+a/b)}{(1+a/b)^{3/2}}$
Two plates (h, h_o, b, t)	$\dfrac{3}{2}\dfrac{h_o^2}{bt}$ (h,b ≫ t)	—	$\dfrac{3}{\sqrt{2}}\dfrac{h_o}{\sqrt{bt}}$	—	$\sqrt{2}\dfrac{h_o}{\sqrt{bt}}$
I-section (h, b, 2t, t)	$\dfrac{1}{2}\dfrac{h}{t}\dfrac{(1+3b/h)}{(1+b/h)^2}$ (h,b ≫ t)	$1.19\left(\dfrac{t}{b}\right)\dfrac{(1+4h/b)}{(1+h/b)^2}$	$\dfrac{1}{\sqrt{2}}\sqrt{\dfrac{h}{t}}\dfrac{(1+3b/h)}{(1+b/h)^{3/2}}$	$1.13\sqrt{\dfrac{t}{b}}\dfrac{(1+4h/b)}{(1+h/b)^{3/2}}$	$\sqrt{2}\sqrt{\dfrac{h^2}{bt}}\dfrac{(1+h/2b)}{(1+h/b)^{3/2}}$
T-section (h, 2t)	$\dfrac{1}{2}\dfrac{h}{t}\dfrac{(1+4bt^2/h^3)}{(1+b/h)^2}$ (h,b ≫ t)	$0.595\left(\dfrac{t}{h}\right)\dfrac{(1+8b/h)}{(1+b/h)^2}$	$\dfrac{3}{4}\sqrt{\dfrac{h}{t}}\dfrac{(1+4bt^2/h^3)}{(1+b/h)^{3/2}}$	$0.565\sqrt{\dfrac{t}{h}}\dfrac{(1+8b/h)}{(1+b/h)^{3/2}}$	$\dfrac{1}{\sqrt{2}}\sqrt{\dfrac{h}{t}}\sqrt{t}\left(\dfrac{h}{(h+b)}\right)^{3/2}\dfrac{1}{\left[1+\dfrac{2t(b-2t)}{h^2}\right]}$
I-section (h, b, 2t, t)	$\dfrac{1}{2}\dfrac{h}{t}\dfrac{(1+4bt^2/h^3)}{(1+b/h)^2}$ (h,b ≫ t)	$1.19\left(\dfrac{t}{h}\right)\dfrac{(1+4b/h)}{(1+b/h)^2}$	$\dfrac{3}{4}\sqrt{\dfrac{h}{t}}\dfrac{(1+4bt^2/h^3)}{(1+b/h)^{3/2}}$	$1.13\sqrt{\dfrac{t}{h}}\dfrac{(1+4b/h)}{(1+b/h)^{3/2}}$	$\dfrac{1}{\sqrt{2}}\sqrt{\dfrac{h}{t}}\sqrt{t}\left(\dfrac{h}{(h+b)}\right)^{3/2}\dfrac{1}{\left[1+\dfrac{2t(b-2t)}{h^2}\right]}$

where dA is the differential element of area at the radial distance r, measured from the center of the section. For non-circular sections, K is less than J; it is defined such that the angle of twist θ is related to the torque T by

$$S_T = \frac{T}{\theta} = \frac{KG}{L} \tag{11.5}$$

where L is length of the shaft and G the shear modulus of the material of which it is made. Approximate expressions for K are listed in Table 11.2.

The shape factor for elastic twisting is defined, as before, by the ratio of the torsional stiffness of the shaped section, S_T, to that, S_{T_o}, of a solid square shaft of the same length L and cross-section A, which, using equation (11.5), is:

$$\phi_T^e = \frac{S_T}{S_{T_o}} = \frac{K}{K_o} \tag{11.6}$$

The torsional constant K_o for a solid square section (Table 11.1, top row with $b = h$) is

$$K_o = 0.14A^2$$

giving

$$\phi_T^e = 7.14 \frac{K}{A^2} \tag{11.7}$$

It, too, has the value 1 for a solid circular square section, and values near 1 for any solid, equiaxed section; but for thin-walled shapes, particularly tubes, it can be large. As before, sections with the same value of ϕ_T^e differ in size but not shape. Values derived from the expressions for K and A in Table 11.2 are listed in Table 11.3.

Onset of failure in bending and twisting

Plasticity starts when the stress, somewhere, first reaches the yield strength, σ_y; fracture occurs when this stress first exceeds the fracture strength, σ_{fr}; fatigue failure if it exceeds the endurance limit σ_e. Any one of these constitutes failure. As in earlier chapters, we use the symbol σ_f for the failure stress, meaning "the local stress that will first cause yielding or fracture or fatigue failure".

In bending, the stress σ is largest at the point y_m in the surface of the beam that lies furthest from the neutral axis; it is:

$$\sigma = \frac{My_m}{I} = \frac{M}{Z} \tag{11.8}$$

where M is the bending moment. Failure occurs when this stress first exceeds σ_f. Thus, in problems of failure of beams, shape enters through the section modulus, $Z = I/y_m$. The strength-efficiency of the shaped beam, ϕ_B^f, is measured by the ratio Z/Z_o, where Z_o is the section modulus of a reference beam of square section with the same cross-sectional area, A:

$$Z_o = \frac{b_o^3}{6} = \frac{A^{3/2}}{6} \qquad (11.9)$$

Thus:

$$\phi_B^f = \frac{Z}{Z_o} = \frac{6Z}{A^{3/2}} \qquad (11.10)$$

Like the other shape-efficiency factor, it is dimensionless and therefore independent of scale, and its value for a beam with a solid square section is unity. Table 11.3 gives expressions for other shapes derived from the values of the section modulus Z, which can be found in Table 11.2. A beam with an failure shape-efficiency factor of 10 is 10 times stronger in bending than a solid square section of the same weight. Figure 11.6 is a plot of Z against A for values of ϕ_B^f (equation (11.10)). As before, the contour for $\phi_B^f = 1$ describes the square-section reference beam. The other contours describe shapes that are more or less efficient, as suggested by the icons.

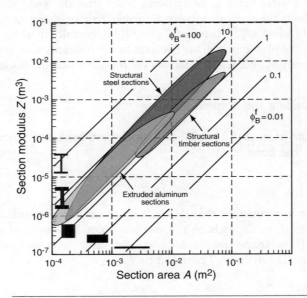

Figure 11.6   The section modulus, Z, plotted against section area A. Efficient structures have high values of the ratio $Z/A^{3/2}$; inefficient structures (ones that bend easily) have low values. Real structural sections have values of Z and A that lie in the shaded zones. Note that there are limits on A and on the maximum shape efficiency ϕ_B^f that depend on material.

In torsion the problem is more complicated. For circular rods or tubes subjected to a torque T (as in Figure 11.3c) the shear stress τ is a maximum at the outer surface, at the radial distance r_m from the axis of bending:

$$\tau = \frac{Tr_m}{J} \tag{11.11}$$

The quantity in J/r_m twisting has the same character as I/y_m in bending. For non-circular sections with ends that are free to warp, the maximum surface stress is given instead by

$$\tau = \frac{T}{Q} \tag{11.12}$$

where Q, with units of m^3 now plays the role of J/r_m or Z. This allows the definition of a shape factor, ϕ_T^f, for failure in torsion, following the same pattern as before:

$$\phi_T^f = \frac{Q}{Q_o} = 4.8 \frac{Q}{A^{3/2}} \tag{11.13}$$

Values of Q and ϕ_T^f are listed in Tables 11.2 and 11.3. Shafts with a solid equiaxed sections all have values of ϕ_T^f close to 1.

Fully plastic bending or twisting (such that the yield strength is exceeded throughout the section) involve a further pair of shape factors. That for fully plastic bending, ϕ_B^{pl}, is listed in Table 11.3. Generally speaking, shapes that resist the onset of plasticity well are resistant to full plasticity also, so ϕ_B^{pl} does not differ much from ϕ_B^f. New shape factors for these are not, at this stage, necessary.

Axial loading and column buckling

A column of length L, loaded in compression, buckles elastically when the load exceeds the Euler load

$$F_c = \frac{n^2 \pi^2 E I_{min}}{L^2} \tag{11.14}$$

where n is a constant that depends on the end-constraints. The resistance to buckling, then, depends on the smallest second moment of area, I_{min}, and the appropriate shape factor (ϕ_B^e) is the same as that for elastic bending (equation (11.4)) with I replaced by I_{min}.

11.3 Microscopic or micro-structural shape factors

Survival, in nature, is closely linked to structural efficiency. The tree that, with a given resource of cellulose, grows the tallest captures the most sunlight.

The creature that, with a given allocation of hydroxyapatite, develops the strongest bone-structure wins the most fights; or — if prey rather than predator — runs the fastest. Structural efficiency means survival. It is worth asking how nature does it.

Microscopic shape

The shapes listed in Tables 11.2 and 11.3 achieve efficiency through their *macroscopic* shape. Structural efficiency can be achieved in another way: through shape on a small scale; *microscopic* or "micro-structural" shape (Figure 11.7). Wood is an example. The solid component of wood (a composite of cellulose, lignin, and other polymers) is shaped into small prismatic cells, dispersing the solid further from the axis of bending or twisting of the branch or trunk of the tree, increasing both stiffness and strength. The added efficiency is characterized by a set of *microscopic shape factors*, ψ, with definitions exactly like those of ϕ. The characteristic of microscopic shape is that the structure repeats itself: it is *extensive*. The micro-structured solid can be thought of as a "material" in its own right: it has a modulus, a density, a strength, and so forth. Shapes can be cut from it which — provided they are large compared with the size of the cells — inherit its properties. It is possible, for instance, to fabricate an I-section from wood, and such a section has macroscopic shape (as defined earlier) as well as microscopic shape as suggested by Figure 11.8. It is shown in a moment that the total shape factor for a wooden I-beam is the product of the shape factor for the wood structure and that for the I-beam, and this can be large.

Many natural materials have microscopic shape. Wood is just one example. Bone, the stalks and leaves of plants and the cuttle-bone of a squid all have structures that give high stiffness at low weight (Figure 11.9). It is harder to think of man-made examples, though it would appear possible to make them.

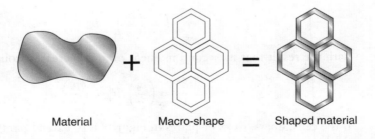

Material　　　　+　　　　Macro-shape　　　　=　　　　Shaped material

Figure 11.7　Mechanical efficiency can be obtained by combining material with microscopic, or internal, shape, which repeats itself to give an extensive structure. The shape is characterized by microscopic shape factors, ψ.

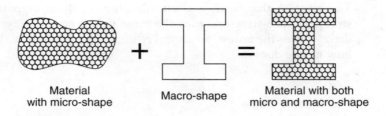

Material
with micro-shape   Macro-shape   Material with both
                                            micro and macro-shape

Figure 11.8 Micro-structural shape can be combined with macroscopic shape to give efficient structures. The overall shape factor is the product of the microscopic and macroscopic shape factors.

Figure 11.10 shows four extensive structures with microscopic shape, all of them found in nature. The first is a wood-like structure of hexagonal-prismatic cells; it is isotropic in the plane of the section when the cells are regular hexagons. The second is an array of fibers separated by a foamed matrix typical of palm wood; it too is isotropic in-plane. The third is an axisymmetric structure of concentric cylindrical shells separated by a foamed matrix, like the stem of some plants. And the fourth is a layered structure, a sort of multiple sandwich-panel, like the shell of the cuttle fish.

Microscopic shape factors. Consider the gain in bending stiffness when a solid square beam like that shown as a black square of side b_o in Figure 11.10 is expanded, at constant mass, to a larger square section with any one of the structures that surround it in the figure. The bending stiffness S_s of the original solid beam is proportional to the product of its modulus E_s and its second moment of area I_s:

$$S_s \propto E_s I_s \tag{11.15}$$

where the subscript "s" means "a property of the solid beam" and $I_s = b_o^4/12$. When the beam is expanded at constant mass its density falls from ρ_s to ρ and its edge-length increases from b_o to b where

$$b = \left(\frac{\rho_s}{\rho}\right)^{1/2} b_o \tag{11.16}$$

with the result that its second moment of area increases from I_s to

$$I = \frac{b^4}{12} = \left(\frac{\rho_s}{\rho}\right)^2 b_o^4 = \left(\frac{\rho_s}{\rho}\right)^2 I_s \tag{11.17}$$

If the cells, fibers, rings or plates in Figure 11.10 extend parallel to the axis of the beam, the modulus parallel to this axis falls from that of the solid, E_s, to

$$E = \left(\frac{\rho}{\rho_s}\right) E_s \tag{11.18}$$

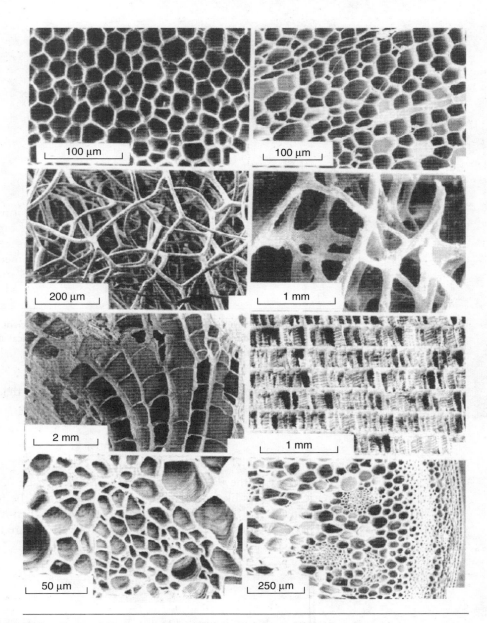

Figure 11.9   Natural materials with internal, or microscopic, shape. Reading from the top left: cork, balsa wood, sponge, cancellous bone, coral, cuttle-bone, and palm plant stalk.

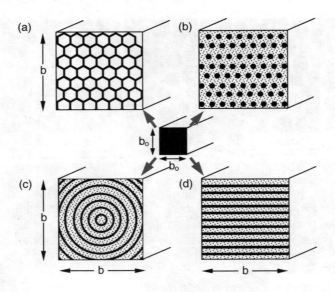

Figure 11.10 Four extensive micro-structured materials that are mechanically efficient: (a) prismatic cells, (b) fibers embedded in a foamed matrix, (c) concentric cylindrical shells with foam between, and (d) parallel plates separated by foamed spacers.

The bending stiffness of the expanded beam scales as EI, so that it is stiffer than the original solid one by the factor

$$\psi_{\mathrm{B}}^{\mathrm{e}} = \frac{S}{S_{\mathrm{s}}} = \frac{EI}{E_{\mathrm{s}}I_{\mathrm{s}}} = \frac{\rho_{\mathrm{s}}}{\rho} \tag{11.19}$$

We refer to $\psi_{\mathrm{B}}^{\mathrm{e}}$ as the *microscopic shape factor for elastic bending*. That for prismatic structures like those of Figure 11.10 is simply the reciprocal of the relative density, ρ/ρ_{s}. Note that, in the limit of a solid (when ρ/ρ_{s}) $\psi_{\mathrm{B}}^{\mathrm{e}}$ takes the value 1, as it obviously should. A similar analysis for failure in bending gives the shape factor

$$\psi_{\mathrm{B}}^{\mathrm{f}} = \left(\frac{\rho_{\mathrm{s}}}{\rho}\right)^{1/2} \tag{11.20}$$

Torsion, as always, is more difficult. When the structure of Figure 11.10(c), which has circular symmetry, is twisted, its rings act like concentric tubes and for these

$$\psi_{\mathrm{T}}^{\mathrm{e}} = \frac{\rho_{\mathrm{s}}}{\rho} \quad \text{and} \quad \psi_{\mathrm{T}}^{\mathrm{f}} = \left(\frac{\rho_{\mathrm{s}}}{\rho}\right)^{1/2} \tag{11.21}$$

The others have lower torsion stiffness and strength (and thus lower micro-shape factors) for the same reason that I-sections, good in bending, perform poorly in torsion.

Structuring, then, converts a solid with modulus E_s and strength $\sigma_{f,s}$ to a new solid with properties E and σ_f. If this new solid is formed to an efficient macroscopic shape (a tube, say, or and I-section) its bending stiffness, to take an example, increases by a further factor of ϕ_B^e. Then the stiffness of the beam, expressed in terms of that of the solid of which it is made, is

$$S = \psi_B^e \phi_B^e S_s \tag{11.22}$$

that is, the shape factors multiply. The same is true for strength.

This is an example of structural hierarchy and the benefits it brings. It is possible to extend it further: the individual cell walls or layers could, for instance, be structured, giving a third multiplier to the overall shape factor, and these units, too could be structured. Nature does this to good effect, but for man-made structures there are difficulties. There is the obvious difficulty of manufacture, imposing economic limits on the levels of structuring. And there is the less obvious one of reliability. If the structure is optimized at all levels of structure, then a failure of a member at any one level can trigger failure at the level above, causing a cascade that ends with the failure of the structure as a whole. The more complex the structure, the harder it becomes to ensure the integrity at all levels. This might be overcome by incorporating redundancy (or a safety factor) at each level, but this implies a cumulative loss of efficiency. Two levels of structure are usually as far as it is practical to go.

As pointed out earlier, a micro-structured material can be thought of as a new material. It has a density, a strength, a thermal conductivity, and so on; difficulties arise only if the sample size is comparable to the cell size, when "properties" become size dependent. This means that micro-structured materials can be plotted on the material charts—indeed, wood appears on them already—and that all the selection criteria developed in Chapter 5 apply, unchanged, to the micro-structured materials. This line of thinking is developed further in Chapter 13, which includes material charts for a range of natural materials.

11.4 Limits to shape efficiency

The conclusions so far: if you wish to make stiff, strong structures that are efficient (using as little material as possible) then make the shape-efficiency factors as large as possible. It would seem, then, that the bigger the value of ϕ the better. True, but there are limits. We examine these next.

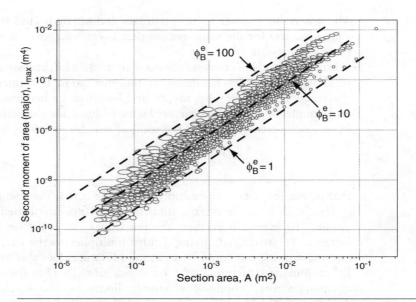

Figure 11.11 *Log(I) plotted against log(A) for standard sections of steel, aluminum, pultruded GFRP and wood. Contours of ϕ_B^e are shown, illustrating that there is an upper limit. A similar plot for log(Z) plotted against log(A) reveals an upper limit for ϕ_B^f. The limits are material-dependent.*

Empirical limits. There are practical limits for the slenderness of sections, and these determine, for a given material, the maximum attainable efficiencies. These limits may be imposed by manufacturing constraints: the difficulty or expense of making an efficient shape may simply be too great. More often they are imposed by the properties of the material itself because these determine the failure mode of the section. We explore these limits in two ways. The first is empirical: by examining the shapes in which real materials — steel, aluminum, etc. — are actually made, recording the limiting efficiency of available sections. The second is by the analysis of the mechanical stability of shaped sections.

Standard sections for beams, shafts, and columns are generally prismatic. Prismatic shapes are easily made by rolling, extrusion, drawing, pultrusion, or sawing. The section may be solid, closed-hollow (like a tube or box) or open-hollow (an I-, U-, or L-section, for instance). Each class of shape can be made in a range of materials. Some are available as standard, off-the-shelf, sections, notably structural steel, extruded aluminum alloy, pultruded GFRP (glass fiber reinforced polyester or epoxy), and structural timber. Figure 11.11 shows values for I and A (the axes of Figure 11.5) for 1880 standard sections made from these four materials, with contours of the shape factor ϕ_B^e superimposed. Some of the sections have $\phi_B^e \approx 1$; they are the ones with solid cylindrical or square sections. More interesting is that none have values of ϕ_B^e greater than about 65; there is an *upper limit* for the shape. A similar plot for Z and A (the axes of Figure 11.6) indicates an upper limit for ϕ_B^f of about 15. When these data are segregated by material[4] it is found that each has its own upper limit of shape, and that they differ greatly. Similar limits hold for the torsional shape factors. They are listed in Table 11.4 and plotted as shaded bands on Figures 11.5 and 11.6. The point: there is an upper limit for shape efficiency, and it depends on material. The ranges of ϕ_B^e and ϕ_B^f of standard sections appear on Figures 11.5 and 11.6 as shaded bands.

The upper limits for shape efficiency are important. They are central to the design of structures that are light, or which, for other reasons (cost, perhaps) the material content should be minimized. Two questions then arise. What sets the upper limit on shape efficiency of Table 11.4? And why does the limit depend on material? One explanation is simply the difficulty of making them. Steel, for example, can be drawn to thin-walled tubes or formed (by rolling, folding or welding) into efficient I-sections; shape factors as high as 50 are common. Wood cannot so easily be shaped; plywood technology could, in principle, be used to make thin tubes or I-sections, but in practice shapes with values of ϕ_B^e greater than 5 are uncommon. That is a manufacturing constraint. Composites, too, can be limited by the present difficulty in making them into thin-walled shapes, though the technology for doing this now exists.

But there is a more fundamental constraint on shape efficiency. It has to do with local buckling.

Limits imposed by local buckling. When efficient shapes *can* be fabricated, the limits of the efficiency are set by the competition between failure modes. Inefficient sections fail in a simple way: they yield, they fracture, or they suffer large-scale buckling. In seeking greater efficiency, a shape is chosen that raises the load required for the simple failure modes, but in doing so the structure is pushed nearer the load at which new modes—particularly those involving *local* buckling—become dominant. It is a characteristic of shapes that approach their limiting efficiency that two or more failure modes occur at almost the same load.

Why? Here is a simple-minded explanation. If failure by one mechanism occurs at a lower load than all others, the section shape can be adjusted to suppress it; but this pushes the load upwards until another mechanism becomes dominant. If the shape is described by a single variable (ϕ) then when two mechanisms occur at the same load you have to stop—no further shape adjustment can improve things. Adding webs, ribs or other stiffeners, gives additional variables, allowing shape to be optimized further, but we shall not pursue that here.

The best way to illustrate this is with a simple example. Think of a drinking straw—it is a thin-walled hollow tube about 5 mm in diameter. It is made of polystyrene, but there is not much of it. If the straw were collapsed into a solid cylinder, the cylinder would have a diameter of less than 1 mm, and because

---

[4] Birmingham and Jobling, 1996; Weaver and Ashby, 1998.

Table 11.4   Empirical upper limits for the shape factors $\phi_B^e, \phi_T^e, \phi_B^f$ and ϕ_T^f

Material	$(\phi_B^e)_{max}$	$(\phi_T^e)_{max}$	$(\phi_B^f)_{max}$	$(\phi_T^f)_{max}$
Structural steel	65	25	13	7
6061 aluminum alloy	44	31	10	8
GFRP and CFRP	39	26	9	7
Polymers (e.g. nylons)	12	8	5	4
Woods (solid sections)	5	1	3	1
Elastomers	< 6	3	–	–

polystyrene has a low modulus, it would have a low bending stiffness. If bent sufficiently it would fail by plastic yielding, and if bent further, by fracturing. Now restore it to its straw-shape and bend it. It is much stiffer than before, but as it bends it ovalizes and then fails suddenly and unpredictably by kinking — a form of local buckling (try it).

A fuller analysis[5] indicates that the maximum practical shape efficiency — when not limited by manufacturing constraints — is indeed dictated by the onset of local buckling. A thick-walled section, loaded in bending, yields before it buckles locally. It can be made more efficient by increasing its slenderness in ways that increase I and Z, thereby increasing both its stiffness and the load it can carry before it yields, but reducing the load at which the increasingly slender walls of the section start to buckle. When the load for local buckling falls below that for yield, it fails by buckling — and this is undesirable because buckling is strongly defect-dependent and can lead to sudden, unpredictable, collapse. The implication drawn from detailed versions of plots like that of Figure 11.11 is that real sections are designed to avoid local buckling, setting the upper limit on shape efficiency. Not surprisingly, the limit depends on the material — those with low strength and high modulus yield easily but do not easily buckle, and vice versa. A rough rule-of-thumb follows from this: it is that

$$(\phi_B^e)_{max} \approx 2.3 \left(\frac{E}{\sigma_f} \right)^{1/2} \tag{11.23a}$$

and[6]

$$(\phi_B^f)_{max} \approx \sqrt{(\phi_B^e)_{max}} \tag{11.23b}$$

allowing approximate estimates for the maximum shape efficiency of materials. Wood, according to these equations, is capable of far greater shape efficiency than that of standard timber sections. This high efficiency can be realized by plywood technology, but such sections are not standard.

Much higher efficiencies are possible when precise loading conditions are known, allowing customized application of stiffeners and webs to suppress

---

[5] See Gerard, G. (1956) and Weaver and Ashby (1998) in Further reading.
[6] A consequence of the fact that $I / Zh \approx 0.5$ where h is the depth of the section.

local buckling. This allows a further increase in the ϕ's until failure or new, localized, buckling modes appear. These, too, can be suppressed by a further hierarchy of structuring; ultimately, the ϕ's are limited only by manufacturing constraints. But this is getting more sophisticated than we need for a general selection of material and shape. Equations (11.23) will do.

11.5 Exploring and comparing structural sections

Stiffness-limited design. The E-ρ material property chart displays properties of materials from which sections can be made. The shape-efficiency chart of Figure 11.5 captures information about the influence of shape on stiffness. By linking them[7] the performance of the section can be explored. In Figure 11.12 the two charts are placed in opposite corners of a square. The material property chart (here much simplified, with only a few materials) is at the top left. The shape-efficiency chart is at the lower right, with its axes interchanged so that I is along the bottom and A is up the side; shaded bands on it show the populated areas, derived from plots like that of Figure 11.11. The remaining two quadrants automatically form two more charts, each sharing axes with the first two. That at the upper right has axes of E and I; the diagonal lines show the bending stiffness of the section, EI. That at the lower left has axes of A and ρ; the diagonal contours show the metric of performance: the mass per unit length, $m_l = \rho A$, of the section.

This set of charts enables the assessment and comparison of stiffness-limited sections. It can be used in several ways, of which the following is typical. It is illustrated in Figures 11.13 and 11.14.

- Chose a material for the section and mark its modulus E and density ρ onto the *materials chart* in the first quadrant of the figure.
- Choose the desired section stiffness (EI); it is a constraint that must be met by the section. Extend a horizontal line from the value of E for the material to the appropriate contour in the *constraint chart* in the second quadrant.
- Drop a vertical from this point onto the *shape chart* in the third quadrant to the line describing the shape factor ϕ_B^e for the section. Values of I and A outside the shaded bands are forbidden.
- Extend a horizontal line from this point to the *performance chart* in the final quadrant (the one on the bottom left). Drop a vertical from the material density ρ in the material chart. The intersection shows the mass per unit length of the section.

The example of Figure 11.14 compares the mass of rolled steel and extruded aluminum sections both with $\phi_B^e = 10$, and timber sections with $\phi_B^e = 2$, with the constraint of a bending stiffness of 10^6 N.m^2. The extruded aluminum

---

[7] Birmingham and Jobling (1996) in Further reading.

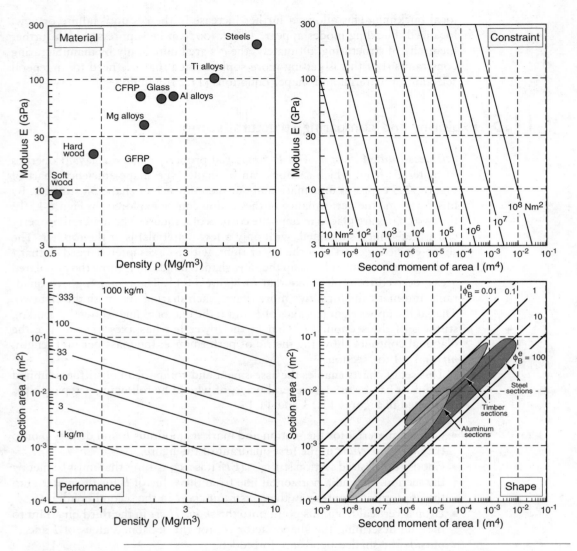

Figure 11.12   The 4-quadrant chart assembly for exploring structural sections for stiffness-limited design. Each chart shares its axes with its neighbors.

section gives the lightest beam. Remarkably, an efficiently shaped steel beam is nearly as light — for a given bending stiffness — as is one made of timber, even though the density of steel is 12 times greater than that of wood. It is because of the higher shape factor possible with steel.

Strength-limited design. The reasoning follows a similar path. In Figure 11.15 the strength-density $(\sigma_f - \rho)$ *material property chart* is placed at

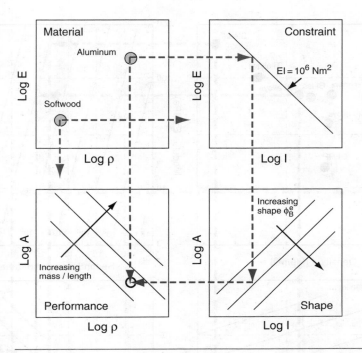

Figure 11.13 A schematic showing how the 4-quadrant chart is used.

the top left. The *Z-A shape chart* (Figure 11.6 with the axes interchanged) is at the bottom right. As before, the remaining two quadrants generate two more charts. This time the *constraint chart* at the upper right has axes of σ_f and Z; the diagonal lines show the bending strength of the section, $Z\sigma_f$. The *performance chart* at the lower left has the same axes as before — A and ρ — and the diagonal contours again show the metric of performance: the mass per unit length, $m_l = \rho A$, of the section.

It is used in the same way as that for stiffness-limited design. Try it for steel with $\phi_B^f = 15$, aluminum with $\phi_B^f = 10$ and GFRP with $\phi_B^f = 5$, for a required bending strength of $10^4 \, \text{N m}$. You will find that GFRP offers the lightest solution.

11.6 Material indices that include shape

The chart-arrays of Figures 11.12 and 11.15 link material, shape, constraint, and performance objective in a graphical, though rather clumsy, way. There is a neater way — it is to build them into the material indices of Chapter 5. Remember that most indices do not need this refinement — the performance

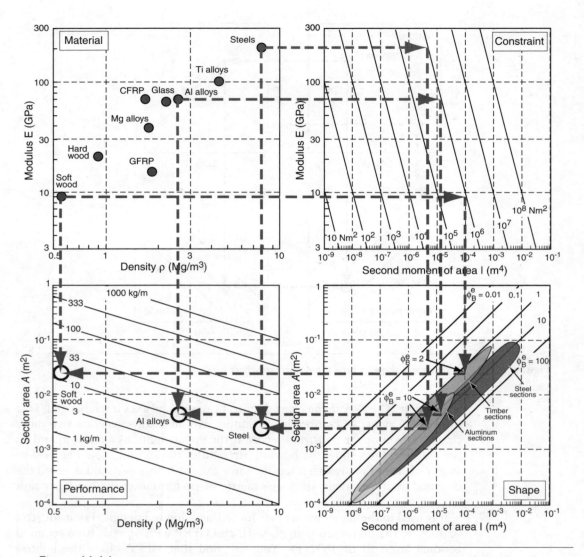

Figure 11.14 A comparison of steel, aluminum and wood section for stiffness-limited design with $EI = 10^6 \, \text{N.m}^2$. Aluminum gives the lightest section.

they characterize does not depend on shape. But stiffness and strength-limited design do. The indices for these can be adapted to include the relevant shape factor, such that they characterize material-shape combinations.

The method is illustrated below for minimum weight design. It can be adapted to other objectives in obvious ways. It follows the derivations of Chapter 5, with one extra step to bring in the shape.

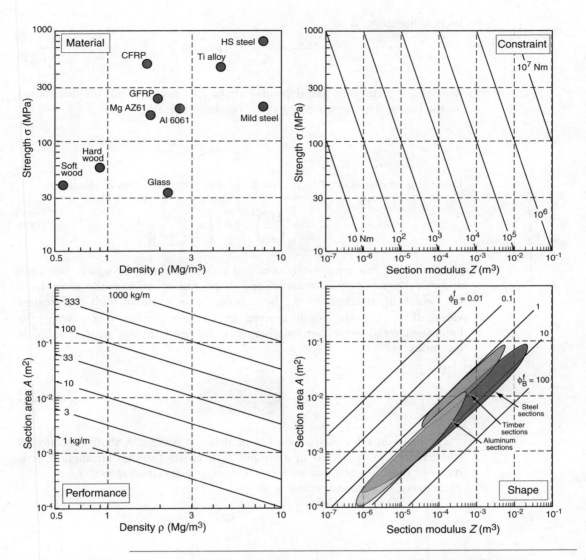

Figure 11.15 The 4-quadrant chart assembly for exploring structural sections for strength-limited design. Like those for stiffness, each chart shares its axes with its neighbors.

Elastic bending of beams and twisting of shafts. Consider the selection of a material for a beam of specified bending stiffness S_B and length L (the constraints), to have minimum mass, m (the objective). The mass m of a beam of length L and section area A is given, as before, by

$$m = AL\rho \qquad (11.24)$$

Its bending stiffness is

$$S_B = C_1 \frac{EI}{L^3} \tag{11.25}$$

where C_1 is a constant that depends only on the way the loads are distributed on the beam. Replacing I by ϕ_B^e using equation (11.3) gives

$$S_B = \frac{C_1}{12} \frac{E}{L^3} \phi_B^e A^2 \tag{11.26}$$

Using this to eliminate A in equation (11.24) gives the mass of the beam.

$$m = \left(\frac{12S_B}{C_1}\right)^{1/2} L^{5/2} \left[\frac{\rho}{(\phi_B^e E)^{1/2}}\right] \tag{11.27}$$

Everything in this equation is specified except the term in square brackets, and this depends only on material and shape. For beams with the *same* shape (for which ϕ_B^e is constant) the best choice is the material with the greatest value of $E^{1/2}/\rho$—the result derived in Chapter 5. But if we want the lightest material-shape combination, it is the one with the greatest value of the index

$$M_1 = \frac{(\phi_B^e E)^{1/2}}{\rho} \tag{11.28}$$

The procedure for elastic twisting of shafts is similar. A shaft of section A and length L is subjected to a torque T. It twists through an angle θ. It is required that the torsional stiffness, T/θ, meet a specified target, S_T, at minimum mass. Its torsional stiffness is

$$S_T = \frac{KG}{L} \tag{11.29}$$

where G is the shear modulus. Replacing K by ϕ_T^e using equation (11.7) gives

$$S_T = \frac{G}{7.14L} \phi_T^e A^2 \tag{11.30}$$

Using this to eliminate A in equation (11.24) gives

$$m = \left(7.14 \frac{S_T}{L^3}\right)^{1/2} L^{3/2} \left[\frac{\rho}{(\phi_T^e G)^{1/2}}\right] \tag{11.31}$$

The best material-and-shape combination is that with the greatest value of $(\phi_T^e G)/\rho^{1/2}$. The shear modulus, G, is closely related to Young's modulus E. For the practical purposes we approximate G by $3/8E$; when the index becomes

$$M_2 = \frac{\left(\phi_T^e E\right)^{1/2}}{\rho} \qquad (11.32)$$

For shafts of the same shape, this reduces to $E^{1/2}/\rho$ again. When shafts differ in both material and shape, the material index (equation (11.32)) is the one to use.

Equations (11.27) and (11.31) give a way of calculating shape factors for complex structures like bridges and trusses. Inverting equation (11.27), for example, gives

$$\phi_B^e = \frac{12 S_B}{C_1} \frac{L^5}{m^2} \left[\frac{\rho^2}{E}\right]$$

Thus if the mass of the structure, its length and its bending stiffness are known (as they are for large bridge-spans) and the density and modulus of the material of which existing bridge spans gives values between 50 and 200. This are large values, larger than the maximum values in Table 11.4. It is because the bridges are "structured-structures" with two or more levels of structure. The high values of ϕ are examples of the way in which shape efficiency can be increased by a hierarchy of structuring.

Failure of beams and shafts. The procedure is the same. A beam of length L, loaded in bending, must support a specified load F without failing and be as light as possible. When section-shape is a variable the best choice is found as follows. Failure occurs if the load exceeds the moment

$$M = Z\sigma_f$$

where Z is the section modulus and σ_f is the stress at which failure occurs. Replacing Z by the shape-factor ϕ_B^f of equation (11.10) gives

$$M = \frac{\sigma_f}{6} \phi_B^f A^{3/2} \qquad (11.33)$$

Substituting this into equation (11.24) for the mass of the beam gives

$$m = (6M)^{2/3} L \left[\frac{\rho^{3/2}}{\phi_B^f \sigma_f}\right]^{2/3} \qquad (11.34)$$

The best material-and-shape combination is that with the greatest value of the index

$$M_3 = \frac{\left(\phi_B^f \sigma_f\right)^{2/3}}{\rho} \qquad (11.35)$$

A similar analysis for torsional failure gives:

$$M_4 = \frac{\left(\phi_T^f \sigma_f\right)^{2/3}}{\rho} \qquad (11.36)$$

At constant shape both indices reduce to the familiar $\sigma_f^{2/3}/\rho$ of Chapter 5; but when shape as well as material are to be compared, the full index must be used.

11.7 Co-selecting material and shape

The indices allow comparison and selection of material-shape combinations. We conclude the chapter by exploring these.

Co-selection by calculation. Consider as an example the selection of a material for a stiff *shaped* beam of minimum mass. Four materials are available, listed in Table 11.5, each with their properties and typical (modest) values for shapes, characterized by ϕ_B^e. We seek the combination with the largest value of the index M_1 of equation (11.28) which, repeated, is

$$M_1 = \frac{\left(\phi_B^e E\right)^{1/2}}{\rho}$$

It identifies the material-shape combinations with the lowest mass for a given stiffness. The second last column of the table shows the simple "fixed shape" index $E^{1/2}/\rho$: wood has the greatest value, more than twice as great as steel. But when each material is shaped efficiently (last column) wood has the *lowest* value of M_1 — even steel is better; the aluminum alloy wins, surpassing steel and GFRP.

Selection for strength follows a similar routine, using the index M_3 of equation (11.35).

Graphical co-selection. The material index for elastic bending (equation (11.28)) can be rewritten as

$$M_1 = \frac{\left(\phi_B^e E\right)^{1/2}}{\rho} = \frac{\left(E/\phi_B^e\right)^{1/2}}{\rho/\phi_B^e} = \frac{E^{*1/2}}{\rho^*} \qquad (11.37)$$

Table 11.5   The selection of material and shape for a light, stiff, beam

Material	$\rho(\mathrm{Mg/m^3})$	$E(\mathrm{GPa})$	ϕ_B^e	$E^{1/2}/\rho$	$\left(\phi_B^e E\right)^{1/2}/\rho$
1020 Steel	7.85	205	20	1.8	8.2
6061-T4 Al	2.7	70	15	3.1	**12.0**
GFRP (isotropic)	1.75	28	8	2.9	8.5
Wood (oak)	0.9	13.5	2	**4.1**	5.8

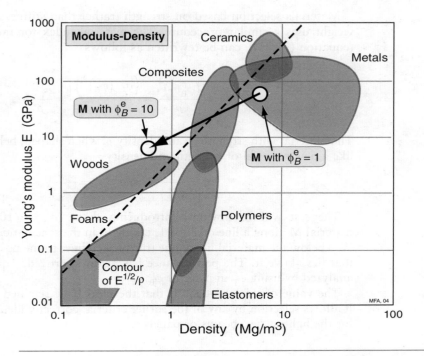

Figure 11.16   The structured material behaves like a new material with a modulus $E^* = E/\phi_B^e$ and a density $\rho^* = \rho/\phi_B^e$, moving it from a position below the broken selection line to one above. A similar procedure can be applied for bending strength, as described in the text.

The equation says: a material with modulus E and density ρ, when structured, can be thought of as a new material with a modulus and density of

$$E^* = \frac{E}{\phi_B^e} \quad \text{and} \quad \rho^* = \frac{\rho}{\phi_B^e} \qquad (11.38)$$

The E-ρ chart is shown schematically in Figure 11.16. The "new" material properties E^* and ρ^* can be plotted onto it. Introducing shape (e.g. $\phi_B^e = 10$) moves the material **M** to the lower left along a line of slope 1, from the position E, ρ to the position $E/10$, $\rho/10$ as shown in the figure. The selection criteria are plotted onto the figure as before: a constant value of the index of $E^{1/2}/\rho$ for instance, plots as a straight line of slope 2; it is shown, for one value of $E^{1/2}/\rho$ as a broken line. The introduction of shape has moved the material from a position below this line to one above; its performance has improved. Elastic twisting of shafts is treated in the same way.

Materials selection based on strength (rather than stiffness) at a minimum weight uses a similar procedure. The material index for failure in bending (equation (11.35)), can be rewritten as follows

$$M_3 = \frac{(\phi_B^f \sigma_f)^{\frac{2}{3}}}{\rho} = \frac{\left(\sigma_f / (\phi_B^f)^2\right)^{\frac{2}{3}}}{\rho / (\phi_B^f)^2} = \frac{\sigma_f^{*2/3}}{\rho^*} \tag{11.39}$$

The material with strength σ_f and density ρ, when shaped, behaves in bending like a new material of strength and density

$$\sigma_f^* = \frac{\sigma_f}{(\phi_B^f)^2} \quad \text{and} \quad \rho^* = \frac{\rho}{(\phi_B^f)^2} \tag{11.40}$$

The rest will be obvious. Introducing shape ($\phi_B^f = \sqrt{10}$, say) moves a material **M** along a line of slope 1, taking it, in the schematic, from a position σ_f, ρ below the material index line (the broken line) to the position $\sigma_f/10$, $\rho/10$ that lies above it. The performance has again improved. Torsional failure is analyzed by using ϕ_T^f in place of ϕ_B^f.

The value of this approach is that the plots have retained their generality. It allows selection by any of the earlier criteria, correctly identifying materials for the lightest ties, beams or panels.

11.8 Summary and conclusions

The designer has two groups of variables with which to optimize the performance of a load-bearing component: the material properties and the shape of the section. They are not independent. The best choice of material, in a given application, depends on the shapes in which it is available or to which it could potentially be formed.

The contribution of shape is isolated by defining four shape factors. The first, ϕ_B^e, is for the elastic bending and buckling of beams; the second, ϕ_T^e, is for the elastic twisting of shafts; the third, ϕ_B^f is for the plastic failure of beams loading in bending; and the last, ϕ_T^f, is for the plastic failure of twisted shafts (Table 11.6). The shape factors are dimensionless numbers that characterize the efficiency of use of the material in each mode of loading. They are defined

Table 11.6   Definitions of shape factors*

Design constraint*	Bending	Torsion
Stiffness	$\phi_B^e = \dfrac{12I}{A^2}$	$\phi_T^e = \dfrac{7.14K}{A^2}$
Strength	$\phi_B^f = \dfrac{6Z}{A^{3/2}}$	$\phi_T^f = \dfrac{4.8Q}{A^{3/2}}$

* A; I, K, Z, and Q are defined in the text and tabulated in Table 11.2.

such that all four have the value 1 for a solid square section. With this definition, all equiaxed solid sections (solid cylinders and hexagonal and other polygonal sections) have shape factors of close to 1, but efficient shapes that disperse the material far from the axis of bending or twisting (I-beams, hollow tubes, box-sections, etc.) have values that are much larger. They are tabulated for common shapes in Table 11.3.

The idea of micro-structural shape factors (symbol ψ) is introduced to characterize the efficiency, in bending and torsion, of cellular, layered and other small-scale structures, common in nature. They are defined in the same way as the ϕs. The difference is that microscopic shape is repeated; structures with microscopic shape are *extensive*. They can be thought of either as a solid with properties E_s, $\sigma_{f,s}$ and ρ_s, with a microscopic shape factor of ψ, or as a new material, with a new set of properties, E_s/ψ, ρ_s/ψ, etc., with a shape-factor of 1. Wood is an example: it can be seen as solid cellulose and lignin shaped to the cells of wood, or as wood itself, with a lower density, modulus and strength then cellulose, but with greater values of indices $E^{1/2}/\rho$ and $\sigma_f^{2/3}/\rho$ that characterize structural efficiency. When micro-structured materials (ψ) are given macroscopic shape (ϕ) the total shape factor is then the product $\phi\,\psi$, and this can be large.

The shapes to which a material can, in practice, be made are limited by manufacturing constraints and by the restriction that the section should yield before it suffers local buckling. These limits can be plotted onto a "shape chart" which, when combined with a material property chart in a four-chart array (Figures 11.12 and 11.15) allow the potential of alternative material-shape combinations to be explored. While this is educational, there is a more efficient alternative: that of developing indices that include the shape factors. The best material-shape combination for a light beam with a prescribed bending stiffness is that which maximizes the material index

$$M_1 = \frac{\left(E\phi_B^e\right)^{1/2}}{\rho}$$

The material-shape combination for a light beam with a prescribed strength is that which maximizes the material index

$$M_3 = \frac{\left(\phi_B^f \sigma_f\right)^{2/3}}{\rho}$$

Similar combinations involving ϕ_T^e and ϕ_T^f gives the lightest stiff or strong shaft. Here, the criterion of "performance" was that of meeting a design specification at minimum weight. Other such material-shape combinations maximize other performance criteria: minimizing cost rather than weight, for example, or maximizing energy storage.

The procedure for selecting material-shape combinations is best illustrated by examples. These can be found in the next chapter.

11.9 Further reading

Ashby, M.F. (1991) Material and shape, *Acta Metall. Mater.* **39**, 1025–1039. (*The paper in which the ideas on which this chapter is based were first developed.*)

Birmingham, R.W. and Jobling, B. (1996) Material selection: comparative procedures and the significance of form, International Conference on Lightweight Materials in Naval Architecture, The Royal Institution of Naval Architects, London. (*The article in which the 4-quadrant charts like those of Figures 11.12 and 11.15 were introduced.*)

Gerard, G. (1956) *Minimum Weight Analysis of Compression Structures*, New York University Press, New York, USA. Library of Congress Catalog Number 55–10052. (*This and the book by Shanley, cited below, establish the principles of minimum weight design. Both, unfortunately, are out of print but can be found in libraries.*)

Gere, J.M. and Timoshenko, S.P. (1985) *Mechanics of Materials*, Wadsworth International, London, U.K. (*An introduction to the mechanics of elastic solids.*)

Gibson, L.J. and Ashby, M.F. (1997) *Cellular Solids*, 2nd edition, Cambridge University Press, Cambridge, UK. ISBN 0-521-49560-1. (*A broad-based introduction to the structure and properties of foams and cellular solids of all types.*)

Parkhouse, J.G. (1984) Structuring: A Process of Material Dilution, in Nooshin (ed.), Third International Conference on Space Structures, Elsevier, London, UK, p. 367. (*Parkhouse develops an unusual approach to analysing material efficiency in structures.*)

Shanley, F.R. (1960) *Weight-Strength Analysis of Aircraft Structures*, 2nd edition, Dover Publications, New York, U.S.A. Library of Congress Catalog Number 60-501011. (*This and the book by Gerard, cited above, establish the principles of minimum weight design. Both, unfortunately, are out of print but can be found in libraries.*)

Timoshenko, S.P. and Gere, J.M. (1961) *Theory of Elastic Stability*, McGraw-Hill Koga Kusha Ltd., London, U.K. Library of Congress Catalog Number 59-8568. (*The definitive text on elastic buckling.*)

Weaver, P.M. and Ashby, M.F. (1998) Material limits for shape efficiency, *Prog. Mater. Sci.* **41**, 61–128. (*A review of the shape-efficiency of standard sections, and the analyses leading to the results used in this chapter for the maximum practical shape factors for bending and twisting.*)

Young, W.C. (1989) *Roark's Formulas for Stress and Strain*, 6th edition, McGraw-Hill, New York, USA. ISBN 0-07-100373-8. (*A sort of Yellow Pages of formulae for stress and strain, cataloging the solutions to thousands of standard mechanics problems.*)

Chapter contents

12.1 Introduction and synopsis

This chapter, like Chapters 6 and 10, is a collection of case studies. They illustrate the use of the 4-quadrant chart and of material indices that include shape. Remember: they are only necessary for the restricted class of problems in which section shape directly influences performance, that is, when the prime function of a component is to carry loads that cause it to bend, twist, or buckle.

Indices that include shape provide a tool for optimizing the co-selection of material-and-shape. The important ones are summarized in Table 12.1. The selection procedure is, first, to identify candidate materials and the section shapes in which each is available, or could be made. The relevant material properties and shape factors for each are tabulated and the relevant index is evaluated. The best material-and-shape combination is that with the greatest value of the index. The same information can be plotted onto material selection charts, allowing a graphical solution to the problem — one that often suggests further possibilities.

Table 12.1a   Indices with shape: stiffness and strength-limited design at minimum weight

Component shape, loading and constraints	Stiffness-limited design	Strength-limited design
Tie (tensile member) Load, stiffness and length specified, section-area free	$\dfrac{E}{\rho}$	$\dfrac{\sigma_f}{\rho}$
Beam (loaded in bending) Loaded externally or by self weight, stiffness, strength and length specified, section area and shape free	$\dfrac{\left(\phi_B^e E\right)^{1/2}}{\rho}$	$\dfrac{\left(\phi_B^f \sigma_f\right)^{2/3}}{\rho}$
Torsion bar or tube Loaded externally, stiffness, strength and length specified, section area and shape free	$\dfrac{\left(\phi_T^e E\right)^{1/2}}{\rho}$	$\dfrac{\left(\phi_T^f \sigma_f\right)^{2/3}}{\rho}$
Column (compression strut) Collapse load by buckling or plastic crushing and length specified, section area and shape free	$\dfrac{\left(\phi_B^e E\right)^{1/2}}{\rho}$	$\dfrac{\sigma_f}{\rho}$

Table 12.1b   Indices with shape: springs, specified energy storage at minimum volume or weight

Component shape, loading and constraints	Flexural springs	Torsion springs
Spring: Specified energy storage, volume to be minimized	$\dfrac{\left(\phi_B^f\right)^2}{\phi_B^e}\dfrac{\sigma_f^2}{E}$	$\dfrac{\left(\phi_T^f\right)^2}{\phi_T^e}\dfrac{\sigma_f^2}{E}$
Spring: Specified energy storage, mass to be minimized	$\dfrac{\left(\phi_B^f\right)^2}{\phi_B^e}\dfrac{\sigma_f^2}{E\rho}$	$\dfrac{\left(\phi_T^f\right)^2}{\phi_T^e}\dfrac{\sigma_f^2}{E\rho}$

For minimum cost design, replace ρ by $C_m\rho$ in the indices.

The method has other uses. It gives insight into the way in which natural materials — many of which are very efficient — have evolved. Bamboo is an example: it has both internal or microscopic shape and a tubular, macroscopic shape, giving it very attractive properties. This and other aspects are brought out in the case studies that follow.

12.2 Spars for man-powered planes

Most engineering design is a difficult compromise: it must meet, as best it can, the conflicting demands of multiple objectives and constraints. But in designing a spar for a man-powered plane the objective is simple: the spar must be as light as possible, and still be stiff enough to maintain the aerodynamic efficiency of the wings (Table 12.2). Strength, safety, even cost, hardly matter when records are to be broken. The plane (Figure 12.1) has two main spars: the transverse spar supporting the wings, and the longitudinal spar carrying the tail assembly. Both are loaded primarily in bending (torsion cannot, in reality, be neglected, although we shall do so here).

Some 60 man-powered planes have flown successfully. Planes of the first generation were built of balsa wood and spruce. The second generation relied on aluminum tubing for the load-bearing structure. The present, third,

Table 12.2 Design requirements for wing spars

Function	Wing spar
Constraints	• Specified stiffness
	• Length specified
Objective	Minimum mass
Free variables	• Choice of material
	• Section shape and scale

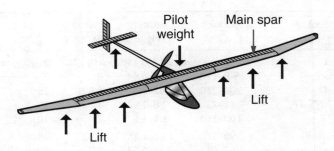

Figure 12.1 The loading on a man-powered plane is carried by two spars, one spanning the wings and the other linking the wings to the tail. Both are designed for stiffness at minimum weight.

generation uses carbon–fiber/epoxy spars, molded to appropriate shapes. How has this evolution come about? And how much further can it go?

The model and the selection. We seek a material-and-shape combination that minimizes mass for a given bending stiffness. The index to be maximized, read from Table 12.1, is

$$M_1 = \frac{\left(\phi_B^e E\right)^{1/2}}{\rho} \tag{12.1}$$

Data for seven materials are assembled in Table 12.3. If all have the same shape, M_1 reduces to the familiar $E^{1/2}/\rho$ and the ranking is that of the fourth column. Balsa and spruce are significantly better than the competition. Woods are extraordinarily efficient; that is why model aircraft builders use them now and the builders of real aircraft relied so heavily on them in the past. The effect of shaping the section, to a rectangle for the woods, to a box-section for aluminum and CFRP, gives the results in the last column. The shape factors listed here are typical of commercially available sections, and are well below the maximum for each material. Aluminum is now marginally better than the woods; CFRP is best of all.

The same information is shown graphically in Figure 12.2, using the method of Section 11.7. Each shape is treated as a new material with modulus $E^* = E/\phi_B^e$ and $\rho^* = \rho/\phi_B^e$. The values of E^* and ρ^* are plotted on the chart. The superiority of both the aluminum tubing with $\phi_B^e = 20$ and the CFRP box-sections with $\phi_B^e = 10$ are clearly demonstrated.

Postscript. Why is wood so good? With no shape it does as well as heavily-shaped steel. It is because wood *is* shaped: its cellular structure gives it internal micro-shape, increasing the performance of the material in bending; it is nature's answer to the I-beam. Advances in the technology of drawing thin-walled

Table 12.3   Materials for wing spars

Material	Modulus E (GPa)	Density ρ (Mg/m³)	Index* $E^{1/2}/\rho$	Shape Factor ϕ_B^e	Index* M_1 $\left(\phi_B^e E\right)^{1/2}/\rho$
Balsa	4.2–5.2	0.17–0.24	10	2	15
Spruce	9.8–11.9	0.36–0.44	8	2	12
Steel	200–210	7.82–7.84	1.8	25	9
Al 7075 T6	71–73	2.8–2.82	3	20	14
CFRP	100–160	1.5–1.6	7	10	23
Beryllium	290–310	1.82–1.86	9.3	15	36
Borosilicate glass	62–64	2.21–2.23	3.7	10	11

*The values of the index are based on mean values of the material properties. The units of the indices are (GPa)$^{1/2}$/(Mg/m³).

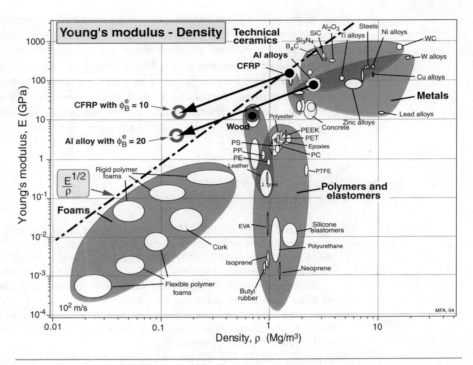

Figure 12.2 The materials-and-shapes for wing-spars, plotted on the modulus-density chart. A spar made of CFRP with a shape factor of 10 out-performs spars made of aluminum with a shape factor of 20, and wood with shape factor close to 1.

aluminum tubes allowed a shape factor that cannot be reproduced in wood, giving aluminum a performance edge — a fact that did not escape the designers of the second generation of man-powered planes. There is a limit of course: tubes that are too thin will kink (as described in Chapter 11), setting an upper limit to the shape factor for aluminum at about 40. Further advance required a new material with lower density and higher modulus, conditions met by CFRP. Add shape and CFRP out-performs them all.

Now lets be inventive. Can we do better? Not easily, but if it is really worth it, maybe. Chapter 13 develops methods for designing material combinations that out-perform anything that one material could do by itself, but we will leave these for later What can be done with a single material? If we rank them by $E^{1/2}/\rho$ we get a list headed by diamond, boron, and — ah! — beryllium, a metal used for space structures, and it can be shaped. After that come more ceramics, not viable because of their brittleness and the difficulty of forming them into any useful shape. And, far below, magnesium and aluminum alloys. But above these, in the middle of the ceramics, is ... glass. A glider with glass wing-spars? Sounds crazy, but think for a moment. Toughed glass, bullet-proof glass, glass floors; glass can be used as a structural material. And it is easy to shape. It is not even that expensive.

So suppose beryllium or glass were given shape — could either one out-perform CFRP? Again, not easily, but maybe. Here we have to guess. In theory (equation (11.23)) beryllium could be given a shape factor of 60, glass of 30. Being more realistic, factors of 15 and 10 are possible. The lower part of the table shows what all this means. Beryllium out-performs CFRP, and by a large margin. Glass doesn't. Well, it was a thought. For more, see Chapter 14.

Further reading

Man-powered flight

There is a large literature. Try:

Drela, M. and Langford, J.D. (1985) Man-powered flight, *Scientific American*, p. 122. (*A concise history of man powered flight up to 1985.*)

Nadel, E.R. and Bussolari, S.R. (1988) The Daedalus project: physiological problems and solutions, *American Scientist*, July–August. (*An account of the Daedalus project, a competition to for man-powered flight from Crete and mainland Greece — the mythical route of Daedalus and his father.*)

Grosser, M (1981) *Gossamer Odyssey*, Dover Publications, New York. ISBN 0-486-26645-1. (*Accounts of the Gossamer Condor and Albatross, attempts to capture the world record for man-powered flight.*)

Bliesner, W. (1991) The design and construction details of the marathon eagle, in *Technology for Human Powered Aircraft*, Proceedings of the Human-powered Aircraft Group Half Day Conference. The Royal Aeronautical Society, London, 30 January 1991. (*Details of another attempt to build a record-breaking plane.*)

Related case studies

6.5   Cost: structural materials for buildings
12.4   Forks for a racing bicycle
12.5   Floor joists: wood, bamboo or steel?

12.3 Ultra-efficient springs

Springs, we deduced in Section 6.7, store energy. They are best made of a material with a high value of σ_f^2/E, or, if mass is more important than volume, then of $\sigma_f^2/\rho E$. Springs can be made more efficient still by shaping their section. Just how much more is revealed below.

We take as a measure of performance the energy stored per unit volume of solid of which the spring is made; we wish to maximize this energy. Energy per unit weight and per unit cost are maximized by similar procedures (Table 12.4).

The model and the selection. Consider a leaf spring first (Figure 12.3a). A leaf spring is an elastically bent beam. The energy stored in a bent beam, loaded by

Table 12.4   Design requirements for ultra-efficient springs

Function	Material-efficient spring
Constraints	Must remain elastic under design loads
Objective	Maximum stored energy per unit volume (or mass, or cost)
Free variables	• Choice of material
	• Section shape

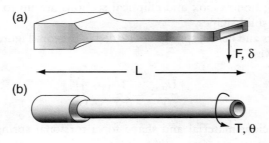

Figure 12.3   Hollow springs use material more efficiently than solid springs. Best in bending is the hollow rectangular or elliptical section; best in torsion is the tube.

a force F, is

$$U = \frac{1}{2}\frac{F^2}{S_B} \tag{12.2}$$

where S_B, the bending stiffness of the spring, is given by equation (11.25), or, after replacing I by ϕ_B^e, by equation (11.26), which, repeated, is:

$$S_B = \frac{C_1}{12}\frac{E}{L^3}\phi_B^e A^2 \tag{12.3}$$

The force F in equation (12.2) is limited by the onset of yield; its maximum value is

$$F_f = C_2 Z \frac{\sigma_f}{L} = \frac{C_2}{6L}\sigma_f\phi_B^f A^{3/2} \tag{12.4}$$

(The constants C_1 and C_2 are tabulated in Appendix A, Sections A.3 and A.4). Assembling these gives the maximum energy the spring can store:

$$\frac{U_{max}}{V} = \frac{C_2^2}{6C_1}\left(\frac{(\phi_B^f\sigma_f)^2}{\phi_B^e E}\right) \tag{12.5}$$

where $V = AL$ is the volume of solid in the spring. The best material and shape for the spring—the one that uses the least material—is that with the greatest value of the quantity

$$M_1 = \frac{\left(\phi_B^f \sigma_f\right)^2}{\phi_B^e E} \qquad (12.6)$$

For a fixed section shape, the ratio involving the two ϕ's is a constant: then the best choice of material is that with the greatest value of σ_f^2/E — the same result as before. When shape is a variable, the most efficient shapes are those with large $\left(\phi_B^f\right)^2/\phi_B^e$. Values for these ratios are tabulated for common section shapes in Table 12.5; hollow box and elliptical sections are up to three times more efficient than solid shapes.

Torsion bars and helical springs are loaded in torsion (Figure 12.3b). A similar calculation gives

$$\frac{U_{max}}{V} = \frac{1}{6.5} \frac{\left(\phi_T^f \sigma_f\right)^2}{\phi_T^e G} \qquad (12.7)$$

The most efficient material and shape for a torsional spring is that with the largest value of

$$M_2 = \frac{\left(\phi_T^f \sigma_f\right)^2}{\phi_T^e E} \qquad (12.8)$$

(where G has been replaced by $3E/8$). The criteria are the same: when shape is not a variable, the best torsion-bar materials are those with high values of σ_f^2/E. Table 12.5 shows that the best shapes are hollow tubes, with a ratio of $\left(\phi_T^f\right)^2/\phi_T^e$ that is twice that of a solid cylinder; all other shapes are less efficient. Springs that store the maximum energy per unit weight (instead of unit volume) are selected with indices given by replacing E by $E\rho$ in equations (12.6) and (12.8). For maximum energy per unit cost, replace E by $EC_m\rho$ where C_m is the material cost per kg.

Postscript. Hollow springs are common in vibrating and oscillating devices and for instruments in which inertial forces must be minimized. The hollow elliptical section is widely used for springs loaded in bending; the hollow tube for those loaded in torsion. More about this problem can be found in the classic paper by Boiten (1963).

Further reading   Design of efficient springs
Boiten, R.G. (1963) Mechanics of instrumentation, *Proc. Int. Mech. Eng.* **177**, 269. (*A definitive analysis of the mechanical design of precision instruments.*)

Related case   6.7   Materials for springs
studies

Table 12.5 Shape factors for the efficiency of springs

Section shape	$(\phi_B^f)^2/\phi_B^e$	$(\phi_T^f)^2/\phi_T^e$
	1	1
	1.33	0.63
	1	$\dfrac{(1 + a^2/b^2)}{2}$ $(a < b)$
	1.33	$\dfrac{2}{3}\left(1 - 0.6\left(\dfrac{b}{h} - \dfrac{b^2}{h^2}\right)\right)$ $(h < b)$
	0.5	0.53
	2 $(t \ll r)$	2 $(t \ll r)$
	2.67 $(t \ll b)$	2
	$\dfrac{1 + 3b/a}{1 + b/a}$ $(a < b)$	$\dfrac{2b^3(a + a^2/b^2)}{(ab)^{3/2}(1 + a/b)}$ $(a < b)$
	4	—
	$\dfrac{4}{3}\dfrac{(1 + 3b/h)}{(1 + b/h)}$	$\dfrac{2}{3}\dfrac{(1 + 4h/b)}{(1 + h/b)}$
	$\dfrac{3}{4}\dfrac{(1 + 4bt^2/h^3)}{(1 + b/h)}$	$\dfrac{2}{3}\dfrac{(1 + 4b/h)}{(1 + b/h)}$
	2	—

12.4 Forks for a racing bicycle

The first consideration in bicycle design (Figure 12.4) is strength (Table 12.6). Stiffness matters, of course, but the initial design criterion is that the frame and forks should not yield or fracture in normal use. The loading on the forks is predominantly *bending*. If the bicycle is for racing, then the mass is a primary consideration: the forks should be as light as possible. What is the best choice of material and shape?

The model and the selection. We model the forks as beams of length L that must carry a maximum load P (both fixed by the design) without plastic collapse or fracture. The forks are tubular, of radius r and fixed wall-thickness, t. The mass is to be minimized. The fork is a light, strong beam. Further details of load and geometry are unnecessary: the best material and shape, read from Table 12.1, is that with the greatest value of

$$M_3 = \frac{\left(\phi_B^f \sigma_f\right)^{2/3}}{\rho} \tag{12.9}$$

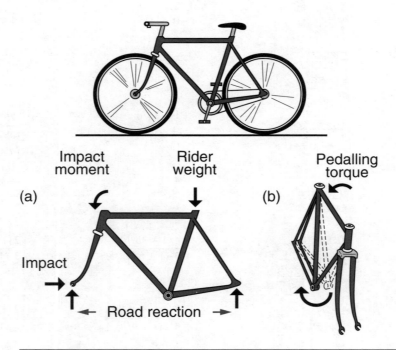

Figure 12.4   The bicycle. The forks are loaded in bending. The lightest forks that will not collapse plastically under a specified design load are those made of the material and shape with the greatest value of $\left(\phi_B^f \sigma_f\right)^{2/3}/\rho$.

Table 12.6   Design requirements for bicycle forks

Function	Bicycle forks
Constraints	Must not fail under design loads — a strength constraint
	• Length specified
Objective	Minimize mass
Free variables	• Choice of material
	• Section shape

Table 12.7   Material for bicycle forks

Material	Strength, σ_f (MPa)	Density, ρ (Mg/m³)	Shape factor, ϕ_B^f	Index*, $\sigma_f^{2/3}/\rho$	Index, $M_3^* \left(\phi_B^f \sigma_f\right)^{2/3}/\rho$
Spruce (Norwegian)	70–80	0.46–0.56	1	<u>36</u>	36
Bamboo	80–160	0.6–0.8	2.2	34	59
Steel (Reynolds 531)	770–990	7.82–7.83	7.5	12	48
Alu (6061–T6)	240–260	2.69–2.71	5.9	15	48
Titanium 6–4	930–980	4.42–4.43	5.9	22	72
Magnesium AZ 61	160–170	1.80–1.81	4.25	17	46
CFRP	300–450	1.5–1.6	4.25	33	<u>88</u>

*The values of the indices are based on mean values of the material properties. The units of the indices are $(MPa)^{2/3}/(Mg/m^3)$.

Table 12.7 lists seven candidate materials with their properties. If the forks were solid, meaning that $\phi_B^f = 1$, spruce wins (see the second last column of the table). Bamboo is special because it grows as a hollow tube with a macroscopic shape factor ϕ_B^f of about 2.2, giving it a bending strength that is much higher than solid spruce (last column). When shape is added to the other materials, however, the ranking changes. The shape factors listed in the table are achievable using normal production methods. Steel is good; CFRP is better; Titanium 6-4 is better still. In strength-limited applications magnesium is poor despite its low density.

Postscript. Bicycles have been made of all seven of the materials listed in the table — you can still buy bicycles made of six of them. Early bicycles were made of wood; present-day racing bicycles of steel, aluminum or CFRP, sometimes interleaving the carbon fibers with layers of glass or Kevlar to improve the fracture-resistance. Mountain bicycles, for which strength and impact resistance are particularly important, have steel or titanium forks.

The reader may be perturbed by the cavalier manner in which theory for a straight beam with an end load acting normal to it is applied to a curved beam loaded at an acute angle. No alarm is necessary. When (as explained in Chapter 5, Section 5.4) the variables describing the functional requirements (F), the geometry (G) and the materials (M) in the performance equation are separable, the details of loading and geometry affect the terms F and G but not M. This is an example: beam curvature and angle of application of load do not change the material index, which depends only on the design requirement of strength in bending at minimum weight.

Further reading   **Bicycle design**

Oliver, T. (1992) *Touring Bikes, a Practical Guide*, Crowood Press, Wiltshire, UK. ISBN 1-85223-339-7. (*A good source of information about bike materials and construction, with tables of data for the steels used in tube sets.*)

Sharp, A. (1979) *Bicycles and Tricycles, an Elementary Treatise on Their Design and Construction*, The MIT Press, Cambridge, MA, USA. ISBN 0-262-69066-7. (*A far from elementary treatise, despite its title, first published in 1977. It is the place to look if you need the mechanics of bicycles.*)

Watson, R. and Gray, M. (1978) *The Penguin Book of the Bicycle*, Penguin Books, Harmondsworth, UK. ISBN 0-1400-4297-0. (*Watson and Gray describe the history and use of bicyles. Not much on design, mechanics or materials.*)

Whitt, F.R. and Wilson, D.G. (1982) *Bicycling Science*, 2nd edition, The MIT Press, Cambridge, MA, USA. ISBN 0-262-73060-X. (*A book by two MIT professorial bike enthusiasts, more easily digestible than Sharp, and with a good chapter on materials.*)

Wilson, D.G. (1986) A short history of human powered vehicles, *The American Scientist*, **74**, 350. (*Typical Scientific American article: good content, balance and presentation. A good starting point.*)

Bike magazines such as *Mountain bike, Which bike?* and *Cycling and mountain biking* carry extensive tables of available bikes and their characteristics — type, maker, cost, weight, gear-set, etc.

Related case
studies

12.5 Floor joists: wood, bamboo or steel?

Floors are supported on *joists*: beams that span the space between the walls. Let us suppose that a joist is required to support a specified bending load (the "floor loading") without sagging excessively or failing; and it must be cheap. Traditionally, joists in the United States and Europe are made of wood with a rectangular section of aspect ratio 2:1, giving an elastic shape factor (Table 11.3) of $\phi_B^e = 2$, or of steel (Figures 6.7 and 12.5). In Asian countries bamboo, with a "natural" shape factor of about $\phi_B^e = 3.2$ (Figure 12.6), replaces wood in smaller buildings. But as wood becomes scarcer and buildings larger, steel replaces wood and bamboo as the primary structural material. Standard steel I-section joists have shape factors in the range $5 < \phi_B^e < 25$ (special I-sections can have much larger values). Are steel I-joists a better choice than wooden ones? Table 12.8 summarizes the design requirements.

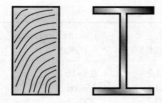

Figure 12.5   The cross-sections of a wooden beam ($\phi_B^e = 2$) and a steel I-beam ($\phi_B^e = 10$). The values of ϕ_B^e are calculated from the ratios of dimensions of each beam, using the formulae of Table 11.3.

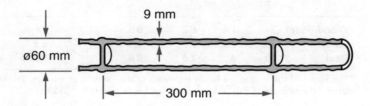

Figure 12.6   The cross-section of a typical bamboo cum. The tubular shape of that shown here gives in "natural" shape factors of $\phi_B^e = 3.2$ and $\phi_B^f = 2.2$ Because of this (and good torsional shape factors also) it is widely used for oars, masts, scaffolding and construction. Several bamboo bicycles have been marketed.

Table 12.8   Design requirements for floor joists

Function	Floor joist
Constraints	• Length specified
	• Stiffness specified
	• Strength specified
Objective	Minimum material cost
Free variables	• Choice of material
	• Section shape

The model and the selection. Consider stiffness first. The cheapest beam, for a given stiffness, is that with the largest value of the index (read from Table 12.1 with ρ replaced by $C_m\rho$ to minimize cost):

$$M_1 = \frac{(\phi_B^e E)^{1/2}}{C_m\rho} \tag{12.10}$$

Data for the modulus E, the density ρ, the material cost C_m and the shape factor ϕ_B^e are listed in Table 12.9, together with the values of the index M_1 with

Table 12.9a   Materials for floor joists

Material	Density, ρ (Mg/m^3)	Cost, C_m (US\$/kg)	Flexural modulus, E (GPa)	Flexural strength, σ_f (MPa)
Wood (Pine)	0.44–0.54	0.8–1.2	8.4–10	37–45
Bamboo	0.6–0.8	1.8–2.1	16–18	38–42
Steel	7.9–7.91	0.6–0.7	200–210	350–360

Table 12.9b   Indices for floor joists

Material	Shape factor, ϕ_B^e	Shape factor, ϕ_B^f	Index, $E^{1/2}/C_m\rho$	Index*, M_1 $\left(\phi_B^e E\right)^{1/2}/C_m\rho$	Index, $\sigma_f^{2/3}/C_m\rho$	Index*, M_2 $\left(\phi_B^f \sigma_f\right)^{2/3}/C_m\rho$
Wood (Pine)	2	1.4	6.2	8.8	24	38
Bamboo	3.2	2.2	3.0	5.4	8.6	14.5
Steel	10	4	2.8	8.8	9.7	24

*The values of the indices are based on means of the material properties. The units of the indices for elastic deflection are (GPa)$^{1/2}$/k\$/m^3; those for failure are (MPa)$^{2/3}$/k\$/m^3.

and without shape. The steel beam with $\phi_B^e = 25$ has a slightly larger value M_1 than wood, meaning that it is a little cheaper for the same stiffness.

But what about strength? The best choice for a light beam of specified strength is that which maximizes the material index :

$$M_2 = \frac{\left(\phi_B^f \sigma_f\right)^{2/3}}{C_m\rho} \qquad (12.11)$$

The quantities of failure strength σ_f shape factor ϕ_B^f and index M_2 are also given in the table. Wood performs better than even the most efficient steel I-beam.

Postscript. So the conclusion: as far as performance per unit material-cost is concerned, there is not much to choose between the standard wood and the standard steel sections used for joists. As a general statement, this is no surprise — if one were much better than the other, the other would no longer exist. But — looking a little deeper — wood dominates in certain market sectors, bamboo in others, and, in others still, steel. Why?

Wood and bamboo are indigenous to some countries and grow locally; steel has to come further, with associated transport costs. Assembling wood structures is easier than those of steel; it is more forgiving of mismatches of dimensions, it can be trimmed on site, you can hammer nails into it anywhere. It is a user-friendly material.

But wood is a variable material, and, like us, is vulnerable to the ravishes of time, prey to savage fungi, insects and small mammals with sharp teeth. The problems they create in a small building — a family home, say — are easily overcome, but in a large commercial building — an office block, for instance — they create greater risks, and are harder to fix. Here, steel wins.

Further reading   Cowan, H.J. and Smith, P.R. (1988) *The Science and Technology of Building Materials*, Van Nostrand Reinhold, New York, USA. ISBN 0-442-21799-4. (*A broad survey of materials for the structure, cladding, insulation and interior surfacing of buildings — and excellent introduction to the subject.*)

Farrelly, D. (1984) *The Book of Bamboo*, Sierra Club Books. ISBN 0-87156-825-X. (*An introduction to bamboo and its many varieties.*)

Janssen, J.J. (1995) *Building with Bamboo: A Handbook* ISBN 1-85339-203-0. (*Bamboo remains a building material of great importance, as well as the material of flooring, matting and basket-making. The techniques of building with bamboo have a long history, documented here.*)

12.6 Increasing the stiffness of steel sheet

How could you make steel sheet stiffer? There are many reasons you might wish to do so. The most obvious: to enable stiffness-limited sheet structures to be lighter than they are now; to allow panels to carry larger in-plane compressive loads without buckling; and to raise the natural vibration frequencies of sheet structures. Bending stiffness is proportional to $E \cdot I$ (E is Young's modulus, I is the second moment of area of the sheet, equal to $t^3/12$ per unit width where t is the sheet thickness). There is nothing much you can do to change the modulus of steel, which is always close to 210 GPa. The answer is to add a bit of shape, increasing I. So consider the design brief of Table 12.10.

The model and the selection. The age-old way to make sheet steel stiffer is to corrugate it, giving it a roughly sinusoidal profile. The corrugations increase the second moment of area of the sheet about an axis normal to the corrugations themselves. The resistance to bending in one direction is thereby increased, but in the cross-direction it is not changed at all.

Corrugations are the clue, but — to be useful — they must stiffen the sheet in all directions, not just one. A hexagonal grid of dimples (Figure 12.7) achieves

Table 12.10   Design requirements for stiffened steel sheet

Function	Steel sheet for stiffness-limited structures
Constraints	• Profile limited to a maximum deviation ± 5 times the sheet thickness from flatness • Cheap to manufacture
Objective	Maximize bending stiffness of sheet
Free variables	Section profile

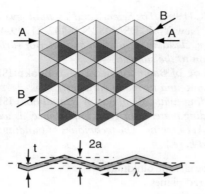

Figure 12.7   A sheet with a profile of adjacent hexagonal dimples that increases its bending stiffness and strength. Shape factors for the section **A–A** are calculated in the text. Those along other trajectories are lower but still significantly greater than 1.

this. There is now no direction of bending that is not dimpled. The dimples need not be hexagons; any pattern arranged in such a way that you cannot draw a straight line across it without intersecting dimples will do. But hexagons are probably about the best.

Consider an idealized cross-section as in the lower part of Figure 12.7, which shows the section A–A, enlarged. As before, we define the shape factor as the ratio of the stiffness of the dimpled sheet to that of the flat sheet from which it originated. The second moment of area of the flat sheet per unit width is

$$I_0 = \frac{t^3}{12} \tag{12.12}$$

That of the dimpled sheet with amplitude a is

$$I \approx \frac{1}{12}(2a + t)^2 t \tag{12.13}$$

giving a shape factor, defined as before as the ratio of the stiffness of the sheet before and after corrugating

$$\phi_B^e = \frac{I}{I_0} \approx \frac{(2a + t)^2}{t^2} \tag{12.14}$$

Note that the shape factor has the value unity when the amplitude is zero, but increases as the amplitude increases. The equivalent shape factor for failure in bending is

$$\phi_B^f = \frac{Z}{Z_0} \approx \frac{(2a + t)}{t} \tag{12.15}$$

These equations predict large gains in stiffness and strength. The reality is a little less rosy. This is because, while all cross-section of the sheet are dimpled, only those that cut through the peaks of the dimples have an amplitude equal to the peak height (all others have less) and, even among these, only some have adjacent dimples; the section **B–B**, for example does not. Despite this, and limits set by the onset of local buckling, the gain is real.

Postscript. Dimpling can be applied to most rolled-sheet products. It is done by making the final roll-pass through mating rolls with meshing dimples, adding little to the cost. It is most commonly applied to sheet steel. Here it finds applications in the automobile industry for bumper armatures, seat frames, side impact bars, allowing weight-saving without compromising mechanical performance. Stiffening sheet also raises its natural vibration frequencies, making them harder to excite, thus helping to suppress vibration in panels.

But a final word of warning: stiffening the sheet may change its failure mechanism. Flat sheet yields when bent; dimpled sheet, if thin, could fail by a local buckling mode. It is this that ultimately limits the useful extent of dimpling.

Further reading Stiffening sheet by dimpling
Fletcher, M. (1998) Cold-rolled dimples gauge strength, *Eureka*, May issue, p. 212. (*A brief account of the shaping of steel panels to improve bending stiffness and strength.*)

Related case studies 12.8 Shapes that flex: leaf and strand structures

12.7 Table legs again: thin or light?

Luigi Tavolino's table (Section 6.4 and Figure 6.5) is a great success. He decides to develop a range of less-expensive tubular-legged furniture. Some are to have thin legs, others have to be light. He needs a more general way of exploring material and shape.

The model and the selection. Tubes can be made with almost any radius r and wall thickness t, though (as we know from Chapter 11) it is impractical to make them with r/t too large because they buckle locally, and that is bad. Luigi chooses GFRP as the material for his legs, and for this material the maximum available shape factor is $\phi_B^e = 10$ corresponding to a value of $r/t = 10.5$ (Table 11.3). The cross-sectional area A and the second moment of area I of a thin-walled tube are

$$A = 2\pi rt$$

and

$$I = \pi r^3 t$$

from which

$$r = \left(\frac{2I}{A}\right)^{1/2} \tag{12.16}$$

This allows contours of constant r to be plotted on the $A–I$ quadrant of the four quadrant diagram shown in Figure 12.8 (they form a family of lines of slope 1), which can now be used to select either to minimize mass or to minimize thickness.

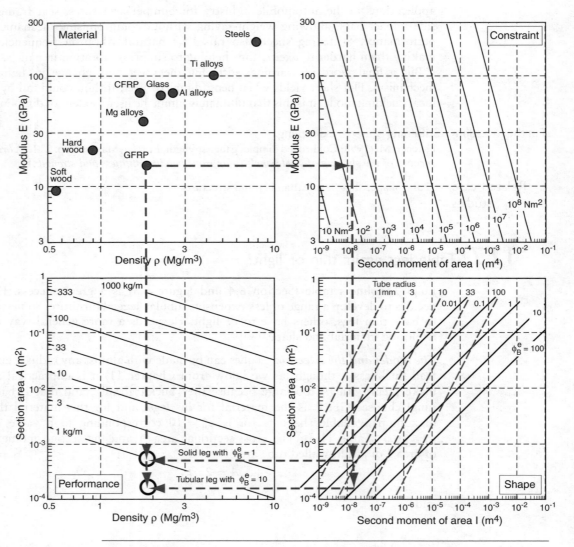

Figure 12.8   The chart-assembly for exploring stiffness-limited design with tubes. The $I – A$ quadrant now has contours of tube radius r. The thinnest GFRP leg is one with $\phi_B^e = 1$. The lightest one is that with $\phi_B^e = 10$, the maximum value for GFRP.

Luigi might use it in the following way. He designs a table with cylindrical, unbraced, legs each with length $L = 1$ m. For safety, each leg must support 50 kg without buckling, requiring that

$$\frac{\pi^2}{4} \frac{EI}{\lambda^3} \geq 500 \, \text{N}$$

from which

$$EI \geq 200 \, \text{N.m}^2$$

Luigi now plots a rectangular selection path on the four-quadrant diagram in the way shown in Figure 12.8. The lightest leg is that with the largest allowable value of ϕ_B^e (10 in the case of GFRP) but this gives a fat leg, one nearly 20 mm in radius, as read from the radius contours. The thinnest leg is a solid one, and that has $\phi_B^e = 1$. It is much thinner — only 5 mm in radius — but it is also almost three times heavier.

Postscript. So Luigi will have to compromise: fat and light or thin and heavy — or somewhere in between. Or he could use another material. The strength of the four-quadrant method is that he can explore other materials with ease. Repeating the construction for aluminum, putting an upper limit on ϕ_B^e of, say, 15, or for CFRP with the same upper limit as GFRP is only a moment's work.

Related case studies   6.4   Materials for table legs

12.8 Shapes that flex: leaf and strand structures

Flexible cables, leaf and helical springs, and flexural hinges require low, not high, structural efficiency. Here the requirement is for low flexural or torsional stiffness about one or two axes while retaining high stiffness and strength in other directions. The simplest example is the torsion bar (Figure 12.9a), with a shape factor $\phi_T^e \approx 1$. The single-leaf spring (Figure 12.9b) allows much lower values of ϕ_B^e ($\phi_B^e = h/w$; the dimensions are defined on the figure).

Multi-strand cables and multi-leaf assemblies do much better. Consider the change in efficiency when a solid square beam of section $A = b^2$ is subdivided into n cylindrical strands each of radius r (Figure 12.9c and d), such that

$$n\pi r^2 = b^2 \tag{12.17}$$

The axial stiffness of the original bar is proportional to EA, and its bending stiffness, S_o, to EI_o where E is the modulus of the material and I_o the second moments of area. When made into a cable of n parallel cylindrical strands,

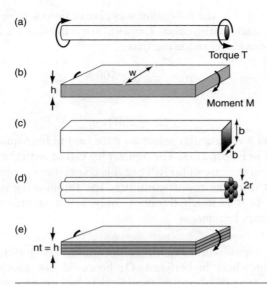

Figure 12.9 (a) a simple torsion bar; (b) a flat panel of width w used in the definition of ϕ_B^e for the multi-leaf structure; (c) the standard shape used in the definition of ϕ_B^e for the cable; (d) a cable with the same section area as the standard shape of (c), but made up of n cylindrical strands; (e) the multi-leaf structure with the same width and cross-sectional area as the panel of (b).

the axial stiffness is unchanged, but the bending stiffness S falls to a value proportional to nEI, where I is the second moment of a single strand. Thus

$$S \propto n E \frac{\pi r^4}{4} = \frac{1}{4\pi} E \frac{b^4}{n} \qquad (12.18)$$

Thus the structural efficiency of the cable is

$$\phi_B^e = \frac{S}{S_0} = \frac{nEI}{EI_0} = \frac{3}{n\pi} \qquad (12.19)$$

The number of strands, n, can be very large, allowing the flexural stiffness to be adjusted over a large range while leaving the axial stiffness unchanged.

Multi-leaf assemblies allow even more dramatic anisotropy. If the thick leaf shown at (b) in the figure is divided into the stack of thin ones of the same width shown at (e), the bending stiffness changes from

$$S_o = Ewh^3/12$$

to

$$S = nEt^3/12$$

where $t = h/n$. The shape efficiency of the stack is

$$\phi_B^e = \frac{S}{S_0} = \frac{nt^3}{h^3} = \frac{1}{n^2} \qquad (12.20)$$

Thus a stack of 10 layers is 100 time less stiff than a simple leaf of the same total thickness, while the in-plane stiffness is unchanged.

Related case studies

12.6   Increasing the stiffness of steel sheet

12.9   Summary and conclusions

In designing components that are loaded such that they bend, twist or buckle, the designer has two groups of variables with which to optimize performance: the choice of *material* and of *section shape*. The best choice of material depends on the shapes in which it is available, or to which it could potentially be formed. The procedure of Chapter 11 gives a method for optimizing the coupled choice of material and shape.

Its use is illustrated in this chapter. Often the designer has available certain stock materials in certain shapes. Then that with the greatest value of the appropriate material index (of which a number were listed in Table 12.1) maximizes performance. Sometimes sections can be specially designed; then material properties and design loads determine a maximum practical value for the shape factor above which local buckling leads to failure; again, the procedure gives an optimal choice of material and shape. Further gains in efficiency are possible by combining microscopic with macroscopic shape.

Chapter 13

Designing hybrid materials

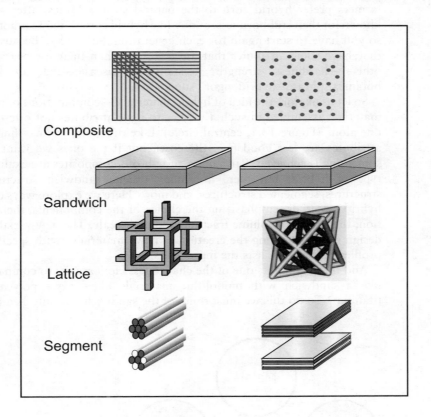

Composite

Sandwich

Lattice

Segment

13.1 Introduction and synopsis

Why do horse breeders cross a horse with a donkey, delivering a mule? Why do farmers prefer hybrid corn to the natural strain? Mules, after all, are best known for their stubbornness, and—like hybrid corn—they cannot reproduce, so you have to start again for each generation. So—why? Because, although they have some attributes that are less good than their forebears, they have others—hardiness, strength, resistance to disease—that are better. The botanical phrase "hybrid vigor" sums it up.

So let us explore the idea of hybrid materials—combinations of two or more materials assembled in such a way as to have attributes not offered by either one alone (Figure 13.1, central circle). Like the mule, we may find that some attributes are less good (e.g. the cost), but if the ones we want are better, something is achieved. Particulate and fibrous composites are examples of one type of hybrid, but there are many others: sandwich structures, lattice structures, segmented structures, and more. Here we explore ways of designing hybrid materials, emphasizing the choice of the components, their configuration, their relative volume fraction, and their scale. The new variables expand design space, allowing the creation of new "materials" with specific property profiles. Table 13.1 lists the ingredients.

And that highlights one of the challenges. How are we to compare a hybrid like a sandwich with monolithic materials like—say—polycarbonate or titanium? To do this we must think of the sandwich not only as a hybrid with

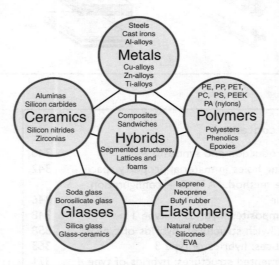

Figure 13.1   Hybrid materials combine the properties of two (or more) monolithic materials, or of one material and space. They include fibrous and particulate composites, foams and lattices, sandwiches, and almost all natural materials.

Table 13.1   Ingredients of hybrid design

Components	The choice of materials to be combined
Configuration	The shape and connectivity of the components
Relative volumes	The volume fraction of each component
Scale	The length-scale of the structural unit

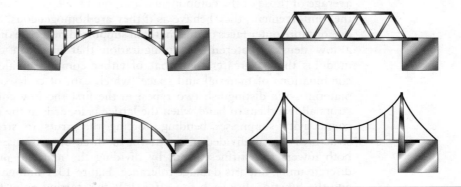

Figure 13.2   Four configurations for a bridge. The design variables describing the performance of each differ. Optimization of performance becomes possible only when a configuration has been chosen.

faces on one material bonded to a core of another, but as a "material" in its own right, with its own set of effective properties; it is these that allow the comparison.

The approach adopted here is one of breadth rather than precision. The aim is to assemble methods to allow the properties of alternative hybrids to be scanned and compared, seeking those that best meet a given set of design requirements. Once materials and configuration have been chosen, standard methods — optimization routines, finite-element analyses — can be used to refine them. But what the standard methods are *not* good at is the quick scan of alternative material–configuration combinations. That is where the approximate methods developed below, in which material and configuration become the variables, pay off.

The word "configuration" requires elaboration. Figure 13.2 shows four different configurations of a bridge. In the first all members are loaded in compression. In the second, members carry both tension and compression, depending on how the bridge is loaded. In the third and fourth, the suspension cables are loaded purely in tension. Any one of these can be optimized, but no amount of optimization will cause one to evolve into another because this involves a discrete jump in configuration, each characterized by its own set of variables.[1]

Hybrid design has the same feature: the classes of hybrid are distinguished by their configuration. Here we focus on four classes, each with a number if discrete members. Figure 13.3 suggests what they look like; they are our bridges. To avoid a mouthful of words every time we refer to one, we use the shorthand on the left of the figure: *composite, sandwich, lattice*, and *segment*. Composites combine two solid components, one (the reinforcement) as fibers or particles, contained in the other (the matrix). Their properties are some average of those of the components, and, on a scale large compared to that of the reinforcement, they behave as if they are homogeneous. Sandwiches have one or more outer faces of one material supported by a core of another, usually a low density material — a configuration that gives an effective flexural modulus that is greater than that of either component alone. Lattices are combinations of material and space (which can, of course, contain another material). We distinguish two types: in the first the low connectivity of the struts allows them to bend when the lattice is loaded; in the other, the higher connectivity suppresses bending, forcing the struts to stretch. Segmented structures are subdivided in one, two, or three dimensions; the subdivisions both lowers the stiffness, and, by dividing the material into a number of discrete units, imparts damage tolerance. Figure 13.4 summarizes the families and the functionality each can offer. It is the starting point for configuration selection and — since each configuration offers more than one type of functionality — for multi-functional design.

The approach we adopt is to use bounding methods to estimate the properties of each configuration. With these, the properties of a given pair of materials in a given configuration can be calculated. These can then be plotted on material selection charts, which become tools for selecting both configuration and material.

13.2 Filling holes in material-property space

All the charts of Chapter 4 have one thing in common: parts of them are populated with materials and parts are not. Some parts are inaccessible for fundamental reasons that relate to the size of atoms and the nature of the forces that bind their atoms together. But other parts are empty even though,

---

[1] Numerical tools are emerging that allow a degree of *topological optimization*, meaning the development of a configuration. They work like this. Start with an envelope — a set of boundaries — and fill it with a homogenous "material" with a relative density initially set at 0.5, with properties that depend linearly on relative density. Impose constraints, meaning the mechanical, thermal, and other loads the structure must support, give it a criterion of excellence (more of this below) and let it condense out into regions of relative density 1 and regions where it is zero, retaining only changes that increase the measure of excellence. The method is computationally intensive but has had some success in suggesting configurations that use a material efficiently (see Further reading for more information).

Family	Examples	Potential functions

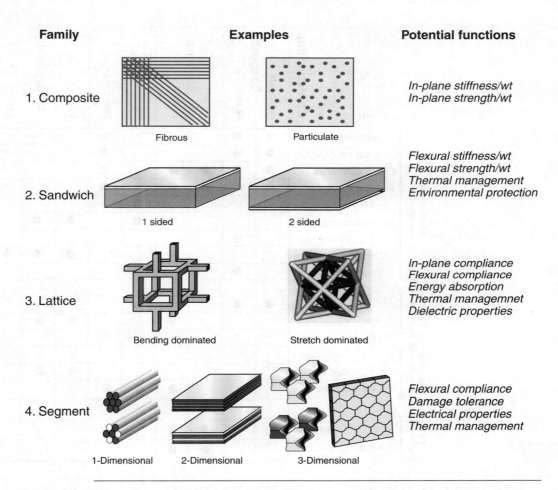

1. Composite — Fibrous / Particulate

In-plane stiffness/wt
In-plane strength/wt

2. Sandwich — 1 sided / 2 sided

Flexural stiffness/wt
Flexural strength/wt
Thermal management
Environmental protection

3. Lattice — Bending dominated / Stretch dominated

In-plane compliance
Flexural compliance
Energy absorption
Thermal managemnet
Dielectric properties

4. Segment — 1-Dimensional / 2-Dimensional / 3-Dimensional

Flexural compliance
Damage tolerance
Electrical properties
Thermal management

Figure 13.3   Four families of configurations of hybrid materials: composites, sandwiches, lattices, and segmented structures.

in principle, they are accessible. If these holes could be filled, the new materials that lay there could allow novel design possibilities.

One approach to this — the traditional one — is that of developing new metal alloys, new polymer chemistries and new compositions of glass and ceramic so as to extend the populated areas of the property charts, but this can be an expensive and uncertain process. An alternative is to combine two or more existing materials so as to allow a superposition of their properties — in short, to create a hybrid. The spectacular success of carbon and glass-fiber reinforced composites at one extreme, and of foamed materials at another (hybrids of material and space) in filling previously empty areas of the property charts is encouragement enough to explore ways in which such hybrids can be designed.

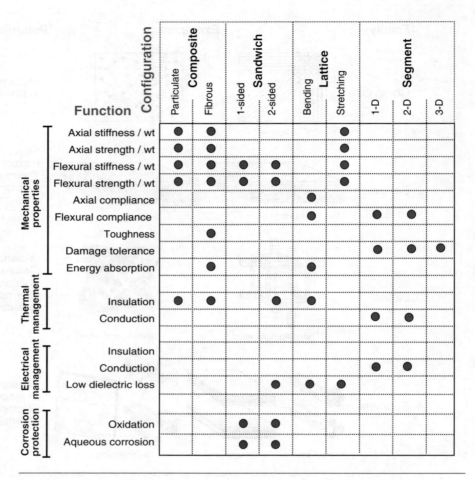

344 Chapter 13 *Designing hybrid materials*

Figure 13.4 The functionality provided by each configuration.

What might we hope to achieve? Figure 13.5 shows schematically the fields occupied by two families of materials, plotted on a chart with properties P_1 and P_2 as axes. Within each field a single member of that family is identified (materials M_1 and M_2). What might be achieved by making a hybrid of the two? The figure shows four scenarios, each typical of a certain class of hybrid. Depending on the shapes of the materials and the way they are combined, we may find any one of the following.

- "The best of both" scenario (point A). The ideal, often, is the creation of a hybrid with the best properties of both components. There are examples, most commonly when a bulk property of one material is combined with the surface properties of another. Zinc coated steel has the strength and toughness of steel with the corrosion resistance of zinc. Glazed pottery

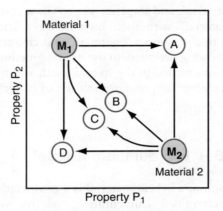

Figure 13.5 The possibilities of hybridization. The properties of the hybrid reflect those of its component materials, combined in one of several possible ways.

exploits the formability and low cost of clay with the impermeability and durability of glass.

- "The rule of mixtures" scenario (point B). When bulk properties are combined in a hybrid, as in structural composites, the best that can be obtained is often the arithmetic average of the properties of the components, weighted by their volume fractions. Thus unidirectional fiber composites have an axial modulus (the one parallel to the fibers) that lies close to the rule of mixtures.

- "The weaker link dominates" scenario (point C). Sometimes we have to live with a lesser compromise, typified by the stiffness of particulate composites, in which the hybrid properties fall below those of a rule of mixtures, lying closer to the harmonic than the arithmetic mean of the properties. Although the gains are less spectacular, they are still useful.

- "The least of both" scenario (point D). Sprinkler systems use a wax–metal hybrid designed to fail, releasing the spray, when the melting point of the lower-melting material (the wax) is exceeded.

These set certain fixed points, but the list is not exhaustive. Other combinations are possible, some giving performance that exceeds even that of point A. These will emerge below.

When is a hybrid a "material"? There is a certain duality about the way in which hybrids are thought about and discussed. Some, like filled polymers, composites or wood are treated as materials in their own right, each characterized by its own set of material properties. Others — like galvanized steel — are seen as one material (steel) to which a coating of a second (zinc) has been applied, even though if could be regarded as a new material with the strength of steel but the surface properties of zinc ("stinc", perhaps?). Sandwich panels illustrate the duality, sometimes viewed as two sheets of face-material separated by a core material, and sometimes—to allow comparison with bulk materials—as a "material" with their own density, flexural stiffness and strength. To call any one of these a "material" and characterize it as such is a useful shorthand, allowing designers to use existing methods when designing with them. But if we are to design the hybrid itself, we must deconstruct it, and think of it as a combination of materials (or of material and space) in a chosen configuration.

13.3 The method: "A + B + configuration + scale"

First, a working definition: a hybrid material is a combination of two or more materials in a predetermined configuration, relative volume and scale, optimally serving a specific engineering purpose", which we paraphrase as "A + B + configuration + scale. Here we allow for the widest possible choice of A and B, including the possibility that one of them is a gas or simply space. These new design variables expand design-space, allowing an optimization of properties that is not possible if choice is limited to single, monolithic materials.

The basic idea, illustrated in Figure 13.6, is this. Monolithic materials offer a certain portfolio of properties on which much engineering design is based. Design requirements isolate a sector of material–property space. If this sector contains materials the requirements can be met by a single-material solution. But if the design requirements are exceptionally demanding, no single material may be found that can meet them all: the requirements lie in a hole in property

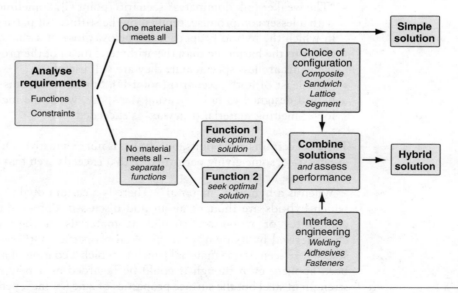

Figure 13.6 The steps in designing a hybrid to meet given design requirements.

space. Then the way forward is to identify and separate the conflicting requirements, seeking optimal material solutions for each, and then combine them in ways that retain the desirable attributes of both. The best choice is the one that ranks most highly when measured by the performance metrics that motivate the design: minimizing mass or cost, or maximizing some aspect of performance (the *criteria of excellence*). The alternative combinations are examined and assessed, using the criteria of excellence to rank them. The output is a specification of a hybrid in terms of its component materials and configurations.

Consider a simple example: materials for long-span power cables. The objectives are to minimize the electrical resistance, but at the same time to maximize the strength since this allows the greatest span. This is an example of multi-objective optimization, discussed in Chapter 9. There we mentioned the convention that each objective is expressed such that a minimum for it is sought. We thus seek materials with the lowest values of resistivity, ρ_e and the reciprocal of yield strength, $1/\sigma_y$. Figure 13.7 shows the result: materials that best meet the design requirements lie near towards the bottom left. But here there is a hole: all 1700 metals and alloys plotted here have properties that lie above the broken red trade-off line. Those with the lowest resistance — copper, aluminum, and certain of their alloys — are not very strong; the materials that are strongest — drawn carbon and low-alloy steel — do not conduct very well. Now consider a cable made by interleaving strands of copper and steel such that each occupies half the cross-section. Assuming that the steel carries no current and the copper no load (the most pessimistic scenario) the performance of the cable will lie at the point shown on the figure — it has twice the resistivity of the copper and half the strength of the steel. It lies in a part of property space that was empty, offering performance that was not previously possible. Other ratios of copper-to-steel fill other parts of the space; by varying the ratio the area shaded in gray on the figure is covered. Similar hybrids of aluminum and steel fill a different area, as is easily seen by repeating the construction using "1000 series Al alloys" in place of "OFHC copper, hard" in the combination. Their combinations of ρ_e and σ_y are less good, but they are lighter and cheaper and for this reason are widely used.

Note the method: decompose the requirements, seek good solutions for each, combine them in a chosen configuration, assess its performance, and chose the combination that offers the best.[2] But while some conflicting requirements can be met in this way, others need a more inventive approach. So the question arises: are there ways in which material hybridization can be explored systematically?

---

[2] The configuration we chose here (the member "multi-strand cable" from the family "segmented structures") did not require that the materials be bonded together. Other families — composites and sandwiches for example — require such bonding, and this is not always easy. Interface engineering is beyond the scope of this book, but is must be recognized that converting a hybrid concept into a reality requires the ability to create an adequate bond between the components.

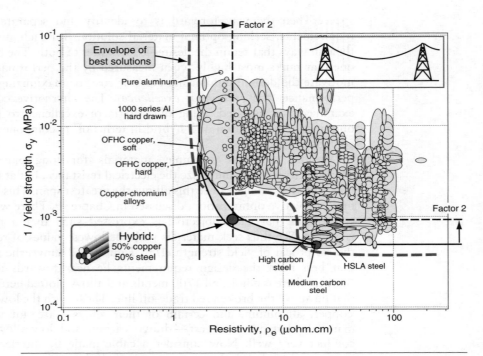

Figure 13.7 Designing a hybrid—here, one with high strength and high electrical conductivity. The figure shows the resistivity and reciprocal of tensile strength for 1700 metals and alloys. We seek materials with the lowest values of both. The construction is for a hybrid of hard-drawn OFHC copper and drawn low alloy steel, but the figure itself allows many hybrids to be investigated.

13.4 Composites: hybrids of type 1

Aircraft engineers, automobile makers, and designers of sports equipment all have one thing in common: they want materials that are stiff, strong, tough and light. The single-material choices that best achieve this are the *light alloys*: alloys based on magnesium, aluminum and titanium. Much research aims at improving their properties. But they are not all that light—polymers have much lower densities. Nor are they all that stiff—ceramics are much stiffer and, specially in the form of small particles or thin fibers, much stronger. These facts are exploited in the family of hybrids that we usually refer to as *particulate and fibrous composites*.

Any two materials can, in principle, be combined to make a composite, and they can be mixed in many geometries (Figure 13.8). In this section we restrict the discussion to *fully dense, strongly bonded, composites* such that there is no tendency for the components to separate at their interfaces when the composite

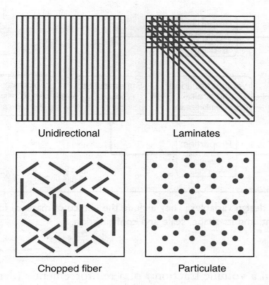

Unidirectional   Laminates

Chopped fiber   Particulate

Figure 13.8   Schematic of hybrids of the composite type: unidirectional fibrous, laminated fiber, chopped fiber and particulate composites. Bounds and limits, described in the text, bracket the properties of all of these.

is loaded, and to those in which the scale of the reinforcement is large compared to that of the atom or molecule size and the dislocation spacing, allowing the use of continuum methods. Figure 13.9 shows the design variables.

On a macroscopic scale—one which is large compared to that of the components—a composite behaves like a homogeneous solid with its own set of thermo-mechanical properties. Calculating these precisely can be done, but is difficult. It is much easier to bracket them by *bounds* or *limits*: upper and lower values between which the properties lie. The term "bound" will be used to describe a rigorous boundary, one which the value of the property *cannot*—subject to certain assumptions—exceed or fall below. It is not always possible to derive bounds; then the best that can be done is to derive "limits" outside which it is *unlikely* that the value of the property will lie. The important point is that the bounds or limits bracket the properties of *all* the configurations of matrix and reinforcement shown in Figure 13.8; by using them we escape from the need to model individual geometries.

Criteria of excellence. We need criteria of excellence to assess the merit of any given hybrid. These are provided by the material indices of Chapter 5. If a possible hybrid has a value of any one of these that exceed those of existing materials, it achieves our goal.

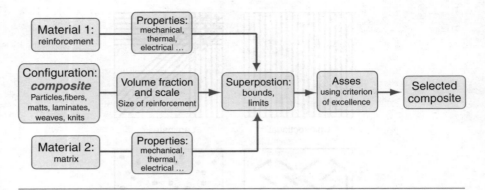

Figure 13.9   Composites: the design variables. They include the choice of material for the matrix and the reinforcement, and its shape, scale and configuration.

Density. When a volume fraction f of a reinforcement r (density ρ_r) is mixed with a volume fraction $(1-f)$ of a matrix m (density ρ_m) to form a composite with no residual porosity, the composite density $\tilde{\rho}$ is given exactly by a rule of mixtures (an arithmetic mean, weighted by volume fraction):

$$\tilde{\rho} = f\rho_r + (1-f)\rho_m \tag{13.1}$$

The geometry or shape of the reinforcement does not matter except in determining the maximum packing-fraction of reinforcement and thus the upper limit for f.

Modulus. The modulus of a composite is bracketed by the well-known Voigt and Reuss bounds. The upper bound, $\tilde{E}_u$, is obtained by postulating that, on loading, the two components suffer the same strain; the stress is then the volume-average of the local stresses and the composite modulus follows a rule of mixtures:

$$\tilde{E}_u = fE_r + (1-f)E_m \tag{13.2}$$

Here E_r is the Young's modulus of the reinforcement and E_m that of the matrix. The lower bound, $\tilde{E}_L$, is found by postulating instead that the two components carry the same stress; the strain is the volume-average of the local strains and the composite modulus is

$$\tilde{E}_L = \frac{E_m E_r}{fE_m + (1-f)E_r} \tag{13.3}$$

More precise bounds are possible but the simple ones are adequate to illustrate the method.

Strength. As the load on a composite is increased, load is redistributed between the components until general yield or fracture of one component occurs. Beyond this point the composite has suffered permanent deformation or damage; we define it as strength of the composite. The composite is strongest if both components reach their failure state simultaneously since if one fails before the other the weaker determines the strength. Thus the upper bound is, as with modulus, a rule of mixtures

$$(\tilde{\sigma}_f)_u = f(\sigma_f)_r + (1-f)(\sigma_f)_m \tag{13.4}$$

where $(\sigma_f)_m$ is the strength of the matrix and $(\sigma_f)_r$ is that of the reinforcement.

A lower bound is more difficult. The literature contains many calculations for special cases: reinforcement by unidirectional fibers, or by a dilute dispersion of spheres. We wish to avoid models which require detailed knowledge of how a particular architecture behaves, and seek a less restrictive lower limit. One, consistent with the requirement that the components of the composite do not separate at their interfaces, is developed in the references listed under Further reading. It describes a "worst case": a continuous, ductile, matrix containing strong reinforcing particles; the lower limit for the composite strength is then the yield strength of the matrix enhanced slightly by the plastic constraint imposed by the reinforcement

$$(\tilde{\sigma}_f)_L = (\sigma_f)_m \left(1 + \frac{1}{16} \left(\frac{f^{1/2}}{1 - f^{1/2}} \right) \right) \tag{13.5}$$

The consequence of this choice is that the bounds are wide, but — as the examples of Chapter 14 will show — they still allow useful conclusions to be reached.

Specific heat. The specific heats of solids at constant pressure, C_p, are almost the same as those at constant volume, C_v. If they were identical, the heat capacity per unit volume of a composite would, like the density, be given exactly by a rule-of-mixtures

$$\rho \tilde{C}_p = f\rho_r(C_p)_r + (1-f)\rho_m(C_p)_m \tag{13.6}$$

where $(C_p)_r$ is the specific heat of the reinforcement and $(C_p)_m$ is that of the matrix (the densities enter because the units of C_p are J/kg.K). A slight difference appears because thermal expansion generates a misfit between the components when the composite is heated; the misfit creates local pressures on the components and thus changes the specific heat. The effect is very small and need not concern us further.

Thermal expansion coefficient. The thermal expansion of a composite can, in some directions, be greater than that of either component, in others, less. This is because an elastic constant — Poisson's ratio — couples the principal elastic strains; if the matrix is prevented from expanding in one direction (by embedded fibers, for instance) then it expands more in the transverse directions. For simplicity we shall use the approximate lower bound

$$\tilde{\alpha}_L = \frac{E_r \alpha_r f + E_m \alpha_m (1-f)}{E_r f + E_m (1-f)} \tag{13.7}$$

(it reduces to the rule of mixtures when the moduli are the same) and the upper bound

$$\tilde{\alpha}_u = f \alpha_r (1 + v_r) + (1-f) \alpha_m (1 + v_m) - \alpha_L [f v_r + (1-f) v_m] \tag{13.8}$$

where α_r and α_m are the two expansion coefficients and v_r and v_m the Poisson's ratios.

Thermal conductivity. The thermal conductivity determines heat flow at steady rate. A composite of two materials, bonded to give good thermal contact, has a thermal conductivity λ that lies between those of the individual components, λ_m and λ_r. Not surprisingly, a composite containing parallel continuous fibers has a conductivity, parallel to the fibers, given by a rule-of-mixtures

$$\tilde{\lambda}_u = f \lambda_r + (1-f) \lambda_m \tag{13.9}$$

This is an upper bound: in any other direction the conductivity is lower. The transverse conductivity of a parallel-fiber composite (again assuming good bonding and thermal contact) lies near the lower bound first derived by Maxwell

$$\tilde{\lambda}_L = \lambda_m \left(\frac{\lambda_r + 2\lambda_m - 2f(\lambda_m - \lambda_r)}{\lambda_r + 2\lambda_m + f(\lambda_m - \lambda_r)} \right) \tag{13.10}$$

Particulate composites, too, have a conductivity near this lower bound. Poor interface conductivity can make λ drop below it. Debonding or an interfacial layer between reinforcement and matrix can cause this; so, too, can a large difference of modulus between reinforcement and matrix (because this reflects phonons, creating an interface impedance) or a structural scale which is shorter than the phonon wavelengths.

Thermal diffusivity. The thermal diffusivity

$$a = \frac{\lambda}{\rho C_p}$$

determines heat flow when conditions are transient, that is, when the temperature field changes with time. It is formed from three of the earlier properties: λ, ρ, and C_p. The second and third of these are given exactly by equations (13.1) and (13.6), allowing the diffusivity to be expressed as

$$\tilde{a} = \frac{\tilde{\lambda}}{f\rho_r(C_p)_r + (1-f)\rho_m(C_p)_m} \qquad (13.11)$$

Its upper and lower bounds are found by substituting those for $\tilde{\lambda}$ (equations (13.9) and (13.10)) into this equation.

Dielectric constant. The dielectric constant $\tilde{\varepsilon}$ is given by a rule of mixtures

$$\tilde{\varepsilon} = f\varepsilon_r + (1-f)\varepsilon_m \qquad (13.12)$$

where ε_r is the dielectric constant of the reinforcement and ε_m that of the matrix.

Electrical conductivity. When the electrical conductivities κ of the components of a composite are of comparable magnitude, bounds for the electrical conductivity are given by those for thermal conductivity with λ replaced by the κ. When, instead, they differ by many orders of magnitude (a metallic powder dispersed in an insulating polymer, for instance) questions of percolation arise. They are considered below.

To see how the bounds are used, consider the following three examples.

Composite design for stiffness at minimum mass

We seek a composite offering high stiffness-to-weight in structures subjected to bending loads. The performance of a material as a light, stiff, beam is measured by the index $E^{1/2}/\rho$ derived in Chapter 5. Imagine, as an example, that the beam is at present made of an aluminum alloy. Beryllium is both lighter and stiffer than aluminum; ceramics are stiffer, but not all are lighter. What can hybrids of aluminum with one of these offer?

Figure 13.10 is a small part of the $E - \rho$ property chart. Three groups of materials are shown: aluminum and its alloys, alumina (Al_2O_3) and beryllium (Be). Composites made by mixing them have densities given exactly by equation (13.1) and moduli that are bracketed by the bounds of equations (13.2) and (13.3). Both of these moduli depend on volume fraction of reinforcement, and through this, on density. Upper and lower bounds for the modulus–density relationship can thus be plotted onto the $E - \rho$ chart using volume fraction f as a parameter, as shown in Figure 13.10. Any composite made by combining aluminum with alumina will have a modulus contained in the envelope for Al–Al_2O_3; the same for Al–Be. Fibrous reinforcement gives a longitudinal modulus (that parallel to the fibers) near the upper bound; particulate reinforcement or transversely loaded fibers give moduli near the lower one.

Superimposed on Figure 13.10 is a grid showing the criterion of excellence $E^{1/2}/\rho$. The bound-envelope for Al–Be composites extends almost normal to the

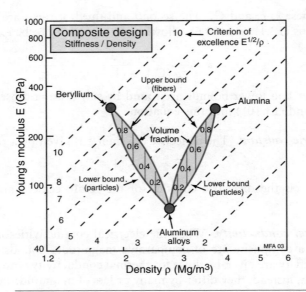

Figure 13.10   Part of the $E - \rho$ property chart, showing aluminum alloys, beryllium and alumina (Al_2O_3). Bounds for the moduli of hybrids made by mixing them are shown. The diagonal contours plot the criterion of excellence, $E^{1/2}/\rho$.

grid, while that for Al–Al$_2$O$_3$ lies at a shallow angle to it. Beryllium fibers improve performance (as measured by $E^{1/2}/\rho$) roughly four times as much as alumina fibers do, for the same volume fraction. The difference for particulate reinforcement is even more dramatic. The lower bound for Al–Be lies normal to the contours: 30% of particulate beryllium increases $E^{1/2}/\rho$ by a factor of 1.5. The lower bound for Al–Al$_2$O$_3$ is, initially, parallel to the $E^{1/2}/\rho$ grid: 30% of particulate Al$_2$O$_3$ gives almost no gain. The underlying reason is clear: both beryllium and Al$_2$O$_3$ increases the modulus, but only beryllium decreases the density; the criterion of excellence is more sensitive to density than to modulus.

Composite design for controlled thermal response

Thermo-mechanical design involves the specific heat, C_p, the thermal expansion, α, the conductivity, λ, and the diffusivity, a. These composite properties are bounded by equations (13.6)–(13.11). They are involved in a number of indices. One is the criterion for minimizing thermal distortion derived in Chapter 6, section 6.16: it is that of maximizing the index λ/α.

Figure 13.11 shows a small part of the $\alpha - \lambda$ materials selection chart, with a grid of lines of the index λ/α superimposed on it. Three groups of materials are shown: aluminum alloys, boron nitride (BN) and silicon carbide (SiC). The thermal properties of Al–BN and Al–SiC are bracketed by lines which

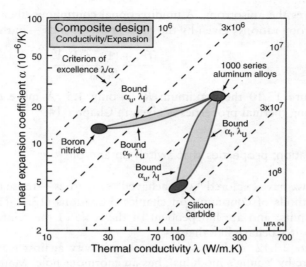

Figure 13.11   A part of expansion-coefficient/conductivity space showing aluminum alloys, boron nitride and silicon carbide. The properties of Al–BN and Al–SiC composites are bracketed by the bounds of equations (13.7)–(13.10). The Al–SiC composites enhance performance, the Al–BN composites reduce it.

show the bounding equations.[3] The plot reveals immediately that SiC reinforcement in aluminum increases performance (as measured by λ/α); reinforcement with BN decreases it.

Similar methods can be used to select materials for optimum strength, and for tailored values of thermal properties. The properties of specific composites can, of course, be computed in conventional ways. The advantage of this graphical approach is the breadth and freedom of conceptual thinking that it allows and the ease of comparison of possible new hybrids with the population of existing materials.

Creating anisotropy

The elastic and plastic properties of bulk monolithic solids are frequently anisotropic, but weakly so — the properties do not depend strongly on direction. Hybridization gives a way of creating controlled anisotropy, and it can be large. We have already seen an example in Figure 13.10, which shows the upper and lower bounds for the moduli of composites. The longitudinal properties of unidirectional long-fiber composites lie near the upper bound, the transverse properties near the lower one. The vertical width of the band in between them

---

[3] Both α and λ have upper and lower bounds (unlike ρ in the previous example) so there are four possible combinations for each material pair. Those shown in the figure are the outermost pair of the four.

measures the anisotropy. A unidirectional continuous-fiber composite has an anisotropy ratio R_a given by the ratio of the bounds — in this example

$$R_a = \frac{(fE_r + (1-f)E_m)^2}{E_r E_m} \tag{13.13}$$

In Figure 13.10 the maximum R_a is only 1.5. A more dramatic example involving thermal properties is given in Chapter 14.

Percolation: properties that switch on and off

So far we have explored "well-behaved" properties — those that can be treated by methods of continuum mechanics. Equations (13.1)–(13.11) are continuum limits, and are independent of the scale of the reinforcement. Not all composites behave like that.

Figure 13.12, a chart of electrical resistivity against elastic stiffness (here measure by Young's modulus), has an enormous hole. Materials that conduct well are stiff; those that are flexible are insulators. Consider designing materials to fill the hole; to be more specific, consider designing one that has low modulus, can be molded like a polymer, and is a good electrical conductor. Such materials find application in anti-static clothing and mats, as pressure sensing elements, even as solder-less connections.

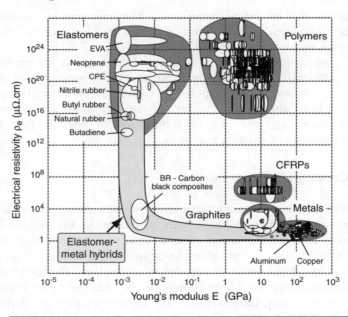

Figure 13.12 When conducting particles or fibers are mixed into an insulating elastomer, a hole in material property space is filled. Carbon-filled butyl rubbers lie in this part of the space.

Metals, carbon, and some carbides and intermetallics are good conductors, but they are stiff and cannot be molded. Thermoplastic and thermosetting elastomers can be molded and are flexible, but they do not conduct. How are they to be combined? Metal coating of polymers is workable if the product will be used in a protected environment, but coatings are easily damaged. If a robust, flexible, product is needed, bulk rather than surface conduction is essential. This can be achieved by mixing conducting particles into the polymer.

To understand how to optimize this we need the concept of *percolation*. Think of mixing conducting and insulating spheres of the same size to give a large array. If there are too few conducting spheres for them to touch, the array is obviously an insulator. If each conducting sphere contacts just one other, there is still no connecting path. If, on average, each touched two, there is still no path. Adding more spheres gives larger clusters, but they can be large yet still discrete. For bulk conduction we need *connectivity*: the array first becomes a conductor when a single trail of contacts links one surface to the other, that is, when the fraction p of conducting spheres reaches the *percolation threshold, p_c*. Percolation problems are easy to describe but difficult to solve. Research since 1960 has provided approximate solutions to most of the percolation problems associated with the design of hybrids (see Further reading for a review). For simple cubic packing $p_c = 0.248$, for close packing $p_c = 0.180$. For a random array it is somewhere in between — approximately 0.2.[4]

Make the spheres smaller and the transition is smeared out. The percolation threshold is still 0.2, but the first connecting path is now thin and extremely devious — it is the only one, out of the vast number of almost complete paths, that actually connects. Increase the volume fraction and the number of conducting paths increases initially as $(p - p_c)^2$, then linearly, reverting to a rule of mixtures. If the particles are very small, as much as 40% may be needed to give good conduction. But a loading of 40% seriously degrades the moldability and compliance of the polymer.

Shape gives a way out. If the spheres are replaced by fibers, they touch more easily and the percolation threshold falls. If their aspect ratio is $\beta = L/d$ (where L is the fiber length, d the diameter) then, empirically, the percolation threshold falls from p_c to roughly $p_c/\beta^{1/2}$. Figure 13.12 shows the area of the property chart where metal–elastomer hybrids lie. With sufficient aspect ratio the percolation threshold falls to a few percentage.

The concept of percolation is a necessary tool in designing hybrids. Electrical conductivity works that way; so too does the passage of liquids through foams or porous media — no connected paths, and no fluid flows; just one (out of a million possibilities) and there is a leak. Add a few more connections and there is a flood. Percolation ideas are particularly important in understanding the transport properties of hybrids: properties that determine the flow of

---

[4] These results are for infinite, or at least very large, arrays. Experiments generally give values in the range 0.19–0.22, with some variability because of the finite size of the samples.

electricity or heat, of fluid, or of flow by diffusion, specially when the differences in properties of the components are extreme. Most polymers differ from metals in their electrical conductivity by a factor of about 10^{20}. It is then that single connections really matter.

Percolation influences mechanical properties too, particularly when mechanical connection is important, as in arrays of loose powders or fibers. If there are no bonds between the particles or fibers, the array has no tensile stiffness or strength. If each particle is bonded to one other, or to several forming discrete clusters, there is still no tensile stiffness or strength. These only appear when there are connected paths running completely through the array. The plasticity of 2-phase hybrids, too, can be viewed as a percolation problem. Plasticity may start in one phase at a low stress, allowing patches of slip to form, but full plasticity requires that the slip patches link to give connected paths through the entire cross section of the sample.

13.5 Sandwich structures: hybrids of type 2

A sandwich panel epitomizes the concept of a hybrid. It combines two materials in a specified geometry and scale, configured such that one forms the faces, the other the core, to give a structure of high bending stiffness and strength at low weight (Figure 13.13). The separation of the faces by the core increases the moment of inertia of the section, *I,* and its section modulus, *Z,* producing a structure that resists bending and buckling loads well. Sandwiches are used where weight-saving is critical: in aircraft, trains, trucks and cars, in portable structures, and in sports equipment. Nature, too, makes use of sandwich designs: sections through the human skull, the wing of a bird and the stalk and leaves of many plants show a low-density foam-like core separating solid faces.

The faces, each of thickness *t,* carry most of the load, so they must be stiff and strong; and they form the exterior surfaces of the panel so they must tolerate the environment in which it operates. The core, of thickness *c,* occupies most of the volume, it must be light, and stiff and strong enough to carry the shear stresses necessary to make the whole panel behave as a load bearing

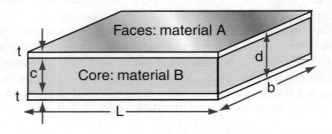

Figure 13.13   The ultimate light, stiff hybrid: the sandwich panel.

unit, but if the core is much thicker than the faces these stresses are small. Figure 13.14 indicates the design variables.

A sandwich as a "material". So far we have spoken of the sandwich as a structure: faces of material A supported on a core of material B, each with its own density and modulus. But we can also think of it as a material with its own set of properties, and this is useful because it allows comparison with more conventional materials. Density is easy — it is given exactly by the rule of mixtures, equation (13.1), with f replaced by $2t$ and $(1-f)$ replaced by c (a substitution we shall use throughout this section):

$$\tilde{\rho} = \frac{2t}{d}\rho_f + \left(1 - \frac{2t}{d}\right)\rho_c \tag{13.14}$$

Mechanical properties. Sandwich panels are designed to be stiff and strong in bending. In thinking of the panel as a "material" we must therefore distinguish the in-plane modulus and strength from those in bending. The in-plane modulus, $\tilde{E}_{//}$, is given exactly by the rule of mixtures, equation (13.2). An upper bound for the in-plane strength is found by postulating that faces and core reach yield or failure together, giving once more the rule of mixtures of equation (13.4). A lower bound is found by proposing that the first yield or failure of either faces or core constitutes failure, giving

$$\tilde{\sigma}_{f,L} = \text{Least of} \left[\left(\frac{\tilde{E}_{//}}{E_f}(\sigma_y)_f\right), \left(\frac{\tilde{E}_{//}}{E_c}(\sigma_y)_c\right)\right] \tag{13.15}$$

Here $(\sigma_y)_f$ is the elastic limit of the material of the faces, and $(\sigma_y)_c$ that of the core.

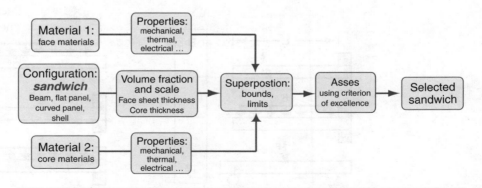

Figure 13.14   Sandwich panels: the design variables. They include the material choice for faces and core, their thicknesses, and the curvature of the panel itself.

The flexural properties are quite different. The bending stiffness of the panel per unit width, S_w is given by

$$S_w = (EI)_{sand} = \left[\frac{1}{12}\left(d^3 - c^3\right)E_f\right]\left\{\frac{1}{1 + BE_f tc/2G_c L^2}\right\}$$

$$\text{or} \quad S_w = (EI)_{sand} = \left[\frac{1}{12}\left(d^3 - c^3\right)E_f\right]K_s \tag{13.16}$$

where the dimensions, d, c, t, and L are identified in Figure 13.13, E_f is Young's modulus of the face sheets and G_c is the shear modulus of the core. The numerical constant B depends only on the way the panel is loaded — Figure 13.15 gives values. Equation (13.16) has two terms. The first, in square brackets, is the stiffness if bending were the only mode of deformation. The second, in curly brackets, is the knock-down in stiffness (condensed here into the symbol K_s) caused by shear in the core. If the core adequately resists shear, the second term reduces to unity. To think of the sandwich as a material, we define an effective flexural modulus $\widetilde{E}_{flex}$ equal to the modulus of a *homogeneous* material with the same bending stiffness as the sandwich, requiring that

$$(EI)_{sand} = \widetilde{E}_{flex}\frac{d^3}{12}$$

from which

$$\widetilde{E}_{flex} = \left(1 - \frac{c^3}{d^3}\right)E_f K_s = \left(1 - \left(1 - \frac{2t}{d}\right)^3\right)E_f K_s \tag{13.17a}$$

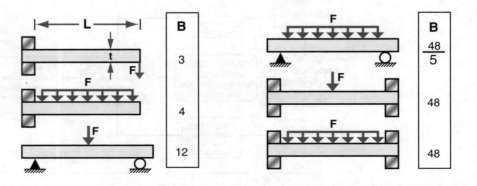

Figure 13.15   Values for the constant B in the equation for the bending stiffness of sandwich panels.

($d^3/12$ is the second moment of area per unit width of a homogeneous panel of thickness d). When $t \ll d$ and $K_s \approx 1$, this reduces to

$$\tilde{E}_{\text{flex}} \approx \frac{6t}{d} E_f \tag{13.17b}$$

It leads to the performance shown in Figure 13.16, which has been constructed in the same way as Figure 13.10. It shows the modulus and density of A+B hybrids. The shaded band is bounded by the upper and lower bounds of equations (13.1)–(13.3), describing particulate and fibrous composites. The flexural performance of the sandwich is shown as a dashed line. At its mid-section the modulus of the sandwich lies a factor 3 above the upper bound rule of mixtures of equation (13.2). The criterion of excellence for a panel of minimum weight with prescribed bending stiffness is that of maximizing $E_{1/3}/\rho$. Contours of this criterion are plotted as diagonal lines on the figure, increasing towards the top left. The sandwich out-performs all alternative hybrids of A + B.

The bending strength or *modulus of rupture* of a sandwich, $\tilde{\sigma}_{\text{MoR}}$, can be bracketed by limits. An upper limit is given by requiring that the faces and core yield simultaneously, when

$$\tilde{Z}\tilde{\sigma}_{\text{MoR}} = \frac{1}{6d}(d^3 - c^3)\sigma_f + \frac{c^2}{6}\sigma_c \tag{13.18}$$

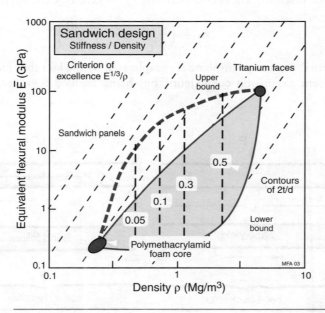

Figure 13.16   The performance of a sandwich structure compared with that of a composite made from the same two materials. The sandwich is more efficient if bending than even the best composites.

But $\tilde{Z}$ is just the section modulus of a homogenous section of thickness d (with value $d^2/6$), so that the upper bound for the panel strength $(\tilde{\sigma}_{MoR})_{ub}$ is given by

$$
\begin{aligned}
(\tilde{\sigma}_{MoR})_u &= \left(1 - \frac{c^3}{d^3}\right)\sigma_f + \frac{c^2}{d^2}\sigma_c \\
&= \left(1 - \left(1 - \frac{2t}{d}\right)^3\right)\sigma_f + \left(1 - \frac{2t}{d}\right)^2\sigma_c
\end{aligned}
\tag{13.19}
$$

The panel can fail in many other ways, illustrated by Figure 13.17, although a well-designed panel should not do so. Each of these can be modeled — such models can be found in the texts listed under Further reading. The bending strength of the panel is determined by the weakest of these competing mechanisms. Some, such a bond failure, can reduce the strength almost to zero. We are forced to accept that there is no lower bound on strength. This is not as serious as it sounds since the upper bound is realizable, and gives a realistic basis for comparing the panel with other materials.

Thermal properties. Thermal properties are treated in a similar way. The specific heat, C_p follows a rule of mixtures, equation (13.6). The in-plane thermal conductivity $\lambda_{//}$, too, follows such a rule — equation (13.9). The through-thickness conductivity, $\lambda_\perp$, is given by the harmonic mean

$$
\tilde{\lambda}_\perp = \left(\frac{2t/d}{\lambda_f} + \frac{(1 - 2t/d)}{\lambda_c}\right)^{-1}
\tag{13.20}
$$

Thermal expansion in-plane is complicated by the fact that faces and core have different expansion coefficients, but being bonded together, they are forced

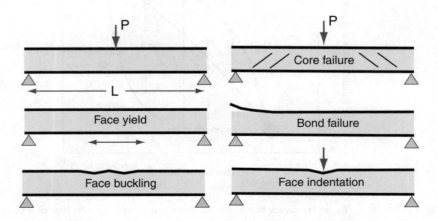

Figure 13.17   Failure modes of sandwich panels.

to suffer the same strain. This constraint leads to an in-plane expansionc coefficient of

$$\tilde{\alpha}_{//} = \frac{2tE_f\alpha_f + cE_c\alpha_c}{2tE_f + cE_c} \quad (13.21)$$

The through-thickness coefficient is simpler; it is given by the weighted mean

$$\tilde{\alpha}_\perp = \frac{2t\alpha_f + c\alpha_c}{2t + c} \quad (13.22)$$

Through-thickness thermal diffusivity is not a single-valued quantity, but depends on time. At short times heat does not penetrate the core and the diffusivity is that of the face, but at longer times the diffusivity tends to the value given by the ratio $\tilde{\lambda}/\tilde{\rho}\tilde{C}_p$.

Electrical properties. The dielectric constant of a sandwich, as with composites, is given by a rule of mixtures, equation (13.12) with $f = 2t/d$. Because the core is usually a foam, this allows the construction of stiff, strong sandwich shells with exceptionally low dielectric loss. In-plane electrical conductivity, too, follows such a rule. Through-thickness conductivity, like that of heat, is described by the harmonic mean (the equivalent of equation (13.20)).

13.6 Lattices: hybrids of type 3

Lattices — foams and other cellular structures — are hybrids of a solid and a gas. The properties of the gas might at first sight seem irrelevant, but this is not so. The thermal conductivity of low-density foams of the sort used for insulation is determined by the conductivity of the gas; and the dielectric constant and breakdown potential, and even the compressibility, depend on the gas properties. We therefore think of lattices as hybrids, and calculate their properties in the way shown in Figure 13.18.

There are two distinct species of cellular solid. The distinction is most obvious in their mechanical properties. The first, typified by foams, are *bending-dominated structures*; the second, typified by triangulated lattice structures, are *stretch dominated* — a distinction explained more fully below. To give an idea of the difference: a foam with a relative density of 0.1 (meaning that the solid cell walls occupy 10% of the volume) is less stiff by a factor of 10 than a triangulated lattice of the same relative density. The word "configuration" has special relevance here.

Bending dominated structures

Figure 13.19 shows an idealized cell of a low-density foam. It consists of solid cell walls or edges surrounding a void space containing a gas or fluid. Cellular solids are characterized by their relative density, which for the structure shown here (with $t \ll L$) is

$$\frac{\tilde{\rho}}{\rho_s} = \left(\frac{t}{L}\right)^2 \tag{13.23}$$

where $\tilde{\rho}$ is the density of the foam, ρ_s is the density of the solid of which it is made, L is the cell size, and t is the thickness of the cell edges.

Mechanical properties. Figure 13.20 shows the compressive stress–strain curve of bending-dominated foams. The material is linear elastic, with modulus $\tilde{E}$ up to its elastic limit, at which point the cell edges yield, buckle or fracture. The foam continues to collapse at a nearly constant stress (the "plateau

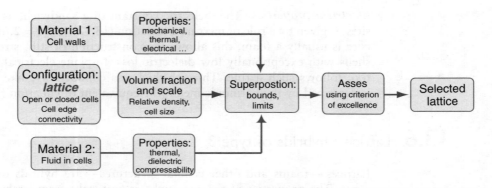

Figure 13.18   Foams and lattice structures: the design variables. They include the choice of material for the cell walls, the gas within the cells, the relative density, the cell size and shape, and — importantly — the connectivity of the cell edges.

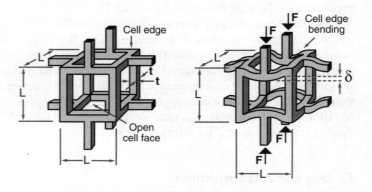

Figure 13.19   A cell in a low-density foam. When the foam is loaded, the cell edges bend, giving a low-modulus structure.

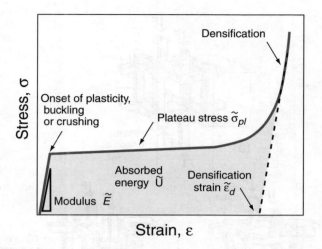

Figure 13.20 The plateau stress is determined by buckling, plastic bending or fracturing of the cell walls.

stress", $\tilde{\sigma}$) until opposite sides of the cells impinge (the "densification strain" $\tilde{\varepsilon}_d$), when the stress rises rapidly. The mechanical properties are calculated in the ways developed below, details of which can be found in the texts listed under Further reading. All are based on an idealized cell shape; real cells are less perfect than this, so the results should be regarded as upper bounds. Experiments show, however, that the results of the models give a good description of real foams.

A remote compressive stress σ exerts a force $F \propto \sigma L^2$ on the cell edges, causing them to bend and leading to a bending deflection δ, as shown in Figure 13.19. For the open-celled structure shown in the figure, the bending deflection is given by

$$\delta \propto \frac{FL^3}{E_s I} \tag{13.24}$$

where E_s is the modulus of the solid of which the foam is made and $I = (t^4/12)$ is the second moment of area of the cell edge of square cross section, $t \times t$. The compressive strain suffered by the cell as a whole is then $\varepsilon = 2\delta/L$. Assembling these results gives the modulus $\tilde{E} = \sigma/\varepsilon$ of the foam as

$$\frac{\tilde{E}}{E_s} \propto \left(\frac{\tilde{\rho}}{\rho_s}\right)^2 \quad \text{(bending-dominated behavior)} \tag{13.25}$$

Since $\tilde{E} = E_s$ when $\tilde{\rho} = \rho_s$ we expect the constant of proportionality to be close to unity—a speculation confirmed both by experiment and by numerical simulation.

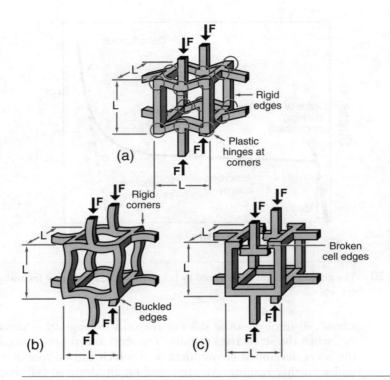

Figure 13.21 Collapse of foams. (a) When a foam made of a plastic materials loaded beyond its elastic limit, the cell edges bend plastically. (b) An elastomeric foam, by contrast, collapses by the elastic buckling of its cell edges. (c) A brittle foam collapses by the successive fracturing of cell edges.

A similar approach can be used to model the collapse, and thus the plateau stress of the foam. The cell walls yield, as shown in Figure 13.21(a) when the force exerted on them exceeds their fully plastic moment

$$M_f = \frac{\sigma_s t^3}{4} \tag{13.26}$$

where σ_s is the yield strength of the solid of which the foam is made. This moment is related to the remote stress by $M \propto FL \propto \sigma L^3$. Assembling these results gives the failure strength $\widetilde{\sigma}_{pl}$:

$$\frac{\widetilde{\sigma}_{pl}}{\sigma_{y,s}} = C\left(\frac{\tilde{\rho}}{\rho_s}\right)^{3/2} \quad \text{(bending-dominated behavior)} \tag{13.27}$$

where the constant of proportionality, $C \approx 0.3$, has been established both by experiment and by numerical computation.

Elastomeric foams collapse not by yielding but by elastic bucking; brittle foams by cell-wall fracture (Figure 13.21(b) and (c)). As with plastic collapse, simple scaling laws describe this behavior well. Collapse by buckling occurs when the stress exceeds $\tilde{\sigma}_{el}$, given by

$$\frac{\tilde{\sigma}_{el}}{E_s} \approx 0.05 \left(\frac{\tilde{\rho}}{\rho_s}\right)^2 \tag{13.28}$$

and by cell wall fracture when it exceeds $\tilde{\sigma}_{cr}$, where

$$\frac{\tilde{\sigma}_{cr}}{\sigma_{cr,s}} \approx 0.3 \left(\frac{\tilde{\rho}}{\rho_s}\right)^{3/2} \tag{13.29}$$

where $\sigma_{cr,s}$ is the modulus of rupture of the cell-wall material. Densification, when the stress rises rapidly, is a purely geometric effect: the opposite sides of the cells are forced into contact and further bending or buckling are not possible. It is found to occur at a strain $\tilde{\varepsilon}_d$ (the densification strain) of

$$\tilde{\varepsilon}_d = 1 - 1.4 \left(\frac{\tilde{\rho}}{\rho_s}\right) \tag{13.30}$$

Foams are often used for cushioning, packaging or to protect against impact. The useful energy that a foam can absorb per unit volume is approximated by

$$\widetilde{U} \approx \tilde{\sigma}\tilde{\varepsilon}_d \tag{13.31}$$

where $\tilde{\sigma}$ is the plateau stress — the yield, buckling or fracturing strength of the foam, whichever is least.

This behavior is not confined to open-cell foams with the structure idealized in Figure 13.19. Most closed-cell foams also follow these scaling laws, at first sight an unexpected result because the cell faces must carry membrane stresses when the foam is loaded, and these should lead to a linear dependence of both stiffness and strength on relative density. The explanation lies in the fact that the cell faces are very thin; they buckle or rupture at stresses so low that their contribution to stiffness and strength is small, leaving the cell edges to carry most of the load.

Thermal properties. The specific heat of foams, when expressed in units of J/m^3 K, is given by a rule of mixtures, summing the contributions from the solid and the gas. The thermal expansion coefficient of an open cell foam is the same as that of the solid from which it is made. The same is true of rigid closed-cell foams but not necessarily of low density elastomeric foams because the expansion of the gas within the cells can expand the foam itself, giving an apparently higher coefficient.

The cells in most foams are sufficiently small that convection of the gas within them is completely suppressed. The thermal conductivity of the foam is thus the sum of that conducted through the cell walls and that through the still air (or other gas) they contain. To an adequate approximation

$$\tilde{\lambda} = \frac{1}{3} \left(\left(\frac{\tilde{\rho}}{\rho_s}\right) + 2 \left(\frac{\tilde{\rho}}{\rho_s}\right)^{3/2} \right) \lambda_s + \left(1 - \left(\frac{\tilde{\rho}}{\rho_s}\right)\right) \lambda_g \tag{13.32}$$

where λ_s is the conductivity of the solid and λ_g that of the gas (for dry air it is 0.025 W/m.K). The term associated with the gas is important: blowing agents for foams intended for thermal insulation are chosen to have a low value of λ_g.

Electrical properties. Insulating foams are attractive for their low dielectric constant, falling towards 1 (the value for air or vacuum) as the relative density decreases:

$$\widetilde{\varepsilon} = 1 + (\varepsilon_s - 1)\left(\frac{\widetilde{\rho}}{\rho_s}\right) \tag{13.33}$$

where ε_s is the dielectric constant of the solid of which the foam is made. The electrical conductivity follows the same scaling law as the thermal conductivity.

Stretch-dominated structures

If conventional foams have low stiffness because the configuration of their cell edges allows them to bend, might it not be possible to devise other configurations in which the cell edges were made to stretch instead? This thinking leads to the idea of *micro-truss lattice-structures*. To understand these we need one of those simple yet profound fundamental laws: it is the Maxwell stability criterion.

The condition that a pin-jointed frame (meaning one that is hinged at its corners) made up of b struts and j frictionless joints, like those in Figure 13.22

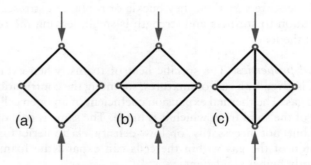

(a)   (b)   (c)

Figure 13.22   The pin-jointed frame at (a) folds up when loaded — it is a mechanism. If its joints are welded together, the cell edges bend (as in Figure 13.16) — it becomes a *bending-dominated structure*. The pin-jointed, triangulated, frame at (b) is stiff when loaded because the transverse bar carries tension, preventing collapse. When its joints are welded its stiffness and strength hardly change because it is a *stretch-dominated structure*. The frame at (c) is over-constrained. If the horizontal bar is tightened, the vertical one is put in tension even when there are no external loads: a state of *self-stress* exists.

to be both statically and kinematically determinate (meaning that it is rigid and does not fold up when loaded) in two dimensions, is

$$M = b - 2j + 3 = 0 \qquad (13.34)$$

In three dimensions the equivalent equation is

$$M = b - 3j + 6 = 0 \qquad (13.35)$$

If $M < 0$, the frame is a *mechanism*. It has no stiffness or strength; it collapses if loaded. If its joints are locked, preventing rotation (as they are in a lattice) the bars of the frame *bend* when the structure is loaded, just as in Figure 13.19. If, instead, $M \geq 0$ the frame ceases to be a mechanism. If it is loaded, its members carry tension or compression (even when pin-jointed), and it becomes a *stretch-dominated* structure. Locking the joints now makes little difference because slender structures are much stiffer when stretched than when bent. There is an underlying principle here: *stretch dominated structures have high structural efficiency; bending dominated structures have low.*

Mechanical properties. These criteria give a basis for the design of efficient micro-truss structures. For the cellular structure of Figure 13.19, $M < 0$, and bending dominates. For the structure shown in Figure 13.23, however, $M > 0$ and it behaves as an almost isotropic, stretch-dominated structure. On average one third of its bars carry tension when the structure is loaded in simple tension, regardless of the loading direction. Thus

$$\frac{\widetilde{E}}{E_s} \approx \frac{1}{3}\left(\frac{\tilde{\rho}}{\rho_s}\right) \quad \text{(isotropic stretch-dominated behavior)} \qquad (13.36)$$

Figure 13.23   A micro-truss structure and its unit cell. This is a stretch dominated structure, and it is over-constrained, meaning that it is possible for it to be in a state of self-stress.

and its collapse stress is

$$\frac{\widetilde{\sigma}}{\sigma_{y,s}} \approx \frac{1}{3}\left(\frac{\tilde{\rho}}{\rho_s}\right) \quad \text{(isotropic stretch-dominated behavior)} \tag{13.37}$$

This is an upper bound since it assumes that the struts yield in tension or compression when the structure is loaded. If the struts are slender, they may buckle before they yield. Then the "strength", like that of a buckling foam (equation (13.28)), is

$$\frac{\widetilde{\sigma}_{el}}{E_s} \approx 0.2\left(\frac{\tilde{\rho}}{\rho_s}\right)^2 \tag{13.38}$$

These results are summarized in Figure 13.24 in which the modulus $\widetilde{E}$ is plotted against the density $\tilde{\rho}$. Stretch dominated, prismatic microstructures have moduli that scale as $\tilde{\rho}/\rho_s$ (slope 1); bending dominated, cellular, microstructures have moduli that scale as $(\tilde{\rho}/\rho_s)^2$ (slope 2). Given that the density can be varied through a wide range, this allows great scope for material design. Note how the use of microscopic shape has expanded the occupied area of $E - \rho$ space.

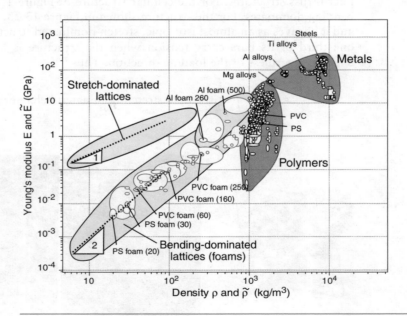

Figure 13.24   Foams and micro-truss structures are hybrids of material and space. Their mechanical response depends on their structure. Foams are usually bending dominated (equations (13.25) and (13.27)), and lie along a line of slope 2 on this chart. Micro-truss structures are stretch-dominated (equations (13.36) and (13.37)) and lie on a line of slope 1. Both extend the occupied area of this chart by many decades.

Thermal and electrical properties. The bending/stretching distinction influences mechanical properties profoundly, but has no effect on thermal or electrical properties, which are adequately described by the equations listed above for foams.

13.7 Segmented structures: hybrids of type 4

"United we stand, divided we fall" as the saying goes. But if "falling" means bending or twisting easily, it is something we sometimes want. And division is the way to get it (Figure 13.25).

Subdivision as a design variable

Shape can be used to reduce flexural stiffness and strength as well as increase them. Springs, suspensions, flexible cables and other structures that must flex yet have high tensile strength, use shape to give a low bending stiffness. This is achieved by shaping the material into strands or leaves, as suggested in Figure 13.26. As explained in Chapter 12, Section 12.8, the slender strands or leaves bend easily but do not stretch when the section is bent: an n-strand cable is less stiff by a factor of $3/\pi n$ than the solid reference section; an n-leaf panel by a factor $1/n^2$.

Subdivision can be used in another way: to impart *damage tolerance*. A glass window, hit by a projectile, will shatter. One made of small glass bricks, laid as bricks usually are, will lose a brick or two but not shatter totally; it is damage-tolerant. By sub-dividing and separating the material, a crack in one segment does not penetrate into its neighbors, allowing local but not global failure. That is the principle of "topological toughening". Builders in stone and brick have exploited the idea for thousands of years: both materials are brittle, but buildings made of them—even those made without cement ("dry-stone

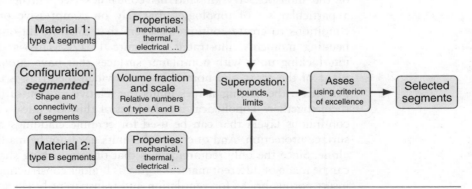

Figure 13.25   Segmented structures: the design variables. They include the choice of materials for the segments, and their shape, size, and stacking.

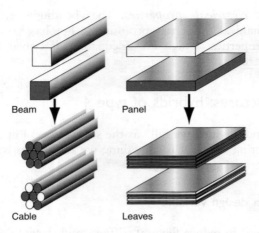

Beam   Panel

Cable   Leaves

Figure 13.26   Making low-efficiency structures. Shape gives the sections a lower flexural stiffness and strength per unit mass than the solid section from which they are made.

building")—survive ground movement, even earthquakes, through their ability to deform with some local failure but without total collapse.

Taking the simplest view, two things are necessary for topological damage tolerance: discreteness of the structural units, and an interlocking of the units in such a way that the array as a whole can carry load. Brick-like arrangements (Figure 13.27a) carry large loads and are damage tolerant in compression and shear, but disintegrate under tension. Strand and layer like structures are damage-tolerant in tension because if one strand fails the crack does not penetrate its neighbors—the principle of multi-strand ropes and cables. The jigsaw puzzle configuration (Figure 13.27b) carries in-plane tension, compression and shear, but at the cost of introducing a stress concentration factor of about, $\sqrt{R/r}$, where R is the approximate radius of a unit and r that of the interlock. Dyskin and his colleagues (see Further reading) explore a particular set of topologies that rely on compressive or rigid boundary conditions to create continuous layers that tolerate out-of-plane forces and bending moments, illustrated in Figure 13.27c. This is done by creating interlocking units with non-planar surfaces that have curvature both in the plane of the array and normal to it. Provided the array is constrained at its periphery, the nesting shapes limits the relative motion of the units, locking them together. Topological interlocking of this sort allows the formation of continuous layers that can be used for ceramic claddings or linings to give surface protection. And of course the units need not be made of one material alone. Since the only requirement is that of interlocking shape, the segments can be made of different materials. Just as builder constructing a brick wall can insert porous bricks for ventilation and transparent bricks to admit light, the designer of a segmented hybrid can add functionality through the choice of material for the units.

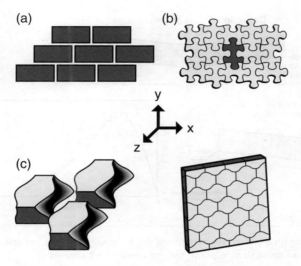

Figure 13.27   Examples of topological interlocking: discrete, unbonded structures that carry load.
(a) Brick-like assemblies of rectangular blocks carry axial compression, but not tension or
shear. (b) The two-dimensional interlocking of a jig-saw puzzle carries in-plane loads.
(c) The units suggested by Dyskin *et al.* (2001), when assembled into a continuous layer and
clamped within a rigid boundary around its edge, can carry out-of-plane loads and
bending moments.

The in-plane compression modulus and strength of any of the arrays of
Figure 13.27 are simply equal to those of the solid of which the segments are
made, combined as a rule of mixtures if more than one material is used to make
them. The in-plain tensile modulus and strength reflect the characteristics of
the peripheral constraint. More interesting are the those for bending.
Figure 13.28 shows schematically what happens when the array (reduced to
just two segments for simplicity) is subjected to a load W/unit depth, causing
a deflection δ. A peripheral constraining force, T per unit depth, holds them
together. Let T have the generalized form

$$T = T_0 + 2\,\Delta d S^* \tag{13.39}$$

Here T_0 is the pre-tension and the constraining stiffness S^* describes how
T increases as the blocks are wedged apart by the distance Δd per block:

$$\Delta d = \frac{b}{d}\delta$$

The lower figure shows a rigid body diagram for one block. Taking moments
about the point A gives the force-displacement response

$$\widetilde{W} = \frac{2(b - \delta)}{d}\left(T_0 + 2S^* \frac{(b - \delta)\delta}{d}\right) \tag{13.40}$$

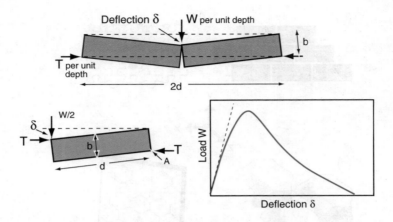

Figure 13.28   The loading on a pair of segments with the free-body diagram for one of them, and the flexural load-deflection response for the constrained segments.

If $S^* = 0$ the array is unstable when $\widetilde{W} > (2b/d)T_0$. If instead $T_0 = 0$ and $S^* > 0$,

$$\widetilde{W} = 4S^* \frac{(b-\delta)\delta}{d}$$

and the array shows linear-elastic response at small δ, becoming unstable at $\delta = b/3$. The behavior is sketched on the right of Figure 13.28.

The damage tolerance can be understood in the following way. We suppose that the units of the structure are all identical, each with a volume V_s, and that they are assembled into a body of volume V_t; there are therefore $n = V_t/V_s$ segments. We describe the probability of failure of a segment under a uniform tensile stress σ by a Weibull probability function:

$$P_f(V, \sigma) = 1 - \exp - \left\{ \frac{V\sigma^m}{V_0\sigma_0^m} \right\} \tag{13.41}$$

where m, V_0 and σ_0 are constants. If the body were made of a single monolithic piece of the brittle solid, this equation, with $V = V_t$ would describe the failure probability. To calculate the design stress σ_t^* we set an acceptable value for P, which we call P^* (say 10^{-6}, meaning that it is acceptable if one in a million fail) and invert the equation to give

$$\sigma_t^* = \sigma_0 \left\{ \frac{V_0}{V_t} \ln(1 - P^*) \right\}^{1/m} \tag{13.42}$$

Now consider the segmented body. A remote stress, if sufficiently large, causes some segments to fail. We refer to the fraction that has failed as the *damage*, D. If loaded such that each segment carried a uniform stress σ the damage is simply $p_f(V_s, \sigma)$. If some segments fail the body as a whole remains intact; global failure requires that a fraction D^*, the critical damage (say, 10%), must fail. Inverting equation (13.41) with $V = V_s$ gives the global failure stress σ_s^* of the segmented body:

$$\sigma_s^* = \sigma_0 \left\{ \frac{V_0}{V_s} \ln(1 - D^*) \right\}^{1/m} \tag{13.43}$$

Thus segmentation increases the allowable design stress from σ_t^* to σ_s^*, a factor of

$$\frac{\sigma_s^*}{\sigma_t^*} = \left\{ n \frac{\ln(1 - D^*)}{\ln(1 - P^*)} \right\}^{1/m} \approx \left\{ n \frac{D^*}{P^*} \right\}^{1/m} \tag{13.44}$$

(expanding the logarithm as a series and retaining the first term — an acceptable approximation for small P^* and D^*). Both n and D^*/P^* are considerably greater than 1, so the equation suggests that segmentation always increases the design stress.

The merit of this argument is its simplicity, illustrating how damage tolerance can arise. But it is a little too simple. What are the underlying assumptions? First, that the number of segments is large; only then can the damage D be reliably equated to the failure probability P_f. Second, that the stress σ is uninfluenced by damage; in reality its average value increases by the factor $1/(1 - D)$. Third, that the stress is uniform; in reality both the interlocking shapes and the load shed by a failed segment onto its immediate neighbors increases the stresses locally by a effective stress concentration factor K_c. And fourth, that segment failure is uncorrelated ("non-associated cracking"), not linked ("associated cracking"). The first of these is simply a pre-condition that must be met for segmentation to be successful. The second and third of these can be included approximately by modifying equation (13.44) to give:

$$\frac{\sigma_s^*}{\sigma_t^*} \approx \left(\frac{1 - D}{K_c} \right) \left\{ n \frac{D^*}{P^*} \right\}^{1/m} \tag{13.45}$$

The last depends on the values of the Weibull constants, and requires numerical modeling to explore the conditions for non-associated cracking. Equation (13.45) however is still useful — it allows for both a gain and a loss in allowable design stress.

Figure 13.29 gives an idea of the messages that it carries. It is constructed using $K_c = 2$, and values of nD^*/P^* between 10^4 and 10^{10}. The Weibull modulus m of brittle solids typically lies in the range 5 (typical of a porous brittle ceramic) and 30 (a value that might describe the variability of a high technical ceramic). When the Weibull modulus is large — more than 25 — segmentation offers little gain, indeed there may be a loss of allowable stress. But when it is low (as it is in building materials like brick and stone, and as it can be for

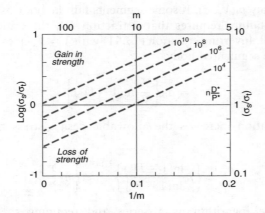

Figure 13.29   The ratio σ_s/σ_t for the listed values of nD^*/P^*, with $K_c = 2$ and $D^* = 0.1$.

certain ceramics and glasses) there is a gain. Segmentation gives a new design variable.

13.8 Summary and conclusions

The properties of engineering materials can be thought of as defining the axes of a multi-dimensional space with each property as a dimension. Sections through this space can be mapped. These maps reveal that some areas of property-space are occupied, others are empty — there are holes. The holes can sometimes be filled by making hybrids: combinations of two (or more) materials in a chosen configuration and scale. Design requirements isolate a small box in a multi-dimensional material–property space. If it is occupied by materials, the requirements can be met. But if it hits a hole in any one of the dimensions, we need a hybrid. A number of families of configuration exist, each offering different combinations of functionality. Each configuration is characterized by a set of bounds, bracketing its effective properties. The methods developed in this chapter provide tools for exploring alternative material–configuration combinations. Chapter 14 gives examples of their use.

13.9 Further reading

Allen, H.G. (1969) *Analysis and Design of Structural Sandwich Panels*, Pergamon Press, Oxford, UK. *(The bible: the book that established the principles of sandwich design).*

Ashby, M.F. (1993) Criteria for selecting the components of composites, *Acta Mater.* **41**, 1313–1335. *(A compilation of models for composite properties, introducing the methods developed here.)*

Ashby, M.F., Evans, A.G., Fleck, N.A., Gibson, Lorna J., Hutchinson J.W. and Wadley, H.N.G. (2000) *Metal Foams: A Design Guide*, Butterworth-Heinemann, Oxford. ISBN 0-7506-7219-6. *(A text establishing the experimental and theoretical basis of the properties of metal foams, with data for real foams and examples of their applications.)*

Bendsoe, M.P. and Sigmund, O. (2003) *Topology Optimisation, Theory Methods and Applications*, Springer-Verlag, Berlin, Germany. ISBN 3-540-42992-1. *(The first comprehensive treatment of emerging methods for optimizing configuration and scale.)*

Chamis, C.C. (1987) in *Engineers Guide to Composite Materials*, pp. 3–8 — 3–24. Am. Soc. Metals, Materials Park, Ohio, USA. *(A compilation of models for composite properties.)*

Deshpande, V.S., Ashby, M.F. and Fleck, N.A. (2001) "Foam topology: bending versus stretching dominated architectures", *Acta Mater.* **49**, 1035–1040. *(A discussion of the bending vs. stretching dominated topologies.)*

Dyskin, A.V., Estrin, Y., Kanel-Belov, A.J. and Pasternak, E. (2001) "Toughening by fragmentation: how topology helps", *Advanced Engineering Materials* **3**, 885–888; (2003) "Topological interlocking of platonic solids: a way to new materials and structures", *Phil. Mag.* **83**, 197–203. *(Two papers that introduce interlocking configurations that carry bending loads, yet give damage tolerance.)*

Gibson, L.J. and Ashby, M.F. (1997) *Cellular Solids, Structure and Properties*, 2nd edition, Cambridge University Press, Cambridge, UK. ISBN 0-521-49560-1. *(A monograph analyzing the properties, performance and uses of foams and giving the derivations and experimental verification of the equations used in Section 13.6.)*

Schoutens, J.E. and Zarate, D.A. (1986) *Composites* **17**, 188. *(A compilation of models for composite properties.)*

Stauffer, D. and Aharony, A. (1994) *Introduction to Percolation Theory*, 2nd edition, Taylor and Francis, London UK. *(A personal but very readable introduction to percolation theory.)*

Watt, J.P., Davies, G.F. and O'Connell, R.J. (1976) *Rev. Geophys. Space Phys.* **14**, 541. *(A compilation of models for composite properties.)*

Weibull, W. (1951) *J. Appl. Mech.* **18**, 293. *(The orginator or the "weakest –link" model of a brittle solid.)*

Zenkert, D. (1995) *An Introduction to Sandwich Construction*, Engineering Advisory Services Ltd., Solihull, UK, published by Chameleon Press Ltd., London, UK. ISBN 0 947817778. *(A primer, aiming to teach the basic analysis of sandwich structures.)*

Hybrid case studies

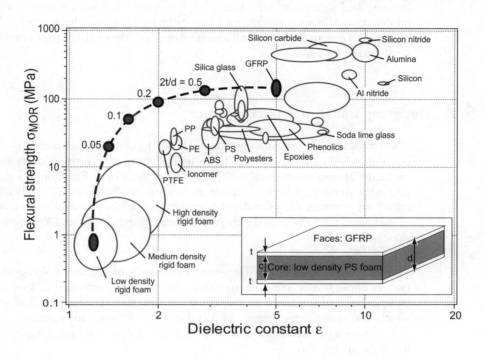

14.1 Introduction and synopsis

In Chapter 13 we explored hybrids of four types: composites, sandwiches, lattices, and segmented structures. Each is associated with a set of property-models that allow its properties to be estimated. In this chapter we illustrate their use to design hybrids to fill specified needs — needs that cannot be filled by single material choices.

14.2 Designing metal matrix composites

The problem. The most common state of loading in structures is that of bending. One measure of excellence in designing materials to carry bending moments at minimum weight is the index $E^{1/2}/\rho$, where E is Young's modulus and ρ the density. Alloys of aluminum and of magnesium rank highly by this criterion; titanium alloys and steels are less good. How could the performance of magnesium (the best of the lot) be enhanced further? Table 14.1 summarizes the challenge.

The method. The configuration–function matrix of Figure 13.4 suggests that two configurations are able to offer high flexural stiffness at low weight: composites and sandwiches. The choice of shape for a sandwich is limited, so we choose composites. Figure 14.1 is a chart of E and ρ for metals and fibers. The criterion of excellence is shown as a set of diagonal contours, increasing towards the top left.

Table 14.2 summarizes the superposition rules for density and modulus. They are plotted as envelopes of attainable performance for four magnesium-based composites; the upper edge of each envelope is the upper bound, the lower edge the lower one. There is an upper limit for the volume fraction, which we will set at 0.5. It is shown as a vertical bar within each envelope. Only the shaded part of the envelope below the bar is accessible. The diagonal lines plot the criterion of excellence, $E^{1/2}/\rho$. The combinations with the highest values of this quantity offer the greatest gain in stiffness per unit weight.

Table 14.1   Design requirements for the panel

Function	Light stiff beam
Constraints	Magnesium matrix
Objective	Maximize stiffness to weight in bending (index $E^{1/2}/\rho$)
Configuration	Fiber composite
Free variables	Choice of reinforcement and volume fraction

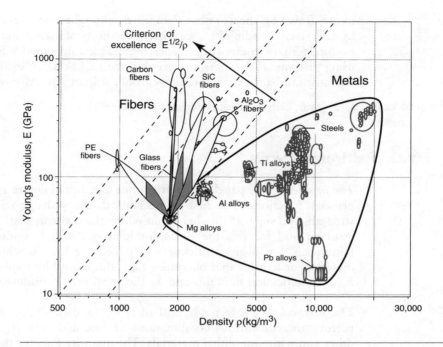

Figure 14.1 Possible magnesium–matrix composites. The lozenges show the areas bracketed by the upper and lower bounds of Table 14.2. The red shaded areas within them extend upto a volume fraction of 0.5.

Table 14.2 Superposition rules for composite density and modulus[*]

Property	Lower bound	Upper bound
Density	$\tilde{\rho} = f\rho_r + (1 - f)\rho_m$ **(exact)**	
Modulus	$\tilde{E}_L = \dfrac{E_m E_r}{f E_m + (1 - f)E_r}$	$\tilde{E}_u = f E_r + (1 - f)E_m$

[*]Subscripts m and r mean "matrix" and "reinforcement"; f = volume fraction.

The results. It is immediately obvious that the greatest promise is shown by the composites of magnesium with drawn polyethylene (PE) fibers. Magnesium–carbon is slightly less good; magnesium–SiC and magnesium Al₂O₃ considerably so. The method allows the potential to be explored rapidly.

Postscript. Magnesium-PE composites look good, but there remains the challenge of actually making them. Polyethylene fibers are already used in ropes and cables because of their high stiffness and strength and low weight. But they are destroyed by temperatures much above 120°C, so casting or sintering of the magnesium around the fibers is not an option. One possibility is to use drawn PE sheets rather than fibers, and fabricate a multi-layer laminate by adhesively bonding PE sheets between sheets of magnesium. A second is to explore ternary composites: dispersing magnesium powder in an epoxy and using this mix as the matrix to contain the PE fibers, for example. Otherwise we must fall back on magnesium–carbon, still an attractive option.

Related case studies   6.12   Stiff, high damping materials for shaker tables

14.3 Refrigerator walls

The problem. The panels of a refrigerator like that in Figure 14.2 perform two primary functions. The first is mechanical: the walls provide stiffness and strength, and support the shelves on which the contents rest. The second is to insulate, and for this the through-thickness thermal conductivity must be minimized. For a given thickness of panel, the first is achieved by seeking materials or hybrids that maximize E_{flex}, the second by minimizing $\lambda_\perp$ where E_{flex} is the flexural modulus and $\lambda_\perp$ the appropriate conductivity.

The method. We select a hybrid of type "sandwich" and explore how the performance of various combinations of face and core compare with each other and with monolithic materials. The quantity E_{flex} for the hybrid is given

Figure 14.2   A refrigerator. The panels of the container unit must insulate, protect against the external environment, and be stiff and strong in bending.

Table 14.3   Design requirements for the insulating panel

Function	Insulating panel
Constraints	Sufficient stiffness to suppress vibration and support internal loads
Objective	Thermal insulation
Configuration	Hybrid of type "sandwich"
Free variables	Material for faces and core; their relative thicknesses

Table 14.4   Superposition rules for sandwich stiffness and conductivity

Property	Lower bound	Upper bound
Flexural modulus	$\tilde{E}_{flex} = \left(1 - \left(1 - \dfrac{2t}{d}\right)^3\right) E_f K_s$	$\tilde{E}_{flex} = \left(1 - (1 - 2t/d)^3\right) E_f$
Through-thickness conductivity	$\tilde{\lambda}_\perp = \left(\dfrac{2t/d}{\lambda_f} + \dfrac{(1 - 2t/d)}{\lambda_c}\right)^{-1}$ (exact)	

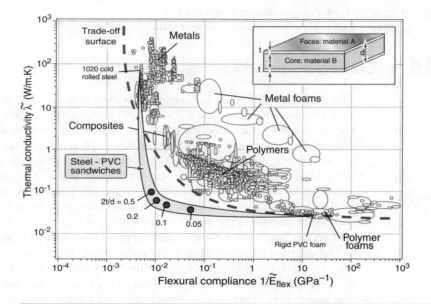

Figure 14.3   The broken line tracks the performance of steel–polystyrene foam bonded sandwiches. The thermal performance is plotted on the vertical axis, the mechanical performance on the horizontal one. Both are to be minimized.

approximately by equation (13.17a) and its through-thickness thermal conductivity is given by equation (13.20). The superposition rules are assembled in Table 14.4, in which t is the face thickness, d the panel thickness, E_f the modulus of the face material, λ_f and λ_c the conductivities of face and core, and K_s is the knock-down factor for core shear, ideally equal to 1 (no shear) but potentially as low as 0.5.

The results. Figure 14.3 shows the appropriate plot, using the thermal conductivity $\tilde{\lambda}_\perp$ and the flexural compliance $1/\tilde{E}_{flex}$ instead of its inverse so that a minimum is sought for both quantities. The approximate performance of a sandwich with mild steel faces and a polystyrene core is plotted using two equations in the table. The panel offers combinations of stiffness and insulation that cannot be matched by monolithic metals, composites, polymers, or foams. Other combinations of face and core (e.g. aluminum or SMC with polystyrene) can be evaluated efficiently using the diagram. Neither perform as well as the steel-polystyrene combination, though both come close.

Postscript. Adhesive technology has advanced rapidly over the last two decades. Adhesives are now available to bond almost any two material, and to do so with high bond-strength (though some adhesives are expensive). Fabricating the sandwich should not be a problem.

Related case studies

6.14 Energy-efficient kiln walls
14.6 Materials for microwave-transparent enclosures

14.4 Connectors that do not relax their grip

The problem. There are kilometers of wiring in a car. The transition, already begun, to drive-by-wire control systems will increase this further. Wires have ends; they do not do much unless the ends are connected to something. The connectors are the problem: they loosen with time until, eventually, connection is lost.

Car makers, responding to market forces, now produce cars designed to run for at least 300,000 km and last, on average, 10 years. The electrical system is expected to operate *without servicing* for the life-time of the car. Its integrity is vital: who would wish to drive-by-wire in a car with loose connectors? With increasing instrumentation on engine and exhaust system many of the connectors get hot; some have to maintain good electrical contact at up to 200°C.

The primary material choice for connectors, today, is a copper-beryllium alloy, Cu 2% Be: it has excellent conductivity and the high strength needed to act as a spring to give the required clamping force at the connection. But the maximum long-term service temperature of copper–beryllium alloys is only about 130°C; at higher temperatures, creep-relaxation causes the connector to loose its grip. The challenge: to suggest a way to solve this problem.

The method. The answer is to separate the functions, select the best material for each, check for compatibility, and combine them. So here goes. Function 1: conduct electricity. Copper excels at this; no other affordable material is as good. Its alloys (among them, copper–beryllium) are stronger, but at the cost of some loss of conductivity. We choose copper to provide the conduction.

Table 14.5   Design requirements for the connector

Function	Conductor for electrical connector
Constraints	Maintain clamping force at 200°C for life of vehicle
Objective	Minimize cost
Configuration	Bi-layer sandwich
Free variables	Materials 1 and 2; their relative thicknesses

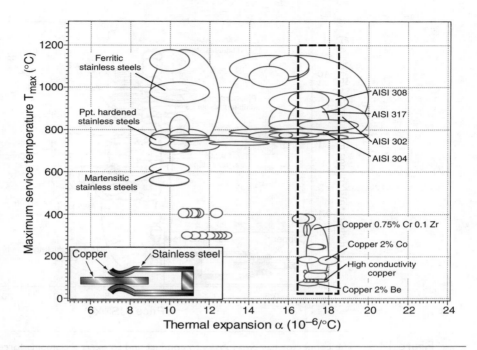

Figure 14.4 A hybrid connector. As in Figure 12.1, we seek materials with matching thermal expansion, but this time with high temperature performance as the other variable. Here copper, because of its superb thermal conduction, is a prerequisite for Material 1. Type 302 or 304 stainless steel is a good choice for Material 2.

Function 2: provide clamping force for the life-time of the vehicle. The material chosen to fill function 2 will have to be bonded to the copper, and if the combination is not to distort when heated, it must have the same expansion coefficient. Figures 14.4 and 14.5 guide the choice. The first shows the maximum service temperature and expansion coefficient for copper, Cu–2% Be, and a range of steels. The box encloses materials with the same expansion coefficient as copper. Type 302 and 304 austenitic stainless steels match copper in expansion coefficient, and can be used at much higher temperatures. But do they make good springs? And are they affordable? The second chart answers these questions. Good materials for springs (Chapter 6, Section 6.7) are those with high values of σ_f^2/E — this appears as one axis of the chart. The other axis is approximate price/kg. The chart shows that both 302 and 304 stainless steels, in the wrought condition, are almost as good as Cu 2% Be as a spring, and considerably cheaper.

The results. The proposed solution, then: a hybrid of copper and 302 stainless, roll-bonded to form a bi-layer like that shown in the inset on both figures. Further detailed design will, of course, be needed to establish the thicknesses

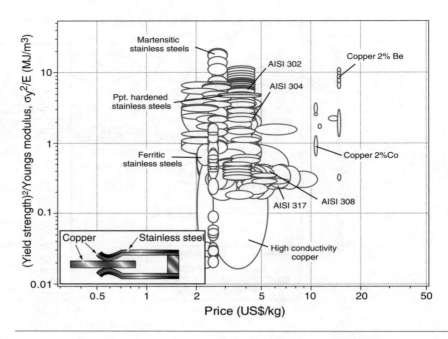

Figure 14.5 The connector has two jobs — to conduct and to exert a clamping force that does not relax. High conductivity copper and 304 stainless steel are both much cheaper that Cu 2% Be. Roll-bonding them will of course add cost, but large volume production could be competitive — and it solves the relaxation problem.

of each layer, the best degree of cold work, the formability and the resistance to the environment in which it will be used. But the method has guided us to a sensible concept, quickly, and efficiently.

Related case studies

6.7   Materials for springs

12.3   Ultra-efficient springs

14.5 Extreme combinations of thermal and electrical conduction

The problem. Materials that are good electrical conductors are always good thermal conductors too. Copper, for example, excels at both (Table 14.6).

Table 14.6   Data for copper and HDPE

Material	Electrical conductivity $(1/\mu\Omega.cm)$	Thermal conductivity (W/m.K)
High conductivity copper	0.6	395
High density polyethylene	1×10^{-25}	0.16

Table 14.7   Requirements for the conducting hybrids of copper and polyethylene

Function	Extreme conduction combinations
Constraints	Materials: copper and polyethylene
Objective	Maximize difference between electrical and thermal conductivities
Configuration	Free choice
Free variables	Configuration, and relative volume fractions of the two materials

Most polymers, by contrast, are electrical insulators (meaning that their conductivity is so low that for practical purposes they do not conduct at all), and as solids go, they are also poor thermal conductors — polyethylene is an example. Thus the "high–high" and the "low–low" combinations of conduction can be met by monolithic materials, and there are plenty of them. The "high–low" and "low–high" combinations are a different matter: nature gives us very few of either one. The challenge: using only copper and polyethylene, devise hybrid materials that achieve both of these combinations (data for both in the table). It is summarized in Table 14.7.

The method. Figure 14.6 shows two possible hybrid configurations. Both are of type "composite", but with very different configuration and volume fractions. The first is a tangle of fine copper wires embedded in a PE matrix. To describe its performance we draw on the bounds of Section 13.4. Electrical conductivity requires percolation. The percolation threshold is minimized by using conducting wires of high aspect ratio. Above the percolation threshold the conductivity tends towards a rule of mixtures (equation (13.9)):

$$\tilde{\kappa}_1 = f\kappa_{Cu} + (1-f)\kappa_{PE} \tag{14.1}$$

The thermal conductivity for a random array like this will lie near the lower bound (equation (13.10)):

$$\tilde{\lambda}_1 = \lambda_{PE}\left(\frac{\lambda_{Cu} + 2\lambda_{PE} - 2f(\lambda_{PE} - \lambda_{Cu})}{\lambda_{Cu} + 2\lambda_{PE} + f(\lambda_{PE} - \lambda_{Cu})}\right) \tag{14.2}$$

The second hybrid is a multi-layered composite with three orthogonal families of PE sheet contained in a discontinuous copper matrix. (We choose a multi-layer in order to create a "material" that, on a scale large compared to the layer spacing, behaves like a homogeneous solid.) We draw on the bounds of Section 13.5. When the PE layers are thin, the through-thickness resistances add, meaning that both the electrical and thermal conductivities are given by the harmonic means (equation 13.20)

$$\tilde{\kappa}_2 = \left(\frac{f}{\kappa_{Cu}} + \frac{(1-f)}{\kappa_{PE}}\right)^{-1} \tag{14.3}$$

$$\tilde{\lambda}_2 = \left(\frac{f}{\lambda_{Cu}} + \frac{(1-f)}{\lambda_{PE}}\right)^{-1} \tag{14.4}$$

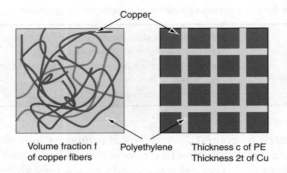

Copper

Volume fraction f
of copper fibers

Polyethylene

Thickness c of PE
Thickness 2t of Cu

Figure 14.6 Two alternative configurations of copper and polyethylene, shown here in two
dimensions, but easily generalized to 3. That on the left has high electrical conductive but
low thermal conductivity; that on the right has the opposite.

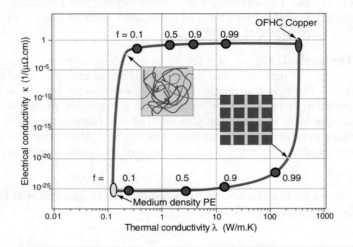

Figure 14.7 Two alternative hybrid-configurations of copper and polyethylene give very different
combinations of thermal and electrical conductivity, and create new "materials" with
properties that are not found in homogeneous materials.

The results. The results are plotted in Figure 14.7. The two hybrids differ
dramatically in their behavior. The hybridization has allowed the creation of
"materials" with extreme combinations of conductivities.

Postscript. Hybrids of the first type are widely used to give electrical screening
to computer and TV cabinets. Those of the second type are less common, but
could find application as heat sinks for power electronics in which large scale
electrical conduction would lead to coupling and eddy current losses.

How would you make it? Perhaps by thermally bonding a stack of copper
sheets interleaved with polyethylene film, dicing the stack normal to the layers

and re-stacking the slices with PE film interlayers, finally dicing and re-stacking a third time to give the last set of layers.

Related case studies 6.18 Materials for heat exchangers
14.4 Connectors that do not relax their grip

14.6 Materials for microwave-transparent enclosures

The problem. Microwave-transparent radomes were introduced in Chapter 6, Section 6.19. The radome is a thin panel or shell, requiring flexural stiffness and strength, but also requiring as low a dielectric constant, ε, as possible. Could hybrids offer better performance than monolithic materials?

The method. The configuration-function matrix of Figure 13.4 suggests that sandwich structures offer control of flexural stiffness and strength, and allow some control of electrical properties. We therefore explore these, seeking to meet the requirements of Table 14.8.

Figure 14.8 shows the flexural strength or modulus of rupture, σ_{MoR}, plotted against the dielectric constant ε. Many polymers have dielectric constants between 2 and 5. Dielectric response is an extensive property—if these polymers are foamed, the dielectric constant falls linearly with the relative density, approaching 1 at low densities (equation (13.33)):

$$\tilde{\varepsilon} = 1 + (\varepsilon_s - 1)\left(\frac{\tilde{\rho}}{\rho_s}\right) \tag{14.5}$$

Foams, however, are not very strong. GFRP, with a dielectric constant of 5, is much stronger. Based on a survey of possible faces and cores from among those plotted in Figure 14.8, we choose to explore a sandwich with faces of GFRP and a core of a low/medium density expanded polymer foam. If it is properly manufactured and the core material has sufficient strength, the flexural strength of the sandwich (equation (13.19)) is

$$(\tilde{\sigma}_{\text{MoR}})_U = \left(1 - \left(1 - \frac{2t}{d}\right)^3\right)\sigma_f + \left(1 - \frac{2t}{d}\right)^2\sigma_c \tag{14.6}$$

Table 14.8 Requirements for low dielectric constant radome skin

Function	Material for protection of microwave detector
Constraints	Must meet constraints on flexural strength
Objective	Minimize dielectric constant
Configuration	Hybrid of type "sandwich" with core of type "foam"
Free variables	• Choice of material for face and core
	• Relative thickness of the two materials

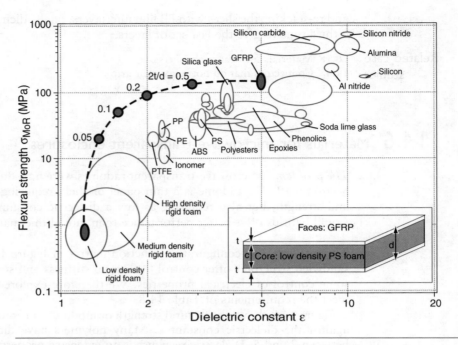

Figure 14.8 A plot of flexural modulus and dielectric constant for low dielectric constant materials. The trajectory shows the possibilities offered by hybrids of GFRP and polymer foam.

Its dielectric constant (equation (13.12)) with $f = 2t/d$ is

$$\tilde{\varepsilon} = \frac{2t}{d}\varepsilon_f + \left(1 - \frac{2t}{d}\right)\varepsilon_c \qquad (14.7)$$

The simplest way to explore them is to plot the two, using $2t/d$ as a parameter to link them.

The results. Figure 14.8 shows the results. The figure is used by identifying the desired flexural strength, σ_{MoR}, and reading off the values of $2t/d$ and dielectric constant. The hybrid of type "sandwich" allows the creation of a set of materials with combinations of flexural strength and dielectric constant that out-perform all the homogeneous materials of Figure 14.8 — indeed it out-performs even the best composites (not shown here).

Further reading Huddleston, G.K. and Bassett, H.L., in Johnson, R.D. and Jasik, H. (eds), *Antenna Engineering Handbook*, 2nd edition, McGraw-Hill, New York, Chapter 44.
Lewis, C.F. (1998) Materials keep a low profile, *Mech. Eng.*, June, 37–41.

Related case studies 6.19 Materials for radomes

14.7 Exploiting anisotropy: heat spreading surfaces

The problem. A saucepan made from a single material, when heated on an open flame, develops hot spots that can locally burn its contents. That is because the saucepan is thin; heat is transmitted through the thickness more quickly than it can be spread transversely to bring the entire pan surface to a uniform temperature. The metals of which saucepans are usually made — cast iron, aluminum, stainless steel, or copper — have thermal conductivities that are isotropic, the same in all directions. What we clearly want is a thermal conductivity that is higher in the transverse direction than in the through-thickness direction. A bi-layer (or multi-layer) hybrid can achieve this. Table 14.9 summarizes the situation.

The method. Heat transmitted in the plane of a bi-layer has two parallel paths; the total heat transmitted is a sum of that in each of the paths. If it is made of a layer of material 1 with thickness t_1 and conductivity λ_1, bonded to a layer of material 2 with thickness t_2 and conductivity λ_2, the conductivity parallel to the layers is

$$\tilde{\lambda}_{//} = f\lambda_1 + (1-f)\lambda_2 \qquad (14.8)$$

(equation (13.9) with $f = t_1/(t_1+t_2)$). Perpendicular to the layers the conductivity is

$$\tilde{\lambda}_\perp = \left(\frac{f}{\lambda_1} + \frac{(1-f)}{\lambda_2}\right)^{-1} \qquad (14.9)$$

(the harmonic mean of equation (13.20)).

If the bi-layer is made of materials that differ greatly in thermal expansion coefficient, the pan will distort when heated. We therefore seek a pair of materials with almost the same expansion coefficient, but that differ as much as possible in conductivity in order to maximize the difference between equations (14.8) and (14.9).

The results. Figures 14.9 and 14.10 enable the selection. Figure 14.9 shows that the choice of copper plus stainless steel meets the requirement that the expansion coefficients match but the thermal conductivities differ. Figure 14.10

Table 14.9   Design requirements for the panel

Function	Heat-spreading surface
Constraints	Temperature up to 200°C without distortion
Objective	Maximize thermal anisotropy while maintaining good conduction
Configuration	Sandwich: Bi-layer or double layer
Free variables	Materials 1 and 2; their relative thicknesses

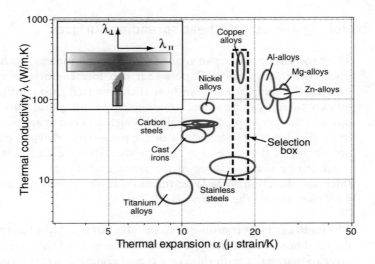

Figure 14.9   To spread heat, $\lambda_{//}$ must be greater than $\lambda_{\perp}$. Monolithic material have almost
isotropic conductivity, to get a bit difference, a hybrid is required, but, to avoid distortion,
the thermal expansion coefficients must match. Copper and austenic stainless steel
have a good match.

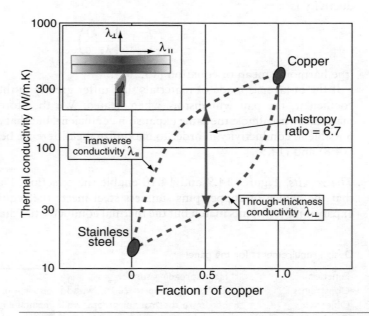

Figure 14.10   Creating anisotropy. A bi-layer of copper and stainless steel creates a "material" with
good conductivity and an anisotropy ratio greater than 6.

shows $\tilde{\lambda}_\perp$ and $\tilde{\lambda}_{//}$ plotted against f for a bi-layer of copper ($\lambda = 390$ W/m.K) and austenitic stainless steel ($\lambda = 16$ W.K). The envelopes are constructed in the same way as those of Figure 13.8, using f as a parameter. The maximum separation between $\tilde{\lambda}_\perp$ and $\tilde{\lambda}_{//}$ occurs broadly at $f = 0.5$ (each material occupies about half the thickness) when the ratio of the two conductivities (the anisotropy ratio) is 6.7. The hybrid has extended the occupied area of property space along an unusual dimension — that of thermal anisotropy.

Related case studies   6.15   Materials for passive solar heating

14.8 The mechanical efficiency of natural materials

"As a general principle natural selection is continually trying to economise every part of the organisation" — Charles Darwin, writing over 100 years ago. Natural materials are remarkably efficient. By efficient we mean that they fulfil the complex requirements posed by the way plants and animals function and that they do so using as little material as possible. Many of these requirements are mechanical in nature: the need to support static and dynamic loads created by the mass of the organism or by wind loading, the need to store and release elastic energy, the need to flex through large angles, and the need to resist buckling and fracture.

Virtually all natural materials are hybrids. They consist of a relatively small number of polymeric and ceramic components or building blocks that often are composites themselves. Plant cell walls, for instance, combine cellulose, hemicellulose and pectin, and can be lignified; animal tissue consists largely of collagen, elastin, keratin, chitin, and minerals such as salts of calcium or silica. From these limited "ingredients" nature fabricates a remarkable range of structured hybrids. Wood, bamboo, and palm consist of cellulose fibers in a lignin–hemicellulose matrix, shaped to hollow prismatic cells of varying wall thickness. Hair, nail, horn, wool, reptilian scales, and hooves are made of keratin while insect cuticle contains chitin in a matrix of protein. The dominant ingredient of mollusc shell is calcium carbonate, bonded with a few percent of protein. Dentine, bone, and antler are formed of "bricks" of hydroxyapatite cemented together with collagen. Collagen is the basic structural element for soft and hard tissues in animals, such as tendon, ligament, skin, blood vessels, muscle, and cartilage, often used in ways that exploit shape.

From a mechanical point of view, there is nothing very special about the individual building blocks. Cellulose fibers have Young's moduli that are about the same as that of nylon fishing-line, but a lot less than steel; and the lignin–hemicellulose matrix in which they are embedded has properties very similar to that of epoxy. Hydroxyapatite has a fracture toughness comparable with that of man-made ceramics, that is to say: it is brittle. It is thus the *configuration* of the components that give rise to the striking efficiency of natural materials.

Like engineering materials, the building blocks of natural materials can be grouped into classes: *natural ceramics* (calcium carbonate, hydroxyapatite), *natural polymers* (the polysaccharides cellulose and the proteins chitin, collagen, silk, and keratin), and *natural elastomers* (elastin, resilin, abductin, skin, artery, and cartilage). These combine to give a range of hybrids, among them and *ceramic composites* and *sandwiches* (bone, antler, enamel, dentine, shell, coral) and *lattices* (natural cellular materials like wood, cork, palm, bamboo, cancellous bone). Their properties, like those of engineering materials, can be explored and compared using material property charts. We conclude this chapter by exploring charts for the properties of these building blocks and of the natural hybrids formed from them.

The Young's modulus — density chart. Figure 14.11 shows data for Young's modulus, E, and density, ρ. Those for the classes of natural materials are circumscribed by a heavy balloon; class members are shown as smaller bubbles within them. The familiar stiffness guidelines E/ρ, $E^{1/2}/\rho$ and $E^{1/3}/\rho$ are shown, each representing the material index for a particular mode of loading.

The natural polymer with the highest efficiency in tension, measured by the index E/ρ, is cellulose; it exceeds that of steel by a factor of about 2.6. The high values for the fibers flax, hemp and cotton derive from this. Wood, palm and bamboo are particularly efficient in bending and resistant to buckling, as indicated by the high values of the flexure index $E^{1/2}/\rho$ when loaded parallel to the grain. That for balsa wood, for example, can be five times greater than that of steel.

The strength — density chart. Data for the strength, σ_f, and density, ρ, of natural materials are shown in Figure 14.12. For natural ceramics, the compressive strength is identified by the symbol (C); the tensile strength (which is much lower) is identified with the modulus of rupture in beam bending, symbol (T). For natural polymers and elastomers, the strengths are tensile strengths. And for natural cellular materials, the compressive strength is the stress plateau, symbol (C), while the tensile stress is either the stress plateau or the modulus of rupture, symbol (T), depending on the nature of the material. Where they differ, the strengths parallel (symbol $\|$) and perpendicular (symbol $\perp$) to the fiber orientation or grain have been plotted separately.

Evolution to provide tensile strength would — we anticipate — lead to materials with high values of σ_f/ρ; where strength in bending or buckling are required we expect to find materials with high $\sigma_f^{2/3}/\rho$. Silk and cellulose have the highest values of σ_f/ρ; that of silk is even higher than that of carbon fibers. The fibers of flax, hemp, and cotton, too, have high values of this index. Bamboo, palm and wood have high values of $\sigma_f^{2/3}/\rho$ giving resistance to flexural failure.

The Young's modulus — strength chart. Data for the strength, σ_f, and modulus, E, of natural materials are shown in Figure 14.13. Two combinations are

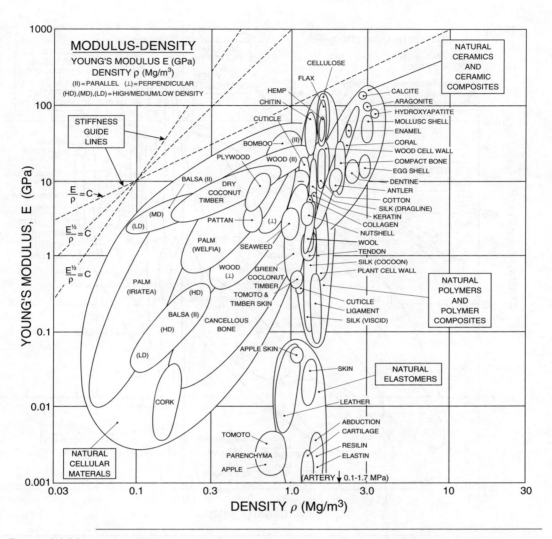

Figure 14.11   A material property chart for natural materials, plotting showing Young's modulus against density. Guidelines identify structurally efficient materials that are light and stiff.

significant. Materials with large values of σ_f^2/E store elastic energy and make good springs; and those with large values of σ_f/E have exceptional resilience. Both are plotted on the figure. Silks (including the silks of spider's webs) stand out as exceptionally efficient, having values of σ_f^2/E that exceed those of spring steel or rubber. High values of the other index, σ_f/E, mean that a material allows large, recoverable deflections and, for this reason, make good elastic hinges. Palm (coconut timber) has a higher value of this index than wood,

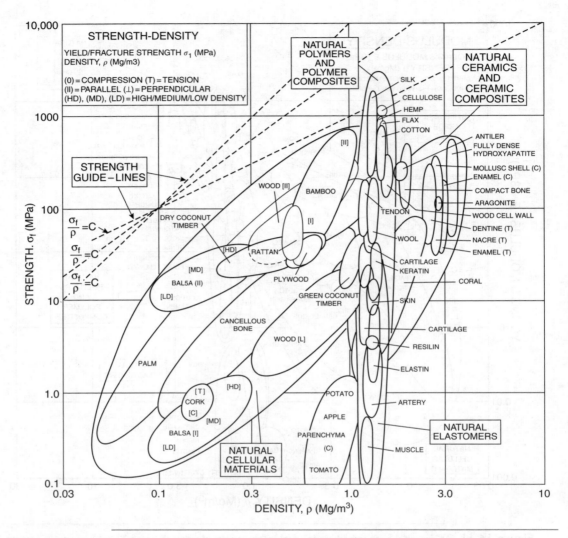

Figure 14.12 A material property chart for natural materials, plotting strength against density. Guidelines identify structurally efficient materials that are light and strong.

allowing palms to flex in a high wind. Nature makes much use of these: skin, leather, and cartilage are all required to act as flexural and torsional hinges.

The toughness—Young's modulus chart. The toughness of a material measures its resistance to the propagation of a crack. The limited data for the toughness, J_c, and Young's modulus, E, of natural materials are shown in Figure 14.14. When the component is required to absorb a given *impact energy*

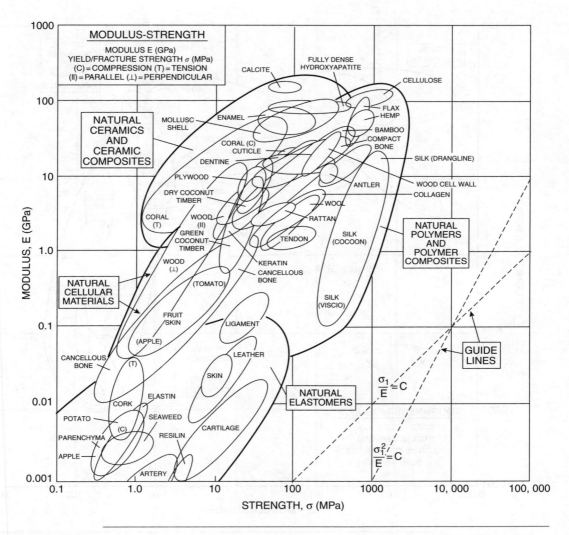

Figure 14.13 A material property chart for natural materials, plotting Young's modulus against strength. Guidelines identify materials that store the most elastic energy per unit volume, and that make good elastic hinges.

without failing, the best material will have the largest value of J_c. These materials lie at the top of Figure 14.14: antler, bamboo and woods stand out. When instead a component containing a crack must carry a given *load* without failing, the safest choice of material is that with the largest values of the fracture toughness $K_{1C} \approx (EJ_c)^{1/2}$. Diagonal contours sloping from the upper left to lower right on Figure 14.14, show values of this index. Nacre and enamel stand out. And when the component must support a given *displacement*

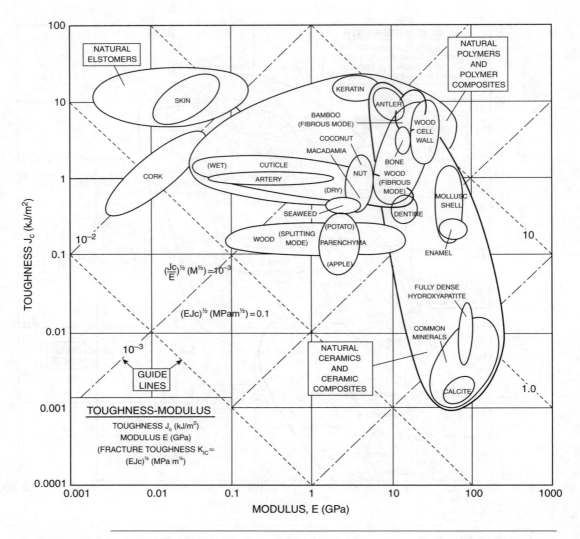

Figure 14.14 A material property chart for natural materials, plotting toughness against Young's modulus. Guidelines show fracture toughness $(EJ_c)^{1/2}(MPa)^{1/2}$ and resilience $(J_c/E)^{1/2}(m)^{1/2}$.

without failure, the material is measured by $(J_c/E)^{1/2}$. This is shown as a second set of diagonal contours sloping from the lower left to upper right on the figure. Skin is identified as particularly good by this criterion.

Many engineering materials — steels, aluminum, alloys — have values of J_c and K_{1C} that are much higher than those of the best natural materials. However, the toughnesses of natural ceramics like nacre, dentine, bone, and enamel are an order of magnitude higher than those of conventional engineering ceramics like alumina. Their toughness derives from their microstructure: platelets of ceramics such as calcite, hydroxyapatite or aragonite, bonded by a small volume fraction of polymer, usually collagen; it increases with decreasing mineral content and increasing collagen content.

Postscript. The charts bear out Darwin's idea that natural materials evolve to make the most of what is available to them — they are efficient, in the sense used earlier. But there is more to it than that. For deeper insight, see the books listed in Section 14.9.

14.9 Further reading: natural materials (see also Appendix D)

Beukers, A. and van Hinte, E. (1998) *Lightness. The Inevitable Renaissance of Minimum Energy Structures*, 010 Publishers, Rotterdam. ISBN 9064503346.

Currey, J.D., Wainwright, S.A. and Biggs, W.D. (1982) *Mechanical Design in Organisms*, Princeton University Press, Princeton, USA. ISBN 0691083088.

McMahon, T. and Bonner, J. (1983) *On Size and Life*, American Books, New York. ISBN 0716750007.

Sarikaya, M. and Aksay, I.A. (eds) (1995) *Biomimetics: Design and Processing of Materials*, AIP Press, Woodbury. ISBN 1563961962.

Thompson, D'A.W. (1992) *On Growth and Form*, Dover Publications. ISBN 0486671356.

Vincent, J.F.V. (1990) *Structural Biomaterials*, revised edition, Princeton University Press, Princeton. ISBN: 0691025134.

Vincent, J.F.V. and Currey, J.D. (1980) The mechanical properties of biological materials, Proceedings of the Symposia of the Society for Experimental Biology; no. 34, Cambridge University Press for the Society for Experimental Biology, Cambridge.

Chapter 15

Information and knowledge sources for design

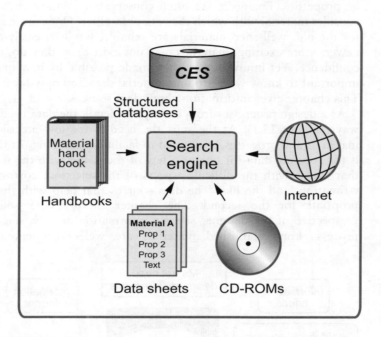

15.1 Introduction and synopsis

The engineer, in selecting a material for a developing design, needs data for its properties. Engineers are often conservative in their choice, reluctant to consider material with which they are unfamiliar. One reason is this: that data for the old, well-tried materials are reliable, familiar, easily-found; data for newer, more exciting, materials may not exist or, if they do, may not inspire confidence. Yet innovation is often made possible by new materials. So it is important to know where to find material data and how far it can be trusted. This chapter gives information about data sources.

As a design progresses from concept to detail, the data needs evolve in two ways (Figure 15.1). At the start the need is for low-precision data for all materials and processes, structured to facilitate screening. At the end the need is for accurate data for one or a few of them, but with the richness of detail that assists with the difficult aspects of the selection: corrosion, wear, cost estimation, and the like. The data sources that help with the first are inappropriate for the second. The chapter surveys data sources from the perspective of the designer seeking information at each stage of the design process. Long-established materials are well documented; less-common

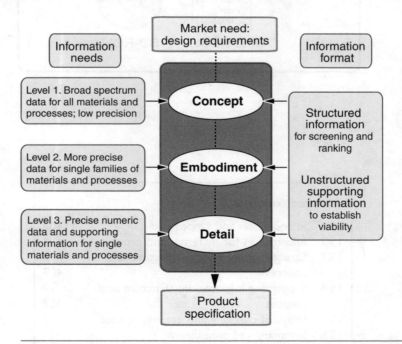

Figure 15.1 Information needs and data structure for screening/ranking and for the supporting information step. The first relies on structured information that is comprehensive and universal, the second makes use of information in any available format.

materials may be less so, posing problems of checking and, sometimes, of estimation. The chapter proper ends with a discussion of how this can be done.

So much for the text. The catalog of data sources, as hard-copy, software and on the internet, are gathered, with brief commentary, in Appendix D at the end of this book. It is intended for reference. When you *really* need data, this is the place to start.

15.2 Information for materials and processes

Materials information for design

If you are going to design something, what sort of materials information do you need? Figure 15.2 draws relevant distinctions. On the left a material is tested and the *data* are captured. But this raw data is, for our purposes, useless. To make it useful requires statistical analysis. What is the mean value of each property when measured on a large batch of samples? What is the standard deviation? Given this it is possible to calculate *allowables*, values of properties that, with a given certainty (say, one part in 10^6) can be guaranteed. Data in design handbooks are allowables. We refer to data with known precision and provenance as *information*. It can generally be reported as tables of numbers, as yes/no statements or as rankings: this is, it can be *structured*. Many attributes that can be structured are common to all materials; all have a density, an elastic modulus, a strength, a thermal conductivity. Thus structured information can be stored in a database and—since all materials have values—it is the starting point for selecting between them. The upper part of Table 15.1 shows structured information for ABS. Each property is stored as

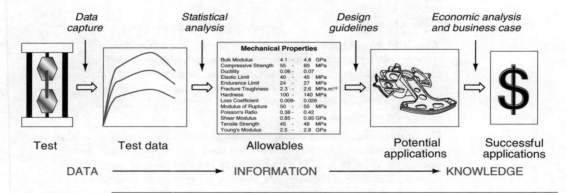

Figure 15.2   Types of material information. We are interested here in the types to be found in the center of this schematic: structured data for design "allowables", and the characteristics of a material that relate to its ability to be formed, joined and finished, records of experience with its use, and design guidelines for its use.

Table 15.1   Part of a record for a material

Acrylonitrile–butadiene–styrene (ABS)

Screening information

General properties

Density	1000–1200 kg/m^3
Price	1.7–3.2 US$/kg

Mechanical properties

Young's modulus	1.1–2.9 GPa
Hardness — Vickers	5.6–15 HV
Elastic limit	19–51 MPa
Tensile strength	28–55 MPa
Elongation	1.5–100%
Endurance limit	11–22 MPa
Fracture toughness	1.2–4.3 MPa.m$^{1/2}$
Loss coefficient	0.014–0.045

Thermal properties

Melting temperature	—
Maximum service temperature	62–77 °C
Thermal conductivity	0.19–0.34 W/m.K
Specific heat	1400–1900 J/kg.K
Thermal expansion	85–230 μ-strain.K

Electrical properties

Resistivity	$3.3 \times 10^{21} - 3 \times 10^{22}$ μΩ.cm
Dielectric constant	2.8–3.2
Breakdown potential	14–22 MV/m
Power factor	$3 \times 10^{-3} - 7 \times 10^{-3}$

Optical properties

Transparent or opaque?	Opaque
Refractive Index	1.5

Supporting Information

What is it? ABS (Acrylonitrile–butadiene–styrene) is tough, resilient, and easily molded. It is usually opaque, although some grades can now be transparent, and it can be given vivid colors. ABS–PVC alloys are tougher than standard ABS and, in self-extinguishing grades, are used for the casings of power tools.

Design guidelines. ABS has the highest impact resistance of all polymers. It takes color well. Integral metallics are possible (as in GE Plastics' Magix.) ABS is UV resistant for outdoor application if stabilizers are added. It is hygroscopic (may need to be oven dried before thermoforming) and can be damaged by petroleum-based machining oils.

ABS can be extruded, compression moulded or formed to sheet that is then vacuum thermo-formed. It can be joined by ultrasonic or hot-plate welding, or bonded with polyester, epoxy, isocyanate or nitrile-phenolic adhesives.

Table 15.1   (*Continued*)

Typical Uses. Safety helmets; camper tops; automotive instrument panels and other interior components; pipe fittings; home-security devices and housings for small appliances; communications equipment; business machines; plumbing hardware; automobile grilles; wheel covers; mirror housings; refrigerator liners; luggage shells; tote trays; mower shrouds; boat hulls; large components for recreational vehicles; weather seals; glass beading; refrigerator breaker strips; conduit; pipe for drain-waste-vent (DWV) systems.

The environment. The acrylonitrile monomer is nasty stuff, almost as poisonous as cyanide. Once polymerized with styrene it becomes harmless. ABS is FDA compliant, can be recycled, and can be incinerated to recover the energy it contains.

a range, here showing the range of properties covered by different grades of ABS (data for a single grade has much narrower ranges).

But this is not enough. To design with a material, you need to know its real character, its strengths, and its weaknesses. How do you shape it? Join it? Who has used it before and for what? Did it fail? Why? This information exists in handbooks, is documented as design guidelines, and is reported in failure analyses and case studies. It consists largely as text, graphs, and images, and while certain bits of it may be available for one material, for another they may not. For this reason it is not useful for screening, but it is crucial for the later steps in reaching a final selection. We refer to this as *specific supporting information* because it relates to a particular material. An example of supporting information for ABS is shown in the lower half of Table 15.1.

There is more. Material uses are subject to standards and codes. These rarely refer to a single material but to classes or subclasses. For a material to be used in contact with food or drugs, it must carry FDA approval or equivalent. Metals and composites for use in US military aircraft must have Military Specification Approval. To qualify for best-practice design for the environment, material usage must confirm to ISO 1400 guidelines. And so forth. We refer to this as *general supporting information*, because it relates to families or classes, not to specific materials. The ensemble of information about a material, structured and unstructured, constitutes *knowledge*.

There is yet more (Figure 15.2, right-hand side). To succeed in the market place, a product must be economically viable and compete successfully, in terms of performance, consumer appeal and cost, with the competition. All of these depend on material choice and the way the material is processed. Much can be said about this, but not here; for now the focus is on structured data and specific and general supporting information.

Data breadth versus data precision

Data needs evolve as a design develops (Figure 15.1). In the conceptual stage, the designer requires approximate data for the widest possible range of materials. At this stage all options are open: a polymer could be the best choice for one concept, a metal for another, even though the function is the same. Breadth is important; precision is less so. Data for this first-level screening is found in wide-spectrum compilations like the charts of this book, ASM Handbooks (1986, 1990, and 2000), and the Chapman and Hall Materials Selector (1997).[1] More effective is software containing these and other information sources such as the CES 4 (2004) selection system. Their ease of access gives the designer the greatest freedom in considering alternatives.

The calculations involved in deciding on the scale and lay-out of the design (the embodiment stage) require more complete information than before, but for fewer candidates. Information for this second-level screening are found in the specialized compilations that include handbooks and computer data-bases, and the data books published by Associations and Federations of material producers. They list, plot and compare properties of closely-related materials, and provide data at a level of precision not usually available in the broad, level 1, compilations. And, if they are doing their job properly, they provide supporting information about processability and possible manufacturing routes. But because they contain much more detail, their breadth (the range of materials and processes they cover) is restricted, and access is more cumbersome.

The final, detailed design, stage requires data at a still higher level of precision and with as much depth as possible, but for only one or a few materials. They are best found in the data-sheets issued by the producers themselves. A given material (ABS, for instance) has a range of properties that derive from differences in the way different producers make it. At the detailed-design stage, a supplier should be identified, and the properties of his product used in the design calculation. But sometimes even this is not good enough. If the component is a critical one (meaning that its failure could be disastrous) then it is prudent to conduct in-house tests, measuring the critical property on a sample of the material that will be used to make the component itself. Parts of power-generating equipment (the turbine disk for instance), of aircraft (the wing spar, the landing-gear) or of nuclear reactors (the pressure vessel) are like this; for these, every new batch of material is tested, and the batch is accepted or rejected on the basis of the test.

Data types

Properties are not all described in the same way. Some, like the atomic number, are described by a single number ("The atomic number of copper $= 29$"); others, like the modulus or the thermal conductivity are characterized by a range ("Young's modulus for low density polyethylene $= 0.1$–0.25 GPa", for instance). Still others can only be described in a qualitative way, or as images.

---

[1] Details in Appendix D.

Table 15.2   Data types

Data type	Example
Numeric point data	Atomic number of magnesium: $N_a = 12$
Numeric range data	Thermal conductivity of polyethylene: $\lambda = 0.28$ to 0.31 W/m.K
Boolean (yes/no) data	304 stainless steel can be welded: Yes
Ranked data	Corrosion resistance of alumina in tap water (scale A to E): A
Text	Supplier for aluminum alloys: Alcan, Canada . . . (Address, phone, fax, e-mail)
Image	Microstructures, charts, pictures of applications, etc.
Hyperlink	URL for supporting information: www.matdata.net URL for material supplier: www.alcan.com

Corrosion resistance is a property too complicated to characterize by a single number; for screening purposes it is ranked on a simple scale: A (very good) to E (very poor), backed up with supporting information stored as text files or graphs. The forming characteristics, similarly, are attributes best described by a list ("mild steel can be rolled, forged, or machined"; "zirconia can be formed by powder methods") with case studies, guidelines and warnings to illustrate how it should be done. The best way to store information about microstructures, or the applications of a material, or the functioning of a process, may be as an image — another data type. Table 15.2 sets out the data types that are typically required for the selection of materials and processes.

15.3 Screening information: structure and sources

Data structure for screening

To "select" means: "to choose". But from what? Behind the concept of selection lies that of a *kingdom of entities* from which the choice is to be made. The kingdom of materials means: all metals, all polymers, all ceramics and glasses, and all composites, as in Figure 5.2. If it is *materials* we mean to select, then the portfolio from which the selection is to be made contains all of these; leave out part, and the selection is no longer one of materials but of some subset of them. If, from the start, the choice is limited to polymers, then the portfolio becomes a single class of materials, that of polymers. A similar statement holds for processes, based on the kingdom of Figures 7.3 and 7.4.

There is a second implication to the concept of selection; it is that all members of the kingdom must be regarded as candidates — they are, after all, *there* — until (by a series of selection stages) they are shown to be otherwise. From this arises the requirement of a data structure that is *comprehensive* (it includes all members of the kingdom) and the need for characterizing attributes that are *universal* (they apply to all members of the kingdom) and *discriminating* (they have recognizably different values for different members of the kingdom). Similar considerations apply to any selection exercise.

In the kingdom of materials, many attributes are universal and discriminating: density, bulk modulus and thermal conductivity are examples. Universal attributes can be used for *screening and ranking*, the initial stage of any selection exercise (Figure 15.3, upper half). But if the values of one or more screening attributes are grossly inaccurate or missing, that material is eliminated by default. It is important, therefore, that the database be *complete* and be of high *quality*, meaning that the data in it can be trusted. This in turn creates the need for data checking and estimation, tackled by methods described in Section 15.5.

The attribute-limits and index methods introduced in Chapters 5 and 7 are examples of the use of attributes to screen, based on design requirements. They provide an efficient way of reducing the vast number of materials in the materials kingdom to a small manageable subset for which supporting information can be sought.

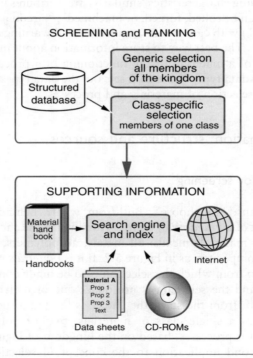

Figure 15.3   Information sources for the screening and ranking step (above), and for the supporting information step (below). The two steps serve different functions, and require different information structures.

Data sources for screening (see also Appendix D)

The traditional source of materials data is the handbook. The more courageous of them span all material classes, providing raw data for generic screening. More specialized handbooks and trade-association publications contain data suitable for second-level screening as well as text and figures that help with supporting information. They are the primary sources, but they are clumsy to use because their data structure is not well-suited to screening. Comparison of materials of different classes is possible but difficult because data are seldom reported in comparable formats; there is too much unstructured information, requiring the user to filter out what he needs; and the data tables are almost always full of holes. Electronic sources for generic screening overcome these problems. If properly structured, they allow direct comparison across classes and selection by multiple criteria, and it is possible (using the methods of Section 15.5) to arrange that they have no holes.

Screening and ranking, as we have seen, identifies a set of viable candidates. We now need their family history. That is the purpose of the "supporting information" step.

15.4 Supporting information: structure and sources

Data structure for supporting information

The data requirements in the supporting information step differ greatly from those for screening (Figure 15.3, lower half). Here we seek additional details about the few candidates that have already been identified by the screening and ranking step. Typically, this is information about availability and pricing; exact values for key properties of the particular version of the material made by one manufacturer; case studies and examples of uses with cautions about unexpected difficulties (e.g. "liable to pitting corrosion in dilute acetic acid" or "material Y is preferred to material X for operation in industrial environments"). It is on this basis that the initial short-list of candidates is narrowed down to one or a few prime choices.

Sources of supporting information (see also Appendix D)

Sources of supporting information typically contain specialist information about a relatively narrow range of materials or processes. The information may be in the form of text, tables, graphs, photographs, computer programs, even video clips. The data can be large in quantity, detailed, and precise in nature, but there is no requirement that it be comprehensive or that the attributes it contains be universal. The most common media are handbooks, trade association publications and manufacturers' leaflets and catalogues. Increasingly such information is becoming available in electronic form on CD-ROMs and on the Internet. The larger materials suppliers (Dow Chemical, GE Plastics, Inco, Corning Glass, etc.) publish design guides and compilations of case studies, and all suppliers have data-sheets describing their products. Because the data is in "free" format, the search strategies differ from the numerical optimization procedures used for the screening step. The simplest approach is to use an index (as in a printed book), or a keyword list, or a computerized full-text search, as implemented in many hyper-media systems.

Data sources on the Internet. The Internet contains a rapidly expanding spectrum of information sources. Some, particularly those on the *world-wide web*, contain information for materials, placed there by standards organizations, trade associations, material suppliers, learned societies, universities, and individuals — some rational, some eccentric — who have something to say. There is no control over the contents of Web pages, so the nature of the information ranges from useful to baffling, and the quality from good to appalling. The Appendix includes WWW sites that offer materials information, but the rate-of-change here is so rapid that it cannot be seen as comprehensive.

The *Material Data Network* (www.matdat.net) is an example of a web-based supporting information source. It enables full-text searching of a number high-quality resources. These include *ASM Handbooks Online* (a 21-volume set of encyclopaedic scope); *ASM Alloy Center Online* (ASM's property data, performance charts, and processing guidelines for specific metals and alloys); *NPL 'MIDAS'* (material property measurement technology and standards from the National Physical Laboratory in the UK); *UKSteel 'SteelSpec'* (contains summaries of British, European and some International steel standards, together with selected proprietary steel grades); *TWI 'Joinit'* (welding and joining processes from TWI, UK); *MIL-HDBK-5* (Granta Design's MIL-Handbook-5 database incorporating data from MIL-HDBK-5H CN1); *IDES ResinSource* (data sheets for more than 40,000 plastics from 390 global suppliers); and *Matweb* (a database of manufacturer's data sheets for more than 31,000 metals, plastics, ceramics, and composites). Matdata.net is evidence of the beginnings of a useful 'supporting information' tool.

Expert systems. The main drawback of the simple, common-or-garden, data-base is the lack of qualification. Some data are valid under all conditions, others are properly used only under certain circumstances. The qualification can be as important as the data itself. Sometimes the question asked of the database is imprecise. The question: "What is the strength of Grade 1020 steel?" could be asking for yield strength or tensile strength or fatigue strength, and in any one of a number of possible states of work-hardening and heat-treatment. If the question were put to a materials expert as part of a larger consultation, he would know from the context what was wanted or could ask if he did not, and would have a shrewd idea of the precision of the information and its limitations. An ordinary database can do none of this.

Expert systems can. They have the potential to solve problems that require reasoning, provided it is based on rules that can be clearly defined: using information about environmental conditions to choose the most corrosion resistant alloy, for instance, or information about material and use to choose an adhesive. It might be argued that a simple check-list or a table in a supplier's data sheet could do most of these things, but the expert system combines qualitative and quantitative information using its rules (the "expertise"), in a way that only someone with experience can. It does more than merely look up data; it qualifies it as well, allowing context-dependent selection of material or process. In the ponderous words of the British Computer Society: "Expert systems offer intelligent advice or take intelligent decisions by embodying in a computer the knowledge-based component of an expert's skill. They must, on demand, justify their line of reasoning in a manner intelligible to the user."

This context-dependent scheme for retrieving data sounds just what we want, but things are not so simple. An expert system is much more complex than a simple data-base: it is a major task to elicit the "knowledge" from the expert; it can require massive programming effort and computer power; and it is difficult to maintain. A full expert system for materials selection is decades away. Success has been achieved in specialized, highly focused applications: guidance in selecting adhesives from a limited set, in choosing a welding technique for a particular steel, or in designing against certain sorts of corrosion. It is only a question of time before more fully developed systems become available. They are something about which to keep informed.

15.5 Ways of checking and estimating data

The value of a database of material properties depends on its precision and its completeness — in short, on its quality. One way of maintaining or enhancing its quality is to subject its contents to validating procedures. The property ranges and dimensionless correlations, described below, provide powerful tools for doing this. The same procedures fill a second function: that of providing estimates for missing data, essential when no direct measurements are available.

Property ranges. Each property of a given class of materials has a characteristic *range*. A convenient way of presenting the information is as a table in which a low (L) and a high (H) value are stored, identified by the material family and class. An example listing Young's modulus, E, is shown in Table 15.3, in which E_L is the lower limit and E_H the upper one.

All properties have characteristic ranges like these. The range becomes narrower if the classes are made more restrictive. For purposes of checking

Table 15.3   Ranges of Young's modulus *E* for broad material classes

Material class	E_L (GPa)	E_H (GPa)
All Solids	0.00001	1000
Classes of Solid		
Metals		
Ferrous	70	220
Non-ferrous	4.6	570
Technical ceramics*	91	1000
Glasses	47	83
Polymers		
Thermoplastic	0.1	4.1
Thermosets	2.5	10
Elastomers	0.0005	0.1
Polymeric foams	0.00001	2
Composites		
Metal-matrix	81	180
Polymer-matrix	2.5	240
Woods		
Parallel to grain	1.8	34
Perpendicular to grain	0.1	18

* Technical ceramics are dense, monolithic ceramics such as SiC, Al_2O_3, ZrO_2, etc.

and estimation, described in a moment, it is helpful to break down the family of *metals* into classes of cast irons, steels, aluminum alloys, magnesium alloys, titanium alloys, copper alloys and so on. Similar subdivisions for polymers (thermoplastics, thermosets, elastomers) and for ceramics and glasses (engineering ceramics, whiteware, silicate glasses, minerals) increases resolution here also.

Correlations between material properties. Materials that are stiff have high melting points. Solids with low densities have high specific heats. Metals with high thermal conductivities have high electrical conductivities. These rules-of-thumb describe correlations between two or more material properties that can be expressed more quantitatively as limits for the values of *dimensionless property groups*. They take the form

$$C_L < P_1 P_2{}^n < C_H \tag{15.1}$$

or

$$C_L < P_1 P_2{}^n P_3{}^m < C_H \tag{15.2}$$

(or larger groupings) where P_1, P_2, P_3 are material properties, n and m are powers (usually -1, $-1/2$, $+1/2$ or $+1$), and C_L and C_H are dimensionless constants—the lower and upper limits between which the values of the property-group lies. The correlations exert tight constraints on the data, giving the "patterns" of property envelopes that appear on the material selection charts. An example is the relationship between expansion coefficient, α (units: K^{-1}), and the melting point, T_m (units: K) or, for amorphous materials, the glass temperature, T_g:

$$C_L \leq \alpha T_m \leq C_H \tag{15.3a}$$
$$C_L \leq \alpha T_g \leq C_H \tag{15.3b}$$

a correlation with the form of equation (15.1). Values for the dimensionless limits C_L and C_H for this group are listed in Table 15.4 for a number of material classes. The values span a factor to 2 to 10 rather than the factor 10 to 100 of the property ranges. There are many such correlations. They form the basis of a hierarchical data-checking and estimating scheme (one used in preparing the charts in this book), described next.

Data checking

The method is shown in Figure 15.4. It proceeds in three steps. Each datum is first associated with a material class, or, at a higher level of checking, with a sub-class. This identifies the values of the property range and correlation

Table 15.4 Limits for the group αT_m and αT_g for broad material classes*

Correlation* $C_L < \alpha T_m < C_H$	C_L ($\times 10^{-3}$)	C_H ($\times 10^{-3}$)
All Solids	0.1	56
Classes of Solid		
Metals		
Ferrous	13	27
Non-ferrous	2	21
Fine Ceramics*	6	24
Glasses	0.3	3
Polymers		
Thermoplastic	18	35
Thermosets	11	41
Elastomers	35	56
Polymeric foams	16	37
Composites		
Metal-matrix	10	20
Polymer-matrix	0.1	10
Woods		
Parallel to grain	2	4
Perpendicular to grain	6	17

*For amorphous solids the melting point T_m is replaced by the glass temperature T_g.

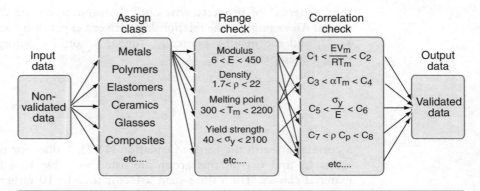

Figure 15.4   A scheme for checking material properties. Class-specific range-checks catch gross errors in all properties. Correlation checks catch subtler errors in certain properties (the Cs are dimensionless constants). The estimation procedure uses the same steps in reverse.

limits against which it will be checked. The datum is then compared with the range-limits L and H for that class and property. If it lies within the range-limits, it is accepted; if it does not, it is flagged for checking.

Why bother with such low-level stuff? Because it is a sanity-check. The commonest error in handbooks and other compilations of material or process properties is that of a value that is expressed in the wrong units, or is, for less obvious reasons, in error by one or more orders of magnitude (slipped decimal point, for instance). Range checks catch errors of this sort. If a demonstration of this is needed, it can be found by applying them to the contents of almost any standard reference data-books; none among those we have tried has passed without errors.

In the third step, each of the dimensionless groups of properties like that of Table 15.4 is formed in turn, and compared with the range bracketed by the limits C_L and C_H. If the value lies within its correlation limits, it is accepted; if not, it is checked. Correlation checks are more discerning than range checks and catch subtler errors, allowing the quality of data to be enhanced further.

Data estimation

The relationships have another, equally useful, function. There remain gaps in our knowledge of material properties. The fracture toughness of many materials has not yet been measured, nor has the electric breakdown potential; even moduli are not always known. The absence of a datum for a material would falsely eliminate it from a screening exercise that used that property, even though the material might be a viable candidate. This difficulty is avoided by using the correlation and range limits to estimate a value for the missing datum, adding a flag to alert the user that it is an estimate.

In estimating property values, the procedure used for checking is reversed: the dimensionless groups are used first because they are the more accurate. They can be surprisingly good. As an example, consider estimating the expansion coefficient, α, of polycarbonate from its glass temperature T_g. Inverting equation (15.3) gives the estimation rule:

$$\frac{C_L}{T_g} \le \alpha \le \frac{C_H}{T_g} \tag{15.4}$$

Inserting values of C_L and C_H from Table 15.3, and the value $T_g = 420\,\text{K}$ for a particular sample of polycarbonate gives the mean estimate

$$\bar{\alpha} = 63 \times 10^{-3}\,\text{K}^{-1} \tag{15.5}$$

The reported value for polycarbonate is

$$\alpha = 54 - 62 \times 10^{-3}\,\text{K}^{-1}$$

The estimate is within 9 percent of the mean of the measured values, perfectly adequate for screening purposes. That it is an estimate must not be forgotten, however: if thermal expansion is crucial to the design, better data or direct measurements are essential.

Only when the potential of the correlations is exhausted are the property ranges invoked. They provide a crude first-estimate of the value of the missing property, far less accurate than that of the correlations, but still useful in providing guide-values for screening.

15.6 Summary and conclusions

Information, by which we mean data with known provenance and pedigree, is central to the selection of materials and processes. It takes several forms. Some is comprehensive and universal. Typically it is numeric; it can be structured into tables using the same format for all members of the population from which a selection is to be made, and it can be managed by standard methods. Structured information is essential for the first stages of selection — those of screening and ranking to give a ranked short-list of candidates.

Much more information is available only in a mix of formats, and is neither comprehensive nor universal. It is often the unique or problematic attributes of a material or process that are documented in this way; information that does not bear on its uses remains undocumented. It takes the form of text, graphs and images detailing design guidelines, case studies, failure analyzes, economics of manufacture and more; it provides a sort of character-sketch. This unstructured information is of no help in screening and ranking, but is vital in making a final choice from the short-list.

Structured and unstructured information must be stored and processed in ways that differ. The first lends itself to the systematic methods that are developed in the earlier chapters of this book. The second, of no less importance, is best accessed by search methods of the sort exemplified by internet search engines: text or association searches based on text strings.

Little can be done to check and validate unstructured data; here the provenance is the only guarantee of quality. For structured data the position is different: there are ways of detecting when data are complete nonsense (range checks) or are dubious (correlation checks). Applying these allows data to be "cleaned", assuring that a minimum level of quality is reached. And the methods allow something else: the ability to make approximate, but still useful, estimates of attributes for which no data are at present available. Why do this? Because the screening stage, designed to reject unsuitable candidates and retain those with promise, should not reject or wrongly retain candidates because one of their attributes is unknown.

Do not leave this chapter without at least glancing at the compilation of data sources in Appendix D. It is probably the most useful bit.

15.7 Further reading

Ashby M.F. (1998) Checks and estimates for material properties, *Proc Roy Soc A* **454**, 1301–1321. (*This and the next reference detail the data-checking and estimation methods, listing the correlations.*)

Bassetti, D., Brechet, Y. and Ashby, M.F. (1998) Estimates for material properties: the method of multiple correlations, *Proc Roy Soc A* **454**, 1323–1336. (*This and the previous reference detail the data-checking and estimation methods, with examples.*)

Cebon D. and Ashby M.F. (1996) Electronic material information systems, I. Mech. E. Conference on Electronic Delivery of Design Information, Oct 1996, London, UK. (*A discussion of data-base design for materials selection.*)

Demerc, M.Y. (1990) *Expert system Applications in Materials Processing and Manufacture*, TMS Publications, Warrendale, USA.

Sargent, P.M. (1991) *Materials Information for CAD/CAM*, Butterworth-Heinemann, Oxford, UK. (*A survey of the way in which materials databases work. No data.*)

Waterman, N.A., Waterman, M. and Poole, M.E. (1992) Computer based materials selection systems. *Metals Materials* **8**, 19–24.

Westbrook, J.H. (1997) Sources of materials property data and information, in *ASM Handbook, Volume 20*, ASM International, Metals Park, OH, USA, pp. 491–506. ISBN 0-87170-386-6. (*A useful compilation of sources for a wide range of materials.*)

Westbrook, J.H. (2003) Materials data on the internet, *Data Sci. J.* **2**, 198–212. (*A review of resources for materials science and engineering available on the internet in 2003, with recommendations for future development.*)

Chapter 16

Materials and the environment

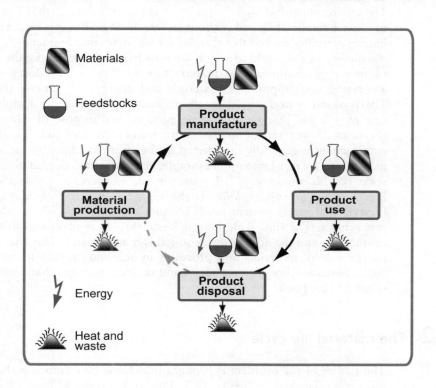

Materials

Feedstocks

Product
manufacture

Material
production

Product
use

Product
disposal

Energy

Heat and
waste

16.1 Introduction and synopsis

All human activity has some impact on the environment in which we live. The environment has some capacity to cope with this, so that a certain level of impact can be absorbed without lasting damage. But it is clear that current human activities exceed this threshold with increasing frequency, diminishing the quality of the world in which we now live and threatening the well-being of future generations. The manufacture and use of products, with their associated consumption of materials and energy, are among the culprits. The position is dramatized by the following statement: at a global growth rate of 3% per year we will mine, process, and dispose of more "stuff" in the next 25 years than in the entire history of mankind. *Design for the environment* is generally interpreted as the effort to adjust our present design methods to correct known, measurable, environmental degradation; the time-scale of this thinking is 10 years or so, an average product's expected life. *Design for sustainability* is the longer view: that of adaptation to a lifestyle that meets present needs without compromising the needs of future generations. The time-scale here is less clear — it is measured in decades or centuries — and the adaptation required is much greater. This chapter focuses on the role of materials and processes in achieving design for the environment. Sustainability requires social and political changes that are beyond the scope of this book.

16.2 The material life cycle

The nature of the problem is brought into focus by examining the materials life cycle, sketched in Figure 16.1. Ore and feedstock, most of them non-renewable, are processed to give materials; these are manufactured into products that are used, and, at the end of their lives, disposed, a fraction perhaps entering a recycling loop, the rest committed to incineration or land-fill. Energy and materials are consumed at each point in this cycle (we shall call them "phases"), with an associated penalty of CO_2 and other emissions — heat, and gaseous, liquid, and solid waste.

The problem, crudely put, is that the sum of these unwanted by-products now often exceeds the capacity of the environment to absorb them. Some of the injury is local, and its origins can be traced and remedial action taken. Some is national, some global, and here remedial action has wider social and organizational prerequisites. Much present environmental legislation aims at modest reductions in the damaging activity; a regulation requiring 20% reduction in — say — the average gasoline consumption of passenger cars is seen by car-makers as a major challenge.

Sustainability requires solutions of a completely different kind. Even conservative estimates of the adjustment needed to restore long-term equilibrium

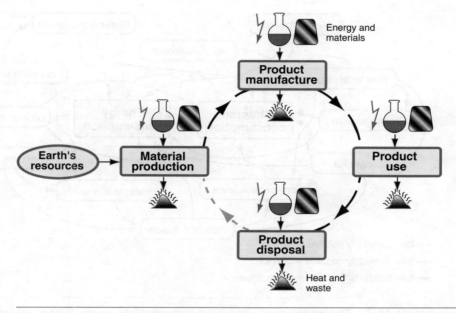

Figure 16.1   The material life cycle. Ore and feedstock are mined and processed to yield a material. These are manufactured into a product that is used and at the end of its life, discarded or recycled. Energy and materials are consumed in each phase, generating waste heat and solid, liquid and gaseous emissions.

with the environment envisage a reduction in the flows of Figure 16.1 by a factor of four or more. Population growth and the growth of the expectations of this population more than cancel any modest savings that the developed nations may achieve. It is here that the challenge is greatest, requiring difficult adaptation, and one for which no generally-agreed solutions yet exist. But it remains the long-term driver of eco-design, to be retained as background to any creative thinking.

16.3 Material and energy-consuming systems

It would seem that the obvious ways to conserve materials is to make products smaller, make them last longer, and recycle them when they finally reach the end of their lives. But the seemingly obvious can sometimes be deceptive. Materials and energy form part of a complex and highly interactive system, of which Figure 16.2 is a cartoon. Here primary catalysts of consumption such as *new technology, planned obsolescence, increasing wealth and education,* and *population growth* influence aspects of product use and through these, the consumption of materials and energy and the by-products that these produce.

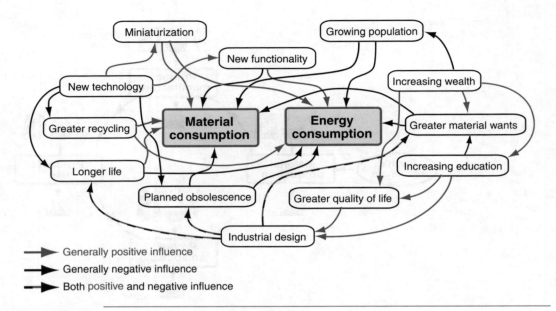

Figure 16.2   The influences on consumption of materials and energy. It is essential to see eco-design as a systems problem, not solved by simple choosing "good" and avoiding "bad" materials, but rather that of matching the material to the system requirements.

The connecting lines indicate influences; a red line suggests positive, broadly desirable influence; black line suggests negative, undesirable influence and red-black suggests that the driver has the capacity for both positive and negative influence.

The diagram brings out the complexity. Follow, for instance the lines of influence of *new technology* and its consequences. It offers more material and energy-efficient products, but by also offering new functionality it creates obsolescence and the desire to replace a product that has useful life left in it. Electronic products are prime examples of this: 80% are discarded while still functional. And observe, even at this simple level, the consequences of *longer life* — a seemingly obvious measure. It can certainly help to conserve materials (a positive influence) but, in an era in which new technology delivers more energy-efficient products (particularly true of cars, electronics and household appliances today), extending the life of old products can have a negative influence on energy consumption.

As a final example, consider the bi-valent influence of industrial design — the subject of Chapter 17. The lasting designs of the past are evidence of its ability to create products that are treasured and conserved. But today it is frequently used as a potent tool to stimulate consumption by deliberate obsolescence, creating the perception that "new" is desirable and that even slightly "old" is unappealing.

The use-patterns of products

Table 16.1 suggests a matrix of product use-patterns. Those in the first row require energy to perform their primary function. Those in the second could function without energy but, for reasons of comfort, convenience or safety, consume energy to provide a secondary function. Those in the last row provide their primary function without any need for energy other than human effort. The load factor across the top is an approximate indicator of the intensity of use — one that will, of course, vary widely.

The choice of materials and processes influences all the phases of Figure 16.1: *production*, through the drainage of resources and the undesired by-products of refinement; *manufacture*, through the level of efficiency and cleanness of the shaping, joining and finishing processes; *use*, through the ability to conserve energy through light-weight design, higher thermal efficiency and lower energy consumption; and (finally) *disposal* through a greater ability to allow reuse, disassembly and recycling.

It is generally true that one of the four phases of Figure 16.1 dominates the picture. Simplifying for a moment, let us take *energy consumption* as a measure of both the inputs and undesired by-products of each phase and use it for a character-appraisal of use-sectors. Figure 16.3 presents the evidence, using this measure. It has two significant features, with important implications. First, one phase almost always dominates, accounting for 80% or more of the energy consumed during life — often much more. If large changes are to be achieved, it is this phase that must be the target; a reduction by a factor of 2, even of 10, in any other makes little significant difference to the total. The second: when differences are as great as those of Figure 16.3, precision is not the issue — an error of a factor of 2 changes very little. It is the nature of people who *measure* things to wish to do so with precision, and precise data must be the ultimate goal. But it is possible to move forward without it: precise judgments can be drawn from imprecise data. This is an important consideration: much information about eco-attributes is imprecise.

Table 16.1 Use matrix of product classes

	High load factor	Modest load factor	Low load factor	
Primary power -consuming	Family car Train set Aircraft	Television Freezer	Coffee maker Vacuum cleaner Washing machine	Energy Intensive
Secondary power -consuming	Housing (heat, light)	Car-park (light)	Household dishes Clothing (washing)	↑
Non power -consuming	Bridges Roads	Furniture Bicycle	Canoe Tent	↓ Material intensive
	High impact ◄———► Low impact			

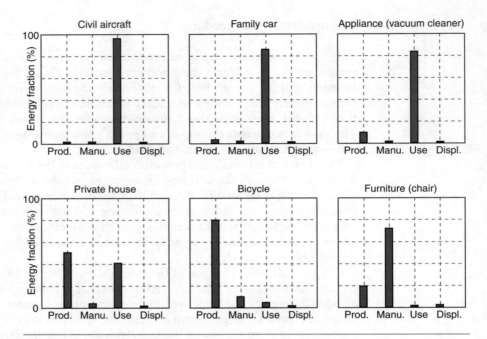

Figure 16.3 Approximate values for the energy consumed at each phase of Figure 16.1 for a range of products.

16.4 The eco-attributes of materials

Material production: energy and emissions

Most of the energy consumed in the four phases of Figure 16.1 is derived from fossil fuels. Some is consumed in that state — as gas, oil, coal, or coke. Much is first converted to electricity at a European average conversion efficiency of about 30%. Not all electricity is generated from fossil fuels — there are contributions from hydroelectric, nuclear and wind/wave generation. But with the exception of Norway (70% hydro) and France (80% nuclear) the predominant energy sources are fossil fuels; and since the national grids of European countries are linked, with power flowing from one to another as needed, it is a reasonable approximation to speak of a European average fossil-fuel energy per kilowatt of delivered electrical power.

The fossil-fuel energy consumed in making one kilogram of material is called its *production energy*. Some of the energy is stored in the created material and can be reused, in one sense or another, at the end of life. Polymers made from oil (as most are) contain energy in another sense — that of the oil that enters the production as a primary feedstock. Natural materials such as wood, similarly, contain "intrinsic" or "contained" energy, this time derived from solar radiation absorbed during growth. Views differ on whether the intrinsic energy should be included in the production energy or not. There is a sense in which not only polymers and woods, but also metals, carry intrinsic energy that could — by chemical reaction or by burning the metal in the form of finely-divided powder — be recovered, so omitting it when reporting production energy for polymers but including it for metals seems inconsistent. For this reason we will include intrinsic energy from non-renewable resources in reporting production energies, which generally lie in the range 25–250 MJ/kg, though a few are much higher. The existence of intrinsic energy has another consequence: that the energy to recycle a material is sometimes much less than that required for its first production, because the intrinsic energy is retained. Typical values lie in the range 10–100 MJ/kg.

The production of 1 kg of material is associated with *undesired gas emissions*, among which CO_2, NO_x, SO_x and CH_4 cause general concern (global warning, acidification, ozone-layer depletion). The quantities can be large — each kilogram of aluminum produced by using energy from fossil fuels creates some 12 kg of CO_2, 40 g of NO_x and 90 g of SO_x. Production is generally associated with other undesirable outputs, particularly toxic wastes and particulates, but these can, in principle, be dealt with at source.

Wood, bamboo, and other plant-based materials, too, contain intrinsic energy, but unlike man-made materials it derives from sunlight, not from non-renewable resources. The production energy data for these materials does not include this intrinsic energy, and the emissions take account of the CO_2 absorbed during their growth. Because of this, woods have a near-neutral energy balance, and a negative value for CO_2 emissions.

Material processing energies at 30% efficiency

Many processes depend on casting, evaporation or deformation. It is helpful to have a feel for the approximate magnitudes of energies required by these.

Melting. To melt a material, it must first be raised to its melting point, requiring a minimum input of the heat $C_p(T_m - T_0)$, and then caused to melt, requiring the latent heat of melting, L_m

$$H_{min} = C_p(T_m - T_0) + L_m \tag{16.1}$$

where H_{min} is the minimum energy per kilogram for melting, C_p is the specific heat, T_m is the melting point, T_0 is the ambient temperature. A close correlation exists between L_m and $C_p T_m$

$$L_m \approx 0.4 C_p T_m \tag{16.2}$$

and for metals and alloys $T_m \gg T_0$ giving

$$H_{min} \approx 1.4 C_p T_m \tag{16.3}$$

Assuming efficiency of 30%, the estimated energy to melt 1 kg, H_m^*, is

$$H_m^* \approx 4.2 C_p T_m \tag{16.4}$$

the asterisk recalling that it is an estimate. For metals and alloys, the quantity H_m^*, lies in the range 0.4–4 MJ/kg.

Vaporization. As a rule of thumb the latent heat of vaporization, L_v, is larger than that for melting, L_m, by a factor of 24 ± 5, and the boiling point T_b is larger than the melting point, T_m, by a factor 2.1 ± 0.5. Using the same assumptions as before, we find an estimate for the energy to evaporate 1 kg of material (as in PVD processing) to be

$$H_v^* \approx 38 C_p T_m \tag{16.5}$$

again assuming an efficiency of 30%. For metals and alloys, the quantity H_v^* lies in the range 3–30 MJ/kg.

Deformation. Deformation processes like rolling or forging generally involve large strains. Assuming an average flow-strength of $(\sigma_y + \sigma_{uts})/2$, a strain of $\varepsilon = 90\%$ and an efficiency factor of 30% we find, for the work of deformation per kilogram to be

$$W_D^* \approx 1.5(\sigma_y + \sigma_{uts})\varepsilon = 1.35(\sigma_y + \sigma_{uts}) \tag{16.6}$$

where σ_y is the yield strength and σ_{uts} is the tensile strength. For metals and alloys, the quantity W_D^* lies in the range 0.01–1 MJ/kg.

We conclude that casting or deformation require processing energies that are small compared to the production energy of the material being processed, but the larger energies required for vapor-phase processing may become comparable with those for material production.

End of life

Quantification of material *recycling* is difficult. Recycling costs energy, and this energy carries its burden of gases. But the *recycle energy* is generally small compared to the initial production energy, making recycling — when it is possible at all — an energy-efficient propositions. It may not, however, be one that is cost efficient; that depends on the degree to which the material has become dispersed. In-house scrap, generated at the point of production or manufacture, is localized and is already recycled efficiently (near 100% recovery). Widely distributed "scrap" — material contained in discarded products — is a much more expensive proposition to collect, separate and clean. Many materials cannot recycled, although they may still be reused in a lower-grade activity; continuous-fiber composites, for instance, cannot be re-separated economically into fiber and polymer in order to recycle them, though they can be chopped and used as fillers. Most other materials require an input of virgin material to avoid build-up of uncontrollable impurities. Thus the fraction of a material production that can ultimately re-enter the cycle of Figure 16.1 depends both on the material itself and on the product into which it has been incorporated. Despite this complexity, some data for the fraction re-entering the cycle of Figure 16.1 are available. More usually, the position is characterized by indicating simply whether the material can or cannot be *recycled, down-cycled, biodegraded, incinerated* or committed to *landfill*.

Bio-data. Some materials are toxic, creating potential problems during production, during use, and particularly in their disposal. Sources of information about this are listed under "Hazardous materials" in Appendix D, Section D.6.

Aggregated measures: eco-indicators

Table 16.2 shows an example of the eco-attributes of a material, here a 1000 series aluminum alloy. A designer, seeking to cope with many interdependent decisions that any design involves, finds it hard to know how best to use these data. How are production energies or CO_2 and SO_x burdens to be balanced against resource depletion, toxicity or ease of recycling? This perception has

Table 16.2 The eco-attributes of one grade of aluminum alloy

WROUGHT 1200 GRADE ALUMINUM			
Material production: energy and emissions			
Production energy		1.9e2–2.1e2	MJ/kg
Carbon dioxide	*	12–13	kg/kg
Nitrogen oxides	*	72–79	g/kg
Sulfur oxides	*	1.2e2–1.4e2	g/kg
Indicators for principal component			
Eco indicator		7.4e2–8.2e2	millipoints/kg
Material processing energy at 30% efficiency			
Min. energy to melt		3.5–3.8	MJ/kg
Min. energy to vaporization		29–32	MJ/kg
Min. energy to 90% deform.		0.04–0.044	MJ/kg
End of life			
Recycle	Yes		
Downcycle	Yes		
Biodegrade	No		
Incinerate	No		
Landfill	Acceptable		
Recycling energy	*	23–26	MJ/kg
Recycle fraction		34–38	%
Bio-data			
Toxicity rating	Non-toxic		

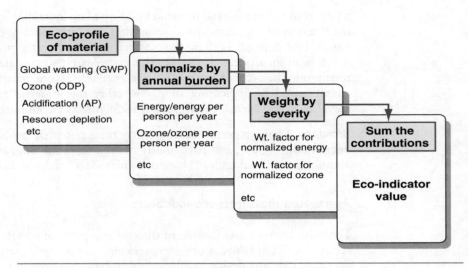

Figure 16.4   The steps in calculating an eco-indicator. Difficulty arises in step 3: there is no agreement on how to choose the weight factors.

lead to efforts to condense the eco-attributes of a material into a single measure or *indicator*, giving the designer a simple, numeric ranking.

To do this, four steps are necessary. They are shown in Figure 16.4. The first is that of *classification* of the data listed in Table 16.2 according to the impact each causes (global warming, ozone depletion, acidification, etc.), giving an eco-profile of the material. The second step is that of *normalization* to remove the units (of which there are several in Table 16.2) and reduce them to a common scale (0–100, for instance). The third step is that of *weighting* to reflect the perceived seriousness of each impact, based on the classification of Step 1: thus global warming might be seen as more serious than resource depletion, giving it a larger weight. In the final step, the weighted, normalized measures are *summed* to give the indicator.

The use of a single-valued indicator is criticized by some. The grounds for criticism are that there is no agreement on normalization or weighting factors, that the method is opaque since the indicator value has no simple physical significance, and that defending design decisions based on a measurable quantity like energy consumption or CO_2 generation carries more conviction than doing so with an indicator. At the time of writing there is no general agreement on how best to use eco-data in design. But on one point there is international agreement (the Kyoto Protocol of 1997): that the developed nations should progressively reduce CO_2 emissions — a considerable challenge in times of industrial growth, increasing affluence and growing population. Thus there is a certain logic in using CO_2 emission as the "indicator", though it is more usual at present to use energy.

Approximate data for production energy H_p and CO_2 burden for materials are listed in Appendix C, Section C.11.

16.5 Eco-selection

To select materials to minimize the impact on the environment we must first ask—as we did in Section 16.2—which phase of the life cycle of the product under consideration makes the largest contribution? The answer guides the choice of strategy to ameliorate it (Figure 16.5). The strategies are described below.

The material production phase

If material production is the dominant phase of life it is this that we must target. Drink containers (Figure 16.6) provide an example: they consume materials and energy during material extraction and production, but (apart from transport, which is minor) not thereafter. We use the energy consumed in extracting and refining the material (the production energy of Table 16.2) as the measure; CO_2, NO_x, and SO_x emissions are related to it, although not in a simple way. The energy associated with the production of one kilogram of a material is H_p, that per unit volume is $H_p\rho$ where ρ is the density of the material. The bar-charts of Figure 16.7(a) and (b) show these two quantities for ceramics, metals, polymers, and composites. On a "per kg" basis (upper chart) glass, the material of the first

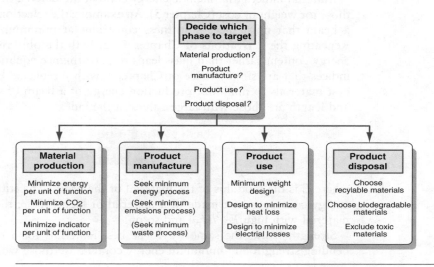

Figure 16.5 Rational design for the environment starts with an analysis of the phase of life to be targeted. This decision then guides the method of selection to minimize the impact of the phase on the environment.

Figure 16.6   Liquid containers: glass, polyethylene, PET, and aluminum. All can be recycled. Which carries the lowest production energy penalty?

container, carries the lowest penalty. Steel is higher. Polymer production carries a much higher burden than does steel. Auminum and the other light alloys carry the highest penalty of all. But if these same materials are compared on a "per m^3" basis (lower chart) the conclusions change: glass is still the lowest, but now commodity polymers such as PE and PP carry a *lower* burden than steel; the composite GFRP is only a little higher. But is comparison "per kg" or "per m^3" the right way to do it? Rarely. To deal with environmental impact at the production phase properly we must seek to minimize the energy, the CO_2 burden or the eco-indicator value *per unit of function*.

Material indices that include energy content are derived in the same way as those for weight or cost (Chapter 5). An example: the selection of a material for a beam that must meet a stiffness constraint at minimum energy content. Repeating the derivations of Chapter 5 but with the objective of minimizing energy content rather than mass leads to performance equations and material indices that are simply those of Chapter 5 with ρ replaced by $H_p\rho$. Thus the best materials to minimize production energy of a beam of specified stiffness and length are those with large values of the index

$$M = \frac{E^{1/2}}{H_p\rho} \tag{16.7}$$

where E is the modulus of the material of the beam. The stiff tie of minimum energy content is best made of a material of high $E/H_p\rho$; the stiff plate, of a material with high $E^{1/3}/H_p\rho$ and so on.

Strength works the same way. The best materials for a beam of specified bending strength and minimum energy content are those with large values of

$$M = \frac{\sigma_f^{2/3}}{H_p\rho} \tag{16.8}$$

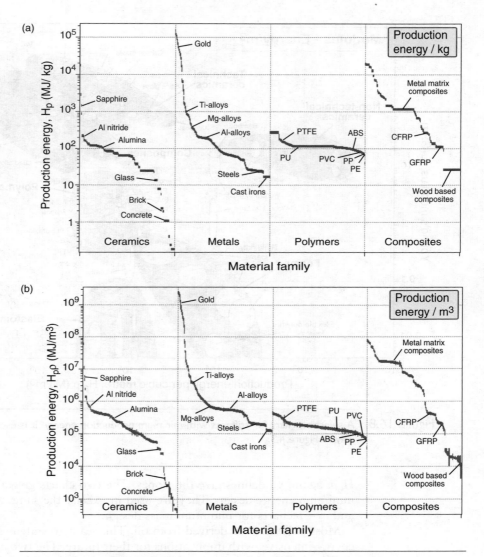

Figure 16.7   (a) and (b). The energy per unit mass and per unit volume associated with material production. Data can be found in Appendix C.

where σ_f is the failure strength of the beam-material. Other indices follow in a similar way.

Figures 16.8 and 16.9 are a pair of materials selection charts for minimizing production energy H_p per unit of function (similar charts for CO_2 burden can be made using the CES 4 software). The first show modulus E plotted against $H_p\rho$; the guidelines give the slopes for three of the commonest performance indices. The second shows strength σ_f (defined as in Chapter 4) plotted against

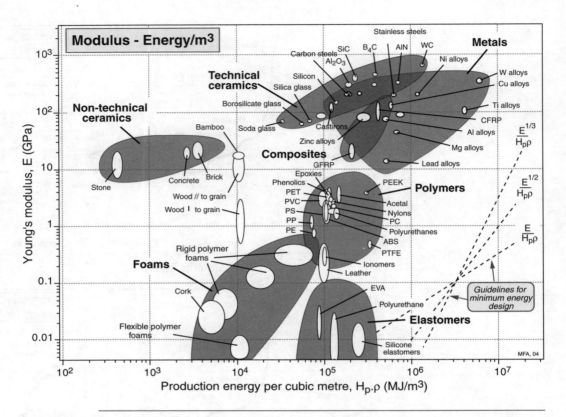

Figure 16.8 A selection chart for stiffness with minimum production energy. It is used in the same way as Figure 4.3.

$H_p\rho$; again, guidelines give the slopes. The two charts give a survey data for minimum energy design. They are used in exactly the same way as the $E - \rho$ and $\sigma_f - \rho$ charts for minimum mass design.

Most polymers are derived from oil. This leads to statements that they are energy-intensive, with implications for their future. The two charts show that, per unit of function in bending (the commonest mode of loading), most polymers carry a lower energy penalty than primary aluminum, magnesium, or titanium, and that several are competitive with steel. Most of the energy consumed in the production of metals such as steel, aluminum or magnesium is used to reduce the ore to the elemental metal, so that these materials, when recycled, require much less energy. Efficient collection and recycling makes important contributions to energy saving.

Table 16.3 contains examples of other indices for eco-selection to minimize impact during the production phase of life. All are derived by adapting the methods of Chapter 5.

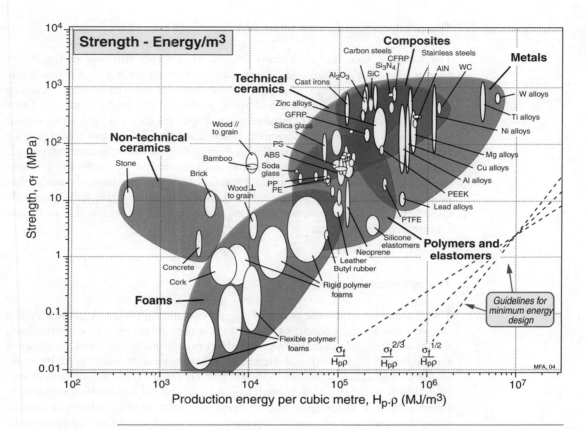

Figure 16.9   A selection chart for strength with minimum production energy. It is used in the same way as Figure 4.4.

Table 16.3   Examples of indices to minimize impact in the production phase

Function	Maximize*
Minimum energy content for given tensile stiffness	$E/H_p\rho$
Minimum CO_2 emissions for given tensile strength	$\sigma_y/[CO_2]\rho$
Minimum energy for given bending stiffness (beam)	$E^{1/2}/H_p\rho$
Minimum energy for given bending stiffness (panel)	$E^{1/3}/H_p\rho$
Minimum CO_2 for given bending strength (beam)	$\sigma_f^{2/3}/[CO_2]\rho$
Minimum CO_2 for given bending strength (panel)	$\sigma_f^{1/2}/[CO_2]\rho$
Minimum eco-indicator points for given thermal conduction	$\lambda/I_e\rho$

*H_p = production energy content per kg; $[CO_2]$ = CO_2 production per kg; I_e = eco-indicator per kg.

The product-manufacture phase

The obvious message of Section 16.3 is that vapor-forming methods are energy-intensive; casting, and deformation processing are less so. Certainly it is important to save energy in production. But higher priority often attaches to the *local* impact of emissions and toxic waste during manufacture, and this depends crucially on local circumstances. Paper-making (to take an example) uses very large quantities of water. Historically the waste water was heavily polluted with alkalis and particulates, devastating the river systems into which it was dumped. Today, the best paper mills discharge water that is as clean and pure as it was when it entered. Production sites of the former communist-block countries are terminally polluted; those producing the same materials elsewhere, using best-practice methods, have no such problems. Clean manufacture is the answer here.

The use phase

The eco-impact of the use phase of energy-consuming products has nothing to do with the energy content of the materials themselves — indeed, minimizing this may frequently have the opposite effect on use-energy. Use-energy depends on mechanical, thermal and electrical efficiencies; it is minimized by maximizing these efficiencies. Fuel efficiency in transport systems (measured, say, by MJ/km) correlates closely with the mass of the vehicle itself; the objective becomes that of minimizing mass. Energy efficiency in refrigeration or heating systems is achieved by minimizing the heat flux into or out of the system; the objective is then that of minimizing thermal conductivity or thermal inertia. Energy efficiency in electrical generation, transmission, and conversion is maximized by minimizing the ohmic losses in the conductor; here the objective is to minimize electrical resistance while meeting necessary constraints on strength, cost, etc. Selection to meet these objectives is exactly what the previous chapters of this book were about.

The product disposal phase

The environmental consequences of the final phase of product life has many aspects. The aggregation of these into the single "indicator" does not appear to be a helpful path to follow. Limited data are available for recycle energies, and for the fraction of current supply currently met by recycling. The simple Boolean (yes/no) classification, shown in Table 16.3, signals that a given material can be recycled, reused in a lower grade activity, bio-degraded, incinerated or committed to landfill.

16.6 Case studies: drink containers and crash barriers

The methods are illustrated below by case studies.

The energy content of containers

The problem. The containers of Figure 16.6 are examples of products for which the first and second phases of life — material production and product manufacture — are ones that consume energy. Thus material selection to minimize energy and consequent gas and particle emissions focuses on these. Table 16.4 summarizes the requirements.

The masses of five competing container-types, the material of which they are made, and the specific energy content of each are listed Tables 16.5a and b. Their production involves molding or deformation; approximate energies for

Table 16.4   Design requirements for the containers

Function	Container for cold drink
Constraints	Must be recyclable
Objective	Minimize production energy per unit capacity
Free variables	Choice of material

Table 16.5a   Details of the containers

Container type	Material	Mass (g)	Mass/liter (g)	Energy/liter (MJ/l)
PET 400 ml bottle	PET	25	62	5.4
PE 1 milk bottle	High density PE	38	38	3.2
Glass 750 ml bottle	Soda glass	325	433	8.2
Al 440 ml can	5000 series Al alloy	20	45	9.0
Steel 440 ml can	Plain carbon steel	45	102	**2.4**

Table 16.5b   Data for the materials of the containers (from Appendix C)

Material	Production energy MJ/kg	Forming method	Forming energy MJ/kg
PET	84	Molding	3.1
PE	80	Molding	3.1
Soda glass	14	Molding	4.9
5000 series Al alloy	200	Deep drawing	0.13
Plain carbon steel	23	Deep drawing	0.15

each are listed. All five of the materials can be recycled. Which container-type carries the lowest overall energy penalty per unit of fluid contained?

The method and results. A comparison of the energies in Tables 16.5(a) and (b) show that the energy to shape the container is always less than that to produce the material in the first place. Only in the case of glass is the forming energy significant. The dominant phase is that of material production. Summing the two energies for each material and multiplying by the container-mass per liter of capacity gives the ranking shown in the second last column of Table 16.5(a). The steel can carries the lowest energy penalty, glass and aluminum the highest.

Crash barriers

The problem. Barriers to protect driver and passengers of road vehicles are of two types: those that are static—the central divider of a freeway, for instance—and those that move—the fender of the vehicle itself (Figure 16.10). The static type line tens of thousands of miles of road. Once in place they consume no energy, create no CO_2 and last a long time. The dominant phases of their life in the sense of Figure 16.1 are those of material production and manufacture. The fender, by contrast, is part of the vehicle; it adds to its weight and thus to its fuel consumption. The dominant phase here is that of use. This means that, if eco-design is the objective, the criteria for selecting materials for the two sorts of barrier will differ. We take as a criterion for the first that of maximizing the energy that the barrier can absorb per unit of production energy; for the second we take absorbed energy per unit mass. Table 16.6 sumarizes.

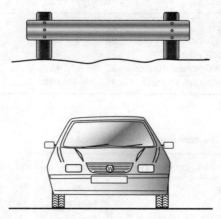

Figure 16.10   Two crash barriers, one static, the other—the fender—attached to something that moves. Different eco-criteria are needed for each.

Table 16.6   Design requirements for the crash barriers

Function	Energy-absorbing crash barriers
Constraints	Must be recyclable
Objective	• Maximize energy absorbed per unit production energy, or • Maximize energy absorbed per unit mass
Free variables	Choice of material

The method and results. The energy absorbed per unit volume by a material when stretched until if fractures is approximately

$$U = \int_0^{\varepsilon_f} \sigma \, d\varepsilon \approx \sigma_y \varepsilon_f \text{ MJ/m}^3$$

Here σ_y is the yield strength and ε_f is the elongation to fracture. Crash barriers are loaded in more complex ways that this, but when the part that is being stretched fractures, the barrier looses effectiveness. The materials with the largest U per unit production energy are those with large values of

$$M = \frac{\sigma_f \varepsilon_f}{H_p \rho} \tag{16.9}$$

Those that do so per unit mass are those with large

$$M = \frac{\sigma_f \varepsilon_f}{\rho} \tag{16.10}$$

Figure 16.11(a) and (b) compare metals and polymers on this basis. The first guides selection for static barriers. Among metals, carbon steels are the best choice. Barriers made of polyethylene or polypropylene could be equally effective. The second figure guides selection for barriers that move. Among metals, titanium alloys offer the lightest solution; steels and light alloys are almost as good and much cheaper. But here polymers become particularly attractive: a polymeric fender, potentially, is much lighter than a metal one.

16.7 Summary and conclusions

Rational selection of materials to meet environmental objectives starts by identifying the phase of product-life that causes greatest concern: production, manufacture, use, or disposal. Dealing with all of these requires data not only for the obvious eco-attributes (energy, CO_2 and other emissions, toxicity, ability to be recycled and the like) but also data for mechanical, thermal, electrical, and chemical properties. Thus if material production is the phase of concern, selection is based on minimizing production energy or the associated emissions (e.g. CO_2 production). But if it is the use-phase that is of concern,

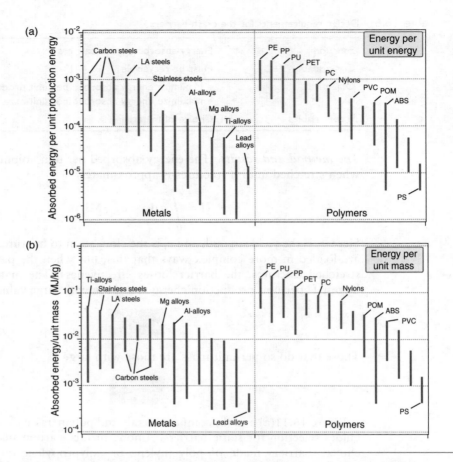

Figure 16.11 (a) The energy absorbed by a material in plastic deformation, per unit of production energy and (b) the energy absorbed per unit mass.

selection is based instead on light weight, excellence as a thermal insulator, or as an electrical conductor (while meeting other constraints on stiffness, strength, cost, etc.). This chapter develops methods to deal with these. The methods are most effective when implemented in software. The Edu version of the CES 4 system, described in Chapters 5 and 6 has limited eco data; the more specialized Eco-selector has much more.

16.8 Further reading

CES'05 (2005) *The Cambridge Engineering Selector*, Granta Design, Cambridge (*The material selection platform now has an optional eco-design module.* www.grantadesign.com)

Eternally Yours (2001)www.eternally-yours.nl *(The Eternally Yours Foundation seeks to analyze why people surround themselves more and more with products they feel less and less attached to, and to explore the design of products that retain their appeal over long periods of time.)*

Fuad-Luke, A. (2002) *The Eco-Design Handbook*, Thames and Hudson, London, UK. ISBN 0-500-28343-5. *(A remarkable source-book of examples, ideas and materials of eco-design.)*

Goedkoop, M.J., Demmers, M. and Collignon, M.X. (1995) *Eco-Indicator '95, Manual*, Pré Consultants, and the Netherlands Agency for Energy and the Environment, Amersfort, Holland. ISBN 90-72130-80-4.

Goedkoop, M., Effting, S. and Collignon, M. (2000) *The Eco-Indicator 99: A Damage Oriented Method for Life Cycle Impact Assessment, Manual for Designers.* *(PRe Consultants market a leading life-cycle analysis tool and are proponents of the eco-indicator method.* http://www.pre.nl)

ISO 14001 (1996) and ISO 14040 (1997, 1998, 1999) International Organisation for Standardisation (ISO), Environmental management system-specification with guidance for use, Geneva, Switzerland.

Kyoto Protocol (1997) United Nations, Framework Convention on Climate Change. Document FCCC/CP1997/7/ADD.1 Environmental management — life cycle assessment and subsections, Geneva, Switzerland. *(An agreement between the developed nations to limit their greenhouse gas emissions relative to the levels emitted in 1990. The US agreed to reduce emissions from 1990 levels by 7% during the period 2008–2012; European nations adopted a more stringent agreement.)*

Lovins, L.H., von Weizsäcker, E. and Lovins, A.B. (1998) *Factor Four: Doubling Wealth, Halving Resource Use*, Earthscan publications, London, UK. ISBN 1-853834-068. *(An influential book arguing that resource productivity can and should grow fourfold, that is, the wealth extracted from one unit of natural resources can quadruple allowing the world to live twice as well yet use half as much.)*

Mackenzie, D. (1997) Green design: design for the environment, 2nd edition, Lawrence King Publishing, London, UK. ISBN 1-85669-096-2. *(A lavishly produced compilation of case studies of eco design in architecture, packaging, and product design.)*

Meadows, D.H., Meadows, D.L., Randers, J. and Behrens, W.W. (1972) *The Limits to Growth — 1st Report of the Club of Rome*, Universe Books, New York, USA. *(A pivotal publication, alerting the world to the possibility of resource depletion, undermined by the questionable qualtiy of the data used for the analysis, but despite that, the catalyst for subsequent studies and for views that are now more widely accepted.)*

Schmidt-Bleek, F. (1997) How much environment does the human being need — factor 10 — the measure for an ecological economy, Deutscher Taschenbuchverlag, Munich, Germany. *(The author argues that true sustainability requires a reduction in energy and resource consumption by the developed nations by a facto of 10.)*

Wenzel, H. Hauschild, M. and Alting, L. (1997) *Environmental Assessment of Products*, Vol. 1, Chapman and Hall, London UK. *(Professor Alting leads Danish research in eco-design.)*

Chapter 17

Materials and industrial design

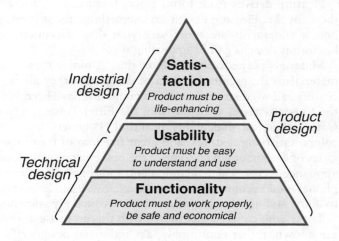

17.1 Introduction and synopsis

Good design works. Excellent design also gives pleasure.

Pleasure derives from form, color, texture, feel, and the associations that these invoke. Pleasing design says something about itself; generally speaking, honest statements are more satisfying than deception, though eccentric or humorous designs can be appealing too.

Materials play a central role in this. A major reason for introducing new materials is the greater freedom of design that they allow. Metals, in the past century, allowed structures which could not have been built before: cast iron, the Crystal Palace; wrought iron, the Eiffel Tower; drawn steel, the Golden Gate Bridge, all undeniably beautiful. Polymers lend themselves to bright colors, satisfying textures, and great freedom of form; they have opened new styles of design, of which some of the best examples are found in the household appliance sector: kitchen equipment, radio and CD-players, hair dryers, telephones, and vacuum cleaners make extensive and imaginative use of materials to allow styling, weight, feel, and form which give pleasure.

Those who concern themselves with this aesthetic dimension of engineering are known, rather confusingly, as "industrial designers". This chapter introduces some of the ideas of industrial design, emphasizing the role of materials. It ends with case-studies. But first a word of caution.

Previous chapters have dealt with systematic ways of choosing material and processes. "Systematic" means that if *you* do it and *I* do it, following the same procedure, we will get the same result, and that the result, next year, will be the same as it is today. Industrial design is not, in this sense, systematic. Success, here, involves sensitivity to fashion, custom and educational background, and is influenced (manipulated, even) by advertising and association. The views of this chapter are partly those of writers who seem to me to say sensible things, and partly my own. You may not agree with them, but if they make you think about designing to give pleasure, the chapter has done what it should.

17.2 The requirements pyramid

The pen with which I am writing this book cost $5 (Figure 17.1, upper image). If you go to the right shop you can find a pen that costs well over $1000 (lower image). Does it write 200 times better than mine? Unlikely; mine writes perfectly well. Yet there is a market for such pens. Why?

A product has a *cost* — the outlay in manufacture and marketing it. It has a *price* — the sum at which it is offered to the consumer. And it has a *value* — a measure of what the consumer thinks it is worth. The expensive pens command the price they do because the consumer perceives their value to justify it. What determines value?

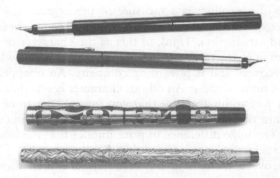

Figure 17.1   Pens, inexpensive and expensive. The chosen material — acrylic in the upper picture, gold, silver and enamel in the lower one — create the aesthetics and the associations of the pens. They are perceived differently, one pair as utilitarian, the other as something rare and crafted. (Lower figure courtesy of David Nishimura of Vintagepens.com). The perceptions have much to do with the materials of which they are made.

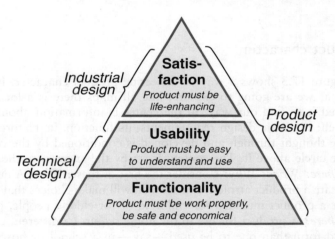

Figure 17.2   The requirements pyramid. The lower part of the pyramid tends to be called "technical design", the upper part, "Industrial design" suggesting that they are separate activities. It is better to think of all three tiers as part of a single process that we shall call "product design".

Functionality, provided by sound technical design, clearly plays a role. The requirements pyramid of Figure 17.2 has this as its base: the product must work properly, be safe and economical. Functionality alone is not enough: the product must be easy to understand and operate, and these are questions of *usability*, the second tier of the figure. The third, completing the pyramid, is the requirement that the product gives *satisfaction*: that it enhances the life of its owner.

The value of a product is a measure of the degree to which it meets (or exceeds) the expectation of the consumer in all three of these — functionality, usability, and satisfaction. Think of this as the character of the product. It is very like human character. An admirable character is one who functions well, interacts effectively and is rewarding company. An unappealing character is one that does none of these. An odious character is one that does one or more of them in a way so unattractive that you cannot bear to be near him.

Products are the same. All the pens in Figure 17.1 function well and are easy to use. The huge difference in price implies that the lower pair provide a degree of satisfaction not offered by the upper ones. The most obvious difference between them is in the materials of which they are made — the upper pair of molded acrylic, the lower pair of gold, silver, and enamel. Acrylic is the material of tooth-brush handles, something you throw away after use. Gold and silver are the materials of precious jewelry, they have associations of craftsmanship, of heirlooms passed from one generation to the next. Well, that's part of the difference, but there is more. So — the obvious question — how do you create product character?

17.3 Product character

Figure 17.3 shows a way of dissecting product character. It is a map of the ideas we are going to explore; like all maps there is a lot of detail, but we need it to find our way. In the center is information about the PRODUCT itself: the basic design requirements, its function, its features. The way these are thought through and developed is conditioned by the *context*, shown in the circle above it. The context is set by the answers to the questions: *Who? Where? When? Why?* Consider the first of these: *Who?* A designer seeking to create a product attractive to women will make choices that differ from those for a product intended for children, or for elderly people, or for sportsmen. *Where?* A product for use in the home requires a different choice of material and form than one to be used — say — in a school or hospital. *When?* One intended for occasional use is designed in a different way than one that is used all the time; one for formal occasions differs from one for informal use. *Why?* A product that is primarily utilitarian involves different design decisions than one that is largely a life-style statement. The context influences and conditions all the decisions that the designer takes in finding a solution. It sets the *mood*.[1]

On the left of PRODUCT lies information about the MATERIALS and the PROCESSES used to shape, join, and finish it. Each illustrates the library, so

---

[1] Many designers, working on a project, assemble a *mood-board* with images of the sort of people for whom the product is intended, the surroundings in which they suppose it will be used, and other products that the intended user group might own, seeking to capture the flavor of their life style.

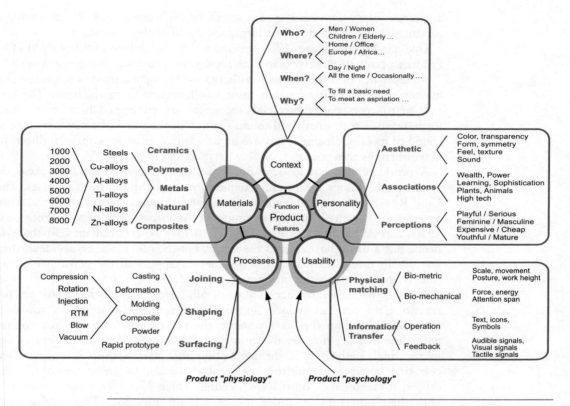

Figure 17.3 The dissection of product character. *Context* defines the intentions or "mood"; *materials* and *processes* create the flesh and bones; the interface with the user determines *usability*, and the aesthetics, associations and perceptions of the product create its *personality*. These terms are explained more fully in the text.

to speak, from which the choices can be made and — or course — they are the families, classes, and members that we have already met. Choosing these to provide functionality — the bottom of the pyramid — was the subject of this book so far. Material and process give the product its tangible form, its flesh, and bones so to speak; they create the *product physiology*.

On the right of Figure 17.3 are two further packages of information. The lower one — USABILITY — characterizes the ways in which the product communicates with the user: the interaction with their sensory, cognitive, and motor functions. Products success requires a mode of operation that, as far as possible, is intuitive, does not require taxing effort, and an interface that communicates the state of the product and its response to user action by visible, acoustic, or tactile response. It is remarkable how many products fail in this, and by doing so, exclude many of their potential users. There is, today, an awareness of this, giving rise to research on *inclusive design*: design to make products that can be used by a larger spectrum of the population.

One circle on Figure 17.3 remains: the one labeled PERSONALITY. Product personality derives from *aesthetics, associations*, and *perceptions*.

Anaesthetics dull the senses. Aesthetics do the opposite: they stimulate the five senses: sight, hearing, touch, taste, smell, and via them, the brain. The first row of the personality box elaborates: we are concerned here with color, form, texture, feel, smell, and sound — think of the smell of a new car and the sound of its door closing. They are no accident. Car makers spend millions to make them as they are.

A products also has associations — the second row of the box. Associations are the things the product reminds you of, the things it suggests. The Land Rover and other SUVs have forms and (often) colors that mimic those of military vehicles. The streamlining of American cars of the 1960s and 1970s carried associations of aerospace. It may be an accident that the VW Beetle has a form that suggests the insect, but the others are no accident; they were deliberately chosen by the designer to appeal to the consumer group (the *Who?*) at which the product was aimed.

Finally, the most abstract quality of all, perceptions. Perceptions are the reactions the product induces in an observer, the way it makes you *feel*. Here there is room for disagreement; the perceptions of a product change with time and depends on the culture and background of the observer. Yet in the final analysis it is the perception that causes the consumer, when choosing between a multitude of similar models, to prefer one above the others; it creates the "must have" feeling. Table 17.1 lists some perceptions with their opposites, in order to sharpen the meaning. They derive from product reviews and magazines specializing in product design; they are a part of a vocabulary, one that is used to communicate views about product character.

Table 17.1   Some perceived attributes of products, with opposites

Perceptions (with opposites)	
Aggressive — Passive	Extravagant — Restrained
Cheap — Expensive	Feminine — Masculine
Classic — Trendy	Formal — Informal
Clinical — Friendly	Hand-made — Mass produced
Clever — Silly	Honest — Deceptive
Common — Exclusive	Humorous — Serious
Decorated — Plain	Informal — Formal
Delicate — Rugged	Irritating — Lovable
Disposable — Lasting	Lasting — Disposable
Dull — Sexy	Mature — Youthful
Elegant — Clumsy	Nostalgic — Futuristic

17.4 Using materials and processes to create product personality

Do materials, of themselves, have a personality? There is a school of thinking that holds as a central tenant that materials must be used "honestly". By this they mean that deception and disguise are unacceptable — each material must be used in ways that expose its intrinsic qualities and natural appearance. It has its roots in the tradition of craftsmanship — the potters' use of clays and glazes, the carpenters' use of woods, the skills of silversmiths and glass makers in crafting beautiful objects that exploit the unique qualities of the materials with which they work, an integrity to craft and material.

This is a view to be respected. But it is not the only one. Design integrity is a quality that consumers value, but they also value other qualities: humor, sympathy, surprise, provocation, even shock. You do not have to look far to find a product that has one of these, and often it is achieved by using materials in ways that deceive. Polymers are frequently used in this way — their adaptability invites it. And, of course, it is partly a question of definition — if you say that a characterizing attribute of polymers is their ability to mimic other materials, then using them in this way *is* honest.

Materials and the senses: aesthetic attributes

Aesthetic attributes are those that relate to the senses: touch, sight, hearing, taste, and smell (Table 17.2). Almost everyone would agree that metals feel 'cold'; that cork feels 'warm'; that a wine glass, when struck, 'rings'; that a

Table 17.2 Some aesthetic attributes of materials

Sense	Attribute	Sense	Attribute
Touch	Warm	Hearing	Muffled
	Cold		Dull
	Soft		Sharp
	Hard		Resonant
	Flexible		Ringing
	Stiff		High pitched
			Low pitched
Sight	Optically		
	Clear	Taste	Bitter
	Transparent	Smell	Sweet
	Translucent		
	Opaque		
	Reflective		
	Glossy		
	Matte		
	Textured		

pewter mug sounds "dull", even "dead". A polystyrene water glass can look indistinguishable from one made of glass, but pick it up and it feels lighter, warmer, less rigid; tap it and it does not sound the same. The impression it leaves is so different from glass that, in an expensive restaurant, it would be completely unacceptable. Materials, then, have certain characterizing aesthetic attributes. Let us see if we can pin these down.

Touch: soft-hard/warm-cold. Steel is "hard"; so is glass; diamond is harder the either of them. Hard materials do not scratch easily; indeed they can be used to scratch other materials. They generally accept a high polish, resist wear, and are durable. The impression that a material is hard is directly related to its Vickers hardness H. Here is an example of a sensory attribute that relates directly to a technical one.

"Soft" sounds like the opposite of "hard" but it has more to do with modulus E than with hardness H. A soft material deflects when handled, it gives a little, it is squashy, but when it is released it returns to its original shape. Elastomers (rubbers) feels soft; so do polymer foams. Both have moduli that are 100 to 10,000 lower than ordinary 'hard' solids; it is this that makes them feel soft. Soft to hard is used as one axis of Figure 17.4. It uses the quantity $\sqrt{EH}$ as a measure.

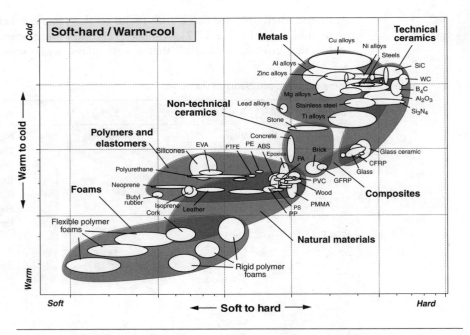

Figure 17.4 Tactile qualities of materials. Foams and many natural materials are soft and warm; metals, ceramics, and glasses are hard and cold. Polymers lie in-between.

A material feels "cold" to the touch if it conducts heat away from the finger quickly; it is "warm" if it does not. This has something to do with its thermal conductivity λ but there is more to it than that—it depends also on its specific heat C_p. A measure of the perceived coldness or warmth of a material is the quantity $\sqrt{\lambda C_p \rho}$. It is shown as the other axis of Figure 17.4, which nicely displays the tactile properties of materials. Polymer foams and low-density woods are warm and soft; so are balsa and cork. Ceramics and metals are cold and hard; so is glass. Polymers and composites lie in between.

Sight: transparency, color, reflectivity. Metals are opaque. Most ceramics, because they are polycrystalline and the crystals scatter light, are either opaque or translucent. Glasses, and single crystals of some ceramics, are transparent. Polymers have the greatest diversity of optical transparency, ranging from transparency of optical quality to completely opaque. Transparency is commonly described by a 4-level ranking that uses easily-understood everyday words : "opaque", "translucent", "transparent", and "water-clear". Figure 17.5 ranks the transparency of common materials. In order to spread the data in a useful way, it is plotted against cost. The cheapest materials offering optical-quality transparency ("water-clarity") are glass, PS, PET, and PMMA. Epoxies can be transparent but not with water clarity. Nylons are, at best, translucent.

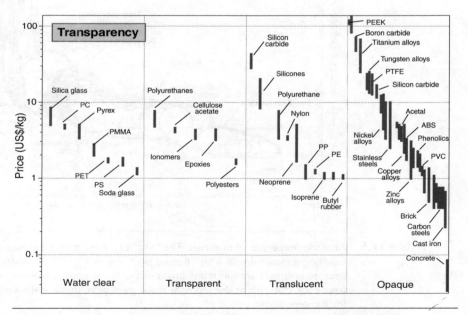

Figure 17.5   Here transparency is ranked on a four-point scale, form water-clear to opaque. Water-clear materials are used for windows, display cases, and lenses. Transparent and translucent materials transmit light but diffuse it in doing so. Opaque materials absorb light.

All metals, most ceramics and all carbon-filled or reinforced polymers are opaque.

Color can be quantified by analyzing spectra but this—from a design standpoint—does not help much. A more effective method is one of color matching, using color charts such as those provided by Pantone[2]; once a match is found it can be described by the code that each color carries. Finally there is reflectivity, an attribute that depends partly on material and partly on the state of its surface. Like transparency, it is commonly described by a ranking: dead matte, eggshell, semi-gloss, gloss, mirror.

Hearing: pitch and brightness. The frequency of sound (pitch) emitted when an object is struck relates to its material properties. A measure of this pitch, $\sqrt{E/\rho}$, is used as one axis of Figure 17.6. Frequency is not the only aspect of acoustic response—the other has to do with the damping or loss coefficient, η. A highly damped material sounds dull and muffled; one with low damping

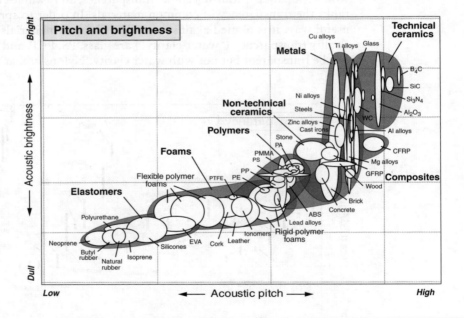

Figure 17.6   Acoustic properties of materials. The "ring" of a wine glass is because glass in an acoustically bright material with a high natural pitch; the dull "ping" of a plastic glass is because polymers are much less bright and—in the same shape—vibrate at a lower frequency. Materials at the top right make good bells; those at the bottom left are good for damping sound.

[2] Pantone (www.pantone.com.) provide detailed advice on color selection, including color-matching chart and good descriptions of the associations and perceptions of color.

rings. Acoustic brightness — the inverse of damping — is used as the other axis of Figure 17.6. It groups materials that have similar acoustic behavior.

Bronze, glass, and steel ring when struck, and the sound they emit has — on a relative scale — a high pitch; they are used to make bells; alumina, on this ranking, has bell-like qualities. Rubber, foams and many polymers sound dull, and, relative to metals, they vibrate at low frequencies; they are used for sound damping. Lead, too, is dull and low-pitched; it is used to clad buildings for sound insulation.

The three figures show that each material class has a certain recognizable aesthetic character. Ceramics are hard, cold, high-pitched and acoustically bright. Metals, too, are relatively hard and cold but although some, like bronze, ring when struck, others — like lead — are dull. Polymers and foams are most nearly like natural materials — warm, soft, low-pitched, and muffled, though some have outstanding optical clarity and almost all can be colored. But their low hardness means that they scratch easily, loosing their gloss.

These qualities of a material contribute to the product personality. The product acquires some of the attributes of the material from which it is made, an effect that designers recognize and use when seeking to create a personality. A stainless steel facia, whether it be in a car or on a hi-fi system, has a different personality than one of polished wood or one of leather, and that in part is because the product has acquired some of the aesthetic qualities of the material.

Materials and the mind: associations and perceptions

So a material certainly has aesthetic qualities — but can it be said to have a personality? At first sight, no — it only acquires one when used in a product. Like an actor, it can assume many different personalities, depending on the role it is asked to play. Wood in fine furniture suggests craftsmanship, but in a packing case, cheap utility. Glass in the lens of a camera has associations of precision engineering, but in a beer bottle, that of disposable packaging. Even gold, so often associated with wealth and power, has different associations when used in microcircuits: that of technical functionality.

But wait. The object in Figure 17.7 has its own sombre association. It appears to be made of polished hardwood — the traditional material for such things. If you had to choose one, you would probably not have any particular aversion to this one — it is a more or less typical example. But suppose I told you it was made of *plastic* — would you feel the same? Suddenly it becomes like a bin, a wastebasket, inappropriate for its dignified purpose. Materials, it seems, *do* have personality.

Expression through material. Think of wood. It is a natural material with a grain that has a surface texture, pattern, color, and feel that other materials do not have. It is tactile — it is perceived as warmer than many other materials, and seemingly softer. It is associated with characteristic sounds and smells.

Figure 17.7 A coffin. Wood is perceived to be appropriate for its sombre, ceremonial function, plastic inappropriate.

It has a tradition; it carries associations of craftsmanship. No two pieces are exactly alike; the wood-worker selects the piece on which he will work for its grain and texture. Wood enhances value: the interior of cheap cars is plastic, that of expensive ones is burr-walnut and calves leather. And it ages well, acquiring additional character with time; objects made of wood are valued more highly when they are old than when they are new. There is more to this than just aesthetics; there are the makings of a personality, to be brought out by the designer, certainly, but there none the less.

And metals . . . Metals are cold, clean, precise. They ring when struck. They reflect light — particularly when polished. They are accepted and trusted: machined metal looks strong, its very nature suggests it has been *engineered*. They are associated with robustness, reliability, permanence. The strength of metals allows slender structures — the cathedral-like space of railway stations or the span of bridges. They can be worked into flowing forms like intricate lace or cast into solid shapes with elaborate, complex detail. The history of man and of metals is intertwined — the titles "bronze age" and "iron age" tell you how important these metals were — and their qualities are so sharply defined that they have become ways of describing human qualities — an iron will, a silvery voice, a golden touch, a leaden look. And, like wood, metals can age well, acquiring a patina that makes them more attractive than when newly polished — think of the bronze of sculptures, the pewter of mugs, the lead of roofs.

Ceramics and glass? They have an exceptionally long tradition — think of Greek pottery and Roman glass. They accept almost any color; this and their total resistance to scratching, abrasion, discoloration and corrosion gives them a certain immortality, threatened only by their brittleness. They are — or were — the materials of great craft-based industries: the glass of Venice, the porcelain of Meissen, the pottery of Wedgwood, valued at certain times more highly than silver. But at the same time they can be robust and functional — think of beer bottles. The transparency of glass gives it an ephemeral quality — sometimes you see it, sometimes you do not. It interacts with light, transmitting it, refracting it, reflecting it. And ceramics today have additional associations — those of advanced technology: kitchen stove-tops, high pressure/high temperature valves, space shuttle tiles...materials for extreme conditions.

And finally polymers. "A cheap, plastic imitation", it used to be said — and that is a hard reputation to live down. It derives from the early use of plastics to simulate the color and gloss of Japanese handmade pottery, much valued in Europe. Commodity polymers *are* cheap. They are easily colored and molded (that is why they are called "plastic"), making imitation easy. Unlike ceramics, their gloss is easily scratched, and their colors fade — they do not age gracefully. You can see where the reputation came from. But is it justified? No other class of material can take on as many characters as polymers: colored, they look like ceramics; printed, they can look like wood or textile; metalized, they look exactly like metal. They can be as transparent as glass or as opaque as lead, as flexible as rubber or as stiff — when reinforced — as aluminum. Plastics emulate precious stones in jewelry, glass in drinking glasses and glazing, wood in counter tops, velvet and fur in clothing, even grass. But despite this chameleon-like behavior they *do* have a certain personality: they feel warm — much warmer than metal or glass. They are adaptable — that is part of their special character; and they lend themselves, particularly, to brightly colored, light-hearted, even humorous, design. But their very cheapness creates problems as well as benefits: our streets, county-side and rivers are littered with discarded plastic bags and packaging that decay only very slowly.

The ways in which material, processes, usability and personality combine to create a product character tuned to the context or "mood" are best illustrated by examples. Figure 17.8 shows the first. The lamp on the left is designed for the office. It is angular, functional, creamy-gray, and it is heavy. Its form and color echo those of computer consoles and keyboards, creating associations of contemporary office technology. Its form and weight transmit the ideas of stability, robustness, efficiency, and fitness for task — but for tasks in the workplace, not in the bedroom. Materials and processes have been chosen to reinforce these associations and perceptions. The enameled frame is pressed and folded sheet steel, the base-weight is cast iron, the reflector is stainless steel set in a high-impact ABS enclosure.

The lamp on the right of Figure 17.8 has the same technical rating of that on the left; the same functionality and usability. But there the resemblance ends. This product is not designed for the busy executive but for children (and adults that still enjoy being children), to be used in the playroom or bedroom. It has a contoured form, contrasting translucent colors, and it is very light. It is made of colored acrylic in translucent and opaque grades, so that the outside of the lamp glows like a neon sign when it is lit. Its form is partly derived from nature, partly from cartoons and comic strips, giving it a light-hearted character. I perceive it as playful, funny, cheerful, and clever — but also as eccentric and easily damaged. You may perceive it in other ways — perception is a personal

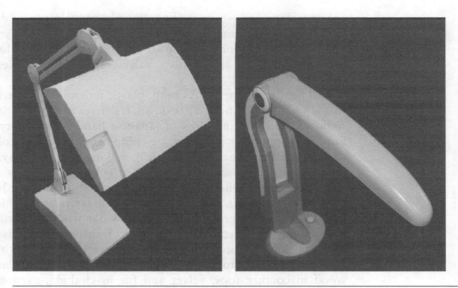

Figure 17.8 Lamps. Both have the same technical rating, but differ completely in their personalities. Materials, processes, form, weight, and color have all contributed to the personality.

thing; it depends where you are coming from. Skilled designers manipulate perception to appeal to the user-group they wish to attract.

Figure 17.9 shows a second example. Here are two contrasting ways of presenting electronic home entertainment systems. On the left: a music center aimed at successful professionals with disposable income, comfortable with (or addicted to) advanced technology, for whom only the best is good enough. The linear form, the use of primitives (rectangles, circles, cylinders, cones) and the matt silver and black proclaim that this product has not just been *made*, it has been *Designed* (big *D*). The formal geometry and finish suggest precision instruments, telescopes, electron microscopes and the shapes resemble those of organ pipes (hence associations of music, of culture). The perception is that of quality cutting-edge technology, a symbol of discriminating taste. The form has much to do with this associations and perceptions, but so too do the materials: brushed aluminum, stainless steel and black enamel—these are not materials you choose for a cuddly toy.

On the right: electronics presented in another way. This is a company that has retained market share, even increased it, by *not* changing, at least as far as appearance is concerned (I had one 40 years ago that looked exactly like this). The context? Clearly, the home, perhaps aimed at consumers who are uncomfortable with modern technology (though the electronics in these radios is modern enough), or who simply feel that it clashes with the home environment. Each radio has a simple form, it is pastel-colored, it is soft and warm to touch. It is the materials that make the difference: these products are available

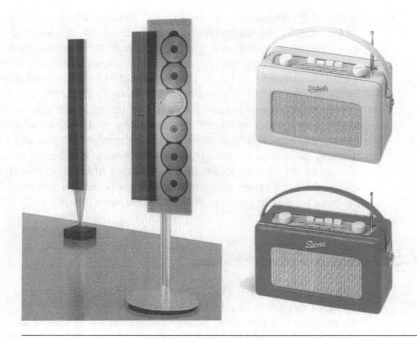

Figure 17.9 Consumer electronics. The products on the left is aimed at a different consumer group than those on the right. The personalities of each (meaning the combination of aesthetics, associations, and perceptions) have been constructed to appeal to the target group. Materials play a central role here in creating personality. (Figure on left courtesy of Bang & Olufsen.)

in suede or leather in six or more colors. The combination of form and material create associations of comfortable furniture, leather purses, and handbags (hence, luxury, comfort, style), the past (hence, stability) and perceptions of solid craftsmanship, reliability, retro-appeal, traditional but durable design.

So there is a character hidden in a material even before it has been made into an recognizable form — a sort of embedded personality, a shy one, not always obvious, easily concealed or disguised, but one that, when appropriately manipulated, imparts its qualities to the design. It is for this reason that certain materials are so closely linked to certain *design styles*. A *style* is a short-hand for a manner of design with a shared set of aesthetics, associations and perceptions. The Early Industrial style (1800–1890)[3] embraced the technologies of the industrial revolution, using cast iron, and steel, often elaborately decorated to give it a historical façade. The Arts and Crafts movement (1860–1910) rejected this, choosing instead natural materials and fabrics to create products

---

[3] The dates are, of course, approximate. Design styles do not switch on an off on specific dates, they emerge as a development of, or reaction to, earlier styles with which they often co-exist, and they merge into the styles that follow.

with the character of traditional hand-crafted quality. Art Nouveau (1890–1918), by contrast, exploited the fluid shapes and durability made possible by wrought iron and cast bronze, the warmth and textures of hard wood and the transparency of glass to create products of flowing, organic character. The Art Deco movement (1918–1935) extended the range of materials to include for the first time plastics (Bakelite and Catalin) allowing production both of luxury products for the rich and also mass-produced products for a wider market. The simplicity and explicit character of Bauhaus designs (1919–1933) is most clearly expressed by the use of chromed steel tubing, glass and moulded plywood. Plastics first reach maturity in product design in the cheeky iconoclastic character of the Pop-Art style (1940–1960). Since then the range of materials has continued to increase, but their role in helping to mould product character remains.

17.5 Summary and conclusions

What do we learn? The element of satisfaction is central to contemporary product design. It is achieved through an integration of good technical design to provide functionality, proper consideration of the needs of the user in the design of the interface, and imaginative industrial design to create a product that will appeal to the consumers at whom it is aimed.

Materials play a central role in this. Functionality is dependant on the choice of proper material and process to meet the technical requirements of the design safely and economically. Usability depends on the visual and tactile properties of materials to convey information and respond to user actions. Above all, the aesthetics, associations, and perceptions of the product are strongly influenced by the choice of the material and its processing, imbuing the product with a personality that, to a greater or lesser extent, reflects that of the material itself.

Consumers look for more than functionality in the products they purchase. In the sophisticated market places of developed nations, the "consumer durable" is a thing of the past. The challenge for the designer no longer lies in meeting the functional requirements alone, but in doing so in a way that also satisfies the aesthetic and emotional needs. The product must carry the image and convey the meaning that the consumer seeks: timeless elegance, perhaps; or racy newness. One Japanese manufacturer goes so far as to say: "*Desire* replaces *need* as the engine of design".

Not everyone, perhaps, would wish to accept that. So we end with simpler words — the same ones with which we started. Good design works. Excellent design also gives pleasure. The imaginative use of materials provides it.

If you found this chapter interesting and would like to read more, you will find the ideas it contains developed more fully in the first book listed in Section 17.6 Further reading.

17.6 Further reading

Ashby, M.F. and Johnson, K. (2002) *Materials and Design—The Art and Science of Materials Selection in Product Design*, Butterworth-Heinemann, Oxford, UK. ISBN0-7506-5554-2. (*A book that develops further the ideas outlined in this chapter.*)

Clark, P. and Freeman, J. (2000) *Design, a Crash Course*, The Ivy Press Ltd, Watson-Guptil Publications, BPI Communications Inc. New York, NY, USA. ISBN 0-8230-0983-1. (*An entertainingly-written scoot through the history of product design from 5000BC to the present day.*)

Dormer, P. (1993) *Design since 1945*, Thames and Hudson, London UK. ISBN 0-500-20269-9. (*A well-illustrated and inexpensive paperback documenting the influence of industrial design in furniture, appliances and textiles—a history of contemporary design that complements the wider-ranging history of Haufe (1998), q.v.*)

Forty, A. (1986) *Objects of Desire—Design in Society Since 1750*, Thames and Hudson, London, UK. ISBN 0-500-27412-6. (*A refreshing survey of the design history of printed fabrics, domestic products, office equipment and transport system. The book is mercifully free of eulogies about designers, and focuses on what industrial design does, rather than who did it. The black and white illustrations are disappointing, mostly drawn from the late 19th or early 20th centuries, with few examples of contemporary design.*)

Haufe, T. (1998) *Design, a Concise History*, Laurence King Publishing, London, UK (originally in German). ISBN 1-5669-134-9. (*An inexpensive soft-cover publication. Probably the best introduction to industrial design for students (and anyone else). Concise, comprehensive, clear and with intelligible layout and good, if small, color illustrations.*)

Jordan, P.S. (2000) *Designing Pleasurable Products*, Taylor and Francis, London, UK. ISBN 0-748-40844-4. (*Jordan, Manager of Aesthetic Research and Philips Design, argues that products today must function properly, must be usable, and must also give pleasure. Much of the book is a description of market-research methods for eliciting reactions to products from users.*)

Julier, G. (1993) *Encyclopedia of 20th Century Design and Designers*, Thames & Hudson, London, UK. ISBN 0-500-20261-3. (*A brief summary of design history with good pictures and discussions of the evolution of product designs.*)

Manzini, E. (1989) *The Material of Invention*, The Design Council, London UK. ISBN 0-85072-247-0 (*Intriguing descriptions of the role of material in design and in inventions. The translation from Italian to English provides interesting—and often inspiring—commentary and vocabulary that is rarely used in traditional writings about materials.*)

McDermott, C. (1999) *The Product Book*, D & AD in association with Rotovison, UK. (*50 essays by respected designers who describe their definition of design, the role of their respective companies and their approach to product design.*)

Norman, D.A. (1988) *The Design of Everyday Things*, Doubleday, New York, USA. ISBN 0-385-26774-6. (*A book that provides insight into the design of products with particular emphasis on ergonomics and ease of use.*)

Chapter 18

Forces for change

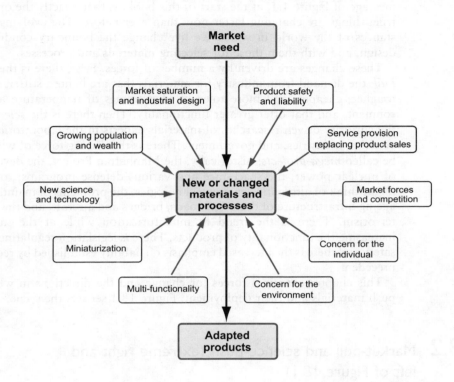

18.1 Introduction and synopsis

If there were no forces for change, everything would stay the same. The clear message of Figure 1.1, at the start of this book, is that exactly the opposite is true: things are changing faster now than ever before. The evolving circumstances of the world in which we live change the boundary conditions for design, and with them those for selecting materials and processes.

These changes are driven by a number of forces. First, there is the *market-pull*: the demand from industry for materials that are lighter, stiffer, stronger, tougher, cheaper, and more tolerant of extremes of temperature and environment, and that offer greater functionality. Then there is the *science-push*: the curiosity-driven researches of materials experts in the laboratories of universities, industries, and government. There is the driving force of what might be called *mega-projects*: historically, the Manhattan Project, the development of nuclear power, the space-race and various defense programs; today, one might think of alternative energy technologies, the problems of maintaining an ageing infrastructure of drainage, roads, bridges and aircraft, and the threat of terrorism. There is the trend to miniaturization while at the same time increasing the functionality of products. There is legislation regulating product safety and there is the increased emphasis on liability established by recent legal precedent.

This chapter examines forces for change and the directions in which they push materials and their deployment. Figure 18.1 set sets the scene.

18.2 Market-pull and science-push (extreme right and left of Figure 18.1)

The power of the market

The end-users of materials are the manufacturing industries. They decide which material they will purchase, and adapt their designs to make best use of them. Their decisions are based on the nature of their products. Materials for large civil structures (which might weigh 10,000 tonnes or more) must be cheap; economy is the overriding consideration. By contrast, the cost of the materials for high-tech products (sports equipment, military hardware, space projects, biomedical applications) plays a less important role: for an artificial heart valve, for instance, material cost is almost irrelevant. Performance, not economy, dictates the choice.

The market price of a product has several contributions. One is the cost of the materials of which the product is made, but there is also the cost of the research and development that went into its design, the cost of manufacture and marketing and the perceived value associated with fashion, scarcity, lack

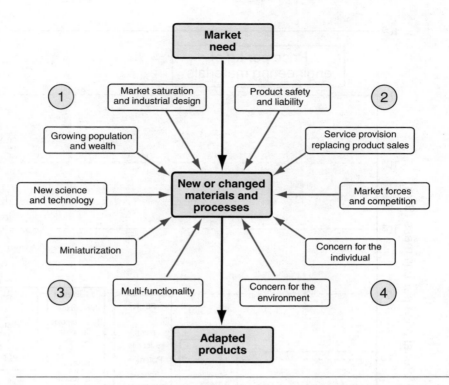

Figure 18.1   Forces for change. Each of the influences exerts pressure to change the choice of material and process, and stimulates efforts to develop new ones.

of competition and such like — they were described in Chapter 7. When the material costs are a large part of the market value (50%, say) — that is, when the value added to the material is small — the manufacturer seeks to economize on material usage to increase profit or market share. When, by contrast, material costs are a tiny fraction of the market value (1%, say), the manufacturer seeks the materials that will most improve the performance of the produce with little concern for their cost.

With this background, examine Figures 18.2 and 18.3. The vertical axis is the price per unit weight ($/kg) of materials and of products: it gives a common measure by which materials and products can be compared. The measure is a crude one but has the great merit that it is unambiguous, easily determined, and bears some relationship to value-added. A product with a price/kg that is only two or three times that of the materials of which it is made is material-intensive and is sensitive to material costs; one with a price/kg that is 100 times that of its materials is insensitive to material costs, and is probably performance-driven rather than cost-driven. On this scale the price per kg of a contact lens differs from that of a glass bottle by a factor of 10^5, even though both are made of almost the same glass; the cost per kg of a heart valve differs from that of

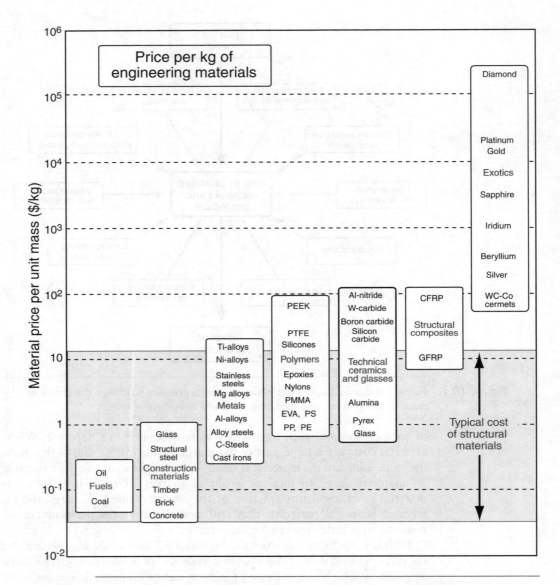

Figure 18.2   The price per unit weight diagram for materials. The shaded band spans the range in which lies the most widely used commodity material of manufacture and construction.

a plastic bottle by a similar factor, even though both are made of polyethylene. There is obviously something to be learned here.

Look first at the price per unit weight of materials (Figure 18.2). The bulk "commodity" materials of construction and manufacture lie in the shaded band; they all cost between $0.05 and $20/kg. Construction materials like

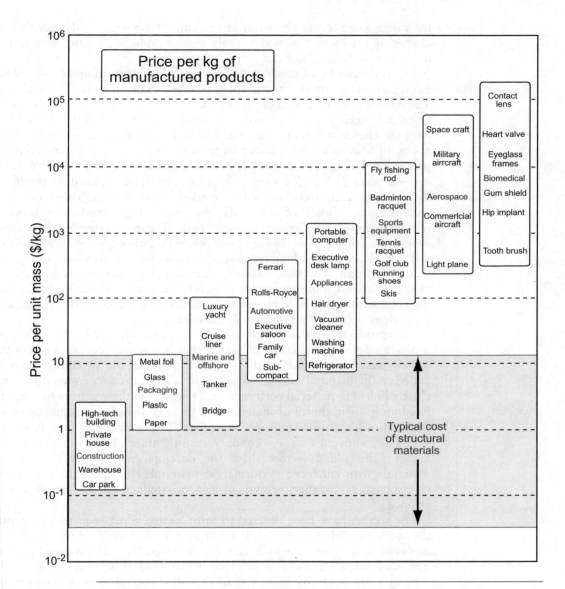

Figure 18.3 The price per unit weight diagram for products. The shaded band spans the range in which lies most of the material of which they are made. Products in the shaded band are material-intensive; those above it are not.

concrete, brick, timber, and structural steel lie at the lower end; high-tech materials like titanium alloys lie at the upper. Polymers span a similar range: polyethylene at the bottom, polytetrafluorethylene (PTFE) near the top. Composites lie higher, with GFRP at the bottom and CFRP at the top of the range Engineering ceramics, at present, lie higher still, though this will change as production increases. Only the low-volume "exotic" materials lie much above the shaded band.

The price per kg of products (Figure 18.3) shows a different distribution. Eight market sectors are shown, covering much of the manufacturing industry. The shaded band on this figure spans the cost of commodity materials, exactly as on the previous figure. Sectors and their products within the shaded band have the characteristic that material cost is a major fraction of product price: about 50% in civil construction, large marine structures and some consumer packaging, falling to perhaps 20% as the top of the band is approached (family car — around 25%). The value added in converting material to product in these sectors is relatively low, but the market volume is large. These constraints condition the choice of materials: they must meet modest performance requirements at the lowest possible cost. The associated market sectors generate a driving force for improved processing of conventional materials in order to reduce cost without loss of performance, or to increase reliability at no increase in cost. For these sectors, incremental improvements in well-tried materials are far more important than revolutionary research-findings. Slight improvements in steels, in precision manufacturing methods, or in lubrication technology are quickly assimilated and used.

The products in the upper half of the diagram are technically more sophisticated. The materials of which they are made account for less than 10% — sometimes less than 1% — of the price of the product. The value added to the material during manufacture is high. Product competitiveness is closely linked to material performance. Designers in these sectors have greater freedom in their choice of material and there is a readier acceptance of new materials with attractive property-profiles. The market-pull here is for performance, with cost as a secondary consideration. These smaller volume, higher value-added sectors drive the development of new or improved materials with enhanced performance: materials that are lighter, or stiffer, or stronger, or tougher, or expand less, or conduct better — or all of these at once.

The sectors have been ordered to form an ascending sequence, prompting the question: what does the horizontal axis measure? Many factors are involved here, one of which can be identified as "information content". The accumulated knowledge involved in the production of a contact lens or a heart valve is clearly greater than that in a beer-glass or a plastic bottle. The sectors on the left make few demands on the materials they employ; those on the right push materials to their limits, and at the same time demand the highest reliability. These features make them information-intensive. But there are also other factors: market size, competition (or lack of it), perceived value, fashion and taste, and so on. For this reason the diagram should not be over-interpreted: it is a help in structuring information, but it is not a quantitative tool.

The manufacturing industry, even in times of recession, has substantial resources; and it is in the interests of governments to support their needs. The market pull is, ultimately, the strongest force for change.

New science: curiosity-driven research

Compulsive curiosity, in children, may be a tribulation, but it is the oxygen of innovative engineering. Technically advanced countries sustain the flow of new ideas by supporting research in three kinds of organization: universities, government laboratories, and industrial research laboratories. Some of the scientists and engineers working in these institutions are encouraged to pursue ideas that may have no immediate economic objective, but that can evolve into the materials and manufacturing methods of future decades. Numerous now-commercial materials started in this way. Aluminum, in the time of Napoleon III, was a scientific wonder — he commissioned a set of aluminum spoons for which he paid more than those of solid silver. Aluminum was not, at that time, a commercial success; now it is. Titanium, more recently, has had a similar history. Amorphous (= non-crystalline) metals, now important in transformer technology and in recording-heads of tape decks, were, for years, of only academic interest. It seems improbable that superconductors or semi-conductors would have been discovered in response to market forces alone; it took long-term curiosity-driven research to carry them to the point that they became commercially attractive. Polyethylene was discovered by chemists studying the effect of pressure on chemical reactions, not by the sales or marketing departments of multi-national corporations. History is dotted with examples of materials and processes that have developed from the inquisitiveness of individuals.

What new ideas are churning in the minds of the materials scientists of today? There are many, some already on the verge of commercialization, others for which the potential is not yet clear. Some, at least, will provide opportunities for innovation; the best may create new markets.

Monolithic ceramics, now produced in commercial quantities, offer high hardness, chemical stability, wear resistance, and resistance to extreme temperatures. Their use as substrates for microcircuits is established; their use in wear-resistant applications is growing, and their use in heat engines is being explored. The emphasis in the development of *composite materials* is shifting towards those that can support loads at higher temperatures. Metal–matrix composites (example: the aluminum containing particles or fibers of silicon–carbide and intermetallic–matrix composites (titanium–aluminide or molybdenum–disilicide containing silicon–carbide, for instance) can do this. So, potentially, can ceramic-matrix composites (alumina with silicon carbide fibers) though the extreme brittleness of these materials requires new design techniques. Metallic foams, up to 90% less dense than the parent metal, promise light, stiff sandwich structures competing with composites.

New *bio-materials*, designed to be implanted in the human body, have structures onto which growing tissue will bond without rejection. New *polymers* that can be used at temperatures up to 350°C allow plastics to replace metals in even more applications — the inlet manifold of the automobile engine is an example. New *elastomers* are flexible but strong and tough; they allow better seals, elastic hinges, and resilient coatings. Techniques for producing *functionally-graded materials* can give tailored gradients of composition and structure through a component so that it could be corrosion resistant on the outer surface, tough in the middle and hard on the inner surface. *Nano-structured materials* promise unique mechanical and electrical properties. *"Intelligent" materials* that can sense and report their condition (via embedded sensors) allow safety margins to be reduced.

Developments in *rapid prototyping* now permit single, complex, parts to be made quickly without dies or moulds, from a wide range of materials. *Micron-scale fabrication methods* create miniature electro-mechanical systems (MEMS). New techniques of *surface engineering* allow the alloying, coating or heat treating of a thin surface layer of a component, modifying its properties to enhance its performance. They include laser hardening, coatings of well-adhering polymers and ceramics, ion implantation, and even the deposition of ultra-hard carbon films with a structure and properties like those of diamond. New *adhesives* displace rivets and spot-welds; the glue-bonded automobile is a real possibility. And new techniques of *mathematical modeling* and *process control* allow much tighter control of composition and structure in manufacture, reducing cost and increasing reliability and safety.

All these and many more are now realities. They have the potential to enable new designs and to stimulate redesign of product that already have a market, increasing their market share. The designer must stay alert.

18.3 Growing population and wealth, and market saturation (sector 1, Figure 18.1)

The world's population is growing in number. Much of this population is growing also in wealth. Whether or not it is wise, they consume more products, and as wealth increases they want still more. But the growth of the world's product-producing capacity has grown faster than either of these, with the result that, in developed and developing countries, product markets are saturated. If you want a product — a mobile phone, a refrigerator, a car — you do not have to stand in line with the hope of getting the only one in the store (as was the case within living memory throughout Europe). Instead, a sales person descends on you to guide you through the task of selecting, from an array of near-identical products with near-identical prices, the one you think you want.

This has certain consequences. One is the massive and continuing growth in consumption of energy and material resources. This, as a force for change in material selection, was the subject of Chapter 16. Another arises because, in a saturated market, the designer must seek now ways to attract the consumer. The more traditional reliance on engineering qualities to sell a product are replaced (or augmented) by visual qualities, associations and carefully managed perceptions created by industrial design. This, too, influences material and process choice, and it, too, was the subject of Chapter 17.

18.4 Product liability and service provision (sector 2, Figure 18.1)

Legislation now requires that, if a product has a fault, it must be recalled and the fault rectified. Product recalls are exceedingly expensive and they damage the company's image. The higher up the chart of Figure 18.3 we go, the more catastrophic a fault becomes. When you buy a packet of six ball-point pens, it is no great tragedy if one does not write properly. The probability ratio 1:6, in a Gaussian distribution, is one standard deviation or "1-sigma". But would you fly in a plane designed on a "1-sigma" basis? Safety critical systems today are designed on a "6-sigma" reliability rating, meaning that the probability that the component or assembly fails to meet specification is less than one part in 10^9.

This impacts the way materials are produced and processed. Reliability and reproducibility requires sophisticated process control and that every aspect of their production is monitored and verified. Clean steels (steels with greatly reduced inclusion-account), aluminum alloys with tight composition control, real-time feedback control of processing, new non-destructive testing for quality and integrity, and random sample testing all help ensure that quality is maintained.

This pressure on manufacturers to assume full-life responsibility for their products, causes them to consider taking maintenance and replacement out of the hands of the consumer and doing it themselves. In this way they can monitor product use, and replace it, taking back the original, not when it is worn out but when it is optimal for them to recondition and re-deploy it. The producer no longer sells the product (it remains his); he sells a service. In common with many others, my university does not own copying machines, though we have many. Instead it has a contract with a provider to guarantee a certain copying-capacity — the provision of "pages per week", you might call it.

This changes the economic boundary conditions for the designer of the copier. The objective is no longer that of building copiers with the most features at the lowest price, but rather those that will, over the long term, provide copies at the lowest cost per page. The result: a priority to design for ease of replacement, reconditioning, and standardization of materials and components between models. The best copier is no longer the one that lasts the longest; it's the one that is the cheapest for the maker to recondition, upgrade, or replace. This is not an isolated example. Aircraft engine makers now adopt a strategy of providing "power by the hour". It is the same as that of the makers of copiers: they own the engines, monitor their use, and replace them on the plane at the point at which the trade-off between cost, reliability and power-providing capacity is optimized.

18.5 Miniaturization and multi-functionality (sector 3, Figure 18.1)

Today we make increasing use of precision mechanical engineering at the micron scale. CD and DVD players require the ability to position the read-head with micron precision. Hard disk drives are yet more impressive: think of storing and retrieving a gigabyte of information on a few square millimeters. (Miniature electro-mechanical systems (MEMS)), now universal as the accelerometers that trigger automobile air-bags, rely on micron-thick cantilever beams that bend under inertial forces in a (very) sudden stop. The MEMS technology promises much more. There are now video projectors that work, not by projecting light through an LCD screen but by reflecting it from arrays of electrostatically actuated mirrors, each a few microns across, allowing much higher light intensities. There are even studies of mechanical micro-processors using bi-stable mechanical switching that could complete with semiconductor switching for information processing.

The smaller you make the components of a device, the greater the functionality you can pack into it. Think (as we already did in Section 10.4) of mobile phones, PDAs, mini-disk players and, above all, portable computers so small that they fit in a jacket pocket without causing a bulge. All these involve components with mechanical functions: protection, positioning, actuation, sensing. Micro-engineering no longer means Swiss watches. It means almost all the devices with which we interact continuously throughout a working day.

Miniaturization imposes new demands on materials. As devices become smaller, it is mechanical failures that become design-limiting. In the past the allowable size and weight of casings, connectors, keypads, motors, actuators gave plenty of margin of safety on bending stiffness, strength, wear-rates, and corrosion-rates. But none of these scales linearly with size. If we measure scale by a characterizing length L, bending stiffness scales as L^3 or L^4, strength as L^2 or L^3, and wear and corrosion, if measured by fractional loss of section per unit time, scales as $1/L$. Thus the smaller the device, the greater are the demands placed on the materials of which it is made.

That is the bad news. The better news is that, since the device is smaller, the amount of material needed to make it is less. Add to that the fact that consumers want small size with powerful functionality, and will pay more for it. The result: expensive, high-performance materials that would not merit consideration for bulky products become viable for this new, miniaturized, generation. Think of the armature of an air-bag actuator: it weighs about 1 mg of almost any material — even gold — costs, in its raw state, almost nothing.

So, it seems, material cost no longer limits choice. It is processing that imposes restrictions. Making micron-scale products with precision presents new processing challenges. The watchmakers of Switzerland perfected tools to meet their needs, but the digital watch decimated the mechanical watch business — their methods were not transferable to the new form of miniaturization, which requires mass production at low cost. The makers of microprocessors, on the other hand, had already coped with this sort of scaling-down in non-mechanical systems — their manufacturing methods, rather than those of the watchmaker, have been adapted to make micro-machines. But in adopting them the menu of materials is drastically curtailed. In their present form these processes can shape silicon, silicon oxides, nitride and carbide, thin films of copper and gold, but little else. The staples of large-scale engineering — carbon and alloy steels, aluminum alloys, polyolefins, glass — do not appear. The current challenge is to broaden the range of micro-fabrication processes to allow a wider choice of materials to be manipulated.

18.6 Concern for the environment and for the individual (sector 4, Figure 18.1)

We live in a carbon-burning economy. Energy from fossil fuels has given the developed countries high standards of living and made them wealthy. Other, much more populous, countries aspire to the same standards of life and wealth and are well advanced in acquiring them. Yet the time is fast approaching when burning hydrocarbons will no longer meet energy needs, and even if it could, the burden it places on the natural environment will force limits to be set. Short-term measures to address this and the implications these have for material selection were the subject of Chapter 16. The changes necessary to allow longer term sustainable development will have to be much further-reaching, changing the ways in which we manufacture, transport ourselves and our goods, and live, with major impact on the way materials, central to all of these, are used.

It is one thing to point out that this force for change is a potent one, but quite another to predict what its consequences will be. It is possible to point to fuel-cell technology, energy from natural sources and safe nuclear energy (all with material challenges) as a way forward, but — in the sense we are discussing here — even these are short term solutions. Remember the figure cited in the introduction to Chapter 16: at a global growth rate of 3% per year we will mine, process, and dispose of more "stuff" in the next 25 years than in the entire history of mankind. The question that has no present answer is: how can a world population growing at about 3% per year, living in countries with economic growth rates of between 2% and 20% per year, continue to meet their aspirations with zero — or even negative — growth in consumption of energy and materials?

Wealth has another dimension: it enables resources to be allocated to health care, for which, with a population with increasing lifespan, there is growing demand. A surprisingly large fraction of the organs in a human body perform predominantly mechanical functions: teeth for cutting and grinding, bone to support structural loads, joints to allow articulation, the heart to pump blood and artery to carry it under pressure to the extremities of the body, muscle to actuate, skin to provide flexible protection. Ageing or accidental damage frequently causes one or more of these to malfunction. Being mechanical, it is possible, in principle, to replace them. One way to do so is to use real body parts, but the shear number of patients requiring such treatment and the ethical and other difficulties of using human replacements drives efforts to develop of artificial substitutes. Man-made teeth and bone implants, hip, and knee joint replacements, artificial artery and skin, even hearts exist already and are widely used. But, at present, they are exceedingly expensive, limiting their availability, and, for the most part, they are only crude substitutes for the real thing. There is major incentive for change here, stimulating research into affordable materials for artificial organs of all types.

Studies show that, in a world that is ageing, the usability of many products and services is unnecessarily challenging to people, whether they are young or old, able-bodied, or less able. Until recently the goal of many designers was to appeal to the youth market (meaning the 15–35 age group) as the mainstay of their business, thereby developing products that, often, could not be used by others. Astute companies have perceived the problem and have sought to redesign their products and services to make them accessible to a wider clientele. The US company OXO has seen sales of its Good Grips range of kitchen and garden tools grow by 50% year on year, and the UK supermarket Tesco has added thousands of internet customers by getting rid of clever but confusing graphics on its website and instead making it quick to down load and simple and intuitive to navigate, thereby including a wider spectrum of people in on-line shopping. There is a powerful business case for *inclusive design* — design to in ensure that the products and services address the needs of the widest possible audience. The social case is equally compelling — at present many people are denied use of electronic and other products because they are unable to understand how to use them or lack the motor skills to do so.

All products exclude some users, sometimes deliberately — the childproof medication bottle for instance — but more usually the exclusion is unintended and unnecessary. Many services rely on the use of products for their delivery, so unusable products deny people access to these too. Video recorders are the standing joke of poor usability, regularly derided in reviews. Installing a VCR is now such a complex task that many retailers offer an installation service. Using them is little better — almost all users have at some point recorded the wrong channel, set the wrong time, recorded the picture but not the sound, or simply given up in despair. A much larger number of contemporary products exclude those with disability—limited sight, hearing, physical mobility or strength, or mental acuity. Countering design exclusion is a growing priority of government and is increasingly seen as important by product makers. This thinking influences material choice, changing the constraints and objectives for their selection. The use of materials that, through color or feel, communicate the function of a control, materials to give good grip, materials to insulate or protect becomes a priority.

18.7 Summary and conclusions

Powerful forces drive the development of new and improved materials, encourage substitution, and modify the way in which materials are produced and used. Market forces and military prerogatives, historically the most influential, remain the strongest. The ingenuity of research scientists, too, drives change by revealing a remarkable spectrum of new materials with exciting possibilities, though the time it takes to develop and commercialize them is long: typically 15 years from laboratory to market place.

Today additional new drivers influence material development and use. Increasing wealth creates markets for ever more sophisticated products. The trend to products that are smaller in size, lighter in weight and have greater functionality makes increasing demands on the mechanical properties of the materials used to make them. Greater insistence on product reliability and safety, with the manufacture held liable for failures or malfunctions, requires materials with consistently reproducible properties and processes that are tightly controlled. Concern for the impact of industrial growth on the natural environment introduces the new objective of selecting materials in such a way as to minimize it. And the perception that many products exclude users by their complexity and difficulty of use drives a reassessment of the way in which they are designed and the choice of materials to make them.

The result is that products that were seen as optimal yesterday are no longer optimal today. There is always scope for reassessing designs and the choice of materials to implement them.

18.8 Further reading

Defence and aerospace materials and structures, National advisory committee (NAC) Annual report 2000. http://www.iom3.org/foresight/nac/html

Delivering the Evidence: Defra's Science and Innovation Strategy (2003–06), Defra. May 2003. http://www.defra.gov.uk/science/S_IS

Energy White Paper: Our energy future—creating a low carbon economy, DTI (Cm 5761). TSO London, February 2003. http://www.dti.gov.uk/energy/whitepaper/index.shtml

Keates, S. and Clarkson, J. (2004) *Countering Design Exclusion — an Introduction to Inclusive Design*, Springer Verlag, London, UK. ISBN 1–85233–769–9. *(An in-depth study of design exclusion and ways to overcome it.)*

van Griethuysen, A.J. (ed.) (1987) *New Applications of Materials*, Scientific, and Technical Publications Ltd, The Hague, The Netherlands. *(A Dutch study of material evolution.)*

Williams, J.C. (1998) Engineering Competitive Materials, *The Bridge* 28(4). *(The Bridge is the journal of the US National Academy of Engineering. Dr. Williams has long experience in the aerospace industry and in academia.)*

Appendix A

Useful solutions to standard problems

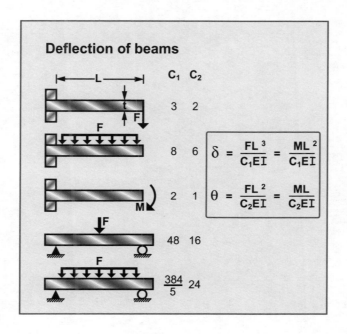

Deflection of beams

$$\delta = \frac{FL^3}{C_1EI} = \frac{ML^2}{C_1EI}$$

$$\theta = \frac{FL^2}{C_2EI} = \frac{ML}{C_2EI}$$

Appendix contents

Introduction and synopsis

Modeling is a key part of design. In the early stage, approximate modeling establishes whether the concept will work at all, and identifies the combination of material properties that maximize performance. At the embodiment stage, more accurate modeling brackets values for the forces, the displacements, the velocities, the heat fluxes, and the dimensions of the components. And in the final stage, modeling gives precise values for stresses, strains, and failure probability in key components; power, speed, efficiency, and so forth.

Many components with simple geometries and loads have been modeled already. Many more-complex components can be modeled approximately by idealizing them as one of these. There is no need to reinvent the beam or the column or the pressure vessel; their behavior under all common types of loading has already been analyzed. The important thing is to know that the results exist and where to find them.

This appendix summarizes the results of modeling a number of standard problems. Their usefulness cannot be overstated. Many problems of conceptual design can be treated, with adequate precision, by patching together the solutions given here; and even the detailed analysis of non-critical components can often be tackled in the same way. Even when this approximate approach is not sufficiently accurate, the insight it gives is valuable.

The appendix contains 15 double page sections that list, with a short commentary, results for constitutive equations; for the loading of beams, columns, and torsion bars; for contact stresses, cracks, and other stress concentrations; for pressure vessels, vibrating beams and plates; and for the flow of heat and matter. They are drawn from numerous sources, listed under "Further reading" in Section A.16.

A.1   Constitutive equations for mechanical response

The behavior of a component when it is loaded depends on the *mechanism* by which it deforms. A beam loaded in bending may deflect elastically; it may yield plastically; it may deform by creep; and it may fracture in a brittle or in a ductile way. The equation that describes the material response is known as a *constitutive equation*. Each mechanism is characterized by a different constitutive equation. The constitutive equation contains one or more than one *material property*: Young's modulus, E, and Poisson's ratio, ν, are the material properties that enter the constitutive equation for linear-elastic deformation; the yield strength, σ_y, is the material property that enters the constitutive equation for plastic flow; creep constants, $\dot{\varepsilon}_0$, σ_0 and n enter the equation for creep; the fracture toughness, K_{1C}, enters that for brittle fracture.

The common constitutive equations for mechanical deformation are listed on the facing page. In each case the equation for uniaxial loading by a tensile stress σ is given first; below it is the equation for multiaxial loading by principal stresses σ_1, σ_2, and σ_3, always chosen so that σ_1 is the most tensile and σ_3 the most compressive (or least tensile) stress. They are the basic equations that determine mechanical response.

A.1   Mechanical constitutive equations

Elastic

Uniaxial

$$\epsilon = \frac{\sigma}{E}$$

General

$$\epsilon_1 = \frac{\sigma_1}{E} - \nu \, (\sigma_2 + \sigma_3) \qquad \text{etc}$$

Plastic

Uniaxial

$$\sigma \geq \sigma_y$$

General

$$\sigma_1 - \sigma_3 \geq \sigma_y \qquad (\sigma_1 > \sigma_2 > \sigma_3) \qquad \text{Tresca}$$

$$\sigma \geq \sigma_y \qquad\qquad\qquad\qquad \text{Von Mises}$$

with $\overline{\sigma} = \left[\frac{1}{2} \left\{ (\sigma_1 - \sigma_2)^2 + (\sigma_2 - \sigma_3)^2 + (\sigma_3 - \sigma_1)^2 \right\} \right]^{\frac{1}{2}}$

Creep

Uniaxial

$$\dot{\epsilon} = \dot{\epsilon}_o \left(\frac{\sigma}{\sigma_o} \right)^n$$

General

$$\dot{\epsilon}_1 = \dot{\epsilon}_o \, \frac{\sigma_o^{\,n-1}}{\sigma_o^{\,n}} \left\{ \sigma_1 - \frac{1}{2} \, (\sigma_2 + \sigma_3) \right\} \text{ etc}$$

Fracture

Uniaxial

$$\sigma \geq CK_{IC} / (\pi a)^{\frac{1}{2}}$$

General

$$\sigma_1 \geq CK_{IC} / (\pi a)^{\frac{1}{2}} \qquad (\sigma_1 > \sigma_2 > \sigma_3)$$

Material properties

E, ν	$=$	Elastic constants
σ_y	$=$	Plastic yield strength
n, ϵ_o, σ_o	$=$	Creep constants
K_{IC}, a	$=$	Fracture toughness, crack length
C	$\approx$	1 Tension
C	$\approx$	15 Compression

A.2 Moments of sections

A beam of uniform section, loaded in simple tension by a force F, carries a stress

$$\sigma = F/A$$

where A is the area of the section. Its response is calculated from the appropriate constitutive equation. Here the important characteristic of the section is its area, A. For other modes of loading, higher moments of the area are involved. Those for various common sections are given on the facing page. They are defined as follows.

The second moment I measures the resistance of the section to bending about a horizontal axis (shown as a broken line). It is

$$I = \int_{\text{section}} y^2 b(y) \; \mathrm{d}y$$

where y is measured vertically and $b(y)$ is the width of the section at y. The moment K measures the resistance of the section to twisting. It is equal to the polar moment J for circular sections, where

$$J = \int_{\text{section}} 2\pi r^3 \; \mathrm{d}r$$

where r is measured radially from the center of the circular section. For non-circular sections K is less than J.

The section modulus $Z = I/y_{\text{m}}$ (where y_{m} is the normal distance from the neutral axis of bending to the outer surface of the beam) measures the surface stress generated by a given bending moment, M:

$$\sigma = \frac{My_{\text{m}}}{I} = \frac{M}{Z}$$

Finally, the moment H, defined by

$$H = \int_{\text{section}} yb(y) \; \mathrm{d}y$$

measures the resistance of the beam to fully-plastic bending. The fully plastic moment for a beam in bending is

$$M_{\text{p}} = H\sigma_{\text{y}}$$

A.2 Moments of sections

Section shape	Area A m	Moment I m^4	Moment K m^4	Moment Z m^4	Moment Q m^4	Moment Z_p m^4
	bh	$\dfrac{bh^3}{12}$	$\dfrac{bh^3}{3}(1-0.58\dfrac{b}{h})$ $(h > b)$	$\dfrac{bh^2}{6}$	$\dfrac{b^2h^2}{(3h+1.8b)}$ $(h>b)$	$\dfrac{bh^2}{4}$
	$\dfrac{\sqrt{3}}{4}a^2$	$\dfrac{a^4}{32\sqrt{3}}$	$\dfrac{\sqrt{3}\,a^4}{80}$	$\dfrac{a^3}{32}$	$\dfrac{a^3}{20}$	$\dfrac{3a^3}{64}$
	πr^2	$\dfrac{\pi}{4}r^4$	$\dfrac{\pi}{2}r^4$	$\dfrac{\pi}{4}r^3$	$\dfrac{\pi}{2}r^3$	$\dfrac{\pi}{3}r^3$
	πab	$\dfrac{\pi}{4}a^3b$	$\dfrac{\pi a^3 b^3}{(a^2+b^2)}$	$\dfrac{\pi}{4}a^2b$	$\dfrac{\pi}{2}a^2b$ $(a<b)$	$\dfrac{\pi}{3}a^2b$
	$\pi(r_o^2 - r_i^2)$ $\approx 2\pi rt$	$\dfrac{\pi}{4}(r_o^4 - r_i^4)$ $\approx \pi r^3 t$	$\dfrac{\pi}{2}(r_o^4 - r_i^4)$ $\approx 2\pi r^3 t$	$\dfrac{\pi}{4r_o}(r_o^4 - r_i^4)$ $\approx \pi r^2 t$	$\dfrac{\pi}{2r_o}(r_o^4 - r_i^4)$ $\approx 2pr^2 t$	$\dfrac{\pi}{3}(r_o^3 - r_i^3)$ $\approx \pi r^2 t$
	$2t(h+b)$ $(h,b \gg t)$	$\dfrac{1}{6}h^3t(1+3\dfrac{b}{h})$	$\dfrac{2tb^2h^2}{(h+b)}(1-\dfrac{t}{h})^4$	$\dfrac{1}{3}h^2t(1+3\dfrac{b}{h})$	$2tbh(1-\dfrac{t}{h})^2$	$bht(1+\dfrac{h}{2b})$
	$\pi(a+b)t$ $(a,b \gg t)$	$\dfrac{\pi}{4}a^3t(1+\dfrac{3b}{a})$	$\dfrac{4\pi(ab)^{5/2}t}{(a^2+b^2)}$	$\dfrac{\pi}{4}a^2t(1+\dfrac{3b}{a})$	$2\pi t(a^3b)^{1/2}$ $(b>a)$	$\pi abt(2+\dfrac{a}{b})$
	$b(h_o - h_i)$ $\approx 2bt$ $(h,b \gg t)$	$\dfrac{b}{12}(h_o^3 - h_i^3)$ $\approx \dfrac{1}{2}bth_o^2$	--	$\dfrac{b}{6h_o}(h_o^3 - h_i^3)$ $\approx bth_o$	--	$\dfrac{b}{4}(h_o^2 - h_i^2)$ $\approx bth_o$
	$2t(h+b)$ $(h,b \gg t)$	$\dfrac{1}{6}h^3t(1+3\dfrac{b}{h})$	$\dfrac{2}{3}bt^3(1+4\dfrac{h}{b})$	$\dfrac{1}{3}h^2t(1+3\dfrac{b}{h})$	$\dfrac{2}{3}bt^2(1+4\dfrac{h}{b})$	$bht(1+\dfrac{h}{2b})$
	$2t(h+b)$ $(h,b \gg t)$	$\dfrac{t}{6}(h^3 + 4bt^2)$	$\dfrac{t^3}{3}(8b + h)$	$\dfrac{t}{3h}(h^3 + 4bt^2)$	$\dfrac{t^2}{3}(8b + h)$	$\dfrac{th^2}{2}\{1+\dfrac{2t(b-2t)}{h^2}\}$

A.3 Elastic bending of beams

When a beam is loaded by a force F or moments M, the initially straight axis is deformed into a curve. If the beam is uniform in section and properties, long in relation to its depth and nowhere stressed beyond the elastic limit, the deflection δ, and the angle of rotation, θ, can be calculated using elastic beam theory (see Further reading in Section A.16). The basic differential equation describing the curvature of the beam at a point x along its length is

$$EI\frac{d^2y}{dx^2} = M$$

where y is the lateral deflection, and M is the bending moment at the point x on the beam. E is Young's modulus and I is the second moment of area (Section A.2). When M is constant this becomes

$$\frac{M}{I} = E\left(\frac{1}{R} - \frac{1}{R_0}\right)$$

where R_0 is the radius of curvature before applying the moment and R the radius after it is applied. Deflections δ and rotations θ are found by integrating these equations along the beam. The stiffness of the beam is defined by

$$S = \frac{F}{\delta} = \frac{C_1 E_1}{L^3}$$

It depends on Young's modulus, E, for the material of the beam, on its length, L, and on the second moment of its section, I. The end-slope of the beam, θ, is given by

$$\theta = \frac{FL^2}{C_2 EI}$$

Equations for the deflection, δ, and end slope, θ, of beams, for various common modes of loading are shown on the facing page together with values of C_1 and C_2.

A.3 Deflection of beams

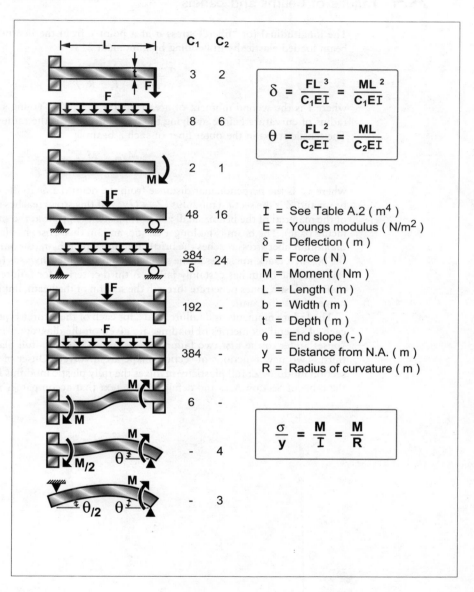

	C_1	C_2
	3	2
	8	6
	2	1
	48	16
	$\dfrac{384}{5}$	24
	192	-
	384	-
	6	-
	-	4
	-	3

$$\delta = \frac{FL^3}{C_1EI} = \frac{ML^2}{C_1EI}$$

$$\theta = \frac{FL^2}{C_2EI} = \frac{ML}{C_2EI}$$

I = See Table A.2 (m^4)
E = Youngs modulus (N/m^2)
δ = Deflection (m)
F = Force (N)
M = Moment (Nm)
L = Length (m)
b = Width (m)
t = Depth (m)
θ = End slope (-)
y = Distance from N.A. (m)
R = Radius of curvature (m)

$$\frac{\sigma}{y} = \frac{M}{I} = \frac{M}{R}$$

A.4   Failure of beams and panels

The longitudinal (or "fiber") stress σ at a point y from the neutral axis of a uniform beam loaded elastically in bending by a moment M is

$$\frac{\sigma}{y} = \frac{M}{I} \, E\left(\frac{1}{R} - \frac{1}{R_0}\right)$$

where I is the second moment of area (Section A.2), E is Young's modulus, R_0 is the radius of curvature before applying the moment and R is the radius after it is applied. The tensile stress in the outer fiber of such a beam is

$$\sigma = \frac{My_m}{I} = \frac{M}{Z}$$

where y_m is the perpendicular distance from the neutral axis to the outer surface of the beam and Z is the section modulus ($Z = I/y_m$). If this stress reaches the yield strength σ_y of the material of the beam, small zones of plasticity appear at the surface (top diagram, facing page). The beam is no longer elastic, and, in this sense, has failed. If, instead, the maximum fiber stress reaches the brittle fracture strength, σ_f (the MR) of the material of the beam, a crack nucleates at the surface and propagates inwards (second diagram); in this case, the beam has certainly failed. A third criterion for failure is often important: that the plastic zones penetrate through the section of the beam, linking to form a plastic hinge (third diagram).

The failure moments and failure loads, for each of these three types of failure, and for each of several geometries of loading, are given on the diagram. The formulae labelled "onset" refer to the first two failure modes; those labelled "full plasticity" refer to the third. Two new functions of section shape are involved. Onset of failure involves the section modulus Z; full plasticity involves the fully plastic modulus H. Both are listed in the table of Section A.2, and defined in the text that accompanies it.

A.4 Failure of beams

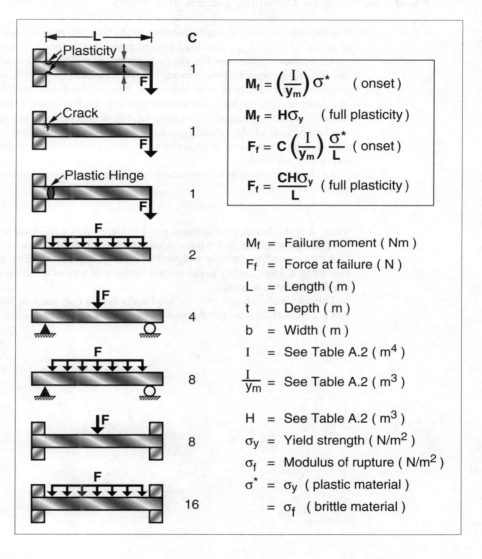

$$M_f = \left(\frac{I}{y_m}\right)\sigma^* \quad \text{(onset)}$$

$$M_f = H\sigma_y \quad \text{(full plasticity)}$$

$$F_f = C\left(\frac{I}{y_m}\right)\frac{\sigma^*}{L} \quad \text{(onset)}$$

$$F_f = \frac{CH\sigma_y}{L} \quad \text{(full plasticity)}$$

M_f = Failure moment (Nm)

F_f = Force at failure (N)

L = Length (m)

t = Depth (m)

b = Width (m)

I = See Table A.2 (m^4)

$\dfrac{I}{y_m}$ = See Table A.2 (m^3)

H = See Table A.2 (m^3)

σ_y = Yield strength (N/m^2)

σ_f = Modulus of rupture (N/m^2)

σ^* = σ_y (plastic material)

  = σ_f (brittle material)

A.5 Buckling of columns, plates, and shells

If sufficiently slender, an elastic column, loaded in compression, fails by elastic buckling at a critical load, F_{crit}. This load is determined by the end constraints, of which five extreme cases are illustrated on the facing page: an end may be constrained in a position and direction; it may be free to rotate but not translate (or "sway"); it may sway without rotation; and it may both sway and rotate. Pairs of these constraints applied to the ends of column lead to the five cases shown opposite. Each is characterized by a value of the constant n that is equal to the number of half-wavelengths of the buckled shape.

The addition of the bending moment M reduces the buckling load by the amount shown in the second box. A negative value of F_{crit} means that a tensile force is necessary to prevent buckling.

An elastic foundation is one that exerts a lateral restoring pressure, p, proportional to the deflection y

$$p = ky$$

where k is the foundation stiffness per unit depth and y the local lateral deflection. Its effect is to increase F_{crit}, by the amount shown in the third box.

A thin-walled elastic tube will buckle inwards under an external pressure p', given in the last box. Here I refers to the second moment of area of a section of the tube wall cut parallel to the tube axis.

Thin or slender shapes may buckle locally before they yield or fracture. It is this that sets a practical limit to the thinness of tube walls and webs.

A.5 Buckling of columns, plates and shells

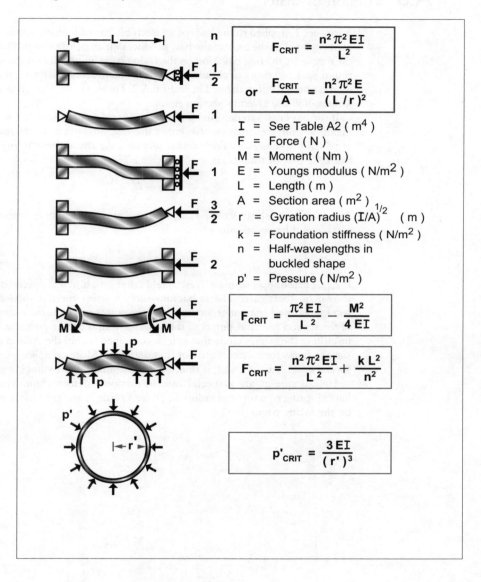

$$F_{CRIT} = \frac{n^2 \pi^2 EI}{L^2}$$

or $\quad \dfrac{F_{CRIT}}{A} = \dfrac{n^2 \pi^2 E}{(L/r)^2}$

I = See Table A2 (m^4)
F = Force (N)
M = Moment (Nm)
E = Youngs modulus (N/m^2)
L = Length (m)
A = Section area (m^2)
r = Gyration radius $(I/A)^{1/2}$ (m)
k = Foundation stiffness (N/m^2)
n = Half-wavelengths in buckled shape
p' = Pressure (N/m^2)

$$F_{CRIT} = \frac{\pi^2 EI}{L^2} - \frac{M^2}{4 EI}$$

$$F_{CRIT} = \frac{n^2 \pi^2 EI}{L^2} + \frac{k L^2}{n^2}$$

$$p'_{CRIT} = \frac{3 EI}{(r')^3}$$

A.6 Torsion of shafts

A torque, T, applied to the ends of an isotropic bar of uniform section, and acting in the plane normal to the axis of the bar, produces an angle of twist θ. The twist is related to the torque by the first equation on the facing page, in which G is the shear modulus. For round bars and tubes of circular section, the factor K is equal to J, the polar moment of inertia of the section, defined in Section A.2. For any other section shape K is less than J. Values of K are given in Section A.2.

If the bar ceases to deform elastically, it is said to have failed. This will happen if the maximum surface stress exceeds either the yield strength σ_y of the material or the stress at which it fractures σ_f. For circular sections, the shear stress at any point a distance r from the axis of rotation is

$$\tau = \frac{Tr}{K} = \frac{G\theta r}{K}$$

The maximum shear stress, τ_{max}, and the maximum tensile stress, σ_{max}, are at the surface and have the values

$$\tau_{max} = \sigma_{max} = \frac{Td_0}{2K} = \frac{G\theta d_0}{2L}$$

If τ_{max} exceeds $\sigma_y/2$ (using a Tresca yield criterion), or if σ_{max} exceeds σ_f, the bar fails, as shown on the figure. The maximum surface stress for the solid ellipsoidal, square, rectangular and triangular sections is at the points on the surface closest to the centroid of the section (the mid-points of the longer sides). It can be estimated approximately by inscribing the largest circle that can be contained within the section and calculating the surface stress for a circular bar of that diameter. More complex section-shapes require special consideration, and, if thin, may additionally fail by buckling.

Helical springs are a special case of torsional deformation. The extension u of a helical spring of n turns of radius R, under a force F, and the failure force F_{crit}, are given on the facing page.

A.6   Torsion of shafts

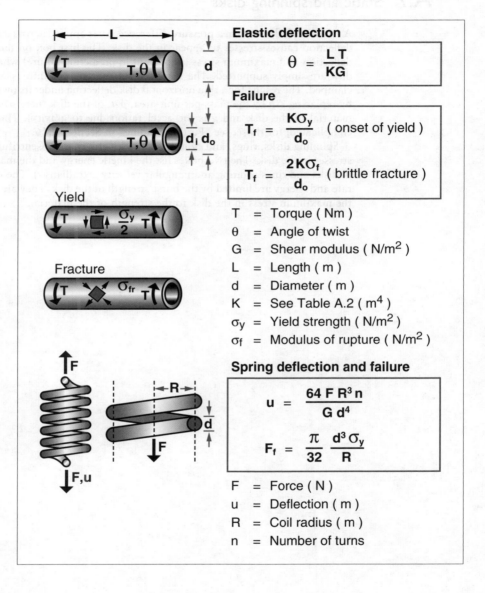

Elastic deflection

$$\theta = \frac{L\,T}{K\,G}$$

Failure

$$T_f = \frac{K\sigma_y}{d_o} \quad (\text{onset of yield})$$

$$T_f = \frac{2K\sigma_f}{d_o} \quad (\text{brittle fracture})$$

T = Torque (Nm)
θ = Angle of twist
G = Shear modulus (N/m^2)
L = Length (m)
d = Diameter (m)
K = See Table A.2 (m^4)
σ_y = Yield strength (N/m^2)
σ_f = Modulus of rupture (N/m^2)

Spring deflection and failure

$$u = \frac{64\,F\,R^3\,n}{G\,d^4}$$

$$F_f = \frac{\pi}{32}\,\frac{d^3\,\sigma_y}{R}$$

F = Force (N)
u = Deflection (m)
R = Coil radius (m)
n = Number of turns

A.7 Static and spinning disks

A thin disk deflects when a pressure difference Δp is applied across its two surfaces. The deflection causes stresses to appear in the disk. The first box on the facing page gives deflection and maximum stress (important in predicting failure) when the edges of the disk are simply supported. The second gives the same quantities when the edges are clamped. The results for a thin horizontal disk deflecting under its own weight are found by replacing Δp by the mass-per-unit-area, $\rho g t$, of the disk (here ρ is the density of the material of the disk and g is the acceleration due to gravity). Thick disks are more complicated; for those, see "Further reading" in Section A.16.

Spinning disks, rings, and cylinders store kinetic energy. Centrifugal forces generate stresses in the disk. The two boxes list the kinetic energy and the maximum stress σ_{max} in disks and rings rotating at an angular velocity ω (radians/s). The maximum rotation rate and energy are limited by the burst-strength of the disk. They are found by equating the maximum stress in the disk to the strength of the material.

A.7 Static and spinning disks

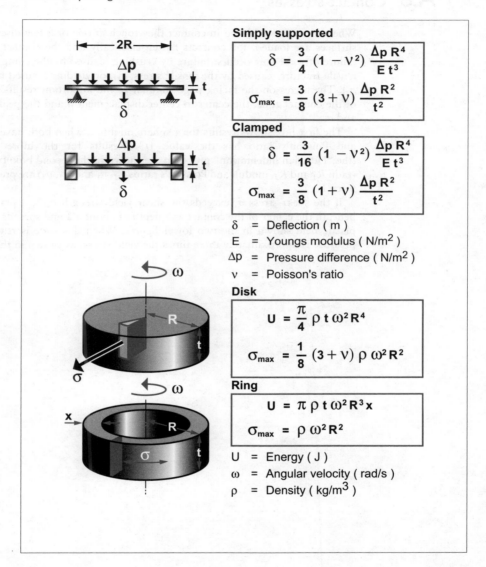

Simply supported

$$\delta = \frac{3}{4}(1 - \nu^2)\frac{\Delta p\, R^4}{E\, t^3}$$

$$\sigma_{max} = \frac{3}{8}(3 + \nu)\frac{\Delta p\, R^2}{t^2}$$

Clamped

$$\delta = \frac{3}{16}(1 - \nu^2)\frac{\Delta p\, R^4}{E\, t^3}$$

$$\sigma_{max} = \frac{3}{8}(1 + \nu)\frac{\Delta p\, R^2}{t^2}$$

δ = Deflection (m)
E = Youngs modulus (N/m^2)
Δp = Pressure difference (N/m^2)
ν = Poisson's ratio

Disk

$$U = \frac{\pi}{4}\rho\, t\, \omega^2 R^4$$

$$\sigma_{max} = \frac{1}{8}(3 + \nu)\rho\, \omega^2 R^2$$

Ring

$$U = \pi \rho\, t\, \omega^2 R^3 x$$

$$\sigma_{max} = \rho\, \omega^2 R^2$$

U = Energy (J)
ω = Angular velocity (rad/s)
ρ = Density (kg/m^3)

A.8 Contact stresses

When surfaces are placed in contact they touch at one or a few discrete points. If the surfaces are loaded, the contacts flatten elastically and the contact areas grow until failure of some sort occurs: failure by crushing (caused by the compressive stress, σ_c), tensile fracture (caused by the tensile stress, σ_t), or yielding (caused by the shear stress σ_s). The boxes on the facing page summarize the important results for the radius, a, of the contact zone, the center-to-center displacement u and the peak values of σ_c, σ_s, and σ_t.

The first box shows results for a sphere on a flat, when both have the same moduli and Poisson's ratio has the value 1/3. Results for the more general problem (the "Hertzian indentation" problem) are shown in the second box: two elastic spheres (radii R_1 and R_2, moduli and Poisson's ratios E_1, ν_1 and E_2, ν_2) are pressed together by a force F.

If the shear stress σ_s exceeds the shear yield strength $\sigma_y/2$, a plastic zone appears beneath the center of the contact at a depth of about $a/2$ and spreads to form the fully-plastic field shown in the two lower figures. When this state is reached, the contact pressure is approximately three times the yield stress, as shown in the bottom box.

A.8   Contact stresses

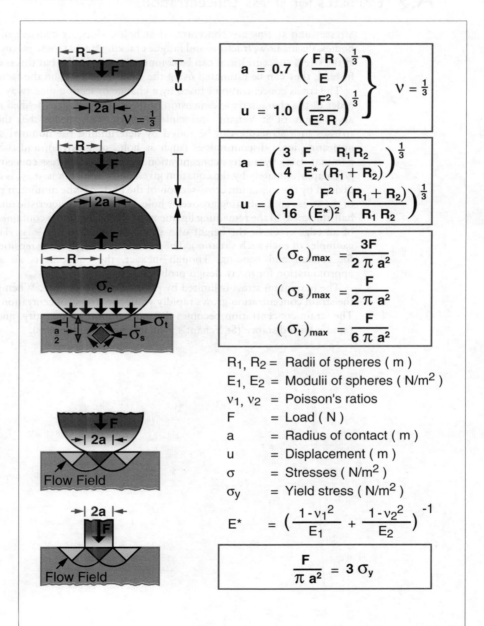

$$a = 0.7 \left(\frac{F R}{E} \right)^{\frac{1}{3}}$$

$$u = 1.0 \left(\frac{F^2}{E^2 R} \right)^{\frac{1}{3}}$$

$\nu = \frac{1}{3}$

$$a = \left(\frac{3}{4} \frac{F}{E^\star} \frac{R_1 R_2}{(R_1 + R_2)} \right)^{\frac{1}{3}}$$

$$u = \left(\frac{9}{16} \frac{F^2}{(E^\star)^2} \frac{(R_1 + R_2)}{R_1 R_2} \right)^{\frac{1}{3}}$$

$$(\sigma_c)_{max} = \frac{3F}{2 \pi a^2}$$

$$(\sigma_s)_{max} = \frac{F}{2 \pi a^2}$$

$$(\sigma_t)_{max} = \frac{F}{6 \pi a^2}$$

$R_1, R_2 =$ Radii of spheres (m)
$E_1, E_2 =$ Modulii of spheres (N/m^2)
$\nu_1, \nu_2 =$ Poisson's ratios
F $\quad=$ Load (N)
a $\quad=$ Radius of contact (m)
u $\quad=$ Displacement (m)
$\sigma \quad=$ Stresses (N/m^2)
$\sigma_y \quad=$ Yield stress (N/m^2)

$$E^\star = \left(\frac{1 - \nu_1^2}{E_1} + \frac{1 - \nu_2^2}{E_2} \right)^{-1}$$

$$\frac{F}{\pi a^2} = 3 \sigma_y$$

A.9 Estimates for stress concentrations

Stresses and strains are concentrated at holes, slots, or changes of section in elastic bodies. Plastic flow, fracture, and fatigue cracking start at these places. The local stresses at the stress concentrations can be computed numerically, but this is often unnecessary. Instead, they can be estimated using the equation shown on the facing page.

The stress concentration caused by a change in section dies away at distances of the order of the characteristic dimension of the section-change (defined more fully below), an example of St Venant's principle at work. This means that the maximum local stresses in a structure can be found by determining the nominal stress distribution, neglecting local discontinuities (such as holes or grooves), and then multiplying the nominal stress by a stress concentration factor. Elastic stress concentration factors are given approximately by the equation given in the table. In it, σ_{nom} is defined as the load divided by the minimum cross-section of the part, ρ is the minimum radius of curvature of the stress-concentrating groove or hole, and c is a characteristic dimension: either the half-thickness of the remaining ligament, the half-length of a contained crack, the length of an edge-crack or the height of a shoulder, whichever is <u>least</u>. The drawings show examples of each such situation. The factor α is roughly 2 for tension, but is nearer 1/2 for torsion and bending. Though inexact, the equation is an adequate working approximation for many design problems.

The maximum stress is limited by plastic flow or fracture. When plastic flow starts, the strain concentration grows rapidly while the stress concentration remains constant. The strain concentration becomes the more important quantity, and may not die out rapidly with distance (St Venant's principle no longer applies).

A.9   Stress concentrations

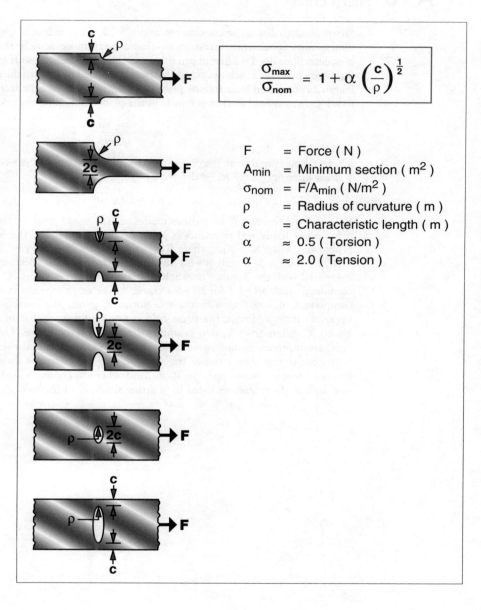

$$\frac{\sigma_{max}}{\sigma_{nom}} = 1 + \alpha \left(\frac{c}{\rho}\right)^{\frac{1}{2}}$$

F	=	Force (N)
A_{min}	=	Minimum section (m^2)
σ_{nom}	=	F/A$_{min}$ (N/m^2)
ρ	=	Radius of curvature (m)
c	=	Characteristic length (m)
α	$\approx$	0.5 (Torsion)
α	$\approx$	2.0 (Tension)

A.10 Sharp cracks

Sharp cracks (that is, stress concentrations with a tip radius of curvature of atomic dimensions) concentrate stress in an elastic body more acutely than rounded stress concentrations do. To a first approximation, the local stress falls off as $1/r^{1/2}$ with radial distance r from the crack tip. A tensile stress σ, applied normal to the plane of a crack of length $2a$ contained in an infinite plate (as in the top figure on the facing page) gives rise to a local stress field σ_e that is tensile in the plane containing the crack and given by

$$\sigma_e = \frac{C\sigma\sqrt{\pi a}}{\sqrt{2\pi r}}$$

where r is measured from the crack tip in the plane $\theta = 0$ and C is a constant. The mode 1 *stress intensity factor* K_1, is defined as

$$K_1 = C\sigma\sqrt{\pi a}$$

Values of the constant C for various modes of loading are given on the figure. The stress σ for point loads and moments is given by the equations at the bottom. The crack propagates when $K_1 > K_{1C}$, where K_{1C} is the *fracture toughness*.

When the crack length is very small compared with all specimen dimensions and compared with the distance over which the applied stress varies, C is equal to 1 for a contained crack and 1.1 for an edge crack. As the crack extends in a uniformly loaded component, it interacts with the free surfaces, giving the correction factors shown opposite. If, in addition, the stress field is non-uniform (as it is in an elastically bent beam), C differs from 1; two examples are given on the figure. The factors, C, given here, are approximate only, good when the crack is short but not when the crack tips are very close to the boundaries of the sample. They are adequate for most design calculations. More accurate approximations, and other less common loading geometries can be found in the references listed in "Further reading" in Section A.16.

A.10 Sharp cracks

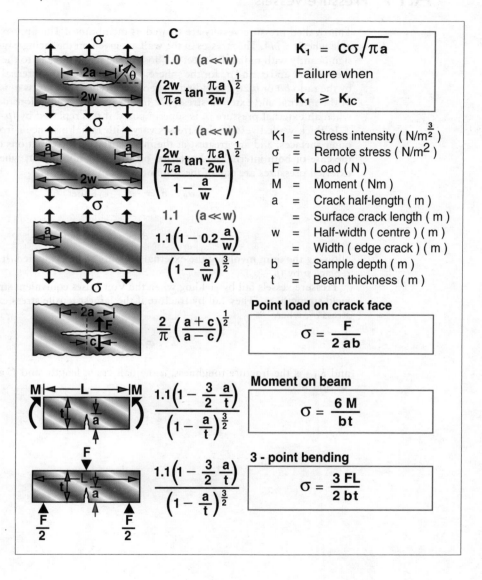

$$K_1 = C\sigma\sqrt{\pi a}$$

Failure when

$$K_1 \geqslant K_{IC}$$

K_1	=	Stress intensity ($N/m^{\frac{3}{2}}$)
σ	=	Remote stress (N/m^2)
F	=	Load (N)
M	=	Moment (Nm)
a	=	Crack half-length (m)
	=	Surface crack length (m)
w	=	Half-width (centre) (m)
	=	Width (edge crack) (m)
b	=	Sample depth (m)
t	=	Beam thickness (m)

Point load on crack face

$$\sigma = \frac{F}{2\,ab}$$

Moment on beam

$$\sigma = \frac{6\,M}{bt}$$

3 - point bending

$$\sigma = \frac{3\,FL}{2\,bt}$$

C

$$1.0 \quad (a \ll w)$$

$$\left(\frac{2w}{\pi a} \tan\frac{\pi a}{2w}\right)^{\frac{1}{2}}$$

$$1.1 \quad (a \ll w)$$

$$\left(\frac{\frac{2w}{\pi a}\tan\frac{\pi a}{2w}}{1-\frac{a}{w}}\right)^{\frac{1}{2}}$$

$$1.1 \quad (a \ll w)$$

$$\frac{1.1\left(1-0.2\frac{a}{w}\right)}{\left(1-\frac{a}{w}\right)^{\frac{3}{2}}}$$

$$\frac{2}{\pi}\left(\frac{a+c}{a-c}\right)^{\frac{1}{2}}$$

$$\frac{1.1\left(1-\frac{3}{2}\frac{a}{t}\right)}{\left(1-\frac{a}{t}\right)^{\frac{3}{2}}}$$

$$\frac{1.1\left(1-\frac{3}{2}\frac{a}{t}\right)}{\left(1-\frac{a}{t}\right)^{\frac{3}{2}}}$$

A.11 Pressure vessels

Thin-walled pressure vessels are treated as membranes. The approximation is reasonable when $t < b/4$. The stresses in the wall are given on the facing page; they do not vary significantly with radial distance, r. Those in the plane tangent to the skin, σ_θ and σ_z for the cylinder and σ_θ and σ_ϕ for the sphere, are just equal to the internal pressure amplified by the ratio b/t or $b/2t$, depending on geometry. The radial stress σ_r is equal to the mean of the internal and external stress, $p/2$ in this case. The equations describe the stresses when an external pressure p_e is superimposed if p is replaced by $(p - p_e)$.

In thick-walled vessels, the stresses vary with radial distance r from the inner to the outer surfaces, and are greatest at the inner surface. The equations can be adapted for the case of both internal and external pressures by noting that when the internal and external pressures are equal, the state of stress in the wall is

$$\sigma_\theta = \sigma_r = -p \quad \text{(cylinder)}$$

or

$$\sigma_\theta = \sigma_\phi = \sigma_r = -p \quad \text{(sphere)}$$

allowing the term involving the external pressure to be evaluated. It is not valid to just replace p by $(p - p_e)$.

Pressure vessels fail by yielding when the Von Mises equivalent stress first exceed the yield strength, σ_y. They fail by fracture if the largest tensile stress exceeds the fracture stress σ_f, where

$$\sigma_f = \frac{CK_{1C}}{\sqrt{\pi a}}$$

and K_{1C} is the fracture toughness, a the half-crack length, and C a constant given in Section A.10.

A.11   Pressure vessels

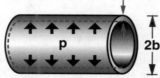

Thin walled   t

p   2b

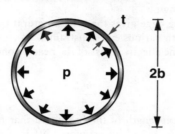

t

p   2b

Cylinder

$$\sigma_\theta = \frac{p\,b}{t}$$

$$\sigma_r = -\frac{p}{2}$$

$$\sigma_z = \frac{p\,b}{2\,t} \quad (\text{Closed ends})$$

Sphere

$$\sigma_\theta = \sigma_\phi = \frac{p\,b}{2\,t}$$

$$\sigma_r = -\frac{p}{2}$$

p = Pressure (N/m^2)
t = Wall thickness (m)
a = Inner radius (m)
b = Outer radius (m)
r = Radial coordinate (m)

Thick walled
2a

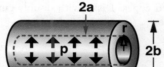

p   r   2b

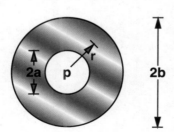

2a   p   2b

$$\sigma_\theta = \frac{p\,a^2}{r^2}\left(\frac{b^2 - r^2}{b^2 - a^2}\right)$$

$$\sigma_r = -\frac{p\,a^2}{r^2}\left(\frac{b^2 + r^2}{b^2 - a^2}\right)$$

$$\sigma_\theta = \sigma_\phi = \frac{p\,a^3}{2\,r^3}\left(\frac{b^3 + 2\,r^3}{b^3 - a^3}\right)$$

$$\sigma_r = -\frac{p\,a^3}{r^3}\left(\frac{b^3 - r^3}{b^3 - a^3}\right)$$

A.12 Vibrating beams, tubes, and disks

Any undamped system vibrating at one of its natural frequencies can be reduced to the simple problem of a mass m attached to a spring of stiffness K. The lowest natural frequency of such a system is

$$f = \frac{1}{2\pi}\sqrt{\frac{K}{m}}$$

Specific cases require specific values for m and K. They can often be estimated with sufficient accuracy to be useful in approximate modelling. Higher natural frequencies are simple multiples of the lowest.

The first box on the facing page gives the lowest natural frequencies of the flexural modes of uniform beams with various end-constraints. As an example, the first can be estimated by assuming that the effective mass of the beam is one quarter of its real mass, so that

$$m = \frac{m_0 L}{4}$$

where m_0 is the mass per unit length of the beam and that K is the bending stiffness (given by F/δ from Section A.3); the estimate differs from the exact value by 2 percent. Vibrations of a tube have a similar form, using I and m_0 for the tube. Circumferential vibrations can be found approximately by "unwrapping" the tube and treating it as a vibrating plate, simply supported at two of its four edges.

The second box gives the lowest natural frequencies for flat circular disks with simply-supported and clamped edges. Disks with doubly-curved faces are stiffer and have higher natural frequencies.

A.12 Vibrating beams, tubes and disks

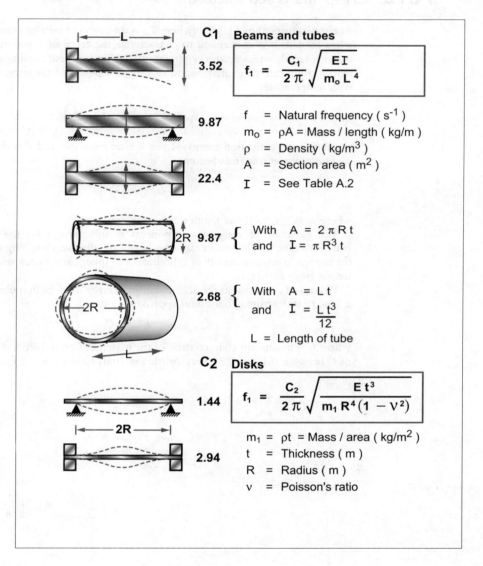

C1 Beams and tubes

3.52

9.87

22.4

$$f_1 = \frac{C_1}{2\pi}\sqrt{\frac{E\,I}{m_o\,L^4}}$$

f = Natural frequency (s^{-1})
$m_o = \rho A$ = Mass / length (kg/m)
ρ = Density (kg/m^3)
A = Section area (m^2)
I = See Table A.2

9.87 $\left\{\begin{array}{l} \text{With} \quad A = 2\pi R t \\ \text{and} \quad I = \pi R^3 t \end{array}\right.$

2.68 $\left\{\begin{array}{l} \text{With} \quad A = L t \\ \text{and} \quad I = \dfrac{L t^3}{12} \end{array}\right.$

L = Length of tube

C2 Disks

1.44

2.94

$$f_1 = \frac{C_2}{2\pi}\sqrt{\frac{E\,t^3}{m_1\,R^4(1-\nu^2)}}$$

$m_1 = \rho t$ = Mass / area (kg/m^2)
t = Thickness (m)
R = Radius (m)
ν = Poisson's ratio

A.13 Creep and creep fracture

At temperatures above $1/3\ T_m$ (where T_m is the absolute melting point), materials creep when loaded. It is convenient to characterize the creep of a material by its behavior under a tensile stress σ, at a temperature T_m. Under these conditions the steady-state tensile strain rate $\dot{\varepsilon}_{ss}$ is often found to vary as a power of the stress and exponentially with temperature:

$$\dot{\varepsilon}_{ss} = A\left(\frac{\sigma}{\sigma_0}\right)^n \exp\left(-\frac{Q}{RT}\right)$$

where Q is an activation energy, A is a kinetic constant, and R is the gas constant. At constant temperature this becomes

$$\dot{\varepsilon}_{ss} = \dot{\varepsilon}_0\left(\frac{\sigma}{\sigma_0}\right)^n$$

where $\dot{\varepsilon}_0\ (\text{s}^{-1})$, $\sigma_0\ (\text{N/m}^2)$, and n are creep constants.

The behavior of creeping components is summarized on the facing page. The equations give the deflection rate of a beam, the displacement rate of an indenter and the change in relative density of cylindrical and spherical pressure vessels in terms of the tensile creep constants.

Prolonged creep causes the accumulation of creep damage that ultimately leads, after a time t_f, to fracture. To a useful approximation

$$t_f\dot{\varepsilon}_{ss} = C$$

where C is a constant characteristic of the material. Creep-ductile material have values of C between 0.1 and 0.5; creep-brittle materials have values of C as low as 0.01.

A.13 Creep

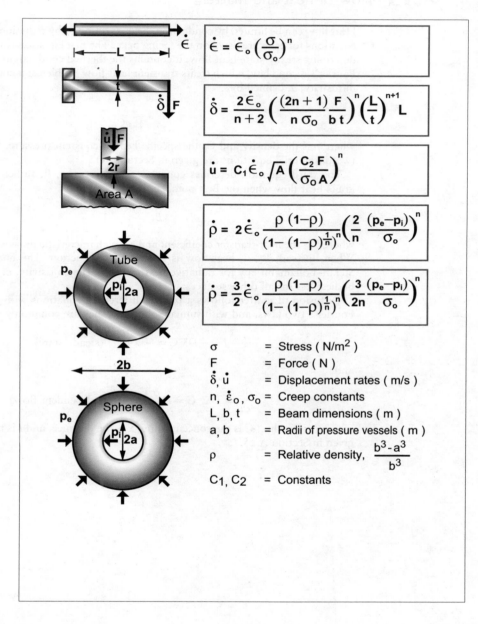

$$\dot{\epsilon} = \dot{\epsilon}_o \left(\frac{\sigma}{\sigma_o} \right)^n$$

$$\dot{\delta} = \frac{2\dot{\epsilon}_o}{n+2} \left(\frac{(2n+1)}{n\,\sigma_o} \frac{F}{b\,t} \right)^n \left(\frac{L}{t} \right)^{n+1} L$$

$$\dot{u} = C_1 \dot{\epsilon}_o \sqrt{A} \left(\frac{C_2 F}{\sigma_o A} \right)^n$$

$$\dot{\rho} = 2\dot{\epsilon}_o \frac{\rho\,(1-\rho)}{\left(1 - (1-\rho)^{\frac{1}{n}}\right)^n} \left(\frac{2}{n} \frac{(p_e - p_i)}{\sigma_o} \right)^n$$

$$\dot{\rho} = \frac{3}{2}\dot{\epsilon}_o \frac{\rho\,(1-\rho)}{\left(1 - (1-\rho)^{\frac{1}{n}}\right)^n} \left(\frac{3}{2n} \frac{(p_e - p_i)}{\sigma_o} \right)^n$$

σ	=	Stress (N/m^2)
F	=	Force (N)
$\dot{\delta}, \dot{u}$	=	Displacement rates (m/s)
$n, \dot{\epsilon}_o, \sigma_o$	=	Creep constants
L, b, t	=	Beam dimensions (m)
a, b	=	Radii of pressure vessels (m)
ρ	=	Relative density, $\dfrac{b^3 - a^3}{b^3}$
C_1, C_2	=	Constants

A.14 Flow of heat and matter

Heat flow can be limited by conduction, by convection or by radiation. The constitutive equations for each are listed on the facing page. The first equation is Fourier's first law, describing steady-state heat flow; it contains the thermal conductivity, λ. The second is Fourier's second law, which treats transient heat-flow problems; it contains the thermal diffusivity, a, defined by

$$a = \frac{\lambda}{\rho C_p}$$

where ρ is the density and C_p the specific heat at constant pressure. Solutions to these two differential equations are given in Section A.15.

The third equation describes convective heat transfer. It, rather than conduction, limits heat flow when the Biot number

$$B_i = \frac{hs}{\lambda} < 1$$

where h is the heat-transfer coefficient and s is a characteristic dimension of the sample. When, instead, $B_i > 1$, heat flow is limited by conduction. The final equation is the Stefan–Boltzmann law for radiative heat transfer. The emissivity, ε, is unity for black bodies; less for all other surfaces.

Diffusion of matter follows a pair of differential equations with the same form as Fourier's two laws, and with similar solutions. They are commonly written

$$J = -D\nabla C = -D\frac{dC}{dx} \quad \text{(steady state)}$$

and

$$\frac{\partial C}{\partial t} = D\nabla^2 C = D\frac{\partial^2 C}{\partial x^2} \quad \text{(time-dependent flow)}$$

where J is the flux, C is the concentration, x is the distance, and t is time. Solutions are given in Section A.15.

A.14 Constitutive equations for heat flow

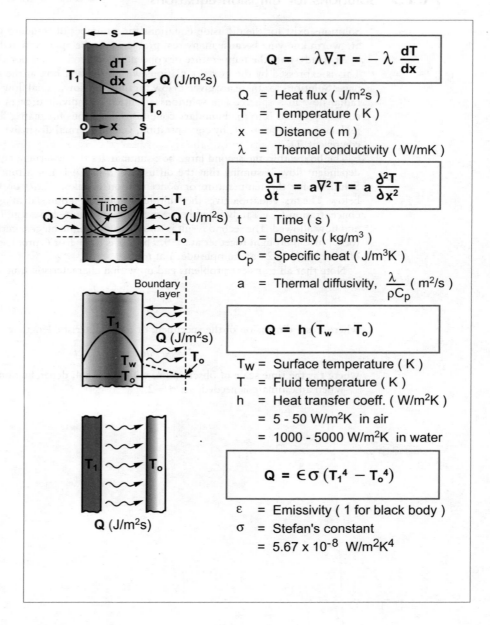

$$Q = - \lambda \nabla.T = - \lambda \frac{dT}{dx}$$

Q = Heat flux (J/m^2s)

T = Temperature (K)

x = Distance (m)

λ = Thermal conductivity (W/mK)

$$\frac{\partial T}{\partial t} = a \nabla^2 T = a \frac{\partial^2 T}{\partial x^2}$$

t = Time (s)

ρ = Density (kg/m^3)

C_p = Specific heat (J/m^3K)

a = Thermal diffusivity, $\dfrac{\lambda}{\rho C_p}$ (m^2/s)

$$Q = h (T_w - T_o)$$

T_w = Surface temperature (K)

T = Fluid temperature (K)

h = Heat transfer coeff. (W/m^2K)

   = 5 - 50 W/m^2K in air

   = 1000 - 5000 W/m^2K in water

$$Q = \epsilon \sigma (T_1{}^4 - T_o{}^4)$$

ε = Emissivity (1 for black body)

σ = Stefan's constant

   = 5.67 x 10^{-8} W/m^2K^4

A.15 Solutions for diffusion equations

Solutions exist for the diffusion equations for a number of standard geometries. They are worth knowing because many real problems can be approximated by one of these.

At steady-state the temperature or concentration profile does not change with time. This is expressed by the boxed equations within the first box at the top of the facing page. Solutions for these are given below for uniaxial flow, radial flow in a cylinder and radial flow in a sphere. The solutions are fitted to individual cases by matching the constants A and B to the boundary conditions. Solutions for matter flow are found by replacing temperature, T, by concentration, C, and thermal dissfusivity, a, by diffusion coefficient, D.

The box within the second large box summarizes the governing equations for time-dependent flow, assuming that the diffusivity (a or D) is not a function of position. Solutions for the temperature or concentration profiles, $T(x,t)$ or $C(x,t)$, are given below. The first equation gives the "thin-film" solution: a thin slab at temperature T_1, or concentration C_1 is sandwiched between two semi-infinite blocks at T_0 or C_0, at $t = 0$, and flow allowed. The second result is for two semi-infinite blocks, initially at T_1 and T_0 (or C_1 or C_0) brought together at $t = 0$. The last is for a T or C profile that is sinusoidal, of wavelength $\lambda/2\pi$ and amplitude A at $t = 0$.

Note that all transient problems end up with a characteristic time constant t^* with

$$t^* = \frac{x^2}{\beta a} \quad \text{or} \quad \frac{x^2}{\beta D}$$

where x is a dimension of the specimen; or a characteristic length x^* with

$$x^* = \sqrt{\beta at} \quad \text{or} \quad \sqrt{\beta Dt}$$

where t is the time scale of observation, with $1 < \beta < 4$, depending on geometry.

When an estimate is needed, set $\beta = 2$.

A.15   Solutions for diffusion equations

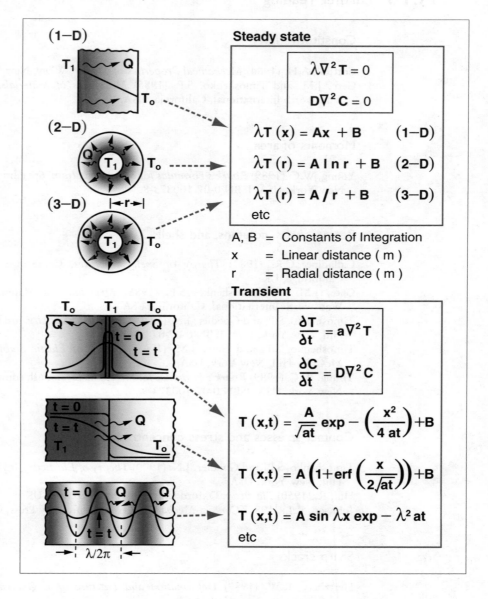

A.16 Further reading

Constitutive laws

Cottrell, A.H. (1964) *Mechanical Properties of Matter*, Wiley, New York, USA.
Gere, J.M. and Timoshenko, S.P. (1985) *Mechanics of Materials*, 2nd SI edition, Wadsworth International, California, USA.

Moments of area

Young, W.C. (1989) *Roark's Formulas for Stress and Strain*, 6th edition, McGraw-Hill, New York, USA. ISBN 0-07-100373-8.

Beams, shafts, columns, and shells

Calladine, C.R. (1983) *Theroy of Shell Structures*, Cambridge University Press, Cambridge, UK.
Gere, J.M. and Timoshenko, S.P. (1985) *Mechanics of Materials*, 2nd edition, Wadsworth International, California, USA.
Timoshenko, S.P. and Goodier, J.N. (1970) *Theory of Elasticity*, 3rd edition, McGraw-Hill, New York, USA. ISBN 0-07-07-2541-1
Timoshenko, S.P. and Gere, J.M. (1961) *Theory of Elastic Stability*, 2nd edition, McGraw-Hill, New York, USA.
Young, W.C. (1989) *Roark's Formulas for Stress and Strain*, 6th edition, McGraw-Hill, New York, USA. ISBN 0-07-100373-8.

Contact stresses and stress concentration

Timoshenko, S.P. and Goodier, J.N. (1970) *Theory of Elasticity*, 3rd edition, McGraw-Hill, New York, USA.
Hill, R. (1950) *Plasticity*, Oxford University Press, Oxford, UK.
Johnson, K.L. (1985) *Contact Mechanics*, Oxford University Press, Oxford, UK.

Sharp cracks

Hertzberg, R.W. (1989) *Deformation and Fracture of Engineering Materials*, 3rd edition, Wiley, New York, USA.
Tada, H., Paris, P.C. and Irwin, G.R. *The Stress Analysis of Cracks Handbook*, 2nd edition, Paris Productions and Del Research Group, Missouri, USA.

Pressure vessels

Timoshenko, S.P. and Goodier, J.N. (1970) *Theory of Elasticity*, 3rd edition, McGraw-Hill, New York, USA.

Hill, R. (1950) *Plasticity*, Oxford University Press, Oxford, UK.

Young, W.C. (1989) *Roark's Formulas for Stress and Strain*, 6th edition, McGraw-Hill, New York, USA.

Vibration

Young, W.C. (1989) *Roark's Formulas for Stress and Strain*, 6th edition, McGraw-Hill, New York, USA. ISBN 0-07-100373-8.

Creep

Finnie, I. and Heller, W.R. (1976) *Creep of Engineering Materials*, McGraw-Hill, New York, USA.

Heat and matter flow

Hollman, J.P. (1981) *Heat Transfer*, 5th edition, McGraw-Hill, New York, USA.

Carslaw, H.S. and Jaeger, J.C. (1959) *Conduction of Heat in Solids*, 2nd edition, Oxford University Press, Oxford, UK.

Shewmon, P.G. (1989) *Diffusion in Solids*, 2nd edition, TMS Warrendale, PA, USA. ISBN 0-87339-05-5.

Appendix B

Material indices

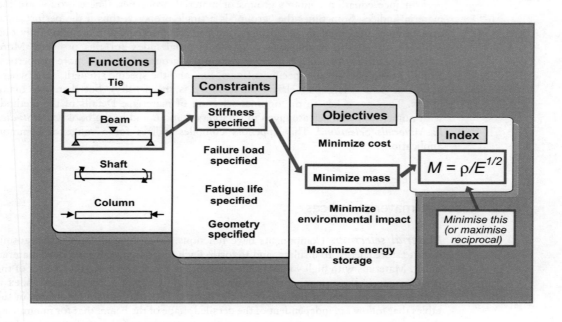

B.1 Introduction and synopsis

The performance, P, of a component is characterized by a performance equation. The performance equation contains groups of material properties. These groups are the material indices. Sometimes the "group" is a single property; thus if the performance of a beam is measured by its stiffness, the performance equation contains only one property, the elastic modulus E. It is the material index for this problem. More commonly the performance equation contains a group of two or more properties. Familiar examples are the specific stiffness, E/ρ, and the specific strength, σ_y/ρ (where σ_y is the yield strength or elastic limit, and ρ is the density), but there are many others. They are a key to the optimal selection of materials. Details of the method, with numerous examples are given in Chapters 5 and 6 and in the book *Case Studies in Materials Selection*.[1] This appendix compiles indices for a range of common applications.

B.2 Uses of material indices

Material selection. Components have functions: to carry loads safely, to transmit heat, to store energy, to insulate, and so forth. Each function has an associated material index. Materials with high values of the appropriate index maximize that aspect of the performance of the component. For reasons given in Chapter 5, the material index is generally independent of the details of the design. Thus the indices for beams in the tables that follow are independent of the detailed shape of the beam; that for minimizing thermal distortion of precision instruments is independent of the configuration of the instrument, and so forth. This gives them great generality.

Material deployment or substitution. A new material will have potential application in functions for which its indices have unusually high values. Fruitful applications for a new material can be identified by evaluating its indices and comparing them with those of existing, established materials. Similar reasoning points the way to identifying viable substitutes for an incumbent material in an established application.

How to read the tables. The indices listed in the Tables B.1–B.7 are, for the most part, based on the objective of minimizing mass. To minimize cost, use the index for minimum mass, replacing the density ρ by the cost per unit volume, $C_m\rho$, where C_m is the cost per kg. To minimize energy content or CO_2 burden, replace ρ by $E_p\rho$ or by $CO_2\rho$ where E_p is the production energy per kg and CO_2 is the CO_2 burden per kg (Appendix C, Table C.11).

---

[1] M. F. Ashby and D. Cebon (1995) *Case Studies in Materials Selection*, Granta Design Ltd, Rustat House, 62 Clifton Road, Cambridge CB1 7EG, UK (www.grantadesign.com).

The symbols. The symbols used in Tables B.1–B.7 are defined below:

Class	Property	Symbol and units
General	Density	ρ (kg/m^3 or Mg/m^3)
	Price	C_m ($/kg)
Mechanical	Elastic moduli (Young's, shear, bulk)	E, G, K (GPa)
	Poisson's ratio	ν (–)
	Failure strength (yield, fracture)	σ_f (MPa)
	Fatigue strength	σ_e (MPa)
	Hardness	H (Vickers)
	Fracture toughness	K_{1C} (MPa.m$^{1/2}$)
	Loss coefficient (damping capacity)	η (–)
Thermal	Thermal conductivity	λ (W/m.K)
	Thermal diffusivity	a (m^2/s)
	Specific heat	C_p (J/kg.K)
	Thermal expansion coefficient	α (K^{-1})
	Difference in thermal conductivity	$\Delta\alpha$ (K^{-1})
Electrical	Electrical resistivity	ρ_e ($\mu\Omega$.cm)
Eco-properties	Energy/kg to produce material per kg	E_p (MJ/kg)
	CO_2/kg burden of material production per kg	CO_2 (kg/kg)

Table B.1 Stiffness-limited design at minimum mass (cost, energy, environmental impact)

Function and constraints	Maximize
Tie (tensile strut)	
Stiffness, length specified; section area free	E/ρ
Shaft (loaded in torsion)	
Stiffness, length, shape specified, section area free	$G^{1/2}/\rho$
Stiffness, length, outer radius specified; wall thickness free	G/ρ
Stiffness, length, wall-thickness specified; outer radius free	$G^{1/3}/\rho$
Beam (loaded in bending)	
Stiffness, length, shape specified; section area free	$E^{1/2}/\rho$
Stiffness, length, height specified; width free	E/ρ
Stiffness, length, width specified; height free	$E^{1/3}/\rho$
Column (compression strut, failure by elastic buckling)	
Buckling load, length, shape specified; section area free	$E^{1/2}/\rho$
Panel (flat plate, loaded in bending)	
Stiffness, length, width specified, thickness free	$E^{1/3}/\rho$
Plate (flat plate, compressed in-plane, buckling failure)	
Collapse load, length and width specified, thickness free	$E^{1/3}/\rho$
Cylinder with internal pressure	
Elastic distortion, pressure and radius specified; wall thickness free	E/ρ
Spherical shell with internal pressure	
Elastic distortion, pressure and radius specified, wall thickness free	$E/(1-\nu)\rho$

Table B.2   Strength-limited design at minimum mass (cost, energy, environmental impact)

Function and constraints	Maximize
Tie (tensile strut)	
Stiffness, length specified; section area free	σ_f/ρ
Shaft (loaded in torsion)	
Load, length, shape specified, section area free	$\sigma_f^{2/3}/\rho$
Load, length, outer radius specified; wall thickness free	σ_f/ρ
Load, length, wall-thickness specified; outer radius free	$\sigma_f^{1/2}/\rho$
Beam (loaded in bending)	
Load, length, shape specified; section area free	$\sigma_f^{2/3}/\rho$
Load length, height specified; width free	σ_f/ρ
Load, length, width specified; height free	$\sigma_f^{1/2}/\rho$
Column (compression strut)	
Load, length, shape specified; section area free	σ_f/ρ
Panel (flat plate, loaded in bending)	
Stiffness, length, width specified, thickness free	$\sigma_f^{1/2}/\rho$
Plate (flat plate, compressed in-plane, buckling failure)	
Collapse load, length and width specified, thickness free	$\sigma_f^{1/2}/\rho$
Cylinder with internal pressure	
Elastic distortion, pressure and radius specified; wall thickness free	σ_f/ρ
Spherical shell with internal pressure	
Elastic distortion, pressure and radius specified, wall thickness free	σ_f/ρ
Flywheels, rotating disks	
Maximum energy storage per unit volume; given velocity	σ
Maximum energy storage per unit mass; no failure	σ_f/ρ

Table B.3   Strength-limited design: springs, hinges, etc. for maximum performance

Function and constraints	Maximize
Springs	
Maximum stored elastic energy per unit volume; no failure	σ_f^2/E
Maximum stored elastic energy per unit mass; no failure	$\sigma_f^2/E\rho$
Elastic hinges	
Radius of bend to be minimized (max flexibility without failure)	σ_f/E
Knife edges, pivots	
Minimum contact area, maximum bearing load	σ_f^3/E^2 and H
Compression seals and gaskets	
Maximum conformability; limit on contact pressure	$\sigma_f^{3/2}/E$ and $1/E$
Diaphragms	
Maximum deflection under specified pressure or force	$\sigma_f^{3/2}/E$
Rotating drums and centrifuges	
Maximum angular velocity; radius fixed; wall thickness free	σ_f/ρ

Table B.4   Vibration-limited design

Function and constraints	Maximize
Ties, columns	
Maximum longitudinal vibration frequencies	E/ρ
Beams, all dimensions prescribed	
Maximum flexural vibration frequencies	E/ρ
Beams, length and stiffness prescribed	
Maximum flexural vibration frequencies	$E^{1/2}/\rho$
Panels, all dimensions prescribed	
Maximum flexural vibration frequencies	E/ρ
Panels, length, width and stiffness prescribed	
Maximum flexural vibration frequencies	$E^{1/3}/\rho$
Ties, columns, beams, panels, stiffness prescribed	
Minimum longitudinal excitation from external drivers, ties	$\eta E/\rho$
Minimum flexural excitation from external drivers, beams	$\eta E^{1/2}/\rho$
Minimum flexural excitation from external drivers, panels	$\eta E^{1/3}/\rho$

Table B.5   Damage-tolerant design

Function and constraints	Maximize
Ties (tensile member)	
Maximum flaw tolerance and strength, load-controlled design	K_{IC} and σ_f
Maximum flaw tolerance and strength, displacement-control	K_{IC}/E and σ_f
Maximum flaw tolerance and strength, energy-control	K_{IC}^2/E and σ_f
Shafts (loaded in torsion)	
Maximum flaw tolerance and strength, load-controlled design	K_{IC} and σ_f
Maximum flaw tolerance and strength, displacement-control	K_{IC}/E and σ_f
Maximum flaw tolerance and strength, energy-control	K_{IC}^2/E and σ_f
Beams (loaded in bending)	
Maximum flaw tolerance and strength, load-controlled design	K_{IC} and σ_f
Maximum flaw tolerance and strength, displacement-control	K_{IC}/E and σ_f
Maximum flaw tolerance and strength, energy-control	K_{IC}^2/E and σ_f
Pressure vessel	
Yield-before-break	K_{IC}/σ_f
Leak-before-break	K_{IC}^2/σ_f

Table B.6   Electro-mechanical design

Function and constraints	Maximize
Bus bars	
Minimum life-cost; high current conductor	$1/\rho_e \rho C_m$
Electro-magnet windings	
Maximum short-pulse field; no mechanical failure	σ_f
Maximize field and pulse-length, limit on temperature rise	$C_p \rho / \rho_e$
Windings, high-speed electric motors	
Maximum rotational speed; no fatigue failure	σ_e / ρ_e
Minimum ohmic losses; no fatigue failure	$1/\rho_e$
Relay arms	
Minimum response time; no fatigue failure	$\sigma_e / E \rho_e$
Minimum ohmic losses; no fatigue failure	$\sigma_e^2 / E \rho_e$

Table B.7   Thermal and thermo-mechanical design

Function and constraints	Maximize
Thermal insulation materials	
Minimum heat flux at steady state; thickness specified	$1/\lambda$
Minimum temp rise in specified time; thickness specified	$1/a = \rho C_p / \lambda$
Minimize total energy consumed in thermal cycle (kilns, etc.)	$\sqrt{a}/\lambda = \sqrt{1/\lambda \rho C_p}$
Thermal storage materials	
Maximum energy stored/unit material cost (storage heaters)	C_p / C_m
Maximize energy stored for given temperature rise and time	$\lambda/\sqrt{a} = \sqrt{\lambda \rho C_p}$
Precision devices	
Minimize thermal distortion for given heat flux	λ/a
Thermal shock resistance	
Maximum change in surface temperature; no failure	$\sigma_f / E \alpha$
Heat sinks	
Maximum heat flux per unit volume; expansion limited	$\lambda / \Delta \alpha$
Maximum heat flux per unit mass; expansion limited	$\lambda / \rho \, \Delta \alpha$
Heat exchangers (pressure-limited)	
Maximum heat flux per unit area; no failure under Δp	$\lambda \sigma_f$
Maximum heat flux per unit mass; no failure under Δp	$\lambda \sigma_f / \rho$

Appendix C

Data and information for engineering materials

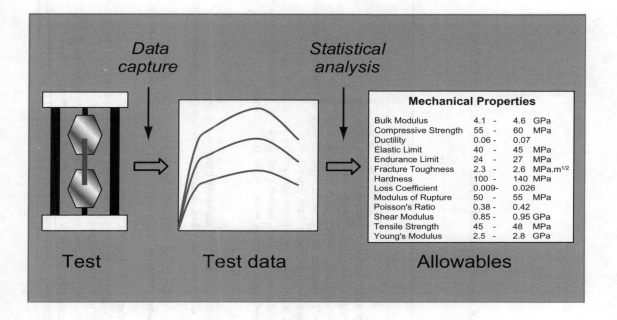

Data capture	Statistical analysis

Mechanical Properties

Bulk Modulus	4.1 -	4.6	GPa
Compressive Strength	55 -	60	MPa
Ductility	0.06 -	0.07	
Elastic Limit	40 -	45	MPa
Endurance Limit	24 -	27	MPa
Fracture Toughness	2.3 -	2.6	MPa.m$^{1/2}$
Hardness	100 -	140	MPa
Loss Coefficient	0.009-	0.026	
Modulus of Rupture	50 -	55	MPa
Poisson's Ratio	0.38 -	0.42	
Shear Modulus	0.85 -	0.95	GPa
Tensile Strength	45 -	48	MPa
Young's Modulus	2.5 -	2.8	GPa

Test   Test data   Allowables

Appendix contents

This appendix lists the names and typical applications of the materials that appear on the charts of Chapter 4, together with the structured data used to construct them.

Table C.1 Names and applications: metals and alloys

Metals	Applications
Ferrous	
Cast irons	Automotive parts, engine blocks, machine tool structural parts, lathe beds
High carbon steels	Cutting tools, springs, bearings, cranks, shafts, railway track
Medium carbon steels	General mechanical engineering (tools, bearings, gears, shafts, bearings)
Low carbon steels	Steel structures ("mild steel") — bridges, oil rigs, ships; reinforcement for concrete; automotive parts, car body panels; galvanized sheet; packaging (cans, drums)
Low alloy steels	Springs, tools, ball bearings, automotive parts (gears connecting rods, etc.)
Stainless steels	Transport, chemical and food processing plant, nuclear plant, domestic ware (cutlery, washing machines, stoves), surgical implements, pipes, pressure vessels, liquid gas containers
Non-ferrous	
Aluminum alloys	
Casting alloys	Automotive parts (cylinder blocks), domestic appliances (irons)
Non-heat-treatable alloys	Electrical conductors, heat exchangers, foil, tubes, saucepans, beverage cans, lightweight ships, architectural panels
Heat-treatable alloys	Aerospace engineering, automotive bodies and panels, lightweight structures and ships
Copper alloys	Electrical conductors and wire, electronic circuit boards, heat exchangers, boilers, cookware, coinage, sculptures
Lead alloys	Roof and wall cladding, solder, X-ray shielding, battery electrodes
Magnesium alloys	Automotive castings, wheels, general lightweight castings for transport, nuclear fuel containers; principal alloying addition to aluminum alloys
Nickel alloys	Gas turbines and jet engines, thermocouples, coinage; alloying addition to austenitic stainless steels
Titanium alloys	Aircraft turbine blades; general structural aerospace applications; biomedical implants.
Zinc alloys	Die castings (automotive, domestic appliances, toys, handles); coating on galvanized steel

Table C.2 Names and applications: polymers and foams

Polymers	Abbreviation	Applications
Elastomer		
Butyl rubber		Tyres, seals, anti-vibration mountings, electrical insulation, tubing
Ethylene-vinyl-acetate	EVA	Bags, films, packaging, gloves, insulation, running shoes
Isoprene	IR	Tyres, inner tubes, insulation, tubing, shoes
Natural rubber	NR	Gloves, tyres, electrical insulation, tubing
Polychloroprene (neoprene)	CR	Wetsuits, O-rings and seals, footware
Polyurethane elastomers	el-PU	Packaging, hoses, adhesives, fabric coating
Silicone elastomers		Electrical insulation, electronic encapsulation, medical implants
Thermoplastic		
Acrylonitrile butadiene styrene	ABS	Communication appliances, automotive interiors, luggage, toys, boats
Cellulose polymers	CA	Tool and cutlery handles, decorative trim, pens
Ionomer	I	Packaging, golf balls, blister packs, bottles
Polyamides (nylons)	PA	Gears; bearings; plumbing, packaging, bottles, fabrics, textiles, ropes
Polycarbonate	PC	Safety goggles, shields, helmets; light fittings, medical components
Polyetheretherketone	PEEK	Electrical connectors, racing car parts, fiber composites
Polyethylene	PE	Packaging, bags, squeeze tubes, toys, artificial joints
Polyethylene terephthalate	PET	Blow molded bottles, film, audio/video tape, sails
Polymethyl methacrylate (acrylic)	PMMA	Aircraft windows, lenses, reflectors, lights, compact discs
Polyoxymethylene (acetal)	POM	Zips, domestic and appliance parts, handles
Polypropylene	PP	Ropes, garden furniture, pipes, kettles, electrical insulation, astroturf
Polystyrene	PS	Toys, packaging, cutlery, audio cassette/CD cases
Polyurethane thermoplastics	tp-PU	Cushioning, seating, shoe soles, hoses, car bumpers, insulation
Polyvinylchloride	PVC	Pipes, gutters, window frames, packaging
Polytetrafluoroethylene (teflon)	PTFE	Non-stick coatings, bearings, skis, electrical insulation, tape
Thermoset		
Epoxies	EP	Adhesives, fiber composites, electronic encapsulation
Phenolics	PHEN	Electrical plugs, sockets, cookware, handles, adhesives
Polyester	PEST	Furniture, boats, sports goods
Polymer foams		
Flexible polymer foam		Packaging, buoyancy, cushioning, sponges, sleeping mats
Rigid polymer foam		Thermal insulation, sandwich panels, packaging, buoyancy

Table C.3   Names and applications: composites, ceramics, glasses, and natural materials

	Applications
Composites	
Metal	
Aluminum/silicon carbide	Automotive parts, sports goods
Polymer	
CFRP	Lightweight structural parts (aerospace, bike frames, sports goods, boat hulls and oars, springs)
GFRP	Boat hulls, automotive parts, chemical plant
Ceramics	
Glasses	
Borosilicate glass	Ovenware, laboratory ware, headlights
Glass ceramic	Cookware, lasers, telescope mirrors
Silica glass	High performance windows, crucibles, high temperature applications
Soda-lime glass	Windows, bottles, tubing, light bulbs, pottery glazes
Porous	
Brick	Buildings
Concrete	General civil engineering construction
Stone	Buildings, architecture, sculpture
Technical	
Alumina	Cutting tools, spark plugs, microcircuit substrates, valves
Aluminum nitride	Microcircuit substrates and heat sinks
Boron carbide	Lightweight armor, nozzles, dies, precision tool parts
Silicon	Microcircuits, semiconductors, precision instruments, IR windows, MEMS
Silicon carbide	High temperature equipment, abrasive polishing grits, bearings, armor
Silicon nitride	Bearings, cutting tools, dies, engine parts
Tungsten carbide	Cutting tools, drills, abrasives
Natural	
Bamboo	Building, scaffolding, paper, ropes, baskets, furniture
Cork	Corks and bungs, seals, floats, packaging, flooring
Leather	Shoes, clothing, bags, drive-belts
Wood	Construction, flooring, doors, furniture, packaging, sports goods

Table C.4   Melting temperature, T_m, and glass temperature, T_g[1]

	$T_m T_g$ (°C)
Metals	
Ferrous	
Cast irons	1130–1250
High carbon steels	1289–1478
Medium carbon steels	1380–1514
Low carbon steels	1480–1526
Low alloy steels	1382–1529
Stainless steels	1375–1450
Non-ferrous	
Aluminum alloys	475–677
Copper alloys	982–1082
Lead alloys	322–328
Magnesium alloys	447–649
Nickel alloys	1435–1466
Titanium alloys	1477–1682
Zinc alloys	375–492
Ceramics	
Glasses	
Borosilicate glass (*)	450–602
Glass ceramic (*)	563–1647
Silica glass (*)	957–1557
Soda-lime glass (*)	442–592
Porous	
Brick	927–1227
Concrete, typical	927–1227
Stone	1227–1427
Technical	
Alumina	2004–2096
Aluminum nitride	2397–2507
Boron carbide	2372–2507
Silicon	1407–1412
Silicon carbide	2152–2500
Silicon nitride	2388–2496
Tungsten carbide	2827–2920
Composites	
Metal	
Aluminum/silicon carbide	525–627
Polymer	
CFRP	n.a.
GFRP	n.a.
Natural	
Bamboo (*)	77–102
Cork (*)	77–102

Table C.4 (*Continued*)

	T_mT_g (°C)
Leather (*)	107–127
Wood, typical (longitudinal) (*)	77–102
Wood, typical (transverse) (*)	77–102
Polymers	
Elastomer	
Butyl rubber (*)	−73−−63
EVA (*)	−73−−23
Isoprene (IR) (*)	−83−−78
Natural rubber (NR) (*)	−78−−63
Neoprene (CR) (*)	−48−−43
Polyurethane elastomers (elPU) (*)	−73−−23
Silicone elastomers (*)	−123−−73
Thermoplastic	
ABS (*)	88–128
Cellulose polymers (CA) (*)	−9–107
Ionomer (I) (*)	27–77
Nylons (PA) (*)	44–56
Polycarbonate (PC) (*)	142–205
PEEK (*)	143–199
Polyethylene (PE) (*)	−25−−15
PET (*)	68–80
Acrylic (PMMA) (*)	85–165
Acetal (POM) (*)	−18−−8
Polypropylene (PP) (*)	−25−−15
Polystyrene (PS) (*)	74–110
Polyurethane thermoplastics (tpPU) (*)	120–160
PVC	75–105
Teflon (PTFE)	107–123
Thermoset	
Epoxies	n/a
Phenolics	n/a
Polyester	n/a
Polymer foams	
Flexible polymer foam (VLD) (*)	112–177
Flexible polymer foam (LD) (*)	112–177
Flexible polymer foam (MD) (*)	112–177
Rigid polymer foam (LD) (*)	67–171
Rigid polymer foam (MD) (*)	67–157
Rigid polymer foam (HD) (*)	67–171

[1.] The table lists the melting point for crystalline solids and the glass temperature for polymeric and inorganic glasses.
(*): glass transition temperature.
n/a: not applicable (materials decompose, rather than melt).

Table C.5   Density, ρ

	ρ (Mg/m^3)
Metals	
Ferrous	
Cast irons	7.05–7.25
High carbon steels	7.8–7.9
Medium carbon steels	7.8–7.9
Low carbon steels	7.8–7.9
Low alloy steels	7.8–7.9
Stainless steels	7.6–8.1
Non-ferrous	
Aluminum alloys	2.5–2.9
Copper alloys	8.93–8.94
Lead alloys	10–11.4
Magnesium alloys	1.74–1.95
Nickel alloys	8.83–8.95
Titanium alloys	4.4–4.8
Zinc alloys	4.95–7
Ceramics	
Glasses	
Borosilicate glass	2.2–2.3
Glass ceramic	2.2–2.8
Silica glass	2.17–2.22
Soda-lime glass	2.44–2.49
Porous	
Brick	1.9–2.1
Concrete, typical	2.2–2.6
Stone	2.5–3
Technical	
Alumina	3.5–3.98
Aluminum nitride	3.26–3.33
Boron carbide	2.35–2.55
Silicon	2.3–2.35
Silicon carbide	3–3.21
Silicon nitride	3–3.29
Tungsten carbide	15.3–15.9
Composites	
Metal	
Aluminum/silicon carbide	2.66–2.9
Polymer	
CFRP	1.5–1.6
GFRP	1.75

Table C.5   *(Continued)*

	ρ (Mg/m^3)
Natural	
Bamboo	0.6–0.8
Cork	0.12–0.24
Leather	0.81–1.05
Wood, typical (longitudinal)	0.6–0.8
Wood, typical (transverse)	0.6–0.8
Polymers	
Elastomer	
Butyl rubber	0.9–0.92
EVA	0.945–0.955
Isoprene (IR)	0.93–0.94
Natural rubber (NR)	0.92–0.93
Neoprene (CR)	1.23–1.25
Polyurethane elastomers (elPU)	1.02–1.25
Silicone elastomers	1.3–1.8
Thermoplastic	
ABS	1.01–1.21
Cellulose polymers (CA)	0.98–1.3
Ionomer (I)	0.93–0.96
Nylons (PA)	1.12–1.14
Polycarbonate (PC)	1.14–1.21
PEEK	1.3–1.32
Polyethylene (PE)	0.939–0.96
PET	1.29–1.4
Acrylic (PMMA)	1.16–1.22
Acetal (POM)	1.39–1.43
Polypropylene (PP)	0.89–0.91
Polystyrene (PS)	1.04–1.05
Polyurethane thermoplastics (tpPU)	1.12–1.24
PVC	1.3–1.58
Teflon (PTFE)	2.14–2.2
Thermoset	
Epoxies	1.11–1.4
Phenolics	1.24–1.32
Polyester	1.04–1.4
Polymer foams	
Flexible polymer foam (VLD)	0.016–0.035
Flexible polymer foam (LD)	0.038–0.07
Flexible polymer foam (MD)	0.07–0.115
Rigid polymer foam (LD)	0.036–0.07
Rigid polymer foam (MD)	0.078–0.165
Rigid polymer foam (HD)	0.17–0.47

Table C.6   Young's modulus, *E*

	E (GPa)
Metals	
Ferrous	
Cast irons	165–180
High carbon steels	200–215
Medium carbon steels	200–216
Low carbon steels	200–215
Low alloy steels	201–217
Stainless steels	189–210
Non-ferrous	
Aluminum alloys	68–82
Copper alloys	112–148
Lead alloys	12.5–15
Magnesium alloys	42–47
Nickel alloys	190–220
Titanium alloys	90–120
Zinc alloys	68–95
Ceramics	
Glasses	
Borosilicate glass	61–64
Glass ceramic	64–110
Silica glass	68–74
Soda-lime glass	68–72
Porous	
Brick	10–50
Concrete, typical	25–38
Stone	120–133
Technical	
Alumina	215–413
Aluminum nitride	302–348
Boron carbide	400–472
Silicon	140–155
Silicon carbide	300–460
Silicon nitride	280–310
Tungsten carbide	600–720
Composites	
Metal	
Aluminum/silicon carbide	81–100
Polymer	
CFRP	69–150
GFRP	15–28

Table C.6   (*Continued*)

	E (GPa)
Natural	
Bamboo	15–20
Cork	0.013–0.05
Leather	0.1–0.5
Wood, typical (longitudinal)	6–20
Wood, typical (transverse)	0.5–3
Polymers	
Elastomer	
Butyl rubber	0.001–0.002
EVA	0.01–0.04
Isoprene (IR)	0.0014–0.004
Natural rubber (NR)	0.0015–0.0025
Neoprene (CR)	0.0007–0.002
Polyurethane elastomers (elPU)	0.002–0.003
Silicone elastomers	0.005–0.02
Thermoplastic	
ABS	1.1–2.9
Cellulose polymers (CA)	1.6–2
Ionomer (I)	0.2–0.424
Nylons (PA)	2.62–3.2
Polycarbonate (PC)	2–2.44
PEEK	3.5–4.2
Polyethylene (PE)	0.621–0.896
PET	2.76–4.14
Acrylic (PMMA)	2.24–3.8
Acetal (POM)	2.5–5
Polypropylene (PP)	0.896–1.55
Polystyrene (PS)	2.28–3.34
Polyurethane thermoplastics (tpPU)	1.31–2.07
PVC	2.14–4.14
Teflon (PTFE)	0.4–0.552
Thermoset	
Epoxies	2.35–3.075
Phenolics	2.76–4.83
Polyester	2.07–4.41
Polymer foams	
Flexible polymer foam (VLD)	0.0003–0.001
Flexible polymer foam (LD)	0.001–0.003
Flexible polymer foam (MD)	0.004–0.012
Rigid polymer foam (LD)	0.023–0.08
Rigid polymer foam (MD)	0.08–0.2
Rigid polymer foam (HD)	0.2–0.48

Table C.7   Yield strength, σ_y, and tensile strength, σ_{ts}

	σ_y (MPa)	σ_{ts} (MPa)
Metals		
Ferrous		
Cast irons	215–790	350–1000
High carbon steels	400–1155	550–1640
Medium carbon steels	305–900	410–1200
Low carbon steels	250–395	345–580
Low alloy steels	400–1100	460–1200
Stainless steels	170–1000	480–2240
Non-ferrous		
Aluminum alloys	30–500	58–550
Copper alloys	30–500	100–550
Lead alloys	8–14	12–20
Magnesium alloys	70–400	185–475
Nickel alloys	70–1100	345–1200
Titanium alloys	250–1245	300–1625
Zinc alloys	80–450	135–520
Ceramics		
Glasses		
Borosilicate glass (*)	264–384	22–32
Glass ceramic (*)	750–2129	62–177
Silica glass (*)	1100–1600	45–155
Soda-lime glass (*)	360–420	31–35
Porous		
Brick (*)	50–140	7–14
Concrete, typical (*)	32–60	2–6
Stone (*)	34–248	5–17
Technical		
Alumina (*)	690–5500	350–665
Aluminum nitride (*)	1970–2700	197–270
Boron carbide (*)	2583–5687	350–560
Silicon (*)	3200–3460	160–180
Silicon carbide (*)	1000–5250	370–680
Silicon nitride (*)	524–5500	690–800
Tungsten carbide (*)	3347–6833	370–550
Composites		
Metal		
Aluminum/silicon carbide	280–324	290–365
Polymer		
CFRP	550–1050	550–1050
GFRP	110–192	138–241

Table C.7 (Continued)

	σ_y (MPa)	σ_{ts} (MPa)
Natural		
Bamboo	35–44	36–45
Cork	0.3–1.5	0.5–2.5
Leather	5–10	20–26
Wood, typical (longitudinal)	30–70	60–100
Wood, typical (transverse)	2–6	4–9
Polymers		
Elastomer		
Butyl rubber	2–3	5–10
EVA	12–18	16–20
Isoprene (IR)	20–25	20–25
Natural rubber (NR)	20–30	22–32
Neoprene (CR)	3.4–24	3.4–24
Polyurethane elastomers (elPU)	25–51	25–51
Silicone elastomers	2.4–5.5	2.4–5.5
Thermoplastic		
ABS	18.5–51	27.6–55.2
Cellulose polymers (CA)	25–45	25–50
Ionomer (I)	8.3–15.9	17.2–37.2
Nylons (PA)	50–94.8	90–165
Polycarbonate (PC)	59–70	60–72.4
PEEK	65–95	70–103
Polyethylene (PE)	17.9–29	20.7–44.8
PET	56.5–62.3	48.3–72.4
Acrylic (PMMA)	53.8–72.4	48.3–79.6
Acetal (POM)	48.6–72.4	60–89.6
Polypropylene (PP)	20.7–37.2	27.6–41.4
Polystyrene (PS)	28.7–56.2	35.9–56.5
Polyurethane thermoplastics (tpPU)	40–53.8	31–62
PVC	35.4–52.1	40.7–65.1
Teflon (PTFE)	15–25	20–30
Thermoset		
Epoxies	36–71.7	45–89.6
Phenolics	27.6–49.7	34.5–62.1
Polyester	33–40	41.4–89.6
Polymer foams		
Flexible polymer foam (VLD)	0.01–0.12	0.24–0.85
Flexible polymer foam (LD)	0.02–0.3	0.24–2.35
Flexible polymer foam (MD)	0.05–0.7	0.43–2.95
Rigid polymer foam (LD)	0.3–1.7	0.45–2.25
Rigid polymer foam (MD)	0.4–3.5	0.65–5.1
Rigid polymer foam (HD)	0.8–12	1.2–12.4

(*): For ceramics, yield stress is replaced by *compressive strength*, which is more relevant in ceramic design. Note that ceramics are of the order of 10 times stronger in compression than in tension.

Table C.8   Fracture toughness (plane-strain), K_{IC}

	K_{IC} (MPa$\sqrt{}$m)
Metals	
Ferrous	
Cast irons	22–54
High carbon steels	27–92
Medium carbon steels	12–92
Low carbon steels	41–82
Low alloy steels	14–200
Stainless steels	62–280
Non-ferrous	
Aluminum alloys	22–35
Copper alloys	30–90
Lead alloys	5–15
Magnesium alloys	12–18
Nickel alloys	80–110
Titanium alloys	14–120
Zinc alloys	10–100
Ceramics	
Glasses	
Borosilicate glass	0.5–0.7
Glass ceramic	1.4–1.7
Silica glass	0.6–0.8
Soda-lime glass	0.55–0.7
Porous	
Brick	1–2
Concrete, typical	0.35–0.45
Stone	0.7–1.5
Technical	
Alumina	3.3– 4.8
Aluminum nitride	2.5–3.4
Boron carbide	2.5–3.5
Silicon	0.83–0.94
Silicon carbide	2.5–5
Silicon nitride	4–6
Tungsten carbide	2–3.8
Composites	
Metal	
Aluminum/silicon carbide	15–24
Polymer	
CFRP	6.1–88
GFRP	7–23

Table C.8   *(Continued)*

	K_{IC} (MPa$\sqrt{m}$)
Natural	
Bamboo	5–7
Cork	0.05–0.1
Leather	3–5
Wood, typical (longitudinal)	5–9
Wood, typical (transverse)	0.5–0.8
Polymers	
Elastomer	
Butyl rubber	0.07–0.1
EVA	0.5–0.7
Isoprene (IR)	0.07–0.1
Natural rubber (NR)	0.15–0.25
Neoprene (CR)	0.1–0.3
Polyurethane elastomers (elPU)	0.2–0.4
Silicone elastomers	0.03–0.5
Thermoplastic	
ABS	1.19–4.30
Cellulose polymers (CA)	1–2.5
Ionomer (I)	1.14–3.43
Nylons (PA)	2.22–5.62
Polycarbonate (PC)	2.1–4.60
PEEK	2.73–4.30
Polyethylene (PE)	1.44–1.72
PET	4.5–5.5
Acrylic (PMMA)	0.7–1.6
Acetal (POM)	1.71–4.2
Polypropylene (PP)	3–4.5
Polystyrene (PS)	0.7–1.1
Polyurethane thermoplastics (tpPU)	1.84–4.97
PVC	1.46–5.12
Teflon (PTFE)	1.32–1.8
Thermoset	
Epoxies	0.4–2.22
Phenolics	0.79–1.21
Polyester	1.09–1.70
Polymer foams	
Flexible polymer foam (VLD)	0.005–0.02
Flexible polymer foam (LD)	0.015–0.05
Flexible polymer foam (MD)	0.03–0.09
Rigid polymer foam (LD)	0.002–0.02
Rigid polymer foam (MD)	0.007–0.049
Rigid polymer foam (HD)	0.024–0.091

Note: K_{IC} is only valid for conditions under which linear elastic fracture mechanics apply (see Chapter 4). The plane-strain toughness, G_{IC}, may be estimated from $K_{IC}^2 = EG_{IC}/(1 - \nu^2) \approx EG_{IC}$ (as $\nu^2 \approx 0.1$).

Table C.9   Thermal conductivity, λ

	λ (W/m.K)
Metals	
Ferrous	
Cast irons	29–44
High carbon steels	47–53
Medium carbon steels	45–55
Low carbon steels	49–54
Low alloy steels	34–55
Stainless steels	11–19
Non-ferrous	
Aluminum alloys	76–235
Copper alloys	160–390
Lead alloys	22–36
Magnesium alloys	50–156
Nickel alloys	67–91
Titanium alloys	5–12
Zinc alloys	100–135
Ceramics	
Glasses	
Borosilicate glass	1–1.3
Glass ceramic	1.3–2.5
Silica glass	1.4–1.5
Soda-lime glass	0.7–1.3
Porous	
Brick	0.46–0.73
Concrete, typical	0.8–2.4
Stone	5.4–6.0
Technical	
Alumina	30–38.5
Aluminum nitride	80–200
Boron carbide	40–90
Silicon	140–150
Silicon carbide	115–200
Silicon nitride	22–30
Tungsten carbide	55–88
Composites	
Metal	
Aluminum/silicon carbide	180–160
Polymer	
CFRP	1.28–2.6
GFRP	0.4–0.55

Table C.9 (*Continued*)

	λ (W/m.K)
Natural	
Bamboo	0.1–0.18
Cork	0.035–0.048
Leather	0.15–0.17
Wood, typical (longitudinal)	0.31–0.38
Wood, typical (transverse)	0.15–0.19
Polymers	
Elastomer	
Butyl rubber	0.08–0.1
EVA	0.3–0.4
Isoprene (IR)	0.08–0.14
Natural rubber (NR)	0.1–0.14
Neoprene (CR)	0.08–0.14
Polyurethane elastomers (elPU)	0.28–0.3
Silicone elastomers	0.3–1.0
Thermoplastic	
ABS	0.19–0.34
Cellulose polymers (CA)	0.13–0.3
Ionomer (I)	0.24–0.28
Nylons (PA)	0.23–0.25
Polycarbonate (PC)	0.19–0.22
PEEK	0.24–0.26
Polyethylene (PE)	0.40–0.44
PET	0.14–0.15
Acrylic (PMMA)	0.08–0.25
Acetal (POM)	0.22–0.35
Polypropylene (PP)	0.11–0.17
Polystyrene (PS)	0.12–0.12
Polyurethane thermoplastics (tpPU)	0.23–0.24
PVC	0.15–0.29
Teflon (PTFE)	0.24–0.26
Thermoset	
Epoxies	0.18–0.5
Phenolics	0.14–0.15
Polyester	0.28–0.3
Polymer foams	
Flexible polymer foam (VLD)	0.036–0.048
Flexible polymer foam (LD)	0.04–0.06
Flexible polymer foam (MD)	0.04–0.08
Rigid polymer foam (LD)	0.023–0.04
Rigid polymer foam (MD)	0.027–0.038
Rigid polymer foam (HD)	0.34–0.06

Table C.10   Thermal expansion, α

	$\alpha \ (10^{-6}/C)$
Metals	
Ferrous	
Cast irons	10–12.5
High carbon steels	11–13.5
Medium carbon steels	10–14
Low carbon steels	11.5–13
Low alloy steels	10.5–13.5
Stainless steels	13–20
Non-ferrous	
Aluminum alloys	21–24
Copper alloys	16.9–18
Lead alloys	18–32
Magnesium alloys	24.6–28
Nickel alloys	12–13.5
Titanium alloys	7.9–11
Zinc alloys	23–28
Ceramics	
Glasses	
Borosilicate glass	3.2–4.0
Glass ceramic	1–5
Silica glass	0.55–0.75
Soda-lime glass	9.1–9.5
Porous	
Brick	5–8
Concrete, typical	6–13
Stone	3.7–6.3
Technical	
Alumina	7–10.9
Aluminum nitride	4.9–6.2
Boron carbide	3.2–3.4
Silicon	2.2–2.7
Silicon carbide	4.0–5.1
Silicon nitride	3.2–3.6
Tungsten carbide	5.2–7.1
Composites	
Metal	
Aluminum/silicon carbide	15–23
Polymer	
CFRP	1–4
GFRP	8.6–33

Table C.10 (*Continued*)

	α $(10^{-6}/C)$
Natural	
Bamboo	2.6–10
Cork	130–230
Leather	40–50
Wood, typical (longitudinal)	2–11
Wood, typical (transverse)	32–42
Polymers	
Elastomer	
Butyl rubber	120–300
EVA	160–190
Isoprene (IR)	150–450
Natural rubber (NR)	150–450
Neoprene (CR)	575–610
Polyurethane elastomers (elPU)	150–165
Silicone elastomers	250–300
Thermoplastic	
ABS	84.6–234
Cellulose polymers (CA)	150–300
Ionomer (I)	180–306
Nylons (PA)	144–150
Polycarbonate (PC)	120–137
PEEK	72–194
Polyethylene (PE)	126–198
PET	114–120
Acrylic (PMMA)	72–162
Acetal (POM)	76–201
Polypropylene (PP)	122–180
Polystyrene (PS)	90–153
Polyurethane thermoplastics (tpPU)	90–144
PVC	100–150
Teflon (PTFE)	126–216
Thermoset	
Epoxies	58–117
Phenolics	120–125
Polyester	99–180
Polymer foams	
Flexible polymer foam (VLD)	120–220
Flexible polymer foam (LD)	115–220
Flexible polymer foam (MD)	115–220
Rigid polymer foam (LD)	20–80
Rigid polymer foam (MD)	20–75
Rigid polymer foam (HD)	22–70

Table C.11   Approximate production energies and CO_2 burden

	Energy (MJ/kg)	CO_2 (kg/kg)
Metals		
Ferrous		
Cast irons	16.4–18.2	1.0–1.1
High carbon steels	24.3–26.9	2.1–2.1
Medium carbon steels	23.4–25.8	2.0–2.2
Low carbon steels	22.4–24.8	1.9–2.1
Low alloy steels	31.0–34.3	1.9–2.2
Stainless steels	77.2–85.3	4.8–5.4
Non-ferrous		
Aluminum alloys	184–203	11.6–12.8
Copper alloys	63.0–69.7	3.9–4.4
Lead alloys	46.6–51.5	2.6–3.8
Magnesium alloys	356–394	22.4–24.8
Nickel alloys	127–140	7.9–8.8
Titanium alloys	885–945	41.7–59.5
Tungsten alloys	313–346	19.7–21.8
Zinc alloys	46.8–51.7	2.5–2.8
Ceramics		
Glasses		
Borosilicate glass	23.8–26.3	1.3–1.4
Glass ceramic	36.2–40.0	1.9–2.2
Silica glass	29.9–33.0	1.6–1.8
Soda-lime glass	13.0–14.4	0.7–0.8
Porous		
Brick	1.9–2.1	0.14–0.16
Concrete, typical	1.1–1.2	0.16–0.18
Stone	0.18–0.2	0.01–0.02
Technical		
Alumina	49.5–54.7	2.7–3.0
Aluminum nitride	209–231	11.3–12.5
Boron carbide	153–169	8.3–9.1
Silicon	56.9–62.9	3.1–3.4
Silicon carbide	70.2–77.6	3.8–4.2
Silicon nitride	70.2–77.6	3.79–4.18
Tungsten carbide	82.4–91.1	4.4–4.9
Composites		
Metal		
Aluminum/silicon carbide	250–300	14–16
Polymer		
CFRP	259–286	21–23
GFRP	107–118	7.5–8.3

Table C.11   (*Continued*)

	Energy (MJ/kg)	CO$_2$ (kg/kg)
Natural		
Bamboo	14.4–15.9	− 1.2− − 1.0
Cork	28.5–31.5	0.19–0.21
Leather	102–113	2.6–2.8
Wood, typical (longitudinal)	14.4–15.9	− 1.2− − 1.0
Wood, typical (transverse)	14.4–15.9	− 1.2− − 1.0
Polymers		
Elastomer		
Butyl rubber	76.5–84.6	2.1–2.4
Eva	86.7–95.8	2.9–3.2
Isoprene (IR)	76.5–84.6	2.2–2.4
Natural rubber (NR)	38.0–42.0	− 0.5− − 0.4
Neoprene (CR)	95.9–106	3.5–3.9
Polyurethane Elastomers (elPU)	109–120	4.5–4.9
Silicone elastomers	152–168	8.2–9.0
Thermoplastic		
ABS	91–102	3.27–3.62
Cellulose polymers (CA)	108–119	4.4–4.87
Ionomer (I)	102–112	3.96–4.38
Nylons (PA)	102–113	4.0–4.41
Polycarbonate (PC)	105–116	3.8–4.2
PEEK	223–246	12.7–14
Polyethylene (PE)	76.9–85	1.95–2.16
PET	79.6–88	2.21–2.45
Acrylic (PMMA)	93.8–104	3.4–3.76
Acetal (POM)	99.5–110	3.8–4.2
Polypropylene (PP)	75.4–83.3	2.07–2.09
Polystyrene (PS)	96–106	2.85–3.13
Polyurethane thermoplastics (tpPU)	113–125	4.77–5.28
PVC	63.5–70.2	1.85–2.04
Teflon (PTFE)	145–160	7.07–7.81
Thermoset		
Epoxies	90–100	3.2–3.6
Phenolics	86–95	2.8–3.2
Polyester	84–90	2.7–3.0
Polymer foams		
Flexible polymer foam (VLD)	113–125	4.78–5.28
Flexible polymer foam (LD)	113–125	4.78–5.28
Flexible polymer foam (MD)	113–125	4.78–5.28
Rigid polymer foam (LD)	138–153	6.59–7.28
Rigid polymer foam (MD)	155–171	7.78–8.8
Rigid polymer foam (HD)	150–165	7.42–8.19

Table C.12   Environmental resistance

	Flammability	Fresh water	Salt water	Sunlight (UV)	Wear resistance
Metals					
Ferrous					
Cast irons	A	B	C	A	A
High carbon steels	A	B	C	A	A
Medium carbon steels	A	B	C	A	A
Low carbon steels	A	B	C	A	A
Low alloy steels	A	B	C	A	A
Stainless steels	A	A	A	A	B
Non-ferrous					
Aluminum alloys	B	A	B	A	C
Copper alloys	A	A	A	A	A
Lead alloys	A	A	A	A	C
Magnesium alloys	A	A	D	A	C
Nickel alloys	A	A	A	A	B
Titanium alloys	A	A	A	A	C
Zinc alloys	A	A	C	A	E
Ceramics					
Glasses					
Borosilicate glass	A	B	B	A	A
Glass ceramic	A	A	A	A	A
Silica glass	A	A	A	A	B
Soda-lime glass	A	A	A	A	A
Porous					
Brick, concrete, stone	A	A	A	A	C
Technical					
Alumina	A	A	A	A	A
Aluminum nitride	A	A	A	A	A
Boron carbide	A	A	A	A	A
Silicon	A	A	B	A	B
Silicon carbide	A	A	A	A	A
Silicon nitride	A	A	A	A	A
Tungsten carbide	A	A	A	A	A
Composites					
Metal					
Aluminum/silicon carbide	A	A	B	A	B
Polymer					
CFRP	B	A	A	B	C
GFRP	B	A	A	B	C

Table C.12   (*Continued*)

	Flammability	Fresh water	Salt water	Sunlight (UV)	Wear resistance
Natural					
Bamboo	D	C	C	B	D
Cork	D	B	B	A	B
Leather	D	B	B	B	B
Wood	D	C	C	B	D
Polymers					
Elastomer					
Butyl rubber	E	A	A	B	B
EVA	E	A	A	B	B
Isoprene (IR)	E	A	A	B	B
Natural rubber (NR)	E	A	A	B	B
Neoprene (CR)	E	A	A	B	B
Polyurethane elastomers (elPU)	E	A	A	B	B
Silicone elastomers	B	A	A	B	B
Thermoplastic					
ABS	D	A	A	C	D
Cellulose polymers (CA)	D	A	A	B	C
Ionomer (I)	D	A	A	B	C
Nylons (PA)	C	A	A	C	C
Polycarbonate (PC)	B	A	A	B	C
PEEK	B	A	A	A	C
Polyethylene (PE)	D	A	A	D	C
PET	D	A	A	B	C
Acrylic (PMMA)	D	A	A	A	C
Acetal (POM)	D	A	A	C	B
Polypropylene (PP)	D	A	A	D	C
Polystyrene (PS)	D	A	A	C	D
Polyurethane thermoplastics (tpPU)	C	A	A	B	C
PVC	A	A	A	A	C
Teflon (PTFE)	A	A	A	B	B
Thermoset					
Epoxies	B	A	A	B	C
Phenolics	B	A	A	A	C
Polyester	D	A	A	A	C
Polymer foams					
Flexible polymer foams	E	A	A	C	D
Rigid polymer foams	C	A	A	B	E

Ranking: A = very good; B = good; C = average; D = poor; E = very poor.

Appendix D

Information and knowledge sources for materials and processes

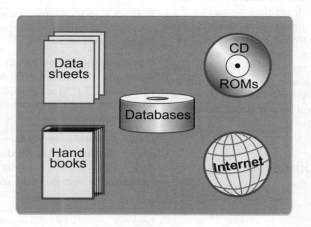

D.1 Introduction

This appendix tells you where to look to find information, both structured and unstructured, for material attributes. The sources, broadly speaking, are of three sorts: hard-copy, software, and the Internet. The hard-copy documents listed below will be found in most engineering libraries. The computer data-bases are harder to find: the supplier is listed, with address and contact number, as well as the hardware required to run the data-base. Internet sites are easy to find but can be frustrating to use.

Section D.2 catalogs information sources for families and classes of material, with a brief commentary where appropriate. Section D.3 provides a starting point for reading on processes. Section D.4 lists software for materials and process data, information and selection. Sections D.5 and D.6 list additional Internet sites on which materials information can be found

D.2 Information sources for materials

All materials

Few hard-copy data-sources span the full spectrum of materials and properties. Six that, in different ways, attempt to do so are listed below.

Materials selector (1997) *Materials Engineering*, Special Issue. Penton Publishing, Cleveland, OH, USA. *Tabular data for a broad range of metals, ceramics, polymers and composites, updated annually. Basic reference work.*

Chapman and Hall, *Materials Selector* (1996) edited by N.A. Waterman and M.F. Ashby. Chapman and Hall, London, UK. *A 3-volume compilation of data for all materials, with selection and design guide. Basic reference work.*

ASM Engineered Materials Reference Book, 2nd Edition (1994) editor: Bauccio, M.L., ASM International, Metals Park, OH, USA. *Compact compilation of numeric data for metals, polymers, ceramics and composites.*

Materials Selector and Design Guide (1974) Design Engineering, Morgan-Grampian Ltd, London. *Resembles the Materials Engineering "Materials Selector", but less detailed and now rather dated.*

Handbook of Industrial Materials, 2nd edition (1992) Elsevier, Oxford, UK. *A compilation of data remarkable for its breadth: metals, ceramics, polymers, composites, fibers, sandwich structures, leather . . . .*

Materials Handbook, 2nd edition (1986) Editors: Brady, G.S. and Clauser, H.R., McGraw-Hill, New York, NY, USA. *A broad survey, covering metals, ceramics, polymers, composites, fibers, sandwich structures and more.*

Handbook of Thermophysical Properties of Solid Materials (1961) Goldsmith, A., Waterman, T.E. and Hirschhorn, J.J. Macmillan, New York, USA. *Thermophysical and thermochemical data for elements and compounds.*

Guide to Engineering Materials Producers (1994) editor: Bittence, J.C. ASM International, Metals Park, OH, USA. *A comprehensive catalog of addresses for material suppliers.*

Internet sources of information on all classes of materials
ASM Handbooks on line, www.asminternational.org/hbk/index.jsp

ASM Alloy center, www.asminternational.org/alloycenter/index.jsp

ASM Materials Information, www.asminternational.org/matinfo/index.jsp

AZOM.com, www.azom.com

Design InSite, www.designinsite.dk

Goodfellow, www.goodfellow.com

K&K Associate's thermal connection, www.tak2000.com

Corrosion source (databases), corrosionsource.com/links.htm — matdbase

Material data network, www.matdata.net

Materials (Research): Alfa Aesar, www.alfa.com

MatWeb, www.matweb.com

MSC datamart, www.mscsoftware.com

NASA Long Duration Exposure Facility, **setas**, www.larc.nasa.gov/setas/PrISMtoHTML/prism_ldef_mat.html

NPL MIDAS, midas.npl.co.uk/midas/index.jsp

All metals

Metals and alloys conform to national and (sometimes) international standards. One consequence is the high quality of data. Hard copy sources for metals data are generally comprehensive, well-structured and easy to use.

ASM Metals Handbook, 9th edition (1986), and 10th edition (1990) ASM International, Metals Park, OH 44073, USA. The 10th edition contains Vol. 1: *Irons and Steels*; Vol. 2: *Non-ferrous Alloys*; Vol. 3: *Heat Treatment*; Vol. 4: *Friction, Lubrication and Wear*; Vol. 5: *Surface Finishing and Coating*; Vol. 6: *Welding and Brazing*; Vol. 7: *Microstructural Analysis Basic reference work, continuously upgraded and expanded.*

ASM Metals Reference Book, 3rd edition, (1993), ed. M.L. Bauccio, ASM International, Metals Park, OH 44073, USA *Consolidates data for metals from a number of ASM publications. Basic reference work.*

Smithells, C.J. (1992), *Metals Reference Book*, 7th edition (Editors: E.A. Brandes and G.B. Brook). Butterworths, London, UK. *A comprehensive compilation of data for metals and alloys. Basic reference work.*

Metals Databook (1990), Colin Robb. The Institute of Metals, 1 Carlton House Terrace, London SW1Y 5DB, UK. *A concise collection of data on metallic materials covered by the UK specifications only.*

Guide to Materials Engineering Data and Information (1986) ASM International, Metals Park, OH, 44073, USA. *A directory of suppliers, trade organizations and publications on metals.*

The Metals Black Book, Vol. 1: *Steels* (1992) editor: J.E. Bringas, Casti Publishing Inc., 14820-29 Street,

Edmonton, Alberta T5Y 2B1, Canada. *A compact book of data for steels.*

The Metals Red Book, Vol. 2: *Nonferrous Metals* (1993), editor: J.E. Bringas, Casti Publishing Inc. 14820-29 Street, Edmonton, Alberta T5Y 2B1, Canada.

Internet sources for two or more metals classes
ASM International Handbooks, www.asminternational.org/hbk/index.jsp

ASM International, Alloy Center, www.asminternational.org/alloycenter/index.jsp

Carpenter Technology Home Page, www.cartech.com

CASTI Publishing Site Catalog, www.casti-publishing.com/intsite.htm

CMW Inc. Home Page, www.cmwinc.com

Engelhard Corporation-Electro Metallics Department, www.engelhard.com

Eurometaux, www.eurometaux.org

Materials (high performance): Mat-Tech, www.mat-tech.com

Metsteel.com, www.metsteel.com

Rare earths: Pacific Industrial Development Corp. pidc.com

Rare Metals: Stanford Materials Inc. www.stanfordmaterials.com

Refractory Metals: Teledyne Wah Chang, www.twca.com

Non-ferrous metals and alloys

In addition to the references listed under the section "All metals", the following sources give data for specific metals and alloys

Pure metals

Most of the sources listed in the previous section contain some information on pure metals. However, the publications listed below are particularly useful in this respect.

Emsley, J. (1989) *The Elements*, Oxford University Press, Oxford, UK. *A book aimed more at chemists and physicists than engineers with good coverage of chemical, thermal, and electrical properties but not mechanical properties. A new edition is expected early in 1997.*

Brandes, E.A. and Brook, G.B. (eds.) (1992) *Smithells Metals Reference Book*, 7th edition, Butterworth-Heinemann, London. *Data for the mechanical, thermal and electrical properties of pure metals.*

Goodfellow Catalogue (1995–96), Goodfellow Cambridge Limited, Cambridge Science Park, Cambridge CB4 4DJ, UK. *Useful though patchy data for mechanical, thermal and electrical properties of pure metals in a tabular format. Free.*

Alfa Aesar Catalog (1995–96) Johnson Matthey Catalog Co. Inc., 30 Bond Street, Ward Hill, MA, 01835-8099, USA. *Coverage similar to that of the Goodfellow Catalogue. Free.*

Samsonov, G.V. (ed.) (1968) Handbook of the Physiochemical Properties of the Elements, Oldbourne, London. *An extensive compilation of data from Western and Eastern sources. Contains a number of inaccuracies, but also contains a large quantity of data on the rarer elements, hard to find elsewhere.*

Gschneidner, K.A. (1964) Physical Properties and Interrelationships of Metallic and Semimetallic Elements, *Solid State Physics* 16, 275–426. *Probably the best source of its time, this reference work is very well referenced, and full explanations are given of estimated or approximate data.*

Internet sources
Winter, M., *WebElements*. Sheffield: University of Sheffield, www.webelements.com. *A comprehensive source of information on all the elements in the Periodic Table. If it has a weakness, it is in the definitions and values of some mechanical properties.*

Martindale's: physics web pages, www.martindalecenter.com/GradPhysics.html

Aluminum alloys

Aluminum Standards and Data (1990) The Aluminum Association Inc., 900, 19th Street N.W., Washington, D.C., USA.

The Properties of Aluminum and its Alloys (1981) The Aluminum Federation, Broadway House, Calthorpe Road, Birmingham, UK.

Technical Data Sheets (1993) ALCAN International Ltd, Kingston Research and Development Center, Box 8400, Kingston, Ontario, Canada KL7 4Z4, and Banbury Laboratory, Southam Road, Banbury, Oxon, UK, X16 7SP.

Technical Data Sheets (1993) ALCOA, Pittsburg, PA, USA.

Technical Data Sheets (1994) Aluminum Pechiney, 23 Bis, rue Balzac, Paris 8, BP 78708, 75360 Paris Cedex 08, France.

Internet sources
Aluminum Federation, www.alfed.org.uk

Aluminum World, www.sovereign-publications. com/index.htm

International Aluminum Institute, www.world-aluminum.org

Babbitt Metal

The term "Babbitt Metal" denotes a series of lead–tin–antimony bearing alloys, the first of which was patented in the USA by Isaac Babbitt in 1839. Subsequent alloys are all variations on his original composition.

ASTM Standard B23-83: "White Metal Bearing Alloys (Known commercially as 'Babbitt Metal')", ASTM Annual Book of Standards, Vol. 02.04.

Beryllium

Designing with Beryllium (1996) Brush Wellman Inc, 1200 Hana Building, Cleveland, OH 44115, USA.

Beryllium Optical Materials (1996) Brush Wellman Inc, 1200 Hana Building, Cleveland, OH 44115, USA.

Internet sources
Brush Wellman, www.brushwellman.com

Cadmium

International Cadmium Association (1991) *Cadmium Production, Properties and Uses*, ICdA, London, UK.

Chromium

ASTM Standard A560–89: *Castings, Chromium-Nickel Alloy*, ASTM Annual Book of Standards, Vol. 01.02.

Cobalt alloys

Betteridge, W. (1982) *Cobalt and its Alloys*, Ellis Horwood, Chichester, UK *A good general introduction to the subject.*

Columbium alloys

See Niobium alloys

Copper alloys

ASM Metals Handbook, 10th edition (1990) ASM International, Metals Park, OH 44073, USA.

The Selection and Use of Copper-based Alloys (1979) E.G. West, Oxford University Press, Oxford, UK.

Copper Development Association Data Sheets, 26 (1988), 27 (1981), 31 (1982), 40 (1979), and Publication 82 (1982), Copper Development Association Inc., Greenwich Office, Park No 2, Box 1840, Greenwich, CT 06836, USA, and The Copper Development Association, Orchard House, Mutton Lane, Potters Bar, Herts, EN6 3AP, UK.

The Copper Development Association (1994) *Megabytes on Coppers*, Orchard House, Mutton Lane, Potters Bar, Herts EN6 3AP, UK; and Granta Design Limited, Rustat House, 62 Clifton Road, Cambridge, CB1 7EG, UK.

Smithells Metals Reference Book, 7th edition, Eds. E.A. Brandes and G.B. Brook, Butterworth-Heinemann Ltd, Oxford UK (1992).

Internet sources
Copper Development Association, www.cda.org.uk

Copper page, www.copper.org

Gold and dental alloys

Gold: art, science and technology (1992) *Interdisciplinary Science Reviews*, **17**(3). ISSN 0308-0188.

Focus on gold (1992) *Interdisciplinary Science Reviews*, **17**(4). ISSN 0308-0188.

ISO Standard 1562:1993, *Dental casting gold alloys*, International Standards Organisation, Switzerland.

ISO Standard 8891:1993, *Dental casting alloys with Noble metal content of 25% up to but not including 75%*, International Standards Organisation, Switzerland.

Internet sources

Goodfellow Metals http://www.goodfellow.com/. *Supplier of pure and precious metals, mainly for laboratory use. Web site contains price and property information for their entire stock.*

Jeneric Pentron Inc., "Casting Alloys", http://www.jeneric.com/casting USA. *An informative commercial site.*

O'Brien, W.J., "Biomaterial Properties Database", http://www.lib.umich.edu/libhome/Dentistry.lib/Dental_tables School of Dentistry, University of Michigan, USA. *An extensive source of information, both for natural biological materials and for metals used in dental treatments.*

Rand Refinery Limited, http://www.bullion.org.za/associates/rr.htm. Chamber of Mines Web-site, SOUTH AFRICA. *Contains useful information on how gold is processed to varying degrees of purity.*

Indium

The Indium Info Center, http://www.indium.com/metalcenter.html, Indium Corp. of America.

Lead

ASTM Standard B29-79: *Pig Lead*, ASTM Annual Book of Standards, Vol. 02.04.

ASTM Standard B102-76: *Lead- and Tin- Alloy Die Castings*, ASTM Annual Book of Standards, Vol. 02.04.

ASTM Standard B749-85: *Lead and Lead Alloy Strip, Sheet, and Plate Products*, ASTM Annual Book of Standards, Vol. 02.04.

Lead Industries Association, *Lead for Corrosion Resistant Applications*, LIA Inc., New York, USA.

ASM Metals Handbook, (1986) 9th edition, Vol. 2, pp. 500–510.

See also Babbitt Metal (above).

Internet sources

Lead Development Association International, www.ldaint.org/default.htm

India Lead Zinc Development Association, www.ilzda.com

Magnesium alloys

Technical Data Sheets (1994) Magnesium Elektron Ltd., PO Box 6, Swinton, Manchester, UK.

Technical Literature (1994) Magnesium Corp. of America, Div. of Renco, Salt Lake City, UT, USA.

Molybdenum

ASTM Standard B386-85: *Molybdenum and Molybdenum Alloy Plate, Sheet, Strip and Foil*, ASTM Annual Book of Standards, Vol. 02.04.

ASTM Standard B387-85: *Molybdenum and Molybdenum Alloy Bar, Rod and Wire*, ASTM Annual Book of Standards, Vol. 02.04.

Nickel

A major data sources for Nickel and its alloys is the Nickel Development Institute (NIDI), a global organization with offices in every continent except Africa. NIDI freely gives away large quantities of technical reports and data compilations, not only for nickel and high-nickel alloys, but also for other nickel-bearing alloys, e.g. stainless steel.

ASTM Standard A297-84, *Steel Castings, Iron-Chromium and Iron-Chromium-Nickel, Heat Resistant, for General Application*, ASTM Annual Book of Standards, Vol. 01.02.

ASTM Standard A344-83, *Drawn or Rolled Nickel-Chromium and Nickel-Chromium-Iron Alloys for Electrical Heating Elements*, ASTM Annual Book of Standards, Vol. 02.04.

ASTM Standard A494-90, *Castings, Nickel and Nickel Alloy*, ASTM Annual Book of Standards, Vol. 02.04.

ASTM Standard A753-85, *Nickel–Iron Soft Magnetic Alloys*, ASTM Annual Book of Standards, Vol. 03.04.

Betteridge, W., *Nickel and its alloys*, Ellis Horwood, Chichester, UK (1984). *A good introduction to the subject.*

INCO Inc., *High-Temperature, High-Strength Nickel Base Alloys*, Nickel Development Institute (1995). *Tabular data for over 80 alloys.*

Elliott, P (1990) *Practical Guide to High-Temperature Alloys*, Nickel Development Institute, Birmingham, UK.

INCO Inc. (1978) *Heat & Corrosion Resistant Castings*, Nickel Development Institute, Birmingham, UK.

INCO Inc. (1969) *Engineering Properties of some Nickel Copper Casting Alloys*, Nickel Development Institute, Birmingham, UK.

INCO Inc. (1968) *Engineering Properties of IN-100 Alloy*, Nickel Development Institute, Birmingham, UK.

INCO Inc. (1969) *Engineering Properties of Nickel-Chromium Alloy 610 and Related Casting Alloys*, Nickel Development Institute, Birmingham, UK.

INCO Inc. (1968) *Alloy 713C: Technical Data*, Nickel Development Institute, Birmingham, UK.

INCO Inc. (1981) *Alloy IN-738: Technical Data*, Nickel Development Institute, Birmingham, UK.

INCO Inc. (1976) *36% Nickel-Iron Alloy for Low Temperature Service*, Nickel Development Institute, Birmingham, UK.

ASTM Standard A658 (Discontinued 1989) *Pressure Vessel Plates, Alloy Steel, 36 Percent Nickel*, ASTM Annual Book of Standards, pre-1989 editions.

ASM Metals Handbook, 9th edition. (1986) Vol. 3, pp. 125–178.

Carpenter Technology Corp. Website, http//www.cartech.com/

Internet sources
Nickel Development Institute, www.nidi.org

Nickel: INCO website, www.incoltd.com

Special Metals Corp., www.specialmetals.com

Steel & nickel based alloys, www.superalloys.co.uk/start.html

Niobium (Columbium) alloys

ASTM Standard B391-89: *Niobium and Niobium Alloy Ingots*, ASTM Annual Book of Standards, Vol. 02.04.

ASTM Standard B392-89: *Niobium and Niobium Alloy Bar, Rod and Wire*, ASTM Annual Book of Standards, Vol. 02.04.

ASTM Standard B393-89: *Niobium and Niobium Alloy Strip, Sheet and Plate*, ASTM Annual Book of Standards, Vol. 02.04.

ASTM Standard B652-85: *Niobium–Hafnium Alloy Ingots*, ASTM Annual Book of Standards, Vol. 02.04.

ASTM Standard B654-79: *Niobium–Hafnium Alloy Foil, Sheet, Strip and Plate*, ASTM Annual Book of Standards, Vol. 02.04.

ASTM Standard B655-85: *Niobium–Hafnium Alloy Bar, Rod and Wire*, ASTM Annual Book of Standards, Vol. 02.04.

Internet sources
Husted, R, http://www-c8.lanl.gov/infosys/html/periodic/41.html Los Alamos National Laboratory, USA. *An overview of Niobium and its uses.*

Palladium

ASTM Standard B540-86: *Palladium Electrical Contact Alloy*, ASTM Annual Book of Standards, Vol. 03.04.

ASTM Standard B563-89: *Palladium–Silver–Copper Electrical Contact Alloy*, ASTM Annual Book of Standards, Vol. 03.04.

ASTM Standard B589-82: *Refined Palladium*, ASTM Annual Book of Standards, Vol. 02.04.

ASTM Standard B683-90: *Pure Palladium Electrical Contact Material*, ASTM Annual Book of Standards, Vol. 03.04.

ASTM Standard B685-90: *Palladium–Copper Electrical Contact Material*, ASTM Annual Book of Standards, Vol. 03.04.

ASTM Standard B731-84: *60% Palladium-40% Silver Electrical Contact Material*, ASTM Annual Book of Standards, Vol. 03.04.

Internet sources
Jeneric Pentron Inc., Casting alloys, http://www.jeneric.com/casting USA. *An informative commercial site, limited to dental alloys.*

Goodfellow Metals http://www.goodfellow.com/. *Supplier of pure and precious metals, mainly for laboratory use. Web site contains price and property information for their entire stock.*

Platinum alloys

ASTM Standard B684-81: *Platinum–Iridium Electrical Contact Material*, ASTM Annual Book of Standards, Vol. 03.04.

Elkonium Series 400 Datasheets, CMW Inc., Indiana, USA.

ASM Metals Handbook, 9th edition, Vol. 2, pp. 688–698 (1986).

Silver alloys

ASTM Standard B413-89: *Refined Silver*, ASTM Annual Book of Standards, Vol. 02.04.

ASTM Standard B617-83: *Coin Silver Electrical Contact Alloy*, ASTM Annual Book of Standards, Vol. 03.04.

ASTM Standard B628-83: *Silver–Copper Eutectic Electrical Contact Alloy*, ASTM Annual Book of Standards, Vol. 03.04.

ASTM Standard B693-87: *Silver–Nickel Electrical Contact Materials*, ASTM Annual Book of Standards, Vol. 03.04.

ASTM Standard B742-90: *Fine Silver Electrical Contact Fabricated Material*, ASTM Annual Book of Standards, Vol. 03.04.

ASTM Standard B780-87: *75% Silver, 24.5% Copper, 0.5% Nickel Electrical Contact Alloy*, ASTM Annual Book of Standards, Vol. 03.04.

Elkonium Series 300 Datasheets, CMW Inc., 70 S. Gray Street, PO Box 2266, Indianapolis, Indiana, USA (1996).

Elkonium Series 400 Datasheets, CMW Inc., 70 S. Gray Street, PO Box 2266, Indianapolis, Indiana, USA (1996).

Internet sources
Silver Institute, www.silverinstitute.org

Jeneric Pentron Inc., Casting Alloys, http://www.jeneric.com/casting, USA. *An informative commercial site, limited to dental alloys.*

Goodfellow Metals http://www.goodfellow.com/. *Supplier of pure and precious metals, mainly for laboratory use. Web site contains price and property information for their entire stock.*

Tantalum alloys

ASTM Standard B365-86: *Tantalum and Tantalum Alloy Rod and Wire*, ASTM Annual Book of Standards, Vol. 02.04.

ASTM Standard B521-86: *Tantalum and Tantalum Alloy Seamless and Welded Tubes*, ASTM Annual Book of Standards, Vol. 02.04.

ASTM Standard B560-86: *Unalloyed Tantalum for Surgical Implant Applications*, ASTM Annual Book of Standards, Vol. 15.01.

ASTM Standard B708-86: *Tantalum and Tantalum Alloy Plate, Sheet and Strip*, ASTM Annual Book of Standards, Vol. 02.04.

Tantalum Data Sheet (1996) The Rembar Company Inc., 67 Main St., Dobbs Ferry, NY 10522, USA.

ASM Handbook, 9th edition, Vol. 3, pp. 323–325 and 343–347 (1986).

Tin alloys

ASTM Standard B32-89: *Solder Metal*, ASTM Annual Book of Standards, Vol. 02.04.

ASTM Standard B339-90: *Pig Tin*, ASTM Annual Book of Standards, Vol. 02.04.

ASTM Standard B560-79: *Modern Pewter Alloys*, ASTM Annual Book of Standards, Vol. 02.04.

Barry, BTK and Thwaites, CJ, *Tin and its Alloys and Compounds*, Ellis Horwood, Chichester, UK (1983).

ASM Metals Handbook, 9th edition, Vol. 2, pp. 613–625.

See also Babbitt Metal (above)
Tin Research Association, www.tintechnology.com

Titanium alloys

Technical Data Sheets (1993) Titanium Development Association, 4141 Arapahoe Ave., Boulder, CO, USA.

Technical Data Sheets (1993) The Titanium Information Group, c/o Inco Engineered Products, Melbourne, UK.

Technical Data Sheets (1995) IMI Titanium Ltd. PO Box 704, Witton, Birmingham B6 7UR, UK.

Internet sources
Titanium Information Group, www.titaniuminfogroup.co.uk

The International Titanium Association http://www.titanium.org/ *This Association site has a large list of member companies and comprehensive information on titanium and its alloys.*

Tungsten alloys

ASTM Standard B777–87, *Tungsten Base, High-Density Metal*, ASTM Annual Book of Standards, Vol. 02.04.

Yih, SWH and Wang, CT, *Tungsten*, Plenum Press, New York (1979).

ASM Metals Handbook, 9th edition, Vol. 7, p. 476 (1986).

Tungsten Data Sheet (1996) The Rembar Company Inc., 67 Main St., Dobbs Ferry, NY 10522, USA.

Royal Ordnance Speciality Metals datasheet (1996) British Aerospace Defence Ltd., PO Box 27, Wolverhampton, West Midlands, WV10 7NX, UK.

CMW Inc. Datasheets (1996) CMW Inc., 70 S. Gray Street, PO Box 2266, Indianapolis, Indiana, USA.

Internet sources
North American Tungsten,
www.northamericantungsten.com

Uranium

Uranium Information Centre, Australia, www.uic.com.au

Ux Jan 96 Uranium Indicator Update, www.uxc.com/review/uxc_g_ind-u.html

Vanadium

Teledyne Wah Chang, *Vanadium Brochure* (1996) TWC, Albany, Oregon, USA.

Zinc

ASTM Standard B6-87: *Zinc*, ASTM Annual Book of Standards, Vol. 02.04, ASTM, USA.

ASTM Standard B69-87: *Rolled Zinc*, ASTM Annual Book of Standards, Vol. 02.04, ASTM, USA.

ASTM Standard B86-88: *Zinc–Alloy Die Castings*, ASTM Annual Book of Standards, Vol. 02.02, ASTM, USA.

ASTM Standard B418-88: *Cast and Wrought Galvanic Zinc Anodes*, ASTM Annual Book of Standards, Vol. 02.04, ASTM, USA.

ASTM Standard B791-88: *Zinc–Aluminum Alloy Foundry and Die Castings*, ASTM Annual Book of Standards, Vol. 02.04, ASTM, USA.

ASTM Standard B792-88: *Zinc Alloys in Ingot Form for Slush Casting*, ASTM Annual Book of Standards, Vol. 02.04, ASTM, USA.

ASTM Standard B793-88: *Zinc Casting Alloy Ingot for Sheet Metal Forming Dies*, ASTM Annual Book of Standards, Vol. 02.04, ASTM, USA.

Goodwin, F.E. and Ponikvar, A.L. (eds) (1989) *Engineering Properties of Zinc Alloys*, 3rd edition, International Lead Zinc Research Organization, North Carolina, USA (1989). *An excellent compilation of data, covering all industrially important zinc alloys.*

Chivers, A.R.L. (1981) *Zinc Diecasting*, Engineering Design Guide no. 41, OUP, Oxford, UK. *A good introduction to the subject.*

ASM Metal Handbook, Properties of Zinc and Zinc Alloys, 9th edition (1986), Vol. 2, pp. 638–645.

Internet sources
International Zinc Association, www.iza.com

Zinc Industrias Nacionales S.A.-Peru, www.zinsa.com/espanol.htm

Zinc: Eastern Alloys, www.eazall.com

India Lead Zinc Development Association, www.ilzda.com

Zirconium

ASTM Standard B350-80: *Zirconium and Zirconium Alloy Ingots for Nuclear Application*, ASTM Annual Book of Standards, Vol. 02.04, ASTM, USA.

ASTM Standard B352-85, B551-83 and B752-85: *Zirconium and Zirconium Alloys*, ASTM Annual Book of Standards, Vol. 02.04, ASTM, USA.

Teledyne Wah Chang (1996) *Zircadyne: Properties & Applications*, TWC, Albany, OR 97231, USA.

ASM Metals Handbook (1986) 9th edition, Vol. 2, pp. 826–831.

Ferrous metals

Ferrous metals are probably the most thoroughly researched and documented class of materials. Nearly every developed country has its own system of standards for irons and steels. Recently, continental and worldwide standards have been developed, which have achieved varying levels of acceptance. There is a large and sometimes confusing literature on the subject. This section is intended to provide the user with a guide to some of the better information sources.

Ferrous metals, general data sources

Bringas, J.E. (ed.) (1995) The Metals Black Book — Ferrous Metals, 2nd edition, CASTI Publishing, Edmonton, Canada (1995). *An excellent short reference work.*

ASM Metals Handbook, 10th edition, Vol. 1 (1990) ASM International, Metals Park, Cleveland, OH, USA. *Authoritative reference work for North American irons and steels.*

ASM Metals Handbook, Desk edition (1985) ASM International, Metals Park, Cleveland, OH, USA. *A summary of the multi-volume ASM Metals Handbook.*

Wegst, C.W. *Stahlschlüssel* (in English: *"Key to Steel"*), Verlag Stahlschlüssel Wegst GmbH, D-1472 Marbach, Germany. *Published every 3 years, in German, French and English. Excellent coverage of European products and manufacturers.*

Woolman, J. and Mottram, R.A. (1966) *The Mechanical and Physical Properties of the British Standard En Steels*, Pergamon Press, Oxford (1966). *Still highly regarded, but is based around a British Standard classification system that has been officially abandoned.*

Brandes, E.A. and Brook, G.R. (eds) (1992) *Smithells Metals Reference Book*, 7th edition, Butterworth-Heinemann, Oxford, UK (1992). *An authoritative reference work, covering all metals.*

Waterman, N.A. and Ashby, M.F. (eds.) (1996) Materials Selector, Chapman and Hall, London, UK *Covers all materials — irons and steels are in Vol. 2.*

Sharpe, C. (ed.) (1993) *Kempe's Engineering Year-Book*, 98th edition, Benn, Tonbridge, Kent, UK. *Updated each year — has good sections on irons and steels.*

Iron and steels standards

Increasingly, national and international standards organizations are providing a complete catalogue of their publications on the world-wide web. Two of the most comprehensive printed sources are listed below:

Iron and Steel Specifications, 9th edition (1998), British Iron and Steel Producers Association (BISPA),

5 Cromwell Road, London, SW7 2HX. *Comprehensive tabulations of data from British Standards on irons and steels, as well as some information on European and North American standards. The same informantion is available on searchable CD.*

ASTM Annual Book of Standards, Vols 01.01 to 01.07. *The most complete set of American iron and steel standards. Summaries of the standards can be found on the WWW at http://www.astm.org/stands.html*

Cross-referencing of similar international standards and grades

It is difficult to match, even approximately, equivalent grades of iron and steel between countries. No coverage of this subject can ever be complete, but the references listed below are helpful:

Gensure, J.G. and Potts, D.L. (1988) *International Metallic Materials Cross Reference*, 3rd edition, Genium Publishing, New York. *Comprehensive worldwide coverage of the subject, well indexed.*

Bringas, J.E. (ed.) (1995) The Metals Black Book — Ferrous Metals, 2nd edition, CASTI Publishing, Edmonton, Canada. *Easy-to-use tables for international cross-referencing. (See General section for more information.)*

Unified Numbering System for Metals and Alloys, 2nd edition (1977), Society of Automotive Engineers, Pennsylvania. *An authoritative reference work, providing a unifying structure for all standards published by US organizations. No coverage of the rest of the world.*

Iron and Steel Specifications, 7th edition (1989), British Steel, 9 Albert Embankment, London, SE1 7SN. *Lists "Related Specifications" for France, Germany, Japan, Sweden, UK and USA.*

Cast irons

Scholes, J.P. (1979) The Selection and Use of Cast Irons, Engineering Design Guides, OUP, Oxford, UK.

Angus, H.T. (1976) *Cast Iron: Physical and Engineering Properties*, Butterworths, London.

Gilbert, G.N.J. (1977) *Engineering Data on Grey Cast Irons.*

Gilbert, G.N.J. (1986) *Engineering Data on Nodular Cast Irons.*

Gilbert, G.N.J. (1983) *Engineering Data on Malleable Cast Irons.*

Smith, L.W.L., Palmer, K.B. and Gilbert, G.N.J. (1986) Properties of modern malleable irons.

Palmer, K.B. (1988) *Mechanical & Physical Properties of Cast Irons at Sub-zero Temperatures.*

Palmer, K.B. (1986) *Mechanical & Physical Properties of Cast Irons up to 500°C.*

Irons, American standards

These can all be found in the Annual Book of ASTM Standards, Vol. 01.02

ASTM A220M-88: *Pearlitic Malleable Iron.*

ASTM A436-84: *Austenitic Gray Iron Castings.*

ASTM A532: *Abrasion-Resistant Cast Irons.*

ASTM A602-70 (Reapproved 1987): *Automotive Malleable Iron Castings.*

Cast irons, international standards

These are available from ISO Central Secretariat, 1, rue de Varembe, Case postale 56, CH-1211 Geneve 20, Switzerland.

ISO 185:1988 *Grey cast iron — classification.*

ISO 2892:1973 *Austenitic Cast Iron.*

ISO 5922:1981 *Malleable Cast Iron.*

Cast irons, British standards

Compared with steels, there are relatively few standards on cast iron, which makes it feasible to list them all. Standards are available from BSI Customer Services, 389 Chiswick High Road, London, W4 4AL, UK.

BS 1452:1990 *Flake Graphite Cast Iron.*

BS 1591:1975 *Specification for Corrosion Resisting High Silicon Castings.*

BS 2789:1985 *Iron Castings with Spheroidal or Nodular Graphite.*

BS 3468:1986 *Austenitic Cast Iron.*

BS 4844:1986 *Abrasion Resisting White Cast Iron.*

BS 6681:1986 *Specification for Malleable Cast Iron.*

Carbon and alloy steels

ASM Metals Handbook, 10th edition, Vol. 1 (1990), ASM International, Metals Park, Cleveland, Ohio, USA. *Authoritative reference work for North American irons and steels.*

Fox, J.H.E. *An Introduction to Steel Selection: Part 1, Carbon and Low-Alloy Steels*, Engineering Design Guide no. 34, Oxford University Press.

Stainless steels

ASM Metals Handbook, 10th edition, Vol. 1 (1990), ASM International, Metals Park, Cleveland, Ohio, USA. *Authoritative reference work for North American irons and steels.*

Elliott, D. and Tupholme, S.M. (1981) *An Introduction to Steel Selection: Part 2, Stainless Steels*, Engineering Design Guide no. 43, Oxford University Press.

Peckner, D. and Bernstein, I.M. (1977) *Handbook of Stainless Steels*, McGraw-Hill, New York.

Design Guidelines for the Selection and Use of Stainless Steel (1991) Designers' Handbook Series no. 9014, Nickel Development Institute.

(The Nickel Development Institute (NIDI) is a worldwide organization that gives away a large variety of free literature about nickel-based alloys, including stainless steels. NIDI European Technical Information Centre, The Holloway, Alvechurch, Birmingham, B48 7QB, UK.)

General internet sites for ferrous metals
British Constructional Steel Work Association, www.steelconstruction.org

International Iron & Steel Institute, www.worldsteel.org

Iron & Steel Trades Confederation, www.istc-tu.org

National Assoc. of Steel Stock Holders, www.nass. org.uk

Steel Manufacturers Association, www.steelnet.org

Steel: Bethlehem Steel's website, www.bethsteel.com/index2.shtml

Steel: Corus home page, www.corusgroup.com/home/index.cfm

Steel: Automotive Steel Library, www.autosteel.org

Steels Construction Institute, www.steel.org.uk

SteelSpec, www.steelspec.org.uk/index.htm

Steelynx, www.steelynx.net

UK Steel, www.uksteel.org.uk

Wire & Wire Rope Employers Association, www.uksteel.org.uk/wwrea.htm

British Stainless Steel Association, www.bssa.org.uk/index.htm

Steel: Great Plains Stainless, www.gpss.com

Polymers and elastomers

Polymers are not subject to the same strict specification as metals. Data tend to be producer-specific. Sources, consequently, are scattered, incomplete and poorly presented. Saechtling is the best; although no single hard-copy source is completely adequate, all those listed here are worth consulting. See also Databases as Software, Section 5; some are good on polymers.

Saechtling: International Plastics Handbook (1983) editor: Dr. Hansjurgen Saechtling, MacMillan Publishing Co (English edition), London, UK. *The most comprehensive of the hard-copy data-sources for polymers.*

Polymers for Engineering Applications (1987) R.B. Seymour. ASM International, Metals Park, OH 44037, USA. *Property data for common polymers. A starting point, but insufficient detail for accurate design or process selection.*

New Horizons in Plastics, a Handbook for Design Engineers (1990), editor: J. Murphy, WEKA Publishing, London, UK.

ASM Engineered Materials Handbook, Vol. 2. *Engineering Plastics* (1989). ASM International, Metals Park, OH 44037, USA.

Handbook of Plastics and Elastomers (1975) Editor: C.A. Harper. McGraw-Hill, New York, USA.

International Plastics Selector, Plastics, 9th edition (1987). Int. Plastics Selector, San Diego, Calif, USA.

Die Kunststoffe und Ihre Eigenschaften (1992) editor: Hans Domininghaus, VDI Verlag, Dusseldorf, Germany.

Properties of Polymers, 3rd edition (1990) D.W. van Krevelen, Elsevier, Amsterdam, Holland. *Correlation of properties with structure; estimation from molecular architecture.*

Handbook of Elastomers (1988) A.K. Bhowmick and H.L. Stephens. Marcel Dekker, New York, USA.

ICI Technical Service Notes (1981) ICI Plastics Division, Engineering Plastics Group, Welwyn Garden City, Herts, UK.

Technical Data Sheets (1995) Malaysian Rubber Producers Research Association, Tun Abdul Razak Laboratory, Brickendonbury, Herts. SG13 8NL. *Data sheets for numerous blends of natural rubber.*

Internet sources
CAMPUS Plastics database, www.campusplastics.com

GE Plastics, www.ge.com/en/company/businesses/ge_plastics.htm

Harboro Rubber Co. Ltd., www.harboro.co.uk

IDES Resin Source, ides.com

MERL, www.merl-ltd.co.uk

Plastics.com, www.plastics.com

Ceramics and glasses

Sources of data for ceramics and glasses, other than the suppliers data-sheets, are limited. Texts and handbooks such as the ASMs (1991) *Engineered*

Materials Handbook, Vol. 4, Morrell's (1985) compilations, Neville's (1996) book on concrete, Boyd and Thompson (1980) *Handbook on Glass* and Sorace's (1996) treatise on stone are useful starting points. The CES Ceramics Database contains recent data for ceramics and glasses. But in the end it is the manufacturer to whom one has to turn: the data sheets for their products are the most reliable source of information.

Ceramics and ceramic–matrix composites

ASM Engineered Materials Handbook, Vol. 4 *Ceramics and Glasses* (1991) ASM International, Metals Park, Ohio 44073, USA.

Waterman, N. and Ashby, M.F. (eds) (1996) *Materials Selector* Chapman and Hall, London, UK.

Brook, R.J. (ed.) (1991) *Concise Encyclopedia of Advanced Ceramic Materials* Pergamon Press, Oxford, UK.

Creyke, W.E.C., Sainsbury, I.E.J. and Morrell, R. (1982) *Design with Non Ductile Materials*, App. Sci., London, UK.

Cheremisinoff, N.P. (ed.) (1990) *Handbook of Ceramics and Composites*, 3 Vols, Marcel Dekker Inc., New York, USA.

Clark, S.P. (ed.) (1966) *Handbook of Physical Constants*, Memoir 97, Geological Society of America.

Schwartz, M.M. (ed.) (1992) *Handbook of Structural Ceramics*, McGraw-Hill, New York, USA. *Lots of data, information on processing and applications.*

Kaye G.W.C. and Laby T.H. (1986) *Tables of Physical & Chemical Constants*, 15th edition, Longman, New York, USA.

Kingery W.D., Bowen H.K. and Uhlmann D.R. (1976) *Introduction to Ceramics*, 2nd edition, Wiley, New York.

Materials Engineering (1992) *Materials Selector*, Penton Press, Cleveland, Ohio, USA.

Morrell, R. (1985) *Handbook of Properties of Technical & Engineering Ceramics*, Parts I and II, National Physical Laboratory, Her Majesty's Stationery Office, London, UK.

Musikant, S. (1991) *What Every Engineer Should Know About Ceramics*, Marcel Dekker, Inc. *Good on data.*

Richerson, D.W. (1992) *Modern Ceramic Engineering*, 2nd edition, Marcel Dekker, New York, USA.

Brandes E.A. and Brook G.B. (eds.) (1992) *Smithells Metals Reference Book*, 7th edition Butterworth-Heinemann, Oxford.

Harper, C.A. (ed.) (2001) *Handbook of Ceramics, Glasses and Diamonds*, McGraw-Hill, New York, NY, USA. ISBN 0-07-026712-X. *A comprehensive compilation of data and design guidelines.*

Richerson, D.W. (2000) *The Magic of Ceramics*, The American Ceramics Society, 735 Ceramic Place, Westerville, Ohio 43081, USA. ISBN 1-57498-050-5. *A readable introduction to ceramics, both old and new.*

Glasses

ASM Engineered Materials Handbook, Vol. 4, *Ceramics and Glasses* (1991) ASM International, Metals Park, Ohio 44073, USA.

Boyd D.C. and Thompson D.A. (1980) *Glass*, Reprinted from Kirk-Othmer: *Encyclopedia of Chemical Techology*, Vol. 11, 3rd edition, pp. 807–880, Wiley, New York.

Oliver, D.S. (1975) *Engineering Design Guide 05: The Use of Glass in Engineering*, Oxford University Press, Oxford, UK.

Bansal, N.P. and Doremus, R.H. (1966) *Handbook of Glass Properties*, Academic Press, New York, USA.

Cement and concrete

Cowan, H.J. and Smith, P.R. (1988) *The Science and Technology of Building Materials*, Van Nostrand-Reinhold, New York, USA.

Illston, J.M., Dinwoodie, J.M. and Smith, A.A. (1979) *Concrete, Timber and Metals*, Van Nostrand-Reinhold, New York, USA.

Neville, A.M. (1996) *Properties of Concrete*, 4th edition, Longman Scientific and Technical. *An excellent introduction to the subject.*

Composites: PMCs, MMCs, and CMCs

The fabrication of composites allows so many variants that no hard-copy data source can capture them all; instead, they list properties of matrix and reinforcement, and of certain generic lay-ups or types. The "Engineers Guide" and the "Composite Materials Handbook", listed first, are particularly recommended.

Composite, general

Weeton, J.W., Peters, D.M. and Thomas, K.L. (eds) (1987) *Engineers Guide to Composite Materials*, ASM International, Metals Park, OH 44073, USA. *The best starting point: data for all classes of composites.*

Schwartz, M.M. (ed.) (1992) *Composite Materials Handbook*, 2nd edition, McGraw-Hill, New York, USA. *Lots of data on PMCs, less on MMCs and CMCs, processing, fabrication, applications and design information.*

ASM Engineered Materials Handbook, Vol. 1: *Composites* (1987) ASM International, Metals Park, Ohio 44073, USA.

Seymour R.B. (1991) *Reinforced Plastics, Properties and Applications*, ASM International, Metals Park, OH 44073, USA.

Cheremisinoff, N.P. (ed.) (1990) *Handbook of Ceramics and Composites*, Vols. 1–3, Marcel Dekker Inc., New York, USA.

Kelly, A. (ed.) (1989) *Concise Encyclopedia of Composited Materials*, Pergamon Press, Oxford, UK.

Middleton, D.H. (1990) *Composite Materials in Aircraft Structures*, Longman Scientific and Technical Publications, John Wiley, New York, NY, USA.

Smith, C.S. (1990) *Design of Marine Structures in Composite Materials*, Elsevier Applied Science, London, UK.

Metal matrix composites

See, first, the sources listed under "All Composite Types", then, for more detail, go to:

ASM Engineered Materials Handbook, Vol. 1: *Composites* (1987) ASM International, Metals Park, Ohio 44073, USA.

Technical Data Sheets, Duralcan USA (1995) 10505 Roselle Street, San Diego, CA 92121, USA.

Technical Data Sheets (1995) 3M Company, 3M Xenter, Building 60-1N-001, St Paul MN 55144-1000, USA.

Foams and cellular solids

Many of the references given under the section "Polymers and elastomers" for polymers and elastomers mention foam. The references given here contains much graphical data, and simple formulae that allow properties of foams to be estimated from its density and the properties of the solid of which it is made, but in the end it is necessary to contact suppliers. See also Data-bases as Software (section "Databases and expert systems in software"); some are good on foams. For Woods and wood-based composites, see section below.

Cellular Polymers (a Journal) (1981–1996) published by RAPRA Technology, Shrewsbury, UK.

Encyclopedia of Chemical Technology (1980) Vol. 2, 3rd edition, pp. 82–126, Wiley, New York, USA.

Encyclopedia of Polymer Science and Engineering (1985) Vol. 3, 2nd edition, Section C, Wiley, New York, USA.

Gibson, L.J. and Ashby, M.F. (1997), *Cellular Solids*, Cambridge University Press, Cambridge, UK. *Basic text on foamed polymers, metals, ceramics and glasses, and natural cellular solids.*

Handbook of Industrial Materials, 2nd edition (1992) Elsevier Advanced Technology, Elsevier, Oxford, UK, pp. 537–556.

Hilyard, N.C. and Cunningham, A. (eds) (1994) *Low Density Cellular Plastics — Physical Basis of Behaviour*, Chapman and Hall, London, UK. *Specialized articles on aspects of polymer-foam production, properties and uses.*

Plascams (1995) Version 6, *Plastics Computer-Aided Materials Selector*, RAPRA Technology Limited, Shawbury, Shrewsbury, Shropshire SY4 4NR, UK.

Hans Jurgen (ed.) (1983) *International Plastics Handbook*, Saechtling, MacMillan Publishing Co. (English edition), London, UK.

Seymour, R.P. (1987) *Polymers for Engineering Applications*, ASM International, Metals Park, OH 44037, USA.

Stone, rocks and minerals

There is an enormous literature on rocks and minerals. Start with the handbooks listed below; then ask a geologist for guidance.

Atkinson, B.K. (1987) *The Fracture Mechanics of Rock*, Academic Press, UK.

Clark, Jr., S.P. (ed.) (1966) *Handbook of Physical Constants*, Memoir 97, The Geological Society of America, 419 West 117 Street, New York, USA (1966). *Old but trusted compilation of property data for rocks and minerals.*

Lama, R.E. and Vutukuri, V.S. (eds) (1978) *Handbook on Mechanical Properties of Rocks*, Vols. 1–4, Trans Tech Publications, Clausthal, Germany.

Griggs, D. and Handin, J. (eds) (1960) *Rock Deformation*, Memoir 79, The Geological Society of America, 419 West 117 Street, New York, USA.

Sorace, S. (1996) Long-term tensile and bending strength of natural building stones, *Mater. Struct.*, **29**, pp. 426–435.

Internet data sources
Building stone, www.wishbone.com

Woods and wood-based composites

Woods, like composites, are anisotropic; useful sources list properties along and perpendicular to the grain. The US Forest Products Laboratory *Handbook of Wood and Wood-based Materials* and Kollmann and Coté *Principles of Wood Science and Technology* are particularly recommended.

Woods, general information

Bodig, J. and Jayne, B.A. (1982) *Mechanics of Wood and Wood Composites*, Van Nostrand-Reinholt Company, New York, USA.

Dinwoodie, J.M. (1989) *Wood, Nature's Cellular Polymeric Fiber Composite*, The Institute of Metals, London, UK.

Dinwoodie, J.M. (1981) *Timber, its Nature and Behaviour*, Van Nostrand-Reinhold, Wokingham, UK. *Basic text on wood structure and properties. Not much data.*

Gibson, L.J. and Ashby, M.F. (1997) *Cellular Solids*, 2nd edition, Cambridge University Press, Cambridge, UK.

Jane, F.W. (1970) *The Structure of Wood*, 2nd edition, A. and C. Black, Publishers, London, UK.

Kollmann, F.F.P. and Côté, W.A. Jr. (1968) *Principles of Wood Science and Technology*, Vol. 1: *Solid Wood*, Springer-Verlag, Berlin, Germany. *The bible.*

Kollmann, F., Kuenzi, E. and Stamm, A. (1968) *Principles of Wood Science and Technology*, Vol. 2: *Wood Based Materials*, Springer-Verlag Berlin.

Schniewind, A.P. (ed.) (1989) *Concise Encyclopedia of Wood and Wood-Based Materials*, Pergamon Press, Oxford, UK.

Woods: data compilations

BRE (1996) *BRE Information Papers*, Building Research Establishment (BRE), Garston, Watford, WD2 7JR, UK.

Forest Products Laboratory (1989), Forest Service, US Department of Agriculture, *Handbook of Wood and Wood-based Materials*, Hemisphere Publishing Corporation, New York. A massive compilation of data for North-American woods.

Informationsdienst Holz (1996) Merkblattreihe Holzarten, Verein Deutscher Holzeinfuhrhäuser e.V., Heimbuder Strabe 22, D-20148 Hamburg, Germany.

TRADA (1978/1979) *Timbers of the World*, Vols 1–9, Timber Research and Development Association, High Wycombe, UK.

TRADA (1991) *Information Sheets*, Timber Research and Development Association, High Wycombe, UK.

Wood and wood-composite standards

Great Britain
British Standards Institution (BSI), 389 Chiswick High Road, GB-London W4 4AL, UK (Tel: +44 181

996 9000; Fax: +44 181 996 7400; e-mail: info@bsi.org.uk).

Germany

Deutsches Institut für Normung (DIN), Burggrafenstrasse 6, D-10772, Berlin, Germany (Tel +49 30 26 01-0; Fax: +49 30 26 01 12 31; e-mail: postmaster@din.de

USA

American Society for Testing and Materials (ASTM), 1916 Race Street, Philadelphia, Pennsylvania 19103-1187 (Tel 215 299 5400; Fax: 215 977 9679).

ASTM European Office, 27–29 Knowl Piece, Wilbury Way, Hitchin, Herts SG4 0SX, UK (Tel: +44 1462 437933; Fax: +44 1462 433678; e-mail: 100533.741@compuserve.com).

Software and Internet data sources

CES woods database, www.grantadesign.com. A database of the engineering properties of softwoods, hardwoods and wood-based composites. PC format, Windows environment.

PROSPECT (Version 1.1), 1995. Oxford: Oxford Forestry Institute, Department of Plant Sciences, Oxford University. A database of the properties of tropical woods of interest to a wood user; includes information about uses, workability, treatments, origins. PC format, DOS environment.

Woods of the World, 1994. Burlington, VT: Tree Talk, Inc. A CD-ROM of woods, with illustrations of structure, information about uses, origins, habitat, etc. PC format, requiring CD drive; Windows environment.

WoodWeb http://www.woodweb.com *Woodwork industry web service. Contains information about woodworking related companies.*

Natural fibers and other natural materials

Houwink, R. (1958) *Elasticity Plasticity and Structure of Matter,* Dover Publications, Inc, New York, USA.

Handbook of Industrial Materials, 2nd edition, (1992) Elsevier, Oxford, UK. *A compilation of data remarkable for its breadth: metals, ceramics, polymers, composites, fibers, sandwich structures, leather... .*

Brady, G.S. and Clauser, H.R. (eds) (1986) *Materials Handbook*, 2nd edition, McGraw-Hill, New York, NY, USA (1986). *A broad survey, covering metals, ceramics, polymers, composites, fibers, sandwich structures, and more.*

Environmental and medical

Biomaterials Properties TOC, www.lib.umich.edu/dentlib/Dental_tables/toc.html

Pentron, www.pentron.com

British Metals Recycling Association, www.britmetrec.org.uk

Green Design Initiative, www.ce.cmu.edu/GreenDesign

IDEMAT, Environmental Materials Database, www.io.tudelft.nl/research/dfs/idemat/index.htm

D.3 Information for manufacturing processes

Alexander, J.M., Brewer, R.C. and Rowe, G.W. (1987) *Manufacturing Technology*, Vol. 2: *Engineering Processes*, Ellis Horwood Ltd., Chichester, UK.

Bralla, J.G. (1986) *Handbook of Product Design for Manufacturing*, McGraw-Hill, New York, USA.

Waterman, N.A. and Ashby, M.F. (eds.) (1996) *Materials Selector*, Chapman and Hall, London, UK.

Dieter, G.E. (1983) *Engineering Design, A Materials and Processing Approach*, McGraw-Hill, New York, USA, Chapter 7.

Kalpakjian, S. (1984) *Manufacturing Processes for Engineering Materials*, Addison Wesley, London, UK.

Lascoe, O.D. (1989) *Handbook of Fabrication Processes*, ASM International, Metals Park, Columbus, OH, USA.

Schey, J.A. (1997) *Introduction to Manufacturing Processes*, McGraw-Hill, New York, USA.

Suh, N.P. (1990) *The Principles of Design*, Oxford University Press, Oxford, UK.

Internet sources
Cast Metals Federation,
www.castmetalsfederation.com

Castings Technologies International,
www.castingstechnology.com

Confederation of British Metalforming,
www.britishmetalforming.com

National Center for Excellence in Metalworking Technology, www.ncemt.ctc.com

PERA, www.pera.com

The British Metallurgical Plant Constructors' Association, www.bmpca.org.uk

TWI (The Welding Institute), www.twi.co.uk

D.4 Databases and expert systems in software

The number and quality of computer-based materials information systems is growing rapidly. A selection of these, with comment and source, is given here. There has been consumer resistance to on-line systems; almost all recent developments are in PC-format. The prices vary widely. Five price groups are given: free, cheap (less than $200 or £125), modest (between $200 or £125 and $2000 or £1250), expensive (between $2000 or £1250 and $10,000 or £6000) and very expensive (more than $10,000 or £6000). The databases are listed in alphabetical order.

ACTIVE LIBRARY ON CORROSION. Materials Park: ASM International. *PC format requiring CD ROM drive. Graphical, numerical and textual information on corrosion of metals. Price modest.*

ASM HANDBOOKS ONLINE. Materials Park: ASM International, www.asminternational.org/hbk/index.jsp. *Annual license fee.*

ALLOYCENTER. Materials Park: ASM International, www.asminternational.org/alloycenter/index.jsp. *Annual license fee.*

ALUSELECT P1.0: Engineering Property Data for Wrought Aluminum Alloys, 1992. Dusseldorf: European Aluminum Association. *PC format, DOS environment. Mechanical, thermal, electrical and environmental properties of wrought aluminum alloys. Price cheap.*

CAMPUS: Computer Aided Material Preselection by Uniform Standards, 1995, www.campusplastics.com. *Polymer data from approx 30 suppliers, measured to set standards.*

CETIM-EQUIST II: Centre Technique des Industries Mécaniques, 1997. Senlis: CETIM. *PC format, DOS environment. Compositions and designations of steels.*

CETIM-Matériaux: Centre Technique des Industries Mécaniques, 1997. Senlis: CETIM. *On-line system. Compositions and mechanical properties of materials.*

CETIM-SICLOP: Centre Technique des Industries Mécaniques, 1997. Senlis: CETIM. *On-line system. Mechanical properties of steels.*

CES: Cambridge Engineering Selector, 1999–2004, www.grantadesign.com. *Comprehensive selection system for all classes of materials and manufacturing processes. Variety of optional reference data sources, connects directly to www.matdata.net. Windows and web format. Modest price.*

CUTDATA: Machining Data System. Cincinnati: Metcut Research Associates Inc, Manufacturing Technology Division. *A PC-based system that guides the choice of machining conditions: tool materials, geometries, feed rates, cutting speeds, and so forth. Modest price.*

EASel: Engineering Adhesives Selector Program, 1986. London: The Design Centre. *PC and Mac formats. A knowledge-based program to select industrial adhesives for joining surfaces. Modest price.*

SF-CD (replacing ELBASE): Metal Finishing/Surface Treatment Technology, 1992. Stevenage: *Metal Finishing Information Services Ltd. PC format. Comprehensive information on published data related to surface treatment technology. Regularly updated. Modest price.*

MATDATA.NET: www.matdata.net. *The Material Data Network provides integrated access to a variety of quality information sources, from ASM International, Granta Design, TWI, NPL, UKSteel, Matweb, IDES, etc.*

M-VISION: 1990: www.mscsoftware.com. *Materials selection and querying, report generation and development of customizable, electronic links to analysis programs. Workstation and web versions. Substantial database of reference materials. Very expensive.*

MEGABYTES ON COPPERS II: Information on Copper and Copper Alloys, www.cda.org.uk. *A CD-ROM with Windows search engine, containing all the current publications as well as interactive programs published by CDA on topics of electrical energy efficiency, cost effectiveness and corrosion resistance. Cheap.*

PAL II: Permabond Adhesives Locator, 1996. Eastleigh: Permabond. *A knowledge-based, PC-system (DOS environment) for adhesive selection among Permabond adhesives. An impressive example of an expert system that works. Modest price.*

PROSPECT. Version 1.1, 1995. Oxford: Oxford Forestry Institute, Department of Plant Sciences, Oxford University. *A database of the properties of tropical woods of interest to a wood user; includes information about uses, workability, treatments, origins. PC format, DOS environment.*

OPS (Optimal Polymer Selector), www.grantadesign. com. *Integrates Granta Design's generic Polymer Universe database, with Chemical resistance data from RAPRA and grade-specific data for approx 6000 polymers from CAMPUS, to produce the only comprehensive polymer selection system available.*

teCal: Steel Heat-Treatment Calculations. Materials Park: ASM International. *PC format, DOS environment. Computes the properties resulting from defined heat-treatments of low-alloy steels, using the composition as input. Modest price.*

WEGST, C.W. 1997. Stahlschlüssel (Key to Steel). 17th ed. Marbach: Verlag Stahlschlüssel Wegst GmbH. *CD ROM, PC format. Excellent coverage of European products and manufacturers.*

SOFINE PLASTICS, 1997. Villeurbanne Cedex: Société CERAP. *Database of polymer properties. Environment and price unknown.*

TAPP 2.0: Thermochemical and Physical Properties, 1994. Hamilton, OH: ES Microware. *PC format, CD ROM, Windows environment. A database of thermochemical and physical properties of solids, liquids and gasses, including phase diagrams neatly packaged with good user manual. Modest price.*

UNSearch: Unified Metals and Alloys Composition Search. Materials Park: ASTM. *PC format, DOS environment. A datbase of information about composition, US designation and specification of common metals and alloys. Modest price.*

WOODS OF THE WORLD, 1994. Burlington, VT: Tree Talk, Inc. A CD-ROM of woods, with illustrations of structure, information about uses, origins, habitat etc. *PC format, requiring CD drive; Windows environment.*

D.5 Additional useful internet sites

Metals prices and economic reports

American Metal Market On-line, www.amm.com

Business Communications Company, www.buscom.com

Daily Economic Indicators, www.bullion.org.za/Level2/Econ&stats.htm

Iron & Steel Statistics Bureau, www.issb.co.uk

Kitco Inc Gold & Precious Metal Prices, www.kitco. com/gold.live.html-ourtable

London Metal Exchange, www.lme.co.uk

Metal Bulletin, www.metalbulletin.plc.uk

Metal Powder Report, www.metal-powder.net

Metallurgia, www.metallurgiaonline.com

Mineral-Resource, minerals.usgs.gov/minerals

Roskill Reports www.roskill.com

The Precious Metal and Gem Connection, www.thebulliondesk.com/default.asp

D.6 Supplier registers, government organizations, standards and professional societies

Supplier registers

Industry and Trade Associations, www.amm.com/index2.htm?/ref/trade.HTM

IndustryLink Homepage, www.industrylink.com

Kellysearch, www.kellysearch.com

Metals Industry Competitive Enterprise, www.metalsindustry.co.uk

Thomas Register-Home Page, www.thomasregister.com

Top 50 US/Canadian Metal Co.s, www.amm.com/ref/top50.HTM

Government organizations

Commonwealth Scientific and Industrial Research Org (Australia), www.csiro.au

National Academies Press, USA, www.nap.edu

National Institute for Standards and Technology (USA), www.nist.gov

The Fraunhofer Institute, www.fhg.de/english.html

Professional societies and trade associations

Aluminum Federation, www.alfed.org.uk

ASM International, www.asminternational.org

ASME International, www.asme.org

British Constructional Steel Work Association, www.steelconstruction.org

Cast Metals Federation, www.castmetalsfederation.com

Confederation of British Metalforming, www.britishmetalforming.com

Copper Development Association, www.cda.org.uk

Institute of Cast Metal Engineers, www.ibf.org.uk

Institute of Spring Technology Ltd, www.ist.org.uk

International Aluminum Institute, www.world-aluminum.org

International Council on Mining and Metals, www.icme.com

International Iron & Steel Institute, www.worldsteel.org

International Titanium Association, www.titanium.org

International Zinc Association, www.iza.com

Iron & Steel Trades Confederation, www.istc-tu.org

Lead Development Association International, www.ldaint.org/default.htm

Materials Information Service, www.iom3.org/mis/index.htm

National Assoc. of Steel Stock Holders, www.nass.org.uk

Nickel Development Institute, www.nidi.org

Pentron, www.pentron.com

Society of Automotive Engineers (SAE), www.sae.org

Steel Manufacturers Association, www.steelnet.org

Steels Construction Institute, www.steel.org.uk

The American Ceramics Society, www.acers.org

The British Metallurgical Plant Constructors' Association, www.bmpca.org.uk

The Institute of Materials, Minerals and Mining, www.instmat.co.uk

The Minerals, Metals & Materials Society, www.tms.org/TMSHome.html

Tin Research Association, www.tintechnology.com

UK Steel, www.uksteel.org.uk

Wire & Wire Rope Employers Association, www.uksteel.org.uk/wwrea.htm

Standards organizations

ASTM, www.astm.org

BSI, www.bsi.org.uk

DIN Deutsches Institut für Normung e.V, www2.din.de/index.php?lang=en

International Standards Organisation, www.iso.org

National Standards Authority of Ireland, www.nsai.ie

Miscellaneous

Acoustic properties of materials, www.ultrasonic.com/tables/index.html

Common unit of measure conversion, www.conweb.com/tblefile/conver.shtml

Dental materials, www.lib.umich.edu.dentlib/Dental_tables/toc.html

Hazardous materials (1), http://hazmat.dot.gov

Hazardous materials (2), www.chemweb.com/databases/brchd

Hazardous materials (3), http://ull.chemistry.uakron.edu/erd

Information for Physical Chemists, www.liv.ac.uk/Chemistry/Links/links.html

K&K associate's thermal connection, www.tak2000.com

Properties of soils, http://homepages.which.net/~fred.moor/soil/links/10101.htm

Spacecraft structural materials, http://esapub.esrin.eas.it/pff/pffv6nl.htm

Temperature dependent elastic and thermal properties, www.jahm.com

Thermal data, www.csn.net/~takinfo/prop-top.html

Unit conversion, http://www.conversion.com

Appendix E

Exercises

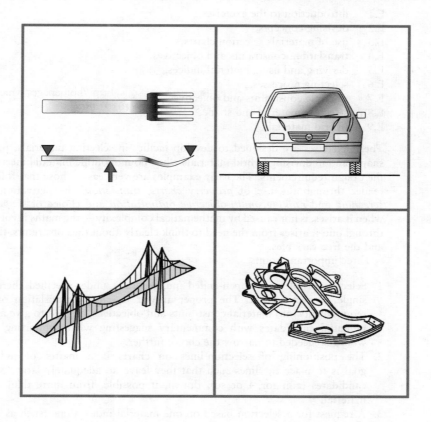

E.1 Introduction to the exercises

The exercises are organized into nine sections:

E.1 introduction to the exercises,
E.2 devising concepts,
E.3 use of materials selection charts,
E.4 translation: constraints and objectives,
E.5 deriving and using material indices,
E.6 selecting processes,
E.7 multiple constraints and objectives,
E.8 selecting material and shape,
E.9 hybrid materials.

These exercises are designed to develop facility in selecting materials, processes and shape, and in devising hybrid materials when no monolithic material meets completely the design requirements. The early examples are very easy. Those that follow lead the reader through the use of *property charts, translation,* the derivation of *indices, screening* and *ranking, multi-objective optimization* and choice of *shape.* Difficulty, when it arises, is not caused by mathematical complexity — the maths involved is simple throughout; it arises from the need to think clearly about the constraints, the objectives and the free variables.

Three important points.

1. Selection problems are open-ended and, generally, under-specified; there is seldom a single, correct answer. The proper answer is sensible translation of the design requirements into material constraints and objectives, applied to give a short-list of potential candidates with commentary suggesting what supporting information would be needed to narrow the choice further.
2. The positioning of selection-lines on charts is a matter of judgment. The goal is to place the lines such that they leave an adequately large "short list" of candidates (aim for 4 or so), drawn, if possible, from more than one class of material.
3. A request for a selection based on one material index alone (such as $M \cdot E^{1/2}/\rho$) is correctly answered by listing the subset of materials that maximize this index. But a request for a selection of materials for a component — a wing spar, for instance (which is a light, stiff beam, for which the index is $M = E^{1/2}/\rho$) — requires more: some materials with high $E^{1/2}/\rho$ such as silicon carbide, are unsuitable for obvious reasons. It is a poor answer that ignores common sense and experience and fails to add further constraints to incorporate them. Students should be encouraged to discuss the implications of their selection and to suggest further selection stages.

The best way to use the charts that are a feature of the book is to make clean copies (or down-load them from http://www.grantadesign.com) on which you can draw, try out alternative selection criteria, write comments, and so forth. Although the book itself is copyrighted, the reader is authorized to make copies of the charts and to reproduce these, with proper reference to their source, as he or she wishes.

All the materials selection problems can be solved using the CES software, which is particularly effective when multiple criteria and unusual indices are involved.

E.2 Devising concepts

These two examples illustrate the way in which concepts are generate. The left-hand part of each diagram describes a physical principle by which the need might be met; the right-hand part elaborates, suggesting how the principle might be used.

Exercise E2.1   *Concepts and embodiments for dust removers.* We met the need for a "device to remove household dust" in Chapter 1, with examples of established solutions. Now it is time for more creative thinking. Devise as many concepts to meet this need as you can. Nothing, at the concept stage, is too far-fetched; decisions about practicality and cost come later, at the detailed stage. So think along the lines of Figure 2.2 and list concepts and outline embodiments as block diagrams like the following:

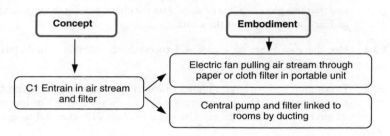

Exercise E2.2   *Cooling power electronics.* Microchips, particularly those for power electronics, get hot. If they get too hot they cease to function. The need: a scheme for removing heat from power microchips. Devise concepts to meet the need and sketch an embodiment of one of them, laying out your ideas in the way suggested in Exercise E2.1.

E.3 Use of material selection charts

The 21 exercises in this section involve the simple use of the charts of Chapter 4 to find materials with given property profiles. They are answered by placing selection lines on the appropriate chart and reading off the materials that lie on the appropriate side of the line. It is a good idea to present the results as a table. All can be solved by using the printed charts.

If the CES EDU Materials Selection software is available the same exercises can be solved by its use. This involves first creating the chart, then applying the appropriate box or line selection. The results, at Level 1 or 2, are the same as those read from the hard copy charts (which were made using the Level 2 database). The software offers links to processes, allows a wider search by using the Level 3 database, and gives access to supporting information via the "Search Web" function.

Exercise E3.1 A component is at present made from a brass, a copper alloy. Use the Young's modulus–density (E–ρ) chart of Figure 4.3 to suggest three other metals that, in the same shape, would be stiffer. "Stiffer" means a higher value of Young's modulus.

Exercise E3.2 Use the Young's modulus–density (E–ρ) chart of Figure 4.3 to identify materials with both a modulus $E > 50\,\text{GPa}$ and a density $\rho < 2\,\text{Mg/m}^3$.

Exercise E3.3 Use the Young's modulus–density (E–ρ) chart of Figure 4.3 to find (a) metals that are stiffer and less dense than steels and (b) materials (not just metals) that are both stiffer and less dense than steel.

Exercise E3.4 Use the E–ρ chart of Figure 4.3 to identify metals with both $E > 100\,\text{GPa}$ and $E/\rho > 23$ GPa/(Mg/m^3)

Exercise E3.5 Use the E–ρ chart of Figure 4.3 to identify materials with both $E > 100\,\text{GPa}$ and $E^{1/3}/\rho > 3$ (GPa)$^{1/3}$/(Mg/m^3). Remember that, on taking logs, the index $M = E^{1/3}/\rho$ becomes

$$\text{Log}(E) = 3\,\text{Log}(\rho) + 3\,\text{Log}(M)$$

and that this plots as a line of slope 3 on the chart, passing through the point $\rho = 1/3 = 0.33$ at $E = 1$ in the units on the chart.

Exercise E3.6 Use the E–ρ chart of Figure 4.3 to establish whether woods have a higher specific stiffness E/ρ than epoxies.

Exercise E3.7 Do titanium alloys have a higher or lower specific strength (strength/density, σ_f/ρ) than tungsten alloys? This is important when you want strength at low weight (landing gear of aircraft, mountain bikes). Use the σ_f/ρ chart of Figure 4.4 to decide.

Exercise E3.8 Use the modulus–strength E–σ_f chart of Figure 4.5 to find materials that have $E > 10\,\text{GPa}$ and $\sigma_f \geq 1000\,\text{MPa}$.

Exercise E3.9 Are the fracture toughnesses, K_{1C}, of the common polymers polycarbonate, ABS, or polystyrene larger or smaller than the engineering ceramic alumina? Are their toughnesses $G_{1C} = K_{1C}^2/E$ larger or smaller? The K_{1C}–E chart, Figure 4.7, will help.

Exercise E3.10 Use the fracture toughness–modulus chart (Figure 4.7) to find materials that have a fracture toughness K_{1C} greater than $100\,\text{MPa.m}^{1/2}$ and a toughness $G_{1C} = K_{1C}^2 E$ (shown as contours on Figure 4.7) greater than $10\,\text{kJ/m}^3$.

Exercise E3.11 The elastic deflection at fracture (the "resilience") of an elastic–brittle solid is proportional to the failure strain, $\varepsilon_{\text{fr}} = \sigma_{\text{fr}}/E$, where is the stress that will cause a crack to propagate:

$$\sigma_{\text{fr}} = \frac{K_{1C}}{\sqrt{\pi c}}$$

where K_{1C} is the fracture toughness and c is the length of the longest crack the materials may contain. Thus

$$\varepsilon_{\text{fr}} = \frac{1}{\sqrt{\pi c}}\left(\frac{K_{1C}}{E}\right)$$

Materials that can deflect elastically without fracturing are therefore those with large values of K_{1C}/E. Use the K_{1C}–E chart of Figure 4.7 to identify the class of materials with $K_{1C} > 1$ MPa.m$^{1/2}$ and high values of K_{1C}/E.

Exercise E3.12 One criterion for design of a safe pressure vessel is that it should leak before it breaks: the leak can be detected and the pressure released. This is achieved by designing the vessel to tolerate a crack of length equal to the thickness t of the pressure vessel wall, without failing by fast fracture. The safe pressure p is then

$$p \leq \frac{4}{\pi} \frac{1}{R} \left(\frac{K_{1C}^2}{\sigma_f} \right)$$

where σ_f is the elastic limit, K_{1C} is the fracture toughness, R is the vessel radius. The pressure is maximized by choosing the material with the greatest value of

$$M = \frac{K_{1C}^2}{\sigma_y}$$

Use the K_{1C}–σ_f chart of Figure 4.8 to identify three alloys that have particularly high values of M.

Exercise E3.13 A material is required for the blade of a rotary lawn-mower. Cost is a consideration. For safety reasons, the designer specified a minimum fracture toughness for the blade: it is $K_{1C} > 30$ MPa.m$^{1/2}$. The other mechanical requirement is for high hardness, H, to minimize blade wear. Hardness, in applications like this one, is related to strength:

$$H \approx 3\sigma_y$$

where σ_f is the strength (Chapter 4 gives a fuller definition). Use the K_{1C}–σ_f chart of Figure 4.8 to identify three materials that have $K_{1C} > 30$ MPa.m$^{1/2}$ and the highest possible strength. To do this, position a "K_{1C}" selection line at 30 MPa.m$^{1/2}$ and then adjust a "strength" selection line such that it just admits three candidates. Use the Cost chart of Figure 4.19 to rank your selection by material cost, hence making a final selection.

Exercise E3.14 Bells ring because they have a low loss (or damping) coefficient, η; a high damping gives a dead sound. Use the loss coefficient–modulus (η–E) chart of Figure 4.9 to identify material that should make good bells.

Exercise E3.15 Use the loss coefficient–modulus (η–E) chart (Figure 4.9) to find metals with the highest possible damping.

Exercise E3.16 The window through which the beam emerges from a high-powered laser must obviously be transparent to light. Even then, some of the energy of the beam is absorbed in the window and can cause it to heat and crack. This problem is minimized by choosing a window material with a high thermal conductivity λ (to conduct the heat away) and a low expansion coefficient α (to reduce thermal strains), that is, by seeking a window material with a high value of

$$M = \lambda/\alpha$$

Use the α–λ chart of Figure 4.12 to identify the best material for an ultra-high powered laser window.

Exercise E3.17 Use the thermal conductivity–electrical resistivity (λ–ρ_e) chart (Figure 4.10) to find three materials with high thermal conductivity, λ, and high electrical resistivity, ρ_e.

Exercise E3.18 Use the strength–maximum service temperature (σ_f–T_{max}) chart (Figure 4.14) to find polymers that can be used above 200°C.

Exercise E3.19 (a) Use the Young's modulus–relative cost (E–C_R) chart (Figure 4.18) to find the cheapest materials with a modulus, E, greater than 100 GPa.
(b) Use the strength–relative cost (σ_f–C_R) chart (Figure 4.19) to find the cheapest materials with a strength, σ_f, above 100 MPa.

Exercise E3.20 Use the friction coefficient chart (Figure 4.15) to find two materials with exceptionally low coefficient of friction.

Exercise E3.21 Identical casings for a power tool could be die-cast in aluminum or molded in ABS or polyester GFRP. Use the appropriate production–energy chart of Figure 16.7 to decide which choice minimizes the material production energy.

E.4 Translation: constraints and objectives

Translation is the task of re-expressing design requirements in terms that enable material and process selection. Tackle the exercises by formulating the answers to the questions in this table. Do not try to model the behavior at this point (that comes in later exercises). Just think out what the component does, and list the constraints that this imposes on material choice, including processing requirements.

Function	• What does component do?
Constraints	• What essential conditions must be met?
Objective	• What is to be maximized or minimized?
Free variables	• What parameters of the problem is the designer free to change?

Here it is important to recognize the distinction between constraints and objectives. As the table says, a constraint is an essential condition that must be met, usually expressed as a limit on a material or process attribute. An objective is an quantity for which an extremum (a maximum or minimum) is sought, frequently cost, mass, or volume, but there are others, several of which appear in the exercises below. Take the example of a bicycle frame. It must have a certain stiffness and strength. If it is not stiff and strong enough it will not work, but it is never required to have infinite stiffness or strength. Stiffness and strength are therefore constraints that become limits on modulus, elastic limit, and shape. If the bicycle is for sprint racing, it should be as light as possible — if you could make it infinitely light, that would be best of all. Minimizing mass, here, is the objective, perhaps with an upper limit (a constraint) on cost. If instead it is a shopping bike to be sold through supermarkets it should be as cheap as possible — the cheaper it is, the more will be sold. This time minimizing cost is the objective, possible with an upper limit (a constraint) on mass. For most bikes, of course, minimizing mass and cost are both objectives, and then trade-off methods are needed. They come later. For now use judgment to choose the single most important objective and make all others into constraints.

Two rules-of-thumb, useful in many "translation" exercises. Many applications require sufficient fracture toughness for the component can survive mishandling and accidental impact during service; a totally brittle material (like un-toughened glass) is unsuitable. Then a necessary constraint is that of "adequate toughness". This is achieved by requiring that the fracture toughness $K_{1C} > 15\,\mathrm{MPa.m^{1/2}}$. Other applications require some ductility, sufficient to allow stress redistribution under loading points, and some ability to bend or shape the material plastically. This is achieved by requiring that the (tensile) ductility $\varepsilon_f > 2\%$. (If the CES software is available it can be used to impose the constraints and to rank the survivors using the objective.)

Exercise E4.1   A material is required for the windings of an electric air-furnace capable of temperatures up to 1000°C. Think out what attributes a material must have if it is to be made into windings and function properly in a furnace. List the function and the constraints; set the objective to "minimize cost" and the free variables to "choice of material".

Exercise E4.2   A material is required to manufacture office scissors. Paper is an abrasive material, and scissors sometimes encounter hard obstacles like staples. List function and constraints; set the objective to "minimize cost" and the free variables to "choice of material".

Exercise E4.3   A material is required for a heat exchanger to extract heat from geo-thermally heated, saline, water at 120°C (and thus under pressure). List function and constraints; set the objective to "minimize cost" and the free variables to "choice of material".

Exercise E4.4   A C-clamp (Figure E.1) is required for processing of electronic components at temperatures up to 450°C. It is essential that the clamp have low thermal inertia so that it reaches temperature quickly, and it must not charge-up when exposed to an electron beam. The time t it takes a component of thickness x to reach thermal equilibrium when the temperature is suddenly changed (a transient heat flow problem) is

$$t \approx \frac{x^2}{2a}$$

where the thermal diffusivity $a = \lambda/\rho C_p$ and λ is the thermal conductivity, ρ the density and C_p the specific heat.

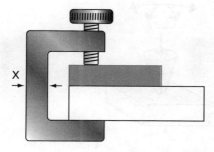

Figure E.1   C-clamp.

List function and constraints; set the objective to "minimize cost" and the free variables to "choice of material".

Exercise E4.5 A furnace is required to sinter powder-metal parts. It operates continuously at 650°C while the parts are fed through on a moving belt. You are asked to select a material for furnace insulation to minimize heat loss and thus to make the furnace as energy-efficient as possible. For reasons of space the insulation is limited to a maximum thickness of $x = 0.2$ m. List the function, constraints, objective and free variable.

Exercise E4.6 Ultra-precise bearings that allow a rocking motion make use of knife-edges or pivots (Figure E.2). As the bearing rocks, it rolls, translating sideways by a distance that depends on the radius of contact. The further it rolls, the less precise is its positioning, so the smaller the radius of contact R the better. But the smaller the radius of contact, the greater is the contact pressure (F/A). If this exceeds the hardness H of either face of the bearing, it will be damaged. Elastic deformation is bad too: it flattens the contact, increasing the contact area and the roll.

A rocking bearing is required to operate in a micro-chip fabrication unit using fluorine gas at 100°C, followed by e-beam processing requiring that all structural parts of the equipment can be earthed to prevent stray charges. Translate the requirements into material selection criteria, listing function, constraints, objective and free variable.

Exercise E4.7 The standard CD ("Jewel" case) cracks easily and, if broken, can scratch the CD. Jewel cases are made of injection molded polystyrene, chosen because it is transparent, cheap, and easy to mould. A material is sought to make CD cases that do not crack so easily. The case must still be transparent, able to be injection molded, and able to compete with polystyrene in cost.

Exercise E4.8 A storage heater captures heat over a period of time, then releases it, usually to an air stream, when required. Those for domestic heating store solar energy or energy from cheap off-peak electricity and release it slowly during the cold part of the day. Those for research release the heat to a supersonic air stream to test system behavior in supersonic flight. What is a good material for the core of a compact storage material capable of temperatures up to 120°C?

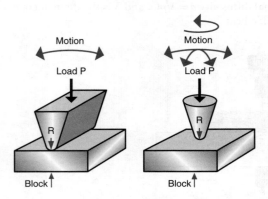

Figure E.2   Ultra-precise bearings with knife-edges and pivots.

E.5 Deriving and using material indices

The exercises in this section give practice in deriving indices.

(a) Start each by listing function, constraints, objectives and free variables; without having those straight, you will get in a mess. Then write down an equation for the objective. Consider whether it contains a free variable other than material choice; if it does, identify the constraint that limits it, substitute, and read off the material index.

(b) If the CES EDU software is available, then use it to apply the constraints and rank the survivors using the index (start with the Level 2 database). Are the results sensible? If not, what constraint has been overlooked or incorrectly formulated?

Exercise E5.1   *Aperture grills for cathode ray tubes* (Figure E.3). Two types of cathode ray tube (CRT) dominate the computer monitor and television marketplace. In the older technology, color separation is achieved by using a *shadow mask*: a thin metal plate with a grid of holes that allow only the correct beam to strike a red, green or blue phosphor. A shadow mask can heat up and distort at high brightness levels ("doming"), causing the beams to miss their targets, and giving a blotchy image. To avoid this, the newest shadow masks are made of Invar, a nickel alloy with a near-zero expansion coefficient between room temperature and 150°C. It is a consequence of shadow-mask technology that the glass screen of the CRT curves inward on all four edges, increasing the probability of reflected glare.

Sony's "Trinitron" technology overcomes this problem and allows greater brightness by replacing the shadow mask by an *aperture grill* of fine vertical wires, each about 200 μm in thickness, that allows the intended beam to strike either the red, the green or the blue phosphor to create the image. The glass face of the Trinitron tube is curved in one plane only, reducing glare.

The wires of the aperture grill are tightly stretched, so that they remain taut even when hot — it is this tension that allows the greater brightness. What

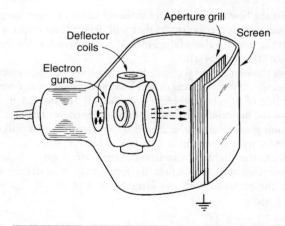

Figure E.3   Aperture grills for cathode ray tubes.

index guides the choice of material to make them? The table summarizes the requirements.

Function	Aperture grill for CRT
Constraints	• Wire thickness and spacing specified
	• Must carry pre-tension without failure
	• Electrically conducting to prevent charging
	• Able to be drawn to wire
Objective	Maximize permitted temperature rise without loss of tension
Free variables	Choice of material

Exercise E5.2 *Material indices for elastic beams with differing constraints* (Figure E.4). Start each of the four parts of this problem by listing the function, the objective, and the constraints. You will need the equations for the deflection of a cantilever beam with a square cross-section $t \times t$, given in Appendix A, Section A.3. The two that matter are that for the deflection δ of a beam of length L under an end load F:

$$\delta = \frac{FL^3}{3EI}$$

and that for the deflection of a beam under a distributed load f per unit length:

$$\delta = \frac{1}{8}\frac{fL^4}{EI}$$

where $I = t^4/12$. For a self-loaded beam $f = \rho A g$ where ρ is the density of the material of the beam, A its cross-sectional area and g the acceleration due to gravity.

(a) Show that the best material for a cantilever beam of given length L and *given* (i.e., fixed) square cross-section $(t \times t)$ that will deflect least under a given end load F is that with the largest value of the index $M = E$, where E is Young's modulus (neglect self-weight) (Figure E.4(a)).

(b) Show that the best material choice for a cantilever beam of given length L and with a given section $(t \times t)$ that will deflect least under its own self-weight is that with the largest value of $M = E/\rho$, where ρ is the density (Figure E.4(b)).

(c) Show that the material index for the lightest cantilever beam of length L and square section (not given, that is, the area is a free variable) that will not deflect by more than δ under its own weight is $M = E/\rho^2$ (Figure E.4(c)).

(d) Show that the lightest cantilever beam of length L and square section (area free) that will not deflect by more than δ under an end load F is that made of the material with the largest value of $M = E^{1/2}/\rho$ (neglect self weight) (Figure E.4(d)).

Exercise E5.3 *Material index for a light, strong beam* (Figure E.5). In stiffness-limited applications, it is elastic deflection that is the active constraint: it limits performance.

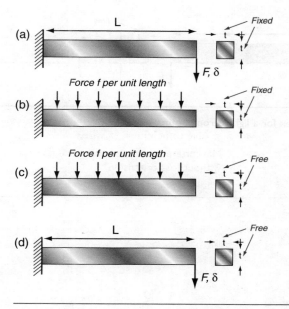

Figure E.4 Material indices for elastic beams with differing constraints.

In strength-limited applications, deflection is acceptable provided the component does not fail; strength is the active constraint. Derive the material index for selecting materials for a beam of length L, specified strength and minimum weight. For simplicity, assume the beam to have a solid square cross-section $t \times t$. You will need the equation for the failure load of a beam (Appendix A, Section A.4). It is

$$F_f = \frac{I \sigma_f}{y_m L}$$

where y_m is the distance between the neutral axis of the beam and its outer filament and $I = t^4/12 = A^2/12$ is the second moment of the cross-section. The table itemizes the design requirements:

Function	Beam
Constraints	• Length L is specified
	• Beam must support a bending load F without yield or fracture
Objective	Minimize the mass of the beam
Free variables	• Cross-section area, A
	• Choice of material

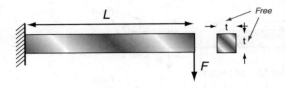

Figure E.5 Material index for a light, strong beam.

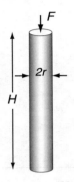

Figure E.6 Material index for a cheap, stiff column.

Exercise E5.4 *Material index for a cheap, stiff column* (Figure E.6). In the last two exercises the objective has been that of minimizing weight. There are many others. In the selection of a material for a spring, the objective is that of maximizing the elastic energy it can store. In seeking materials for thermal-efficient insulation for a furnace, the best are those with the lowest thermal conductivity and heat capacity. And most common of all is the wish to minimize cost. So here is an example involving cost.

Columns support compressive loads: the legs of a table; the pillars of the Parthenon. Derive the index for selecting materials for the cheapest cylindrical column of specified height, H, that will safely support a load F without buckling elastically. You will need the equation for the load F_{crit} at which a slender column buckles. It is

$$F_{crit} = \frac{n\pi^2 EI}{H^2}$$

where n is a constant that depends on the end constraints and $I = \pi r^4/4 = A^2/4\pi$ is the second moment of area of the column (see Appendix A for both). The table lists the requirements:

Function	Cylindrical column
Constraints	• Length L is specified
	• Column must support a compressive load F without buckling
Objective	Minimize the material cost of the column
Free variables	• Cross-section area, A
	• Choice of material

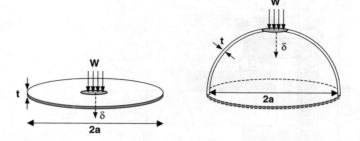

Figure E.7   Indices for stiff plates and shells.

Exercise E5.5   *Indices for stiff plates and shells* (Figure E.7). Aircraft and space structures make use of plates and shells. The index depends on the configuration. Here you are asked to derive the material index for

(a) a circular plate of radius a carrying a central load W with a prescribed stiffness $S = W/\delta$ and of minimum mass,

(b) a hemispherical shell of radius a carrying a central load W with a prescribed stiffness $S = W/\delta$ and of minimum mass, as shown in the figure.

Use the two results listed below for the mid-point deflection δ of a plate or spherical shell under a load W applied over a small central, circular area.

$$\text{Circular plate:} \quad \delta = \frac{3}{4\pi} \frac{Wa^2}{Et^3} (1 - \nu^2) \left(\frac{3 + \nu}{1 + \nu} \right)$$

$$\text{Hemispherical shell:} \quad \delta = A \frac{Wa}{Et^2} (1 - \nu^2)$$

in which $A \approx 0.35$ is a constant. Here E is Young's modulus, t is the thickness of the plate or shell and ν is Poisson's ratio. Poisson's ratio is almost the same for all structural materials and can be treated as a constant. The table summarizes the requirements:

Function	• Stiff circular plate, or • Stiff hemispherical shell
Constraints	• Stiffness S under central load W specified • Radius a of plate or shell specified
Objective	Minimize the mass of the plate or shell
Free variables	• Plate or shell thickness, t • Choice of material

Exercise E5.6   *The C-clamp in more detail* (Figure E.8). Exercise E4.4 introduced the C-clamp for processing of electronic components. The clamp has a square cross-section of width x and given depth b. It is essential that the clamp have low thermal inertia so that it reaches temperature quickly. The time t it takes a component of thickness x to reach

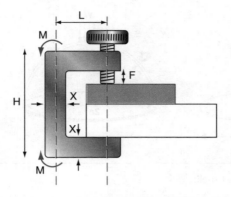

Figure E.8 The C-clamp in more detail.

thermal equilibrium when the temperature is suddenly changed (a transient heat flow problem) is

$$t \approx \frac{x^2}{2a}$$

where the thermal diffusivity $a = \lambda/\rho C_p$ and λ is the thermal conductivity, ρ the density and C_p the specific heat. The time to reach thermal equilibrium is reduced by making the section x thinner, but it must not be so thin that it fails in service. Use this constraint to eliminate x in the equation above, thereby deriving a material index for the clamp. Use the fact that the clamping force F creates a moment on the body of the clamp of $M = FL$, and that the peak stress in the body is given by

$$\sigma = \frac{x}{2} \frac{M}{I}$$

where $I = bx^3/12$ is the second moment of area of the body. The table summarizes the requirements.

Function	C-clamp of low thermal inertia
Constraints	• Depth b specified
	• Must carry clamping load F without failure
Objective	Minimize time to reach thermal equilibrium
Free variables	• Width of clamp body, x
	• Choice of material

Exercise E5.7 *Springs for trucks* (Figure E.9). In vehicle suspension design it is desirable to minimize the mass of all components. You have been asked to select a material and dimensions for a light spring to replace the steel leaf-spring of an existing truck suspension. The existing leaf-spring is a beam, shown schematically in the figure. The new spring must have the same length L and stiffness S as the existing one, and must deflect through a maximum safe displacement δ_{max} without failure. The width b and thickness t are free variables.

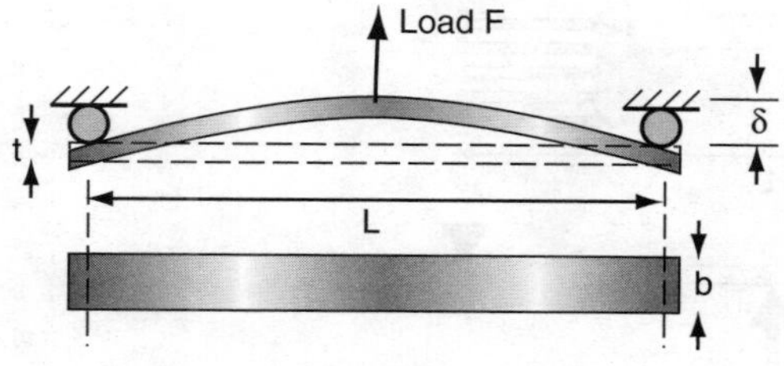

Figure E.9   Springs for trucks.

Derive a material index for the selection of a material for this application. Note that this is a problem with two free variables: b and t; and there are two constraints, one on safe deflection δ_{max} and the other on stiffness S. Use the two constraints to fix free variables. The table catalogs the requirements:

Function	Leaf spring for truck
Constraints	• Length L specified • Stiffness S specified • Maximum displacement δ_{max} specified
Objective	Minimize the mass
Free variables	• Spring thickness t • Spring width b • Choice of material

You will need the equation for the mid-point deflection of an elastic beam of length L loaded in three-point bending by a central load F:

$$\delta = \frac{1}{48}\frac{FL^3}{EI}$$

and that for the deflection at which failure occurs

$$\delta_{max} = \frac{1}{6}\frac{\sigma_f L^2}{tE}$$

where I is the second moment of area; for a beam of rectangular section, $I = bt^3/12$ and E and σ_f are the modulus and failure stress of the material of the beam (both results can be found in Appendix A).

Exercise E5.8   *Disposable knives and forks* (Figure E.10). Disposable knives and forks are ordered by an environmentally-conscious pizza-house. The shape of each (and thus the length, width, and profile) are fixed, but the thickness is free: it is chosen to give enough bending-stiffness to cut and impale the pizza without excessive flexure. The pizzeria-proprietor wishes to enhance the greenness of his image by minimizing the

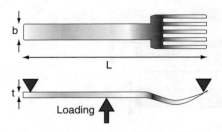

Figure E.10 Disposable knives and forks.

energy-content of his throw-away tableware, which could be molded from polystyrene (PS) or stamped from aluminum sheet.

Establish an appropriate material index for selecting materials for energy-economic forks. Model the eating implement as a beam of fixed length L and width w, but with a thickness t that is free, loaded in bending, as in the figure. The objective-function is the volume of material in the fork times its energy content, $H_p\rho$, per unit volume (H_p is the production energy per kg, and ρ the density). The limit on flexure imposes a stiffness constraint (Appendix A, Section A.3). Use this information to develop the index.

Flexure, in cutlery, is an inconvenience. Failure — whether by plastic deformation or by fracture — is more serious: it causes loss-of-function; it might even cause hunger. Repeat the analysis, deriving an index when a required strength is the constraint.

This is a straightforward application of the method illustrated by Exercise E5.2; the only difference is that energy content, not weight, is minimized. The free variable is the thickness of the shaft of the fork; all other dimensions are fixed. There are two alternative constraints, first, that the fork should not flex too much, second, that it should not fail.

Function	Environmentally friendly disposable forks
Constraints	• Length L specified
	• Width b specified
	• Stiffness S specified, or
	• Failure load F is specified
Objective	Minimize the material energy-content
Free variables	• Shaft thickness, t
	• Choice of material

The selection can be implemented using Figures 16.8 and 16.9. If the CES software is available, make a chart with the stiffness index as one axis and the strength index as the other. The materials that best meet *both* criteria lie at the top right.

Exercise E5.9 *Fin for a rocket* (Figure E.11). A tube-launched rocket has stabilizing fins at its rear. During launch the fins experience hot gas at $T_g = 1700°C$ for a time $0.3\,s$. It is

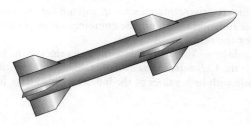

Figure E.11   Fin for a rocket.

important that the fins survive launch without surface melting. Suggest a material index for selecting a material for the fins. The table summarizes the requirements:

Function	High heat-transfer rocket fins
Constraints	• All dimensions specified • Must not suffer surface melting during exposure to gas at 1700°C for 0.3 s
Objective	• Minimize the surface temperature rise during firing • Maximize the melting point of the material
Free variables	Choice of material

This is tricky. Heat enters the surface of the fin by transfer from the gas. If the heat transfer coefficient is h, the heat flux per unit area is

$$q = h(T_g - T_s)$$

where T_s is the surface temperature of the fin—the critical quantity we wish to minimize. Heat diffuses into the fin surface by thermal conduction. If the heating time is small compared with the characteristic time for heat to diffuse through the fin, a quasi steady-state exists in which the surface temperature adjusts itself such that the heat entering from the gas is equal to that diffusing inwards by conduction. This second is equal to

$$q = \lambda \frac{(T_s - T_i)}{x}$$

where λ is the thermal conductivity, T_i is the temperature of the (cold) interior of the fin, and x is a characteristic heat-diffusion length. When the heating time is short (as here) the thermal front, after a time t, has penetrated a distance

$$x \approx (2at)^{1/2}$$

where $a = \lambda/\rho C_p$ is the thermal diffusivity. Substituting this value of x in the previous equation gives

$$q = \left(\lambda \rho C_p\right)^{1/2} \frac{(T_s - T_i)}{x}$$

where ρ is the density and C_p the specific heat of the material of the fin.

Proceed by equating the two equations for q, solving for the surface temperature T_s to give the objective function. Read off the combination of properties that minimizes T_s; it is the index for the problem.

The selection is made by seeking materials with large values of the index and with a high melting point, T_m. If the CES software is available, make a chart with these two as axes and identify materials with high values of the index that also have high melting points.

E.6 Selecting processes

The exercises of this section use the process selection charts of Chapters 7 and 8. They are useful in giving a feel for process attributes and the way in which process choice depends on material and the shape. Here the CES software offers greater advantages: what is cumbersome and of limited resolution with the charts is easy with the software, which offers much greater resolution.

Each exercise has three parts, labeled (a)–(c). The first involves translation. The second uses the selection charts of Chapter 7 (which you are free to copy) in the way that was illustrated in Chapter 8. The third involves the use of the CES software if available.

Exercise E6.1 *Elevator control quadrant* (Figure E.12). The quadrant sketched here is part of the control system for the wing-elevator of a commercial aircraft. It is to be made of a light alloy (aluminum or magnesium) with the shape shown in Figure E.12. It weighs about 5 kg. The minimum section thickness is 5 mm, and — apart from the bearing surfaces — the requirements on surface finish and precision are not strict: surface finish $\leq 10\,\mu m$ and precision $\leq 0.5\,mm$. The bearing surfaces require a surface finish $\leq 1\,\mu m$ and a precision $\leq 0.05\,mm$. A production run of 100–200 is planned.

(a) Itemize the function and constraints, leave the objective blank and enter "choice of process" for the free variable.
(b) Use copies of the charts of Chapter 7 in succession to identify processes to shape the quadrant.
(c) If the CES software is available, apply the constraints and identify in more detail the viable processes.

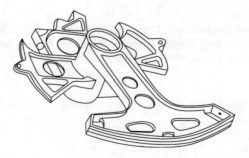

Figure E.12   Elevator control quadrant.

Figure E.13   Casing for an electric plug.

Exercise E6.2   *Casing for an electric plug* (Figure E.13). The electric plug is perhaps the commonest of electrical products. It has a number of components, each performing one or more functions. The most obvious are the casing and the pins, though there are many more (connectors, a cable clamp, fasteners, and, in some plugs, a fuse). The task is to investigate processes for shaping the two-part insulating casing, the thinnest part of which is 2 mm thick. Each part weighs about 30 grams and is to be made in a single step from a thermoplastic or thermosetting polymer with a planned batch size of 5×10^4–2×10^6. The required tolerance of 0.3 mm and surface roughness of 1 μm must be achieved without using secondary operations.

(a) Itemize the function and constraints, leave the objective blank and enter "choice of process" for the free variable.
(b) Use the charts of Chapter 7 successively to identify possible processes to make the casing.
(c) Use the CES software to select materials for the casing.

Exercise E6.3   *Ceramic valves for taps* (Figure E.14). Few things are more irritating than a dripping tap. Taps drip because the rubber washer is worn or the brass seat is pitted by corrosion, or both. Ceramics wear well, and they have excellent corrosion resistance in both pure and salt water. Many household taps now use ceramic valves.

The sketch shows how they work. A ceramic valve consists of two disks mounted one above the other, spring-loaded so that their faces are in contact. Each disk has a diameter of 20 mm, a thickness of 3 mm and weighs about 10 grams. In order to seal well, the mating surfaces of the two disks must be flat and smooth, requiring high levels of precision and surface finish; typically tolerance <0.02 mm and surface roughness <0.1 μm. The outer face of each has a slot that registers it, and allows the upper disk to be rotated through 90° (1/4 turn). In the "off" position the holes in the upper disk are blanked off by the solid part of the lower one; in the "on" position the holes are aligned. A production run of 10^5–10^6 is envisaged.

(a) List the function and constraints, leave the objective blank and enter "choice of process" for the free variable.
(b) Use the charts of Chapter 7 to identify possible processes to make the casing.
(c) Use the CES software to select materials for the casing.

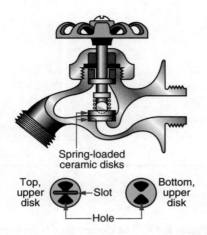

Figure E.14 Ceramic valves for taps.

Figure E.15 Shaping plastic bottles.

Exercise E6.4 *Shaping plastic bottles* (Figure E.15). Plastic bottles are used to contain fluids as various as milk and engine oil. A typical polyethylene bottle weighs about 30 grams and has a wall thickness of about 0.8 mm. The shape is 3-D hollow. The batch size is large (1,000,000 bottles). What process could be used to make them?

    (a) List the function and constraints, leave the objective blank and enter "choice of process" for the free variable.

    (b) Use the charts of Chapter 7 to identify possible processes to make the casing.

    (c) Use the CES software to select materials for the casing.

Exercise E6.5 *Car hood (bonnet)* (Figure E.16). As weight-saving assumes greater importance in automobile design, the replacement of steel parts with polymer-composite substitutes

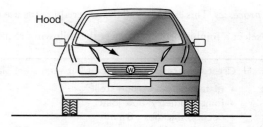

Figure E.16 Car hood (bonnet).

Figure E.17 Complex structural channels.

becomes increasingly attractive. Weight can be saved by replacing a steel hood with one made from a thermosetting composites. The weight of the hood depends on the car model: a typical composite hood weighs is 8–10 kg. The shape is a dished-sheet and the requirements on tolerance and roughness are 0.2 mm and 2 μm, respectively. A production run of 100,000 is envisaged.

(a) List the function and constraints, leave the objective blank and enter "choice of process" for the free variable.
(b) Use the charts of Chapter 7 to identify possible processes to make the casing.
(c) Use the CES software to select materials for the casing.

Exercise E6.6 *Complex structural channels* (Figure E.17). Channel sections for window frames, for slide-together sections for versatile assembly and for ducting for electrical wiring can be complex in shape. The figure shows an example. The order is for 10,000 such sections, each 1 m in length and weighing 1.2 kg, with a minimum section of 4 mm. A tolerance of 0.2 mm and a surface roughness of less than 1 μm must be achieved without any additional finishing operation.

(a) List the function and constraints, leave the objective blank and enter "choice of process" for the free variable.
(b) Use the charts of Chapter 7 to identify possible processes to make the casing.
(c) Use the CES software to select materials for the casing.

Exercise E6.7 *Selecting joining processes.* This exercise and the next require the use of the CES software.

(a) Use CES to select a joining process to meet the following requirements:

Function	Create a permanent butt joint between steel plates
Constraints	• Material class: carbon steel • Joint geometry: butt joint • Section thickness: 8 mm • Permanent • Watertight
Objective	—
Free variables	Choice of process

(b) Use CES to select a joining process to meet the following requirements:

Function	Create a watertight, demountable lap joint between glass and polymer
Constraints	• Material class: glass and polymers • Joint geometry: lap joint • Section thickness: 4 mm • Demountable • Watertight
Objective	—
Free variables	Choice of process

Exercise E6.8 *Selecting surface treatment processes.* This exercise, like the last, requires the use of the CES *software.*

(a) Use CES to select a surface treatment process to meet the following requirements:

Function	Increase the surface hardness and wear resistance of a high carbon steel component
Constraints	• Material class: carbon steel • Purpose of treatment: increase surface hardness and wear resistance
Objective	—
Free variables	Choice of process

(b) Use CES to select a joining process to meet the following requirements:

Function	Apply color and pattern to the curved surface of a polymer molding
Constraints	• Material class: thermoplastic • Purpose of treatment: aesthetics, color • Curved surface coverage: good or very good
Objective	—
Free variables	Choice of process

E.7 Multiple constraints and objectives

Over-constrained problems are normal in materials selection. Often it is just a case of applying each constraint in turn, retaining only those solutions that meet them all. But when constraints are used to eliminate free variables in an objective function (as discussed in Section 9.2) the "active constraint" method must be used. The first three exercises in this section illustrate problems with multiple constraints.

The remaining two concern multiple objectives and trade-off methods. When a problem has two objectives — minimizing both mass m and cost C of a component, for instance — a conflict arises: the cheapest solution is not the lightest and vice versa. The best combination is sought by constructing a trade-off plot using mass as one axis, and cost as the other. The lower envelope of the points on this plot defines the trade-off surface. The solutions that offer the best compromise lie on this surface. To get further we need a penalty function. Define the penalty function

$$Z = C + \alpha m$$

where α is an exchange constant describing the penalty associated with unit increase in mass, or, equivalently, the value associated with a unit decrease. The best solutions are found where the line defined by this equation is tangential to the trade-off surface. (Remember that objectives must be expressed in a form such that a minimum is sought; then a low value of Z is desirable, a high one is not.)

When a substitute is sought for an existing material it is better to work with ratios. Then the penalty function becomes

$$Z^* = \frac{C}{C_o} + \alpha^* \frac{m}{m_o}$$

in which the subscript "o" means properties of the existing material and the asterisk "*" on Z^* and α^* is a reminder that both are now dimensionless. The relative exchange constant α^* measures the fractional gain in value for a given fractional gain in performance.

Exercise E7.1   *Multiple constraints: a light, stiff, strong tie* (Figure E.18). A tie, of length L loaded in tension, is to support a load F, at minimum weight without failing (implying a constraint on strength) or extending elastically by more than δ (implying a constraint on stiffness, F/δ). The table summarizes the requirements:

Function	Tie rod
Constraints	• Must not fail by yielding under force F
	• Must have specified stiffness, F/δ
	• Length L and axial load F specified
Objective	Minimize mass m
Free variables	• Section area A
	• Choice of material

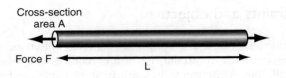

Figure E.18   Multiple constraints: a light stiff, strong tie.

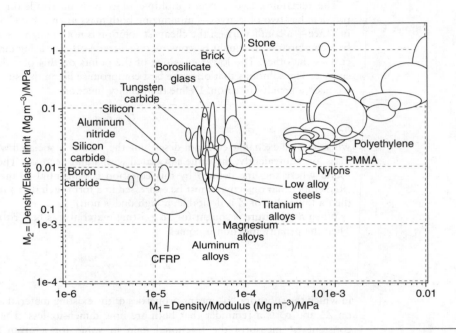

Figure E.19   Material chart.

(a) Follow the method of Chapter 9 to establish two performance equations for the mass, one for each constraint, from which two material indices and one coupling equation linking them are derived. Show that the two indices are

$$M_1 = \frac{\rho}{E} \quad \text{and} \quad M_2 = \frac{\rho}{\sigma_y}$$

and that a minimum is sought for both.

(b) Use these and the material chart of Figure E.19, which has the indices as axes, to identify candidate materials for the tie (i) when $L/\delta = 100$ and (ii) when $L/\delta = 10^3$.

Exercise E7.2   *Multiple constraints: a light, safe, pressure vessel* (Figure E.20). When a pressure vessel has to be mobile; its weight becomes important. Aircraft bodies, rocket casings and liquid-natural gas containers are examples; they must be light, and at the same time they must be safe, and that means that they must not fail by yielding or by fast

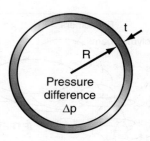

Figure E.20 Multiple constraints: a light, safe, pressure vessel.

fracture. What are the best materials for their construction? The table summarizes the requirements:

Function	Pressure vessel
Constraints	• Must not fail by yielding
	• Must not fail by fast fracture.
	• Diameter $2R$ and pressure difference Δp specified
Objective	Minimize mass m
Free variables	• Wall thickness, t
	• Choice of material

(a) Write, first, a performance equation for the mass m of the pressure vessel. Assume, for simplicity, that it is spherical, of specified radius R, and that the wall thickness, t (the free variable) is small compared with R. Then the tensile stress in the wall is

$$\sigma = \frac{\Delta p R}{2t}$$

where Δp, the pressure difference across this wall, is fixed by the design. The first constraint is that the vessel should not yield, that is, that the tensile stress in the wall should not exceed σ_y. The second is that it should not fail by fast fracture; this requires that the wall-stress be less than $K_{1C}\sqrt{\pi c}$, where K_{1C} is the fracture toughness of the material of which the pressure vessel is made and c is the length of the longest crack that the wall might contain. Use each of these in turn to eliminate t in the equation for m; use the results to identify two material indices

$$M_1 = \frac{\rho}{\sigma_y} \quad \text{and} \quad M_2 = \frac{\rho}{K_{1C}}$$

and a coupling relation between them. It contains the crack length, c.

(b) Figure E.21 shows the chart you will need with the two material indices as axes. Plot the coupling equation onto this figure for two values of c: one of 5 mm, the other 5 μm. Identify the lightest candidate materials for the vessel for each case.

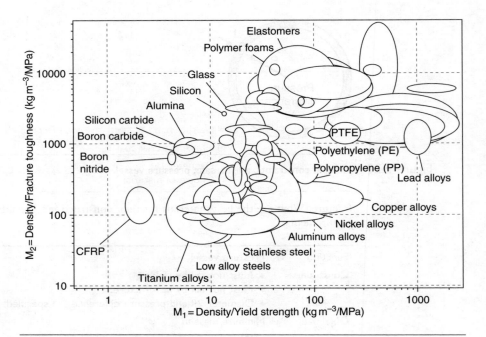

Figure E.21 Density/yield strength and density/tracture toughness chart.

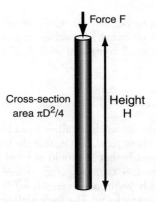

Figure E.22 A cheap column that must not buckle or crush.

Exercise E7.3 *A cheap column that must not buckle or crush* (Figure E.22). The best choice of material for a light strong column depends on its aspect ratio: the ratio of its height H to its diameter D. This is because short, fat columns fail by crushing; tall slender columns buckle instead. Derive two performance equations for the material cost of a column of solid circular section and specified height H, designed to support a load F large compared to its self-load, one using the constraints that the column must not crush, the other that it must not buckle. The table summarizes the needs:

Function	Column
Constraints	• Must not fail by compressive crushing • Must not buckle • Height H and compressive load F specified
Objective	Minimize material cost C
Free variables	• Diameter D • Choice of material

(a) Proceed as follows:
  (1) Write down an expression for the material cost of the column — its mass times its cost per unit mass.
  (2) Express the two constraints as equations, and use them to substitute for the free variable, D, to find the cost of the column that will just support the load without failing by either mechanism
  (3) Identify the material indices M_1 and M_2 that enter the two equations for the mass, showing that they are

$$M_1 = \left(\frac{C_m\rho}{\sigma_c}\right) \quad \text{and} \quad M_2 = \left[\frac{C_m\rho}{E^{1/2}}\right]$$

where C_m is the material cost per kg, the material density, σ_c its crushing strength and E its modulus.
(b) Data for six possible candidates for the column are listed in the table. Use these to identify candidate materials when $F = 10^5$ N and $H = 3$ m. Ceramics are admissible here, because they have high strength in compression.

Data for candidate materials for the column

Material	Density ρ (kg/m^3)	Cost/kg C_m ($/kg)	Modulus E (MPa)	Compression strength, σ_c (MPa)
Wood (spruce)	700	0.5	10,000	25
Brick	2100	0.35	22,000	95
Granite	2600	0.6	20,000	150
Poured concrete	2300	0.08	20,000	13
Cast iron	7150	0.25	130,000	200
Structural steel	7850	0.4	210,000	300
Al-alloy 6061	2700	1.2	69,000	150

(c) Figure E.23 show a material chart with the two indices as axes. Identify and plot coupling lines for selecting materials for a column with $F = 10^5$ N and $H = 3$ m (the same conditions as above), and for a second column with $F = 10^3$ N and $H = 20$ m.

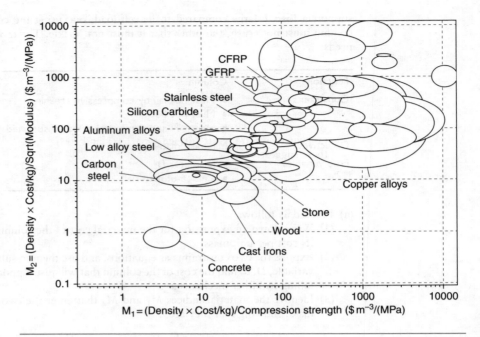

Figure E.23 Material chart.

Exercise E7.4 *An air cylinder for a truck* (Figure E.24). Trucks rely on compressed air for braking and other power-actuated systems. The air is stored in one or a cluster of cylindrical pressure tanks like that shown here (length L, diameter $2R$, hemispherical ends). Most are made of low-carbon steel, and they are heavy. The task: to explore the potential of alternative materials for lighter air tanks, recognizing the there must be a trade-off between mass and cost — if it is too expensive, the truck owner will not want it even if it is lighter. The table summarizes the design requirements.

Function	Air cylinder for truck
Constraints	• Must not fail by yielding
	• Diameter $2R$ and length L specified
Objective	• Minimize mass, m
	• Minimize material cost, C
Free variables	• Wall thickness, t
	• Choice of material

(a) Show that the mass and material cost of the tank relative to one made of low-carbon steel are given by

$$\frac{m}{m_o} = \left(\frac{\rho}{\sigma_y}\frac{\sigma_{y,o}}{\rho_o}\right) \quad \text{and} \quad \frac{C}{C_o} = \left(\frac{C_m\rho}{\sigma_y}\right)\left(\frac{\sigma_{y,o}}{C_{m,o}\rho_o}\right)$$

where ρ is the density, σ_y the yield strength and C_m the cost per kg of the material, and the subscript "o" indicates values for mild steel.

(b) Explore the trade-off between relative cost and relative mass using Figure E.25, which has these quantities as axes. Mild steel lies at the co-ordinates (1, 1). Sketch a trade-off surface. Define a relative penalty function

$$Z^* = \alpha^* \frac{m}{m_o} + \frac{C}{C_o}$$

where α^* is a relative exchange constant, and plot a contour that is approximately tangent to the trade-off surface for $\alpha^* = 1$ and for $\alpha^* = 100$. What selections do these suggest?

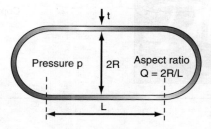

Figure E.24   An air cylinder for a truck.

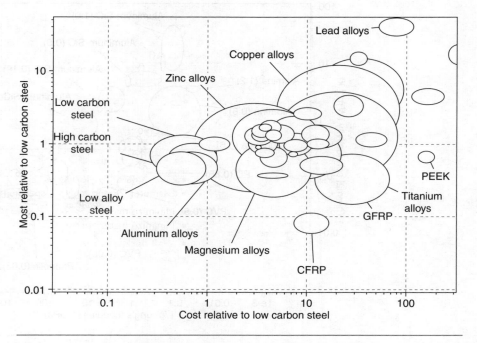

Figure E.25   Relative cost and relative mass.

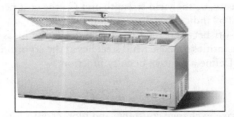

Figure E.26 Insulation walls for freezers.

Exercise E7.5

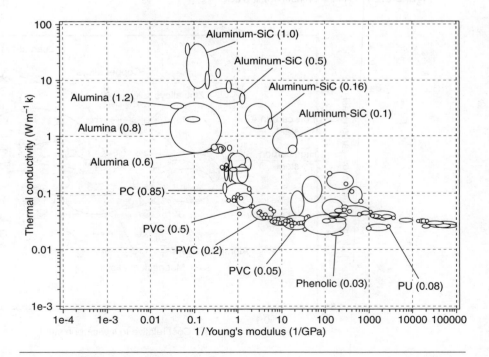

Figure E.27 Thermal conductivity and elastic compliance.

Insulating walls for freezers (Figure E.26). Freezers and refrigerated trucks have panel-walls that provide thermal insulation, and at the same time are stiff, strong and light (stiffness to suppress vibration, strength to tolerate rough usage). To achieve this the panels are usually of sandwich construction, with two skins of steel, aluminum or GFRP (providing the strength) separated by, and bonded to, a low density insulating core. In choosing the core we seek to minimize thermal conductivity, λ, and at the same time to maximize stiffness, because this allows thinner steel faces, and thus a lighter panel, while still maintaining the overall panel stiffness. The table summarizes the design requirements:

Function	Foam for panel-wall insulation
Constraints	Panel wall thickness specified.
Objective	• Minimize foam thermal conductivity, λ • Maximize foam stiffness, meaning Young's modulus, E
Free variables	Choice of material

Figure E.27 shows the thermal conductivity λ of foams plotted against their elastic compliance $1/E$ (the reciprocal of their Young's moduli E, since we must express the objectives in a form that requires minimization). The numbers in brackets are the densities of the foams in Mg/m^3. The foams with the lowest thermal conductivity are the least stiff; the stiffest have the highest conductivity. Explain the reasoning you would use to select a foam for the truck panel using a penalty function.

E.8 Selecting material and shape

The examples in this section relate to the analysis of material and shape of Chapters 11 and 12. They cover the derivation of shape factors, of indices that combine material and shape, and the use of the 4-quadrant chart arrays to explore material and shape combinations. For this last purpose it is useful to have clean copies of the chart arrays of Figures 11.12 and 11.15. Like the material property charts, they can be copied from the text without restriction of copyright.

Exercise E8.1   *Shape factors for tubes* (Figure E.28)

   (a) Evaluate the shape factor ϕ_B^e for stiffness-limited design in bending of a square box section of outer edge-length $h = 100\,mm$ and wall thickness $t = 3\,mm$. Is this shape more efficient than one made of the same material in the form of a tube of diameter

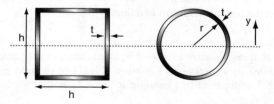

Figure E.28   Shape factors for tubes.

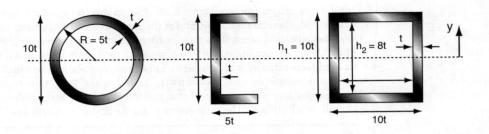

Figure E.29 Deriving shape factors for shiftness-limited design.

$2r = 100$ mm and wall thickness $t = 3.82$ mm (giving it the same mass per unit length, m/L)? Treat both as thin-walled shapes.

(b) Make the same comparison for the shape factor ϕ_B^f for strength-limited design.

Use the expressions given in Table 11.3 for the shape factors ϕ_B^e and ϕ_B^f.

Exercise E8.2 *Deriving shape factors for stiffness-limited design* (Figure E.29). Derive the expression for the shape-efficiency factor ϕ_B^e for stiffness-limited design for a beam loaded in bending with each of the three sections listed below (do not assume that the thin-wall approximations is valid):

(a) a closed circular tube of outer radius $5t$ and wall thickness t,

(b) a channel section of thickness t, overall flange width $5t$ and overall depth $10t$, bent about its major axis;

(c) a box section of wall thickness t, and height and width $h_1 = 10t$.

Exercise E8.3 *Deriving shape factors for strength-limited design.* (a) Determine the shape-efficiency factor ϕ_B^f for strength limited design in bending, for the same three sections shown in Exercise E8.2, Figure E.29. You will need the results for I for the sections derived in Exercise E8.2. (b) A beam of length L, loaded in bending, must support a specified bending moment M without failing and be as light as possible. Show that to minimize the mass of the beam per unit length, m/L, one should select a material and a section-shape to maximize the quantity

$$M = \frac{\left(\phi_B^f \sigma_f\right)^{2/3}}{\rho}$$

where σ_f is the failure stress and ρ the density of the material of the beam, and ϕ_B^f is the shape-efficiency factor for failure in bending.

Exercise E8.4 *Determining shape factors from stiffness data.* The elastic shape factor measures the gain in stiffness by shaping, relative to a solid square section of the same area. Shape factors can be determined by experiment. Equation (11.27) of the text gives the mass m of a beam of length L and prescribed bending stiffness S_B with a section of efficiency ϕ_B^e as

$$m = \left(\frac{12 S_B}{C_1}\right)^{1/2} L^{5/2} \left[\frac{\rho}{\left(\phi_B^e E\right)^{1/2}}\right]$$

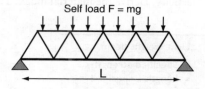

Self load F = mg

L

Figure E.30   Stell truss bridge.

where C_1 is a constant that depends on the distribution of load on the beam. Inverting the equation gives

$$\phi_B^e = \left(\frac{12L^5 S_B}{C_1 m^2}\right)\left(\frac{\rho^2}{E}\right)$$

Thus if S_B and m are measured, and the modulus E and density ρ of the material of the beam are known, ϕ_B^e can be calculated.

(a) Calculate the shape factor ϕ_B^e from the following experimental data, measured on an aluminum alloy beam loaded in 3-point bending (for which $C_1 = 48$, see Appendix A, Section A3) using the data shown in the following table:

Beam stiffness	$S_B = 7.2 \times 10^5$ N/m	Beam material	6061 aluminum alloy
Mass/unit length	$m/L = 1$ kg/m	Material density	$\rho = 2670$ k/gm³
Beam length	$L = 1$ m	Material modulus	$E = 69$ GPa

(b) A steel truss bridge shown in Figure E.30 has a span L and is simply supported at both ends. It weighs m tonnes. As a rule of thumb, bridges are designed with a stiffness S_B such that the central deflection δ of a span under its self-weight is less than 1/300 of the length L (thus $S_B \geq 300\, mg/L$ where g is the acceleration due to gravity, 9.81 m/s²). Use this information to calculate the minimum shape factor ϕ_B^e of the three steel truss bridge spans listed in the table. Take the density ρ of steel to be 7900 kg/m³ and its modulus E to be 205 GPa. ($C_1 = 384/5 = 76.8$ for uniformly distributed load, Appendix A, Section A3).

Bridge and construction date*	Span L (m)	Mass m (tonnes)
Royal Albert bridge, Tamar, Saltash UK (1857)	139	1060
Carquinez Strait bridge, California (1927)	132	650
Chesapeake Bay bridge, Maryland USA (1952)	146	850

*Data from the *Bridges Handbook*.

Exercise E8.5   *Deriving indices for bending and torsion.*

(a) A beam, loaded in bending, must support a specified bending moment M without failing and be as light as possible. Section shape is a variable, and "failure" here means the first onset of plasticity. Derive the material index. The table summarizes the requirements:

Function	Light weight beam
Constraints	• Specified strength
	• Length L specified
Objective	Minimum mass m
Free variables	• Choice of material
	• Section shape and scale

(b) A shaft of length L, loaded in torsion, must support a specified torque T without failing and be as cheap as possible. Section shape is a variable and "failure" again means the first onset of plasticity. Derive the material index. The table summarizes the requirements:

Function	Cheap shaft
Constraints	• Specified strength
	• Length L specified
Objective	Minimum material cost C
Free variables	• Choice of material
	• Section shape and scale

Exercise E8.6 *Use of the four segment chart for stiffness-limited design.*

(a) Use the 4-segment chart for stiffness-limited design of Figure 11.12 to compare the mass per unit length, m/L, of a section with $EI = 10^5 \, \mathrm{Nm^2}$ made from:
  (i) structural steel with a shape factor ϕ_B^e of 20, modulus $E = 210 \, \mathrm{GPa}$ and density $\rho = 7900 \, \mathrm{kg/m^3}$
  (ii) carbon fiber reinforced plastic with a shape factor ϕ_B^e of 10, modulus $E = 70 \, \mathrm{GPa}$ and density $\rho = 1600 \, \mathrm{kg/m^3}$, and
  (iii) structural timber with a shape factor ϕ_B^e of 2, modulus $E = 9 \, \mathrm{GPa}$ and density $= 520 \, \mathrm{kg/m^3}$.
  The schematic Figure E.31 illustrates the method.

(b) Show, by direct calculation, that the conclusions of part (a) are consistent with the idea that to minimize mass for a given stiffness one should maximize $\sqrt{E^*}/\rho$ with $E^* = E/\phi_B^e$ and $\rho^* = \rho/\phi_B^e$.

Exercise E8.7 *Use of the four-segment chart for strength*
Use the four-segment chart for stiffness-limited design of Figure 11.15 to compare the mass per unit length, m/L, of a section with $Z\sigma_f = 10^4 \, \mathrm{Nm}$ (where Z is the section modulus) made from:

  (i) mild steel with a shape factor ϕ_B^f of 10, strength $\sigma_f = 200 \, \mathrm{MPa}$ and density $\rho = 7900 \, \mathrm{kg/m^3}$
  (ii) 6061 grade aluminum alloy with a shape factor ϕ_B^f of 3, strength $\sigma_f = 200 \, \mathrm{MPa}$ and density $\rho = 2700 \, \mathrm{kg/m^3}$, and

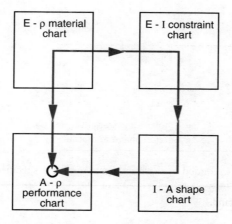

Figure E.31   Use of the four segment chart for shiftness. Limited design.

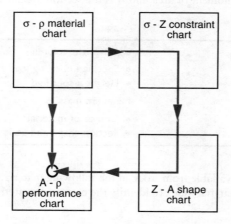

Figure E.32   Use of the four segment chart for strength.

(iii)   a titanium alloy with a shape factor ϕ_B^f of 10, strength $\sigma_f = 480$ MPa and density $\rho = 4420\,\text{kg/m}^3$.

(iv)   The schematic Figure E.32 illustrates the method.

Exercise E8.8   *A light weight display stand* (Figure E.33) The figure shows a concept for a lightweight display stand. The stalk must support a mass m of $100\,\text{kg}$, to be placed on its upper surface at a height h, without failing by elastic buckling. It is to be made of stock tubing and must be as light as possible. Use the methods of Chapter 11 to derive a material index for the tubular material of the stand that meets these requirements, and that includes the shape of the section, described by the shape factor

$$\phi_B^e = \frac{12I}{A^2}$$

Figure E.33 A light weight display stand.

where I is the second moment of area and A is the section area. The table summarizes the requirements:

Function	Light weight column
Constraints	• Specified buckling load F
	• Height h specified
Objective	Minimum mass m
Free variables	• Choice of material
	• Section shape and scale

Cylindrical tubing is available from stock in the following materials and sizes. Use this information and the material index to identify the best stock material for the column of the stand.

Material	Modulus E (GPa)	Tube radius r (mm)	Wall thickness/tube radius, t/r
Aluminum alloys	69	25	0.07–0.25
Steel	210	30	0.045–0.1
Copper alloys	120	20	0.075–0.1
Polycarbonate (PC)	3.0	20	0.15–0.3
Various woods	7–12	40	Solid circular sections only

Exercise E8.9 *Energy-efficient floor joists*. Floor joists are beams loaded in bending. They can be made of wood, of steel, or of steel-reinforced concrete, with the shape factors listed below. For a given bending stiffness and strength, which of these carries the lowest production-energy burden? The relevant data, drawn from the tables of Appendix C, are listed:

Material	Density, ρ (kg/m³)	Modulus, E (GPa)	Strength, σ_f (MPa)	Energy, H_p (MJ/kg)	ϕ_B^e	ϕ_B^f
Soft wood	700	10	40	15	2	1.4
Reinforced concrete	2900	35	10	3.2	2	1.4
Steel	7900	210	200	25	15	4

(a) Start with stiffness. Locate from Table 16.3 of the text the material index for stiffness-limited, shaped beams of minimum production energy. Introduce shape by multiplying the modulus E by the shape factor ϕ_B^e in the index. Use the modified index to rank the three beams.

(b) Repeat the procedure, using the appropriate index for strength at minimum energy content modified by including the shape factor ϕ_B^f.

(c) Compare wood and steel in a second way. Construct a line of slope 1, passing through "carbon steels" on the Modulus/Production energy chart of Figure 16.8, and move steels down both axes by the factor 15, as in Figure 11.16. Move Wood (// to grain) down by a factor of 2. The shaped steel can be compared with other materials — wood, say — using the ordinary guidelines which are shown on the figure. Is the conclusion the same?

(d) Repeat the procedure for strength, using the strength/production energy chart of Figure 16.9.

(e) What do you conclude about the relative energy-penalty of design with wood and with steel?

Exercise E8.10 *Microscopic shape: tube arrays* (Figure E.34). Calculate the gain in bending efficiency, ψ_B^e, when a solid is formed into small, thin-walled tubes of radius r and wall thickness t that are then assembled and bonded into a large array, part of which is shown in the figure. Let the solid of which the tubes are made have modulus E_s and density . Express the result in terms of r and t.

Exercise E8.11 *Microscopic shape: foams.*

(a) In the discussion of microscopic shape the assumption was made that, in converting a solid into a micro-structured material, Young's modulus changed with relative

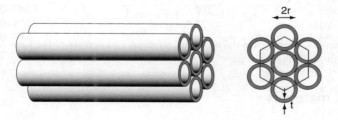

Figure E.34 Tube arrays.

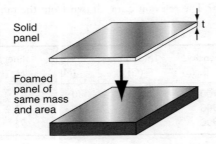

Figure E.35 The structural efficiency of foamed panels.

density as

$$E = \left(\frac{\rho}{\rho_s}\right) E_s$$

where E and ρ are the modulus and density of the microstructured material and E_s and ρ_s are those of the solid from which it was made. This is true for the in-plane moduli of layered structures and for the modulus of a material with prismatic cells parallel to the prism axis (like wood). But for polymeric and other foams with equiaxed cells the modulus scales instead as

$$E = \left(\frac{\rho}{\rho_s}\right)^2 E_s$$

What, then, is the shape-efficiency factor for elastic bending, ψ_B^e?

(b) The strength σ_f of foams, similarly, is not a linear function of relative density but scales as

$$\sigma_f = \left(\frac{\rho}{\rho_s}\right)^{3/2} \sigma_{f,s}$$

What is the value of ψ_f^B for foams?

Exercise E8.12 *The structural efficiency of foamed panels* (Figure E.35). Calculate the change in structural efficiency for both bending stiffness and strength when a solid flat panel of unit area and thickness t is foamed to give a foam panel of unit area and thickness h, at constant mass. The modulus E and strength σ_f of foams scale with relative density ρ/ρ_s as

$$E = \left(\frac{\rho}{\rho_s}\right)^2 E_s \quad \text{and} \quad \sigma_f = \left(\frac{\rho}{\rho_s}\right)^{3/2} \sigma_{f,s}$$

where E, σ_f and ρ are the modulus, strength and density of the foam and and E_s, $\sigma_{f,s}$ and ρ_s those of the solid panel.

E.9 Hybrid materials

The examples in this section relate to the deign of hybrid material described in Chapters 13 and 14. The first two involve the use of bounds for evaluating the potential of composite systems. The third is an example of hybrid design to fill holes in property space. The next three make use of the charts for natural materials. The last is a challenge: to explore the potential of hybridizing two very different materials.

Exercise E9.1 *Concepts for light, stiff composites.* Figure E.36 is a chart for exploring stiff composites with light alloy or polymer matrices. A construction like that of Figure 13.10 of the text allows the potential of any given matrix-reinforcement combination to be assessed.

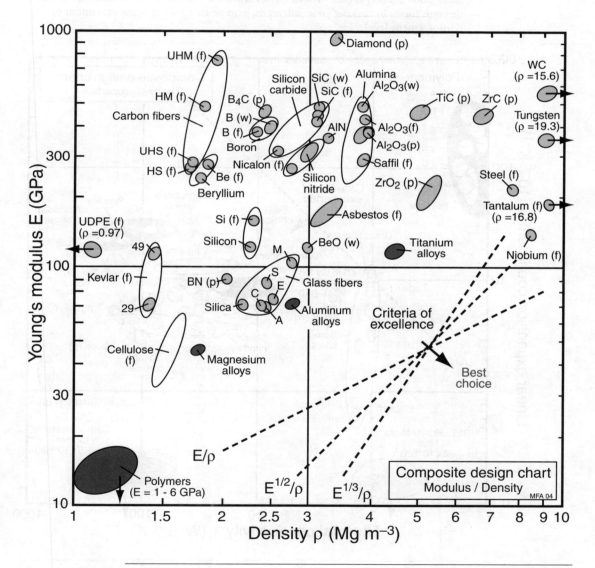

Figure E.36   The density and Young's modulus chart.

Four matrix materials are shown, highlighted in red. The materials shown in gray are available as fibers (f), whiskers (w) or particles (p). The criteria of excellence (the indices E/ρ, $E^{1/2}/\rho$ and $E^{1/3}/\rho$ for light, stiff structures) are shown; they increase in value towards the top left. Use the chart to compare the performance of a titanium-matrix composite reinforced with (a) zirconium carbide, ZrC, (b) Saffil alumina fibers, and (c) Nicalon silicon carbide fibers. Keep it simple: use equations (13.1)–(13.3) to calculate the density and upper and lower bounds for the modulus at a volume fraction of $f = 0.5$ and plot these points. Then sketch arcs of circles from the matrix to the reinforcement to pass through them. In making your judgment, assume that $f = 0.5$ is the maximum practical reinforcement level.

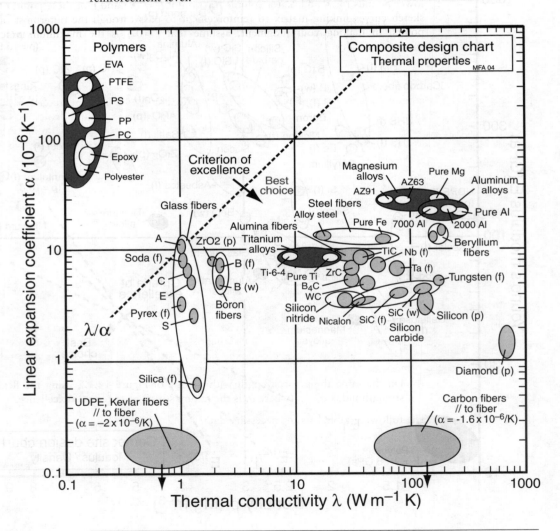

Figure E.37   The thermal conductivity and expansion coefficient chart.

Exercise E9.2 *Concepts for composites with tailored thermal properties.* Figure E.37 is a chart for exploring the design of composites with desired combinations of thermal conductivity and expansion, using light alloy or polymer matrices. A construction like that of Figure 13.11 allows the potential of any given matrix-reinforcement combination to be assessed. One criterion of excellence (the index for materials to minimize thermal distortion, λ/α) is shown; it increases in value towards the bottom right. Use the chart to compare the performance of a magnesium AZ63 alloy-matrix composite reinforced with (a) alloy steel fibers; (b) silicon carbide fibers, SiC (f); and (c) diamond-structured carbon particles. Keep it simple: use equation (13.7)–(13.10) to calculate the upper and lower bounds for α and λ at a volume fraction of $f = 0.5$ and plot these points. Then sketch curves linking matrix to reinforcement to pass through the outermost of the points. In making your judgment, assume that $f = 0.5$ is the maximum practical reinforcement level.

Exercise E9.3 *Hybrids with exceptional combinations of stiffness and damping.* The loss-coefficient–modulus (η–E) chart of Figure 4.9 is populated only along one diagonal band $\delta\eta$. (The loss coefficient measures the fraction of the elastic energy that is dissipated during a loading cycle.) Monolithic materials with low E have high η those with high E have low η. The challenge here is to devise hybrids to fill the holes, with the following applications in mind.

(a) Sheet steel (as used in car body panels, for instance) is prone to lightly damped vibration. Devise a hybrid sheet that combines the high stiffness, E, of steel with high loss coefficient, η.
(b) High loss coefficient means that energy is dissipated on mechanical cycling. This energy appears as heat, sometimes with undesirable consequences. Devise a hybrid with low modulus E and low loss coefficient, η.

 Assume for simplicity that the loss coefficient is independent of shape.

Exercise E9.4 *Natural hybrids that are light, shiff and strong*

(a) Plot aluminum alloys, steels, CFRP and GFRP onto a copy of the E–ρ chart for natural materials (Figure 14.11), where E is Young's modulus and ρ is the density. How do they compare, using the flexural stiffness index $E^{1/2}/\rho$ as criterion of excellence?
(b) Do the same thing for strength with a copy of Figure 14.12, using the flexural strength index $\sigma_f^{2/3}/\rho$ (where σ_f is the failure stress) as criteria of excellence.

The following table lists the necessary data:

Material	Young's modulus E (GPa)	Density ρ (Mg/m^3)	Strength, σ_f (MPa)
Aluminum alloys	68–82	2.5–2.9	30–450
Steels	189–215	7.6–8.1	200–1000
CFRP	69–150	1.5–1.6	550–1000
GFRP	15–28	1.75–2.0	110–192

Exercise E9.5 *Natural hybrids that act as springs.* The table lists the moduli and strengths of spring materials. Plot these onto a copy of the E–σ_f chart for natural materials of Figure 14.13, and compare their energy-storing performance with that of natural materials, using the σ_f^2/E as the criterion of choice. Here σ_f is the failure stress, E is Young's modulus and ρ is the density.

Material	Modulus (GPa)	Strength (MPa)	Density (Mg/m³)
Spring steel	206	1100	7.85
Copper–2% Beryllium	130	980	8.25
CFRP woven fabric laminate	68	840	1.58

Exercise E9.6 *Finding a substitute for bone.* Find an engineering material that most closely resembles compact bone in its strength/weight (σ_f/ρ) characteristics by plotting data for this material, read from Figure 14.12, onto a copy of the σ_f–ρ chart for engineering materials (Figure 4.4). Here σ_f is the failure stress and ρ is the density.

Exercise E9.7 *Creativity: what could you do with X?* The same 68 materials appear on all the charts of Chapter 4. These can be used as the starting point for "what if...?" exercises. As a challenge, use any chart or combination of charts to explore what might be possible by hybridizing any pair of the materials listed below, in any configuration you care to choose.

(a) Cement
(b) Wood
(c) Polypropylene
(d) Steel
(e) Copper.

Index

 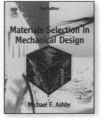